Applied Iterative Methods

Applied Iterative Methods

LOUIS A. HAGEMAN

Westinghouse Electric Corporation
West Mifflin, Pennsylvania

DAVID M. YOUNG

Center for Numerical Analysis
The University of Texas at Austin
Austin, Texas

1981

ACADEMIC PRESS

A Subsidiary of Harcourt Brace Jovanovich, Publishers

New York   London   Toronto   Sydney   San Francisco

ACADEMIC PRESS, INC.
111 Fifth Avenue, New York, New York 10003

United Kingdom Edition published by
ACADEMIC PRESS, INC. (LONDON) LTD.
24/28 Oval Road, London NW1 7DX

Library of Congress Cataloging in Publication Data

Hageman, Louis A.
    Applied iterative methods.

    (Computer science and applied mathematics)
    Bibliography: p.
    Includes index.
    1. Iterative methods (Mathematics) I. Young,
David M., Date   joint author. II. Title.
QA297.8.H34   519.4   80-29546
ISBN 0-12-313340-8

PRINTED IN THE UNITED STATES OF AMERICA

81 82 83 84   9 8 7 6 5 4 3 2 1

089202

Contents

v

Chapter 3   Polynomial Acceleration

Chapter 4   Chebyshev Acceleration

Chapter 5   An Adaptive Chebyshev Procedure Using Special Norms

Chapter 6   Adaptive Chebyshev Acceleration

Chapter 7   Conjugate Gradient Acceleration

Chapter 8   Special Methods for Red/Black Partitionings

Chapter 9   Adaptive Procedures for the Successive Overrelaxation Method

Chapter 10   The Use of Iterative Methods in the Solution of Partial Differential Equations

Chapter 11   Case Studies

Chapter 12   The Nonsymmetrizable Case

Preface

In this book we are primarily concerned with the practical utilization of iterative methods for solving large, sparse systems of linear algebraic equations. Such systems often arise in the numerical solution of partial differential equations by finite difference methods or by finite element methods. For such problems, the number of unknowns may vary from a few hundred to a few million.

Systems of linear algebraic equations can be solved either by direct methods or by iterative methods. For systems of moderate size, the use of direct methods is often advantageous. Iterative methods are used primarily for solving large and complex problems for which, because of storage and arithmetic requirements, it would not be feasible or it would be less efficient to solve by a direct method. For example, iterative methods are usually used to solve problems involving three spatial variables, problems involving nonlinear systems of equations, problems resulting from the discretization of coupled partial differential equations, and time-dependent problems involving more than one spatial variable.

The formulation and use of iterative methods require specialized knowledge and experience. Three of the main stumbling blocks to the efficient use of iterative methods are the following:

(a) uncertainty as to which iterative method should be used and how to implement a given method;
(b) uncertainty about how to select iteration parameters which are required

by certain methods (e.g., the relaxation factor ω for the successive overrelaxation method); and

(c) uncertainty about when the iterative process should be terminated.

Because of the diversity of problems to be solved and because of the large number of iterative procedures available, the complete removal of these uncertainties is not possible.

The choice of an effective iterative solution method for a particular problem depends heavily on the details peculiar to the problem and on the particular architecture of the computer to be used. Thus, no general rules governing the selection of the best solution method can be given. However, knowledge of the relative merits of several general iterative procedures can greatly simplify this task. Our approach to the problem of choice of method is to present the underlying computational and theoretical principles of certain general methods which then can be used as realistic bases upon which to select an effective iterative solution method.

For each general method that we consider, our aim is to present computational procedures for automatically determining good estimates of any iteration parameters required and for automatically deciding when to stop the iterative process. The computational procedures are presented in algorithmic form, using an informal programming language. In almost every case, the description provided is sufficiently complete and self-contained so that the reader could write a code based on the algorithm alone. However, it is strongly recommended that the user study the relevant part of the text before attempting to use any algorithm. Listings of FORTRAN language subroutines that implement some of the algorithms are given in the appendixes. These subroutines are designed for use as software packages to provide required acceleration parameters and to measure the iteration error vectors for certain iterative procedures.

Another aim of this book is to relate our experience in the use of iterative methods in the solution of multidimensional boundary-value problems. Discussions are presented of various problem aspects, such as mesh structure, discretization stencil, and matrix partitioning, which affect the cost-effectiveness of iterative solution procedures. Also discussed is the use of iterative methods to obtain numerical solutions to three particular boundary-value problems. These case studies are given to illustrate the versatility of iterative methods and to examine some problem aspects which must be considered in their use. The important concept of inner–outer or multistage iterations is embodied in each of the three problems studied.

We focus our attention on polynomial acceleration procedures applied to certain basic iterative methods and on the successive overrelaxation (SOR) method. The polynomial acceleration procedures considered are Chebyshev acceleration and conjugate gradient acceleration. It is assumed that the basic methods are "symmetrizable" in the sense defined in Chapter 2. The basic iterative methods considered for illustrative purposes are the RF (Richardson's) method, the Jacobi method, and the symmetric SOR (SSOR) method.

The organization of the material is presented as follows: The first two chapters are introductory in nature. Chapter 1 consists of background material on linear algebra and related topics. Chapter 2 contains descriptions of basic iterative methods which are used for illustrative purposes. Chapter 3 provides a general description of polynomial acceleration of basic iterative methods. Chapters 4–7 are devoted to the development of computational algorithms based on Chebyshev and conjugate gradient acceleration methods. In Chapter 8 we describe special Chebyshev and conjugate gradient acceleration procedures which are applicable when it is feasible to partition the linear system into a "red/black" block form. Chapter 9 contains a description of adaptive computational algorithms for the successive overrelaxation (SOR) method. Computational aspects in the utilization of iterative algorithms for solving multidimensional problems are discussed in Chapters 10 and 11. The iterative procedures discussed in Chapters 3–10 are applicable primarily to symmetric and positive definite matrix problems. A brief discussion of solution methods for nonsymmetric problems is given in Chapter 12. Numerical examples are given at the end of most chapters.

A reader who is interested in one particular method may choose to study only those parts of the text which are relevant. For Chebyshev acceleration, Chapters 2–6 and a part of Chapter 8 should be read. For conjugate gradient acceleration, the reader should refer to Chapters 2, 3, 7, and a part of Chapter 8. The reader interested primarily in the SOR method should study Chapters 2 and 9. Computational and other aspects of these procedures are discussed in Sections 5.6, 6.7, 8.5, 8.6, and 9.9 and in Chapters 10 and 11.

Many of the iterative procedures and algorithms described in this book have been used successfully in production-type computer programs involving a large number of unknowns, often exceeding 10^5. In addition, programs based on some of the algorithms have been developed and are included in the iteration solution package ITPACK (Grimes, Kincaid, MacGregor, and Young [1978]), a research-oriented software package for solving large sparse linear systems iteratively. A principal application of the ITPACK programs is found in connection with the ELLPACK system (Rice [1977]). ELLPACK is a modular system which is designed to serve as a research tool for the development and evaluation of software for the solution of elliptic partial differential equations.

In order to make the book useful to as many readers as possible, a minimal amount of mathematical background has been assumed. The main text presupposes that the reader has some knowledge of computer programming and linear algebra, and some experience with the use of iterative methods. Some of this background material is reviewed and collected in Chapters 1 and 2.

References are given by author and year, e.g., Axelsson [1972]. The list of references is by no means complete. For the most part, we have included only those references that we feel will provide the reader with supplementary information on particular topics.

A decimal notation is used for numbering sections and chapters. For example, the third section of Chapter 2 is referred to as Section 2.3. The eleventh numbered equation in Section 2.3 is (2-3.11). A similar system is used for theorems, tables, figures, etc. For the convenience of the reader, a list of frequently used symbols is given separately.

Acknowledgments

In the preparation of the manuscript, we received encouragement and useful suggestions from many sources. We especially wish to acknowledge the encouragement of the publisher, Academic Press, Inc., and its Consulting Editor, Dr. Werner C. Rheinboldt, who made many constructive suggestions. We are particularly grateful to Dr. Garrett Birkhoff for the stimulation, interest, encouragement, and ideas that he has contributed over the years. We also wish to acknowledge the helpful suggestions and comments concerning the manuscript of Dr. Ibrahim Abu-Shumays, Dr. Thomas Porsching, Dr. Myron Sussman, and Dr. Richard Varga.

Thanks are due to a number of people at the University of Texas at Austin and at Westinghouse who helped with the preparation and testing of the many computer routines which led to the development of some of the algorithms described in this book. At the University of Texas, Dr. David Kincaid, in addition to reviewing portions of the manuscript, supervised most of the computational effort. Others who participated in this effort include Dr. Linda Hayes, Dr. Baker Kearfott, Dr. Minna Chao, Mr. Edward Schleicher, Mrs. Kang Jea, Mr. Roger Grimes, and Mr. William MacGregor. At Westinghouse, Mr. Charles Pfeifer and Mr. Carl Spitz gave invaluable assistance and made many helpful suggestions, especially on the implementation of iterative procedures for large-scale scientific computations.

The preparation of this book required many preliminary drafts in addition to the final manuscript. We wish to give special thanks to Mrs. Darleen Aiken and Mrs. Dorothy Baker for their extremely careful and competent work in the preparation of the manuscript and to Mrs. Marilyn Hageman for indicating how to clarify passages in the final version of the text.

The first-named author is grateful for the opportunity afforded him by Westinghouse to pursue this work. Special thanks are due to Dr. Benjamin Mount, Mr. Milton Galper, and Mr. Richard Smith for their support and encouragement.

The work of the second-named author on the preparation of the book was supported in part by the National Science Foundation through Grant No. MCS76-03141 with The University of Texas at Austin. This support is gratefully acknowledged.

Notation

1. MATRICES

Symbol	Meaning	First Used in Section
$E^{N,N}$	Set of all $N \times N$ matrices	1.2
$A = (a_{i,j})$	Matrix of order N with elements $(a_{i,j})$	1.2
$A_{i,j}$	Submatrix of the partitioned matrix A	1.5
A^{T}	Transpose of A	1.2
I	Identity matrix	1.2
$\det(A)$	Determinant of A	1.3
A^{-1}	Inverse of A	1.2
SPD matrix	Symmetric and positive definite matrix	1.2
$A^{1/2}$	Square root of the SPD matrix A	1.2
$m(A)$	Smallest (algebraically) real eigenvalue of A	1.3
$M(A)$	Largest (algebraically) real eigenvalue of A	1.3
$\mathbf{S}(A)$	Spectral radius of A	1.3
$\left. \|A\|_2, \|A\|_\infty, \atop \|A\|_L \right\}$	Norms of the matrix A	1.4
$\kappa(A)$	Spectral condition number of the matrix A	1.4

2.   VECTORS

Symbol	Meaning	First Used in Section
E^N	Set of all $N \times 1$ column vectors	1.2
$v = (v_i)$	A vector whose ith component is v_i	1.2
v^T	Transpose of v	1.2
v^H	Conjugate transpose of v	1.2
(w, v)	Inner product of w with v	1.2
$\dfrac{(v, Av)}{(v, v)}$	Rayleigh quotient	1.3
$\{v(i)\}_{i=1}^L$	Set of L vectors	1.2
$\{v^{(i)}\}$	Sequence (possibly infinite) of vectors	1.4
$r^{(n)}$	Residual vector	3.3
$\delta^{(n)}$	Pseudoresidual vector	5.2
$\Delta^{(n)}$	Difference vector	8.3, 9.4
$\|v\|_2, \|v\|_\infty,$ $\|v\|_L, \|v\|_{\beta, z}$	Norms of the vector v	1.4, 5.4

3.   BASIC ITERATIVE METHODS

Symbol	Meaning	First Used in Section
$Au = b$	Matrix equation representing system of N equations in N unknowns	2.1
$\bar{u}$	Exact solution to $Au = b$	2.2
G	Iteration matrix of the general basic iterative method $u^{(n+1)} = Gu^{(n)} + k$	2.1
n	Iteration step number	2.1
$\varepsilon^{(n)}$	Error vector $\varepsilon^{(n)} = u^{(n)} - \bar{u}$	2.2
$R_n(G)$	Average rate of convergence	2.2
$R_\infty(G)$	Asymptotic rate of convergence	2.2
W	Symmetrization matrix for the basic method	2.2
$\{\mu_i\}_{i=1}^N$	Set of N eigenvalues of G	5.1
$v(i)$	Eigenvector of G associated with the eigenvalue of μ_i	6.2
γ	Extrapolation factor	2.2
$G_{[\gamma]}$	Iteration matrix associated with extrapolated method	2.2
Q	Splitting matrix	2.2
$B, \mathcal{L}_\omega, \mathcal{S}_\omega$	Iteration matrices corresponding, respectively, to the Jacobi, SOR, and symmetric SOR methods	2.3
ω	Relaxation factor	2.3
ω_b	Optimum relaxation factor	2.3

4. POLYNOMIAL ACCELERATION

Symbol	Meaning	First Used in Section
$Q_n(G)$	General matrix polynomial of degree n	1.3
$\gamma_1, \gamma_2, \ldots$ $\rho_1, \rho_2, \ldots$	Parameters used in three-term form of polynomial acceleration	3.2
$\bar{\mathbf{S}}(Q_n(G))$	Virtual spectral radius of $Q_n(G)$	3.2
$\bar{R}_n(Q_n(G))$	Virtual average rate of convergence	3.2
$\bar{R}_\infty(Q_n(G))$	Virtual asymptotic rate of convergence	3.2
$T_n(x)$	Chebyshev polynomial of degree n	4.2
$P_n(x)$	Normalized Chebyshev polynomial	4.2
m_E	Estimate for m(G)	4.3
M_E	Estimate for M(G)	4.3
$P_{n,E}(x)$	Normalized Chebyshev polynomial based on the estimates m_E, M_E	4.3

5. MISCELLANEOUS

Symbol	Meaning
N	Number of equations in the system $Au = b$
ζ	Iterative stopping criterion number
∎	End of proof
$\equiv$	Identity mark used to define a symbol or a function
$\doteq$	Used to denote approximate equality, i.e., $x \doteq y$ if $\lvert x - y \rvert / \lvert x \rvert$ is small
$\dot{\lessgtr}$	Used to denote approximate inequality, i.e., $x \dot{\lessgtr} y$ if $\lvert x - y \rvert / \lvert x \rvert$ is either negative or small
$\sim$	Used to denote asymptotic behavior, i.e., $x(t) \sim y(t)$ as $t \to a$ if $\lim_{t \to a} \{[y(t) - x(t)]/x(t)\} = 0$

Applied Iterative Methods

CHAPTER

1

Background on Linear Algebra
and Related Topics

1.1 INTRODUCTION

In this chapter we give some background material from linear algebra which will be helpful for our discussion of iterative methods. In Sections 1.2–1.5, we present a brief summary of basic matrix properties and principles which will be used in subsequent chapters. No proofs are given. It is assumed that the reader is already familiar with the general theory of matrices such as presented, for instance, in Noble and Daniel [1977] or in Faddeev and Faddeeva [1963, Chap. 1]. In Sections 1.6 and 1.7, we discuss the matrix problem which is obtained from a simple discretization of the generalized Dirichlet problem. The purpose of this example is to illustrate some of the matrix concepts presented in this chapter and to illustrate the formulations of matrix problems arising from the discretization of boundary-value problems.

1.2   VECTORS AND MATRICES

We let E^N denote the set of all $N \times 1$ column matrices, or vectors, whose components may be real or complex. A typical element v of E^N is given by

$$v = \begin{bmatrix} v_1 \\ v_2 \\ \vdots \\ v_N \end{bmatrix}. \tag{1-2.1}$$

To indicate that v_i, $i = 1, 2, \ldots, N$, are the components of v, we write $v = (v_i)$. A collection of vectors $v(1), v(2), \ldots, v(s)$ is said to be *linearly dependent* if there exist real or complex numbers $c_1, c_2, \ldots, c_s$, not all zero, such that

$$c_1 v(1) + c_2 v(2) + \cdots + c_s v(s) = 0.$$

If this equality holds only when all the constants $c_1, \ldots, c_s$ are zero, then the vectors $v(1), \ldots, v(s)$ are said to be *linearly independent*. A *basis* for E^N is a set of N linearly independent vectors of E^N. Given such a basis, say, $v(1)$, $v(2), \ldots, v(N)$, then any vector w in E^N can be expressed uniquely as a linear combination of basis vectors; i.e., there exists a unique set of numbers $c_1, c_2, \ldots, c_N$ such that

$$w = \sum_{i=1}^{N} c_i v(i). \tag{1-2.2}$$

The *transpose* of the vector v is denoted by v^T and the *conjugate transpose* by v^H. Given two vectors w and v of E^N, the *inner product* (w, v) of the vector w with v is defined by

$$(w, v) \equiv w^H v. \tag{1-2.3}$$

Similarly, $E^{N,N}$ denotes the set of all $N \times N$ square matrices whose elements may be real or complex. A typical element of $E^{N,N}$ is given by

$$A = \begin{bmatrix} a_{1,1} & a_{1,2} & \cdots & a_{1,N} \\ a_{2,1} & a_{2,2} & \cdots & a_{2,N} \\ \vdots & \vdots & & \vdots \\ a_{N,1} & a_{N,2} & \cdots & a_{N,N} \end{bmatrix} \tag{1-2.4}$$

or, equivalently, in abbreviated form by $A = (a_{i,j})$ for $1 \le i, j \le N$. We denote the *transpose* of the matrix A by A^T and the *conjugate transpose* of A by A^H. If the elements of A are real, then $A^H = A^T$. Normally, we shall deal only with real matrices. The matrix A is *symmetric* if $A = A^T$.

The special $N \times N$ matrix $A = (a_{i,j})$, where $a_{i,i} = 1$ and $a_{i,j} = 0$ if $i \neq j$, is called the *identity matrix* and is denoted by I. A matrix A in $E^{N,N}$ is *nonsingular* if there exists a matrix H such that $AH = HA = I$. If such an H exists, it is unique. The matrix H is called the *inverse* of A and is denoted by A^{-1}.

A real matrix A in $E^{N,N}$ is *symmetric and positive definite* (SPD) if A is symmetric and if $(v, Av) > 0$ for any nonzero vector v. If A is SPD, then A is nonsingular. The matrix LL^T is SPD for any real nonsingular matrix L. Also, if A is SPD, there exists a unique SPD matrix J such that $J^2 = A$. The matrix J is called the *square root* of the SPD matrix A and is denoted by $A^{1/2}$.

1.3 EIGENVALUES AND EIGENVECTORS

An *eigenvalue* of the $N \times N$ matrix A is a real or complex number λ which, for some nonzero vector y, satisfies the matrix equation

$$(A - \lambda I)y = 0. \qquad (1\text{-}3.1)$$

Any nonzero vector y which satisfies (1-3.1) is called an *eigenvector* of the matrix A corresponding to the eigenvalue λ.

In order for (1-3.1) to have a nontrivial solution vector y, the determinant of $A - \lambda I$ (denoted by $\det(A - \lambda I)$) must be zero. Hence, any eigenvalue λ must satisfy

$$\det(A - \lambda I) = 0. \qquad (1\text{-}3.2)$$

Equation (1-3.2), which is called the *characteristic equation* of A, is a polynomial of degree N in λ. The eigenvalues of A are the N zeros of the polynomial (1-3.2).

A matrix A in $E^{N,N}$ has precisely N eigenvalues, $\{\lambda_i\}_{i=1}^N$, some of which may be complex. The existence of at least one eigenvector corresponding to each eigenvalue λ_i is assured since (1-3.1) with $\lambda = \lambda_i$ has a nontrivial solution. Eigenvectors corresponding to unequal eigenvalues are linearly independent. Thus, when all the eigenvalues of A are distinct, the set of eigenvectors for A includes a basis for the vector space E^N. However, this is not always the case when some eigenvalues of A are repeated. In this book, we shall be concerned, for the most part, with those matrices whose eigenvalues are real and whose set of eigenvectors includes a basis for E^N. Such eigenproperties are satisfied, for example, by the eigenvalues and eigenvectors of symmetric matrices.

Before discussing symmetric matrices and related matrices, we introduce some notations that will be used repeatedly in this and subsequent chapters.

The *spectral radius* $\mathbf{S}(A)$ of the $N \times N$ matrix A is defined as the maximum of the moduli of the eigenvalues of A; i.e., if $\{\lambda_i\}_{i=1}^{N}$ is the set of eigenvalues of A, then

$$\mathbf{S}(A) \equiv \max_{1 \leq i \leq N} |\lambda_i|. \tag{1-3.3}$$

If the eigenvalues of A are real, we let $m(A)$ *and* $M(A)$ denote, respectively, the algebraically smallest and largest eigenvalues of A, i.e.,

$$m(A) \equiv \min_{1 \leq i \leq N} \lambda_i, \qquad M(A) \equiv \max_{1 \leq i \leq N} \lambda_i. \tag{1-3.4}$$

Eigenproperties of Symmetric and Related Matrices

Important eigenproperties of real symmetric matrices are summarized in the following theorem.

Theorem 1-3.1. If the $N \times N$ matrix A is real and symmetric, then
(1) the eigenvalues λ_i, $i = 1, \ldots, N$, of A are real, and
(2) there exists N real eigenvectors $\{y(i)\}_{i=1}^{N}$ for A such that

(a) $Ay(i) = \lambda_i y(i)$, $i = 1, \ldots, N$,
(b) $\{y(i)\}_{i=1}^{N}$ is a basis for E^N,   and
(c) $(y(i), y(j)) = 0$ if $i \neq j$ and $(y(i), y(j)) = 1$ if $i = j$.

When A is SPD, in addition to the eigenproperties given in Theorem 1-3.1, the eigenvalues of A are also positive. Since the matrix A is nonsingular if and only if no eigenvalue of A equals zero, it follows that a SPD matrix is also nonsingular.

Two matrices A and B are *similar* if $B = WAW^{-1}$ for some nonsingular matrix W. Similar matrices have identical eigenvalues.

Except for (2c), the conclusions of Theorem 1-3.1 also are valid for any real matrix A which is similar to a real symmetric matrix C.

For any real matrix A in $E^{N,N}$ and for any nonzero vector v in E^N (real or complex), the *Rayleigh quotient* of v with respect to A is defined as the quotient of inner products $(v, Av)/(v, v)$. If A is symmetric, then for any nonzero vector v in E^N

$$m(A) \leq (v, Av)/(v, v) \leq M(A). \tag{1-3.5}$$

Here $m(A)$ and $M(A)$ are defined by (1-3.4). Moreover, there exist nonzero vectors w and z such that

$$(w, Aw)/(w, w) = m(A), \qquad (z, Az)/(z, z) = M(A) \tag{1-3.6}$$

and

$$Aw = m(A)w, \qquad Az = M(A)z. \tag{1-3.7}$$

Eigenproperties of Real Nonsymmetric Matrices

The material in this subsection is used primarily in Chapter 9. Since the discussion is somewhat involved, many readers may wish to skip it on a first reading.

A real nonsymmetric matrix A may have complex eigenvalues. Since the coefficients of the characteristic polynomial (1-3.2) are real for this case, any complex eigenvalues of A must occur in complex conjugate pairs; i.e., if λ_i is a complex eigenvalue of the real matrix A, then $\lambda_k = \lambda_i^*$ is also an eigenvalue of A. Here, λ_i^* denotes the complex conjugate of λ_i. Moreover, if $y(i)$ is an eigenvector corresponding to λ_i, then $y(k) = y^*(i)$ is an eigenvector of A corresponding to $\lambda_k = \lambda_i^*$.

For an $N \times N$ nonsymmetric matrix A, it is not always possible to find a basis for E^N from the set of eigenvectors of A. However, it is always possible to form a basis from the independent eigenvectors of A supplemented by other vectors (called principal vectors) which are associated with the eigenvalues and eigenvectors of A. Such a basis can best be described in terms of the Jordan canonical form associated with A. The following is a restatement of the results given in Noble and Daniel [1977].

A square matrix of order ≥ 1 that has the form

$$J = \begin{bmatrix} \lambda & 1 & 0 & \cdots & 0 \\ 0 & \lambda & 1 & \cdots & 0 \\ \vdots & & \ddots & \ddots & \vdots \\ & & & & 1 \\ 0 & & \cdots & 0 & \lambda \end{bmatrix} \qquad (1\text{-}3.8)$$

is called a *Jordan block*. Note that the elements of J are zero except for those on the principal diagonal, which are all equal to λ, and those on the first superdiagonal, which are all equal to unity. Any matrix A can be reduced to a direct sum of Jordan blocks by a similarity transformation. More precisely, we have

Theorem 1-3.2.   For any $N \times N$ matrix A, there exists a nonsingular matrix V such that

$$V^{-1}AV = \begin{bmatrix} J_1 & 0 & \cdots & 0 \\ 0 & J_2 & \cdots & 0 \\ \vdots & & & \vdots \\ 0 & \cdots & 0 & J_k \end{bmatrix} \equiv \mathscr{J}, \qquad (1\text{-}3.9)$$

where each J_i, $1 \leq i \leq k$, is a Jordan block whose constant diagonal element is an eigenvalue of A. The number of linearly independent eigenvectors of A is equal to the number k of Jordan blocks in (1-3.9).

The matrix $\mathscr{J}$ in (1-3.9) is called the *Jordan canonical form* of A and is unique up to a permutation of the diagonal submatrices. Let the column vectors of V be denoted by $\{v(i)\}_{i=1}^{N}$; i.e., $V = [v(1), v(2), \ldots, v(N)]$. Since V is nonsingular, the set of vectors $\{v(i)\}_{i=1}^{N}$ is a basis for E^{N}. For use in later chapters, we now examine the behavior of the matrix–vector products $A^{l}v(i)$, $i = 1, \ldots, N$.

From the relation $AV = V\mathscr{J}$ and the form of $\mathscr{J}$, it follows that the vectors $v(i)$ separate, for each Jordan block J, into equations of the form

$$Av(i) = \lambda_i v(i) + v_i v(i-1), \tag{1-3.10}$$

where λ_i is an eigenvalue of A and v_i is either 0 or 1, depending on $\mathscr{J}$. If $v_i = 0$, then $v(i)$ is an eigenvector of A. When $v(i)$ is an *eigenvector of A*, we have by (1-3.10) that

$$A^{l}v(i) = A^{l-1}(\lambda_i v(i)) = (\lambda_i)^{l}v(i). \tag{1-3.11}$$

If each Jordan block of $\mathscr{J}$ is 1×1, then $v_i = 0$ for all i.† For this case, each column of V is an eigenvector of A and satisfies (1-3.11).

The relationship (1-3.11), however, is not valid for all $v(i)$ when some of the Jordan blocks are of order greater than unity. To illustrate this case, consider the example

$$V^{-1}AV = \begin{bmatrix} J_1 & 0 \\ 0 & J_2 \end{bmatrix}, \quad \text{where} \quad J_1 = [\lambda_1] \quad \text{and} \quad J_2 = \begin{bmatrix} \lambda_2 & 1 \\ 0 & \lambda_2 \end{bmatrix}. \tag{1-3.12}$$

From (1-3.10), the column vectors of V here separate into equations of the form

$$\begin{aligned} Av(1) &= \lambda_1 v(1), \\ Av(2) &= \lambda_2 v(2), \\ Av(3) &= \lambda_2 v(3) + v(2). \end{aligned} \tag{1-3.13}$$

The vectors $v(1)$ and $v(2)$ are eigenvectors of A. The other vector $v(3)$ is known as a *principal vector* (or generalized eigenvector) of grade 2, corresponding to the eigenvalue λ_2. From (1-3.13), the eigenvectors $v(1)$ and $v(2)$ satisfy

$$A^{l}v(1) = (\lambda_1)^{l}v(1) \quad \text{and} \quad A^{l}v(2) = (\lambda_2)^{l}v(2), \tag{1-3.14}$$

while the principal vector of grade 2 satisfies

$$A^{l}v(3) = A^{l-1}(\lambda_2 v(3) + v(2)) = (\lambda_2)^{l}v(3) + l(\lambda_2)^{l-1}v(2). \tag{1-3.15}$$

Note that if λ_2 is less than unity in absolute value, then both the sequences $\{A^{l}v(2)\}_{i=1}^{\infty}$ and $\{A^{l}v(3)\}_{i=1}^{\infty}$ converge to the null vector. However, the sequence involving the principal vector of grade 2 converges at a slower rate.

---

† This is the case, for example, if A is symmetric or if all eigenvalues of A are distinct.

Indeed, from (1-3.14) and (1-3.15), $\{A^l v(2)\}_{l=1}^{\infty}$ converges to the null vector at a rate governed by $|\lambda_2|^l$, while $\{A^l v(3)\}_{l=1}^{\infty}$ converges at the slower rate governed by $l|\lambda_2|^{l-1}$.

In general, the f column vectors of V associated with a Jordan block of order f consist of one eigenvector and $(f - 1)$ principal vectors of grade 2 through f. However, we shall not discuss the more general case since for reasons of simplicity we consider in later chapters only Jordan blocks of order 2 or less. We remark that eigenvectors are often defined as principal vectors of grade 1. A matrix whose set of eigenvectors does not include a basis is said to have an *eigenvector deficiency*.

Matrix Polynomials

If A is an $N \times N$ matrix, an expression of the form

$$Q_n(A) \equiv \alpha_0 I + \alpha_1 A + \alpha_2 A^2 + \cdots + \alpha_n A^n, \qquad (1\text{-}3.16)$$

where $\alpha_1, \ldots, \alpha_n$ are complex numbers, is called a *matrix polynomial*. A matrix polynomial $Q_n(A)$ can be obtained by substitution of the matrix A for the variable x in the associated algebraic polynomial

$$Q_n(x) \equiv \alpha_0 + \alpha_1 x + \alpha_2 x^2 + \cdots + \alpha_n x^n. \qquad (1\text{-}3.17)$$

The eigenvalues of the matrix polynomial $Q_n(A)$ can be obtained by substitution of the eigenvalues of A for the variable x in $Q_n(x)$. That is, if $\{\lambda_i\}_{i=1}^{N}$ is the set of eigenvalues for the matrix A in $E^{N,N}$, then $\{Q_n(\lambda_i)\}_{i=1}^{N}$ is the set of eigenvalues for the $N \times N$ matrix $Q_n(A)$.

1.4   VECTOR AND MATRIX NORMS

We shall consider several different vector and matrix norms. The vector norms that we consider are the following:

$$\|v\|_2 \equiv \sqrt{(v, v)} = \left[\sum_{i=1}^{N} |v_i|^2 \right]^{1/2}, \qquad (1\text{-}4.1)$$

$$\|v\|_\infty \equiv \max_{i=1,2,\ldots,N} |v_i|, \qquad (1\text{-}4.2)$$

$$\|v\|_L \equiv \|Lv\|_2. \qquad (1\text{-}4.3)$$

Here L is any nonsingular matrix. We also consider the corresponding matrix norms

$$\|A\|_2 \equiv \sqrt{\mathbf{S}(AA^{\mathrm{H}})}, \tag{1-4.4}$$

$$\|A\|_\infty \equiv \max_{i=1,2,\ldots,N} \left\{ \sum_{j=1}^{N} |a_{i,j}| \right\}, \tag{1-4.5}$$

$$\|A\|_L \equiv \|LAL^{-1}\|_2. \tag{1-4.6}$$

For $\alpha = 2, \infty$, and L, it can be shown that

$$\|Av\|_\alpha \le \|A\|_\alpha \|v\|_\alpha \tag{1-4.7}$$

and

$$\|A\|_\alpha = \sup_{v \ne 0}(\|Av\|_\alpha/\|v\|_\alpha). \tag{1-4.8}$$

An important property of matrix norms is that

$$\mathbf{S}(A) \le \|A\|_\beta \tag{1-4.9}$$

for $\beta = 2, \infty$, and L. If A is symmetric, then

$$\|A\|_2 = \mathbf{S}(A), \tag{1-4.10}$$

while if LAL^{-1} is symmetric, then

$$\|A\|_L = \mathbf{S}(A). \tag{1-4.11}$$

The sequence of vectors $v^{(0)}, v^{(1)}, \ldots$ converges to the vector v if and only if

$$\lim_{n \to \infty} \|v^{(n)} - v\|_\alpha = 0 \tag{1-4.12}$$

for any vector norm α. Similarly, the sequence of matrices $A^{(0)}, A^{(1)}, \ldots$ converges to A if and only if

$$\lim_{n \to \infty} \|A^{(n)} - A\|_\beta = 0 \tag{1-4.13}$$

for any matrix norm β. It can be shown that

$$\lim_{n \to \infty} A^n = 0 \tag{1-4.14}$$

and

$$\lim_{n \to \infty} A^n v = 0 \tag{1-4.15}$$

for all vectors v if and only if

$$\mathbf{S}(A) < 1. \tag{1-4.16}$$

For any nonsingular matrices A and L, we define the L-condition number $\kappa_L(A)$ of the matrix A by

$$\kappa_L(A) \equiv \|A\|_L \|A^{-1}\|_L. \tag{1-4.17}$$

The *spectral condition number* $\kappa(A)$ is obtained for the special case $L = I$, i.e.,

$$\kappa(A) \equiv \kappa_I(A) = \|A\|_2 \|A^{-1}\|_2. \tag{1-4.18}$$

If A is SPD, then by (1-4.10), the spectral condition number of A is given by

$$\kappa(A) = M(A)/m(A). \tag{1-4.19}$$

1.5   PARTITIONED MATRICES

In writing the matrix equation $Au = b$, an ordering of the unknowns (and equations) is implied. For the iterative methods that we consider, this implied ordering usually determines the sequence in which the unknowns are improved in the iterative process. For *block iterative methods*, blocks or groups of unknowns are improved simultaneously. The blocks of unknowns to be improved simultaneously are determined by an imposed partitioning of the coefficient matrix A. Such a partitioning is defined by the integers $n_1, n_2, \ldots, n_q$, where $n_i \geq 1$ for all i and where

$$n_1 + n_2 + \cdots + n_q = N. \tag{1-5.1}$$

Given the set $\{n_i\}_{i=1}^q$, which satisfies (1-5.1), the $q \times q$ partitioned form of the $N \times N$ matrix A is then given by

$$A = \begin{bmatrix} A_{1,1} & A_{1,2} & \cdots & A_{1,q} \\ A_{2,1} & A_{2,2} & \cdots & A_{2,q} \\ \vdots & \vdots & & \vdots \\ A_{q,1} & A_{q,2} & \cdots & A_{q,q} \end{bmatrix}, \tag{1-5.2}$$

where $A_{i,j}$ is an $n_i \times n_j$ submatrix.

If $q = N$, i.e., if $n_i = 1$ for all i, then we say (1-5.2) is a *point partitioning* of A. When $q = 2$, we obtain the special case

$$A = \begin{bmatrix} A_{1,1} & A_{1,2} \\ A_{2,1} & A_{2,2} \end{bmatrix}, \tag{1-5.3}$$

which is called a *red/black partitioning*. A 2×2 partitioning of the coefficient matrix A is required for some of the iterative methods discussed later in Chapters 8 and 9. Examples of red/black partitionings are given in Sections 1.7 and 9.2, and in Chapters 10 and 11.

If $A_{i,j} = 0$ whenever $i \neq j$, then A is called a *block diagonal matrix*; i.e., A has the form

$$A = \begin{bmatrix} A_{1,1} & & & \\ & A_{2,2} & & 0 \\ & & \ddots & \\ 0 & & & A_{q,q} \end{bmatrix}, \qquad (1\text{-}5.4)$$

where the $A_{i,i}$ submatrices are square. In abbreviated form (1-5.4) is written as $A = \mathrm{diag}(A_{1,1}, \ldots, A_{qq})$.

We always assume that the unknown vector u and the known vector b in the matrix equation $Au = b$ are partitioned in a form consistent with A. Thus if A is given by (1-5.2), then u is assumed to be partitioned as

$$u = \begin{bmatrix} U_1 \\ U_2 \\ \vdots \\ U_q \end{bmatrix}, \qquad (1\text{-}5.5)$$

where U_i is an $n_i \times 1$ matrix (column vector). Conversely, a partitioning for A is implied by a given partitioning (1-5.5) for u. Thus a partitioning for the matrix problem $Au = b$ can be given by specifying a partitioning either for A or for u.

As we shall see in Chapter 2, in order to carry out one iteration of a given block iterative process corresponding to the partitioning (1-5.2), it will be necessary to solve subsystems of the form $A_{i,i} U_i = y_i$, where y_i is some known vector. In general, the larger the sizes of the diagonal blocks $A_{i,i}$, the more difficult it will be to solve these subsystems. On the other hand, the larger the blocks, the faster the convergence of a given method. (This is usually, though not always, true.) Any convergence improvement obtained through the use of larger blocks needs to be balanced against the additional work per iteration. In this book, we shall be concerned primarily with those partitionings that result in diagonal submatrices $A_{i,i}$ whose sizes are considerably larger than unity and whose sparseness structures are such that the subsystems $A_{i,i} U_i = y_i$ are considerably easier to solve than the complete system $Au = b$. In subsequent chapters, submatrices $A_{i,i}$ that satisfy these conditions will be called "easily invertible." Examples of such partitionings for some practical problems are given in Chapters 10 and 11.

1.6   THE GENERALIZED DIRICHLET PROBLEM

In this and the next section, we discuss the matrix problem that is obtained from a discretization of the generalized Dirichlet problem. The purpose of this simple example is to illustrate some of the matrix concepts presented in this chapter and to familiarize the reader with the formulation of matrix equations arising from the discretization of boundary value problems. The examples given of red/black partitionings are referred to in later chapters and will be more meaningful then. Thus on the first reading of this chapter, these examples may be skimmed.

Consider the problem of numerically solving the generalized Dirichlet problem

$$\frac{\partial}{\partial x}\left(B\,\frac{\partial \mathcal{U}}{\partial x}\right) + \frac{\partial}{\partial y}\left(C\,\frac{\partial \mathcal{U}}{\partial y}\right) + F\mathcal{U} = G \tag{1-6.1}$$

in a square region R with boundary S. We assume that B and C are positive functions that are twice continuously differentiable with respect to x and y in R. Moreover, it is assumed that $F \leq 0$ and that F and G are continuous in R. Given a function $g(x, y)$ that is defined and continuous on S, we seek a solution $\mathcal{U}(x, y)$ which satisfies (1-6.1) in R and such that

$$\mathcal{U}(x, y) = g(x, y) \tag{1-6.2}$$

for (x, y) on S.

To solve this problem numerically, we impose a uniform square mesh of size $h = L/(M + 1)$ on R (see Fig. 1-6.1). Here L is the side length of R, and we exclude the trivial case $M = 0$. With i and j integers, the set of mesh points Ω_h is defined by (see Fig. 1-6.1) $\Omega_h = \{(x_i, y_j): x_i = ih, y_j = jh$ for $0 \leq i, j \leq (M + 1)\}$. Moreover, we let $R_h = R \cap \Omega_h$ be the set of interior mesh points and $S_h = S \cap \Omega_h$ be the set of boundary points.

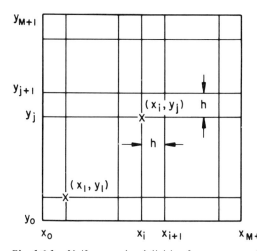

Fig. 1-6.1. Uniform mesh subdivision for a square region.

To discretize the problem defined by (1-6.1)–(1-6.2), we replace (see, e.g., Varga [1962]) the differential equation (1-6.1) at a mesh point (x_i, y_j) in R_h by the finite difference equation

$$h^{-2}\{B(x_i + \tfrac{1}{2}h, y_j)[u_{i+1,j} - u_{i,j}] - B(x_i - \tfrac{1}{2}h, y_j)[u_{i,j} - u_{i-1,j}]$$
$$+ C(x_i, y_j + \tfrac{1}{2}h)[u_{i,j+1} - u_{i,j}] - C(x_i, y_j - \tfrac{1}{2}h)[u_{i,j} - u_{i,j-1}]\}$$
$$+ F(x_i, y_j)u_{i,j}$$
$$= G(x_i, y_j). \tag{1-6.3}$$

Here $u_{i,j}$ is the approximation to $\mathscr{U}(x_i, y_j)$. An equation of the form (1-6.3) holds for each of the M^2 mesh points in R_h.

Equation (1-6.3), which we refer to as the five-point formula, can also be expressed in the form

$$-h^{-2}\{-E_{i,j}u_{i+1,j} - W_{i,j}u_{i-1,j} - N_{i,j}u_{i,j+1} - S_{i,j}u_{i,j-1} + P_{i,j}u_{i,j}\}$$
$$= G(x_i, y_j), \tag{1-6.4}$$

where

$$E_{i,j} = B(x_i + \tfrac{1}{2}h, y_j), \qquad W_{i,j} = B(x_i - \tfrac{1}{2}h, y_j),$$
$$N_{i,j} = C(x_i, y_j + \tfrac{1}{2}h), \qquad S_{i,j} = C(x_i, y_j - \tfrac{1}{2}h),$$

and

$$P_{i,j} = E_{i,j} + W_{i,j} + N_{i,j} + S_{i,j} - h^2 F(x_i, y_j).$$

For (x_i, y_j) on S_h, we require that the $u_{i,j}$ satisfy

$$u_{i,j} = g(x_i, y_j). \tag{1-6.5}$$

If we now multiply both sides of (1-6.4) by $-h^2$ and transfer to the right-hand side those terms involving the known boundary values $u_{i,j}$ on S_h, we obtain a linear system of the form

$$Au = b. \tag{1-6.6}$$

Here A is an $N \times N$ matrix, b a known $N \times 1$ vector, u the $N \times 1$ vector of unknowns, and $N = M^2$ the number of points of R_h.

When a system of linear equations such as (1-6.4) is expressed in matrix form $Au = b$, it is implied that a correspondence between equations and unknowns exists and that an ordering of the unknowns has been chosen. In writing (1-6.6), if the kth unknown in the vector u is $u_{i,j}$, we assume that the kth row of A is obtained from the difference equation (1-6.4), corresponding to the mesh point (x_i, y_j). Independent of the ordering of the unknowns $u_{i,j}$ for u, this correspondence between equations and unknowns determines the

define the *red/black line ordering*, we let U_j be a *red line* of unknowns if
'd and let U_j be a *black line* of unknowns if j *is even*. The red/black line
ing is then any ordering such that every black line of unknowns follows
e red lines. In the next section, we shall use this red/black line ordering
: unknowns to obtain another red/black partitioning for A.

1.7   THE MODEL PROBLEM

r our later discussion and for illustrative purposes, it is convenient to
nt in some detail the following model elliptic differential equation.
ider the discrete approximation of the Poisson equation

$$\frac{\partial^2 \mathscr{U}}{\partial x^2} + \frac{\partial^2 \mathscr{U}}{\partial y^2} = G(x, y), \qquad 0 < x, y < 1, \tag{1-7.1}$$

boundary conditions (1-6.2) on S, the boundary of the unit square R.
the mesh subdivision given by Fig. 1-6.1, the finite difference approxi-
n (1-6.4) to Poisson's equation at a mesh point (x_i, y_j) in R_h may be
:n as

$$4u_{i,j} - u_{i+1,j} - u_{i-1,j} - u_{i,j+1} - u_{i,j-1} = -h^2 G(x_i, y_j). \tag{1-7.2}$$

ow give several point and line partitionings for the corresponding
cient matrix A when $M + 1 = 4$ (see Fig. 1-6.1). For this special case,
are nine unknowns and $h = \frac{1}{4}$. In what follows, we indicate the ordering
e unknown vector u by first numbering the mesh points of the problem
ion region. We then let the kth component u_k of the vector u be the
own corresponding to the mesh point marked k.

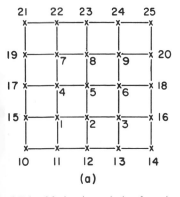
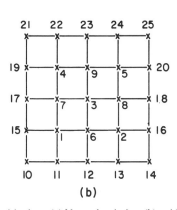

1-7.1.   Mesh point ordering for point partitionings. (a) Natural ordering, (b) red/black
ig.

diagonal elements of A.† To illustrate this, suppose th
unknown $u_{1,1}$. With $A = (a_{l,m})$, this correspondence
where $P_{1,1}$ is defined in (1-6.4). However, if $u_{1,1}$ were
then this correspondence would imply that $a_{2,2} = P$
is a diagonal element of A.

Moreover, with this correspondence between equ
is easy to see that A is symmetric. This follows sin
$u_{i+1,j}$ in the equation corresponding to the unkno
while the coefficient $W_{i+1,j}$ of $u_{i,j}$ in the equation co
$B(x_{i+1} - \frac{1}{2}h, y_j)$, which is equal to $B(x_i + \frac{1}{2}h, y_j)$. Sin
for any other pair of adjacent points (x_i, y_j). It can
difficulty that A also is positive definite and, henc
Varga [1962]).

The above properties of A are independent of t
ordering of unknowns for u. However, the behavior
discussed in Chapter 2 and in later chapters often de
unknowns and on the partitioning imposed. We now
of u for both point and block partitionings, whicl
purposes later.

We consider two orderings for the point parti
natural ordering defined by

$$\{u_{i^*,j^*} \text{ follows } u_{i,j} \text{ if } j^* > j \text{ or if } j^* = j \text{ a}$$

Relative to this ordering for u, the elements of A
uniquely. An example is given in the next section (s

Another ordering for the point partitioning is the
ordering, which is defined as follows: Let the *red unk*
such that $(i + j)$ is *even* and let the *black unknowns* be
$(i + j)$ is *odd*. The red/black ordering is then any ⟨
black unknown follows all the red unknowns. In the r
this ordering of unknowns leads to a red/black part
such that the submatrices $A_{1,1}$ and $A_{2,2}$ are diagon

We now consider natural and red/black orderings
partitioned by lines. Let U_j denote the vector of unk
(see Fig. 1-6.1). The *natural ordering* for this *line par*
vector u is defined by

$$\{U_{j^*} \text{ follows } U_j \text{ if } j^* > j\}.$$

An example is given in the next section (see (1-7.7))

---

† This correspondence usually also ensures that the largest ⟨
diagonal.

Point Partitionings

For point partitionings the natural ordering of unknowns for u is defined by the mesh point numbering given in Fig. 1-7.1a. With the unknown at mesh point k denoted by u_k and with all boundary terms moved to the right-hand side, the system of difference equations for this case can be written as

$$
\begin{bmatrix}
4 & -1 & 0 & -1 & & & & & \\
-1 & 4 & -1 & 0 & -1 & & & 0 & \\
0 & -1 & 4 & 0 & 0 & -1 & & & \\
-1 & 0 & 0 & 4 & -1 & 0 & -1 & & \\
& -1 & 0 & -1 & 4 & -1 & 0 & -1 & \\
& & -1 & 0 & -1 & 4 & 0 & 0 & -1 \\
& & & -1 & 0 & 0 & 4 & -1 & 0 \\
& 0 & & & -1 & 0 & -1 & 4 & -1 \\
& & & & & -1 & 0 & -1 & 4
\end{bmatrix}
\begin{bmatrix}
u_1 \\ u_2 \\ u_3 \\ u_4 \\ u_5 \\ u_6 \\ u_7 \\ u_8 \\ u_9
\end{bmatrix}
=
\begin{bmatrix}
g_{11} + g_{15} - h^2 G_1 \\
g_{12} \qquad\;\; - h^2 G_2 \\
g_{13} + g_{16} - h^2 G_3 \\
g_{17} - h^2 G_4 \\
- h^2 G_5 \\
g_{18} - h^2 G_6 \\
g_{19} + g_{22} - h^2 G_7 \\
g_{23} - h^2 G_8 \\
g_{20} + g_{24} - h^2 G_9
\end{bmatrix}.
$$

$$(1\text{-}7.3)$$

A red/black ordering for the point partitioning is defined by the mesh point numbering in Fig. 1-7.1b. For this ordering for the unknowns, the difference equations (1-7.2) can be expressed in the matrix form

$$
\begin{bmatrix}
4 & & & & & -1 & -1 & 0 & 0 \\
& 4 & & 0 & & -1 & 0 & -1 & 0 \\
& & 4 & & & -1 & -1 & -1 & -1 \\
0 & & & 4 & & 0 & -1 & 0 & -1 \\
& & & & 4 & 0 & 0 & -1 & -1 \\
-1 & -1 & -1 & 0 & 0 & 4 & & & \\
-1 & 0 & -1 & -1 & 0 & & 4 & 0 & \\
0 & -1 & -1 & 0 & -1 & & 0 & 4 & \\
0 & 0 & -1 & -1 & -1 & & & & 4
\end{bmatrix}
\begin{bmatrix}
u_1 \\ u_2 \\ u_3 \\ u_4 \\ u_5 \\ u_6 \\ u_7 \\ u_8 \\ u_9
\end{bmatrix}
=
\begin{bmatrix}
g_{11} + g_{15} - h^2 G_1 \\
g_{13} + g_{16} - h^2 G_2 \\
- h^2 G_3 \\
g_{19} + g_{22} - h^2 G_4 \\
g_{20} + g_{24} - h^2 G_5 \\
g_{12} \qquad - h^2 G_6 \\
g_{17} \qquad - h^2 G_7 \\
g_{18} \qquad - h^2 G_8 \\
g_{23} \qquad - h^2 G_9
\end{bmatrix}.
$$

$$(1\text{-}7.4)$$

The red unknowns are $u_1, u_2, \ldots, u_5$ and the black unknowns are $u_6, \ldots,$ u_9. Note that if u is now partitioned by red unknowns and black unknowns, we obtain the red/black partitioning (1-5.3). Indeed, if we let

$$
u_R =
\begin{bmatrix}
u_1 \\ u_2 \\ \vdots \\ u_5
\end{bmatrix}
\qquad \text{and} \qquad
u_B =
\begin{bmatrix}
u_6 \\ \vdots \\ u_9
\end{bmatrix},
\qquad (1\text{-}7.5)
$$

then from (1-7.4) we obtain

$$\begin{bmatrix} D_R & H \\ H^T & D_B \end{bmatrix} \begin{bmatrix} u_R \\ u_B \end{bmatrix} = \begin{bmatrix} F_R \\ F_B \end{bmatrix},$$
(1-7.6)

where

$$D_R = \begin{bmatrix} 4 & & & \\ & 4 & & 0 \\ & & 4 & \\ & 0 & & 4 \\ & & & & 4 \end{bmatrix},$$

$$D_B = \begin{bmatrix} 4 & & 0 \\ & 4 & \\ & & 4 \\ 0 & & & 4 \end{bmatrix}, \quad \text{and} \quad H = \begin{bmatrix} -1 & -1 & 0 & 0 \\ -1 & 0 & -1 & 0 \\ -1 & -1 & -1 & -1 \\ 0 & -1 & 0 & -1 \\ 0 & 0 & -1 & -1 \end{bmatrix}.$$

Line Partitionings

For line partitionings the natural ordering of unknowns for u can be given, for example, by the mesh point numbering given in Fig. 1-7.2a. With the unknowns on line k denoted by U_k and with all boundary terms moved to the right-hand side, the system of equations (1-7.2) can be written in the partitioned form

$$\begin{bmatrix} A_{1,1} & A_{1,2} & 0 \\ A_{2,1} & A_{2,2} & A_{2,3} \\ 0 & A_{3,2} & A_{3,3} \end{bmatrix} \begin{bmatrix} U_1 \\ U_2 \\ U_3 \end{bmatrix} = \begin{bmatrix} F_1 \\ F_2 \\ F_3 \end{bmatrix},$$
(1-7.7)

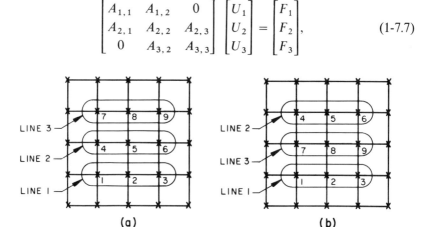

(a)   (b)

Fig. 1-7.2. Mesh point ordering for line partitionings. (a) Natural ordering, (b) red/black ordering.

where the submatrix $A_{k,l}$ gives the couplings of the unknowns from line k to those on line l. Because of matrix symmetry, $A_{k,l} = A_{l,k}^T$. For the mesh point numbering of unknowns† within a line given in Fig. 1-7.2a, we have

$$A_{k,k} = \begin{bmatrix} 4 & -1 & 0 \\ -1 & 4 & -1 \\ 0 & -1 & 4 \end{bmatrix} \quad \text{and} \quad A_{k,k+1} = \begin{bmatrix} -1 & 0 & 0 \\ 0 & -1 & 0 \\ 0 & 0 & -1 \end{bmatrix}. \quad (1\text{-}7.8)$$

A red/black ordering for the line partitioning is defined by the mesh point numbering given in Fig. 1-7.2b. The partitioned matrix resulting from this ordering is

$$\begin{bmatrix} A_{1,1} & 0 & A_{1,3} \\ 0 & A_{2,2} & A_{2,3} \\ A_{3,1} & A_{3,2} & A_{3,3} \end{bmatrix} \begin{bmatrix} U_1 \\ U_2 \\ U_3 \end{bmatrix} = \begin{bmatrix} F_1 \\ F_2 \\ F_3 \end{bmatrix}. \quad (1\text{-}7.9)$$

The red lines of unknowns are U_1 and U_2, while U_3 is the only black line. If we now partition u by red and black lines, i.e., let

$$u_R = \begin{bmatrix} U_1 \\ U_2 \end{bmatrix} \quad \text{and} \quad u_B = U_3, \quad (1\text{-}7.10)$$

then from (7.9) we obtain the red/black partitioning

$$\begin{bmatrix} D_R & H \\ H^T & D_B \end{bmatrix} \begin{bmatrix} u_R \\ u_B \end{bmatrix} = \begin{bmatrix} F_R \\ F_B \end{bmatrix}, \quad (1\text{-}7.11)$$

where

$$D_R = \begin{bmatrix} A_{1,1} & 0 \\ 0 & A_{2,2} \end{bmatrix}, \quad D_B = A_{3,3}, \quad \text{and} \quad H = \begin{bmatrix} A_{1,3} \\ A_{2,3} \end{bmatrix}.$$

Note that here the matrices D_R and D_B are themselves block diagonal matrices. Contrast this with the form of D_R and D_B for the point red/black partitioning (1-7.6). However, in both the point and line cases, D_R and D_B are block diagonal matrices whose diagonal blocks are determined by the diagonal submatrices of the basic point or line partitioning imposed on u.

Iterative methods utilizing the red/black partitioning (discussed in Chapters 8 and 9) require that subsystems of the form $D_R u_R = F_R$ and $D_B u_B = F_B$ be solved for every iteration. The work required to solve these subsystems is reduced significantly when D_R and D_B are block diagonal matrices which are "easily invertible." For some physical problems, careful thought must be given in order to obtain such partitionings. This problem is discussed in more detail in Chapters 9–11.

---

† As we point out later, the ordering of unknowns within a line does not affect the iterative behavior of block methods but can affect the work required per iteration.

2

Background on Basic
Iterative Methods

2.1 INTRODUCTION

In this chapter we give some background material on a class of iterative methods for solving the linear system

$$Au = b, \qquad (2\text{-}1.1)$$

where A is a given real $N \times N$ nonsingular matrix and b is a given $N \times 1$ real column matrix.

All methods considered in this chapter are *linear stationary methods of first degree*. Such methods may be expressed in the form

$$u^{(n+1)} = Gu^{(n)} + k, \qquad n = 0, 1, 2, \ldots, \qquad (2\text{-}1.2)$$

where G is the real $N \times N$ iteration matrix for the method and k is an associated known vector. The method is of first degree since $u^{(n+1)}$ depends explicitly only on $u^{(n)}$ and not on $u^{(n-1)}, \ldots, u^{(0)}$. The method is linear since neither G nor k depends on $u^{(n)}$, and it is stationary since neither G nor k depends on n. In this book we refer to any method of the form (2-1.2) as a *basic iterative method*.

In Section 2.2, we briefly discuss general principles concerning convergence and rates of convergence of basic iterative methods. In Section 2.3, we describe those basic methods which we consider in later chapters. These methods are well known and include the RF (Richardson's) method, the Jacobi method, the Gauss–Seidel method, the successive overrelaxation (SOR) method, and the symmetric successive overrelaxation (SSOR) method. We limit our attention to these basic methods because of the limitations of space and our own experience. However, in Section 2.5 we give a brief introduction to other solution methods which, although useful, will not be considered elsewhere in this book. As in Chapter 1, no proofs are given. It is assumed that the reader already has some familiarity with the use of basic iterative methods, such as that provided by Varga [1962] or Young [1971]. References will be cited only for those results that are not given in either of these basic texts.

2.2   CONVERGENCE AND OTHER PROPERTIES

In this section we discuss convergence and other properties of basic iterative methods that will be used in later portions of the book.

We assume throughout that

$$G = I - Q^{-1}A, \qquad k = Q^{-1}b \tag{2-2.1}$$

for some nonsingular matrix Q. Such a matrix Q is called a *splitting matrix*. The assumptions of (2-2.1) together with the fact that A is nonsingular imply that $\bar{u}$ is a solution to the *related system*

$$(I - G)u = k \tag{2-2.2}$$

if and only if $\bar{u}$ is also the unique solution to (2-1.1), i.e.,

$$\bar{u} = A^{-1}b. \tag{2-2.3}$$

An iterative method (2-1.2) whose related system (2-2.2) has a unique solution $\bar{u}$ which is the same as the solution of (2-1.1) is said to be *completely consistent*.

If $\{u^{(n)}\}$ is the sequence of iterates determined by (2-1.2), then complete consistency implies that (a) if $u^{(n)} = \bar{u}$ for some n, then $u^{(n+1)} = u^{(n+2)} = \cdots = \bar{u}$ and (b) if the sequence $\{u^{(n)}\}$ converges to some vector $\hat{u}$, then $\hat{u} = \bar{u}$.

We always assume that the basic iterative method (2-1.2) is *completely consistent* since this property seems essential for any reasonable method. Another property of basic iterative methods, which we do not always assume, is that of *convergence*. The method (2-1.2) is said to be convergent if for any $u^{(0)}$ the sequence $u^{(1)}, u^{(2)}, \ldots$ defined by (2-1.2) converges to $\bar{u}$. A necessary and sufficient condition for convergence is that

$$\mathbf{S}(G) < 1. \tag{2-2.4}$$

To measure the rapidity of convergence of the linear stationary iterative method (2-1.2), let the error vector $\varepsilon^{(n)}$ be defined by

$$\varepsilon^{(n)} \equiv u^{(n)} - \bar{u}. \tag{2-2.5}$$

Using (2-1.2) together with the fact that $\bar{u}$ also satisfies the related equation (2-2.2), we have

$$\varepsilon^{(n)} = G\varepsilon^{(n-1)} = \cdots = G^n\varepsilon^{(0)}. \tag{2-2.6}$$

Therefore, for any vector norm β and corresponding matrix norm β, we have by (1-4.7) that

$$\|\varepsilon^{(n)}\|_\beta \leq \|G^n\|_\beta\|\varepsilon^{(0)}\|_\beta. \tag{2-2.7}$$

Thus $\|G^n\|_\beta$ gives a measure by which the norm of the error has been reduced after n iterations. We define the *average rate of convergence* of (2-1.2) by†

$$R_n(G) \equiv -n^{-1}\log\|G^n\|_\beta. \tag{2-2.8}$$

It can be shown that if $\mathbf{S}(G) < 1$, then

$$\lim_{n\to\infty}(\|G^n\|_\beta)^{1/n} = \mathbf{S}(G). \tag{2-2.9}$$

Hence we are led to define the *asymptotic rate of convergence* by

$$R_\infty(G) \equiv \lim_{n\to\infty}R_n(G) = -\log\mathbf{S}(G). \tag{2-2.10}$$

We remark that whereas $R_n(G)$ depends on the norm β which is used, $R_\infty(G)$ is independent of β. Frequently we shall refer to $R_\infty(G)$ as the *rate of convergence*.

If $\mathbf{S}(G) < 1$, a rough approximation to the number of iterations n needed to reduce the norm of the initial error vector by a factor ζ can be given by

$$n \doteqdot -(\log\zeta)/R_\infty(G). \tag{2-2.11}$$

The estimate given by (2-2.11) is often much too low if the matrix G has principal vectors of grade two or higher (see Section 1.3) associated with one or more eigenvalues of G. The use of the approximate formula

$$n \doteqdot -(\log\zeta)/R_n(G), \tag{2-2.12}$$

where $R_n(G)$ is the average rate of convergence given by (2-2.8), would give much more accurate results. Unfortunately, however, $R_n(G)$ is seldom available. In any case, the reciprocal of either $R_\infty(G)$ or $R_n(G)$ can be used as a measure of the number of iterations required to reduce the error vector by a factor of $\zeta = e^{-1}$.

---

† In this book, $\log x$ denotes the logarithm of x to the base e.

For most of the acceleration methods considered in this book, it is not necessary that the basic method (2-1.2) be convergent. Normally, it is sufficient that the method be "symmetrizable" in the following sense.

Definition 2-2.1. The iterative method (2-1.2) is *symmetrizable* if for some nonsingular matrix W the matrix $W(I - G)W^{-1}$ is SPD. Such a matrix W is called a *symmetrization matrix*. An iterative method that is not symmetrizable is *nonsymmetrizable*.

With this definition, the following results follow easily from the material given in Section 1.3.

Theorem 2-2.1. If the iterative method (2-1.2) is symmetrizable, then (a) the eigenvalues of G are real, (b) the algebraically largest eigenvalue $M(G)$ of G is less than unity, and (c) the set of eigenvectors for G includes a basis for the associated vector space.

As we shall see in later chapters, properties (a)–(c) of the matrix G given above turn out to suffice for the effective use of polynomial acceleration methods. The difficulties encountered when one or more of these properties is not valid will be discussed briefly in Chapters 6 and 12. For some of the acceleration procedures given in later chapters, a symmetrization matrix must be available for computational purposes.

Many iterative methods are symmetrizable. For example, the basic method (2-1.2) is symmetrizable whenever A and the splitting matrix Q in (2-2.1) are SPD. In such a case, $A^{1/2}$ and $Q^{1/2}$ are symmetrization matrices. Moreover, any matrix W such that $Q = W^T W$ is also a symmetrization matrix. The W obtained from the factorization $Q = W^T W$ is usually the most computationally convenient choice. In the next section, symmetrization matrices W are given for some of the basic iterative methods discussed there. In Section 5.6, computational aspects in the use of symmetrization matrices are discussed.

We remark that the symmetrization property need not imply convergence. If the iterative method (2-1.2) is symmetrizable, then the eigenvalues of G are less than unity but not necessarily less than unity in absolute value. Hence, the convergence condition (2-2.4) need not be satisfied. However, as we now describe, there always exists a so-called *extrapolated method* based on (2-1.2) which is convergent whenever the basic method is symmetrizable.

The extrapolated method applied to (2-1.2) is defined by

$$u^{(n+1)} = \gamma(Gu^{(n)} + k) + (1 - \gamma)u^{(n)} = G_{[\gamma]}u^{(n)} + \gamma k, \qquad (2\text{-}2.13)$$

where

$$G_{[\gamma]} \equiv \gamma G + (1 - \gamma)I. \qquad (2\text{-}2.14)$$

Here γ is a parameter that is often referred to as the "extrapolation factor." If the iterative method is symmetrizable, then the optimum value $\bar{\gamma}$ for γ, in the sense of minimizing $\mathbf{S}(G_{[\gamma]})$, is given by

$$\bar{\gamma} = 2/(2 - M(G) - m(G)), \qquad (2\text{-}2.15)$$

where $m(G)$ and $M(G)$ are the smallest and largest eigenvalues of G, respectively. Moreover, it easily follows that

$$\mathbf{S}(G_{[\bar{\gamma}]}) = (M(G) - m(G))/(2 - M(G) - m(G)) < 1. \qquad (2\text{-}2.16)$$

Thus the *optimum extrapolated method*, which we define by

$$u^{(n+1)} = G_{[\bar{\gamma}]}u^{(n)} + \bar{\gamma}k, \qquad (2\text{-}2.17)$$

is convergent.

2.3   EXAMPLES OF BASIC ITERATIVE METHODS

In this section we describe five well-known basic iterative methods which will be used in subsequent chapters. The methods considered are the RF method, the Jacobi method, the Gauss–Seidel method, the successive overrelaxation (SOR) method, and the symmetric successive overrelaxation (SSOR) method. Primary attention in later chapters is given to the Jacobi and SOR methods. The Jacobi method is used primarily in discussions concerning computational aspects and costs. The SOR method will be studied in detail in Chapter 9.

We assume throughout this section that the matrix A of the system (2-1.1) *is symmetric and positive definite (SPD).* For each method, we describe a computational procedure for carrying out the iterations and discuss convergence properties. Specific rates of convergence are given for the model problem. We also discuss whether or not a particular method is symmetrizable.

The RF Method

The RF method is based on a variant of the method of Richardson [1910] and is defined by

$$u^{(n+1)} = (I - A)u^{(n)} + b. \qquad (2\text{-}3.1)$$

The iteration matrix G for the RF method is simply $G = I - A$. The associated splitting matrix Q is the identity matrix. Since A is SPD, it is clear that the RF method is symmetrizable; for example, $W = I$ is a symmetrization matrix.

Each eigenvalue μ of the iteration matrix $G = I - A$ for the RF method is equal to $1 - v$ for some eigenvalue v of A. Thus we have

$$\mathbf{S}(I - A) = \max(|1 - m(A)|, |1 - M(A)|),$$

where $m(A)$ and $M(A)$ are defined by (1-3.4). From this it follows that the RF method is convergent if and only if $M(A)$ satisfies

$$M(A) < 2. \tag{2-3.2}$$

The optimum extrapolation method (2-2.17) based on the RF method, however, is always convergent. This extrapolation method, which we denote by RF–OE, can be expressed in the form

$$u^{(n+1)} = (I - \bar{\gamma}A)u^{(n)} + \bar{\gamma}b, \tag{2-3.3}$$

where

$$\bar{\gamma} = \frac{2}{2 - M(G) - m(G)} = \frac{2}{M(A) + m(A)}. \tag{2-3.4}$$

The spectral radius of the corresponding iteration matrix $G_{[\bar{\gamma}]} = I - \bar{\gamma}A$ is

$$\mathbf{S}(I - \bar{\gamma}A) = \frac{M(A) - m(A)}{M(A) + m(A)} = \frac{\kappa(A) - 1}{\kappa(A) + 1}, \tag{2-3.5}$$

where $\kappa(A)$ is the spectral condition number (1-4.18) of the matrix A. Therefore, the rate of convergence for the RF–OE method can be given by

$$R_\infty(I - \bar{\gamma}A) = -\log\frac{\kappa(A) - 1}{\kappa(A) + 1} \sim \frac{2}{\kappa(A)} \qquad \text{as} \quad \kappa(A) \to \infty. \tag{2-3.6}$$

For the model problem of Section 1.7, it can easily be shown that

$$m(A) = 8 \sin^2 \tfrac{1}{2}\pi h, \qquad M(A) = 8 \cos^2 \tfrac{1}{2}\pi h. \tag{2-3.7}$$

For this case, using (2-3.5), we have that

$$R_\infty(I - \bar{\gamma}A) = -\log \mathbf{S}(I - \bar{\gamma}A) = -\log\frac{\cot^2 \tfrac{1}{2}\pi h - 1}{\cot^2 \tfrac{1}{2}\pi h + 1}$$

and, after some manipulation, that

$$R_\infty(I - \bar{\gamma}A) \sim \tfrac{1}{2}\pi^2 h^2, \qquad h \to 0. \tag{2-3.8}$$

The Jacobi Method

We assume for the remainder of this section that the system (2-1.1) is partitioned in the form

$$\begin{bmatrix} A_{1,1} & A_{1,2} & \cdots & A_{1,q} \\ A_{2,1} & A_{2,2} & \cdots & A_{2,q} \\ \vdots & \vdots & \cdots & \vdots \\ A_{q,1} & A_{q,2} & \cdots & A_{q,q} \end{bmatrix} \begin{bmatrix} U_1 \\ U_2 \\ \vdots \\ U_q \end{bmatrix} = \begin{bmatrix} F_1 \\ F_2 \\ \vdots \\ F_q \end{bmatrix}, \tag{2-3.9}$$

where $A_{i,j}$ is an $n_i \times n_j$ submatrix and $n_1 + n_2 + \cdots + n_q = N$. Here the U_i and F_i represent subvectors of order n_i. Since A is SPD, each diagonal submatrix $A_{i,i}$ is also SPD. For use throughout this section, we express the matrix A as the matrix sum

$$A = D - C_L - C_U, \tag{2-3.10}$$

where

$$
D = \begin{bmatrix} A_{1,1} & & & \huge{0} \\ & A_{2,2} & & \\ & & \ddots & \\ \huge{0} & & & A_{q,q} \end{bmatrix}, \quad
C_U = - \begin{bmatrix} 0 & A_{1,2} & A_{1,3} & \cdots & A_{1,q} \\ 0 & 0 & A_{2,3} & \cdots & A_{2,q} \\ 0 & 0 & 0 & \cdots & A_{3,q} \\ \vdots & \vdots & \vdots & \cdots & \vdots \\ 0 & 0 & 0 & \cdots & 0 \end{bmatrix},
$$

$$
C_L = - \begin{bmatrix} 0 & 0 & 0 & \cdots & 0 \\ A_{2,1} & 0 & 0 & \cdots & 0 \\ A_{3,1} & A_{3,2} & 0 & \cdots & 0 \\ \vdots & \vdots & \vdots & \cdots & \vdots \\ A_{q,1} & A_{q,2} & A_{q,3} & \cdots & 0 \end{bmatrix}. \tag{2-3.11}
$$

As we shall see, the Jacobi, Gauss–Seidel, SOR, and SSOR iteration methods can be defined uniquely in terms of these D, C_L, and C_U matrices.

Relative to the partitioning (2-3.9), the Jacobi method is defined by

$$A_{i,i} U_i^{(n+1)} = - \sum_{\substack{j=1 \\ j \neq i}}^{q} A_{i,j} U_j^{(n)} + F_i, \quad i = 1, 2, \ldots, q. \tag{2-3.12}$$

If the $A_{i,i}$ are 1×1 matrices, the method is sometimes referred to as the *point Jacobi method*. Otherwise, it will be referred to as the *block Jacobi method*. In order for the block Jacobi method to be practical, it is essential that the subsystems

$$A_{i,i} U_i^{(n+1)} = y_i \tag{2-3.13}$$

can be solved easily for $U_i^{(n+1)}$, given any y_i. Frequently, each $A_{i,i}$ will be a tridiagonal matrix or a matrix with small bandwidth for which special direct methods can be used. Later in this section, we give a systematic procedure which may be used to solve directly a symmetric and positive definite tridiagonal matrix.

In the matrix notation of (2-3.11), the Jacobi method may be expressed as

$$u^{(n+1)} = Bu^{(n)} + k, \tag{2-3.14}$$

where B is the *Jacobi iteration matrix* defined by

$$B \equiv D^{-1}(C_L + C_U) = I - D^{-1}A \tag{2-3.15}$$

and where

$$u^{(n)} \equiv \begin{bmatrix} U_1^{(n)} \\ U_2^{(n)} \\ \vdots \\ U_q^{(n)} \end{bmatrix}, \qquad k \equiv D^{-1} \begin{bmatrix} F_1 \\ F_2 \\ \vdots \\ F_q \end{bmatrix}. \tag{2-3.16}$$

Since A is SPD, it follows from (2-3.11) and (2-3.15) that D is also SPD and that the Jacobi method is symmetrizable with $W = D^{1/2}$. Other choices for the symmetrization matrix include $W = A^{1/2}$ or $W = S$, where S is any matrix such that $S^T S = D$.

The Jacobi method is convergent if and only if $\mathbf{S}(B) < 1$. It can be shown that $\mathbf{S}(B) < 1$, for example, if the SPD matrix A has *property $\mathcal{A}$* or if A is *irreducible* with *weak diagonal dominance*. We refer to Young [1971] for a definition of these properties. Property $\mathcal{A}$ is also defined and discussed in Chapter 9. We remark that if the matrix A of (2-3.9) has property $\mathcal{A}$, then the eigenvalues of B satisfy

$$m(B) = -M(B). \tag{2-3.17}$$

As for the RF method, the optimum extrapolation method (2-2.17) based on the Jacobi method, which we denote by J–OE, is always convergent. The J–OE method can be expressed in the form

$$u^{(n+1)} = \bar{\gamma}(Bu^{(n)} + k - u^{(n)}) + u^{(n)}, \tag{2-3.18}$$

where $\bar{\gamma}$ is given by (2-2.15) with $G = B$. When the eigenvalues of B satisfy (2-3.17), $\bar{\gamma} = 1$. For this case, the J–OE method reduces to the Jacobi method without extrapolation.

For the model problem of Section 1.7, it can be shown that

$$\mathbf{S}(B) = M(B) = -m(B) = \begin{cases} \cos \pi h & \text{for point Jacobi,} \\[2mm] \dfrac{\cos \pi h}{2 - \cos \pi h} & \text{for line Jacobi.} \end{cases} \tag{2-3.19}$$

The values given in (2-3.19) are valid for either the natural ordering or the red/black ordering. From (2-2.10) and (2-3.19), it can be shown that

$$R_\infty(B) \sim \begin{cases} \frac{1}{2}\pi^2 h^2 & \text{for point Jacobi,} \\ \pi^2 h^2 & \text{for line Jacobi,} \end{cases} \qquad h \to 0. \tag{2-3.20}$$

Since $M(B) = -m(B)$ for this problem, the J–OE method here reduces to the unextrapolated Jacobi method.

We now describe a direct solution procedure for solving the subsystem (2-3.13) under the assumption that $A_{i,i}$ is a $p \times p$ tridiagonal matrix of the form

$$A_{i,i} = \begin{bmatrix} b_1 & c_1 & & & \\ c_1 & b_2 & c_2 & & 0 \\ & \ddots & \ddots & \ddots & \\ 0 & & & & c_{p-1} \\ & & & c_{p-1} & b_p \end{bmatrix}. \tag{2-3.21}$$

The procedure we describe is based on a Cholesky decomposition (or factorization) of the matrix $A_{i,i}$. Since $A_{i,i}$ is SPD, there exists an upper bidiagonal matrix S_i of the form

$$S_i = \begin{bmatrix} d_1 & f_1 & & & \\ 0 & d_2 & f_2 & & 0 \\ & & \ddots & \ddots & \\ 0 & & & & f_{p-1} \\ & & 0 & & d_p \end{bmatrix} \tag{2-3.22}$$

such that

$$A_{i,i} = S_i^T S_i. \tag{2-3.23}$$

The d_j and f_j for $j = 1, \ldots, p$ are given by

$$d_1 = \sqrt{b_1}, \qquad f_1 = c_1/d_1,$$
$$d_j = \sqrt{b_j - f_{j-1}^2}, \quad j = 2, 3, \ldots, p, \qquad f_j = c_j/d_j, \quad j = 2, 3, \ldots, p - 1. \tag{2-3.24}$$

From (2-3.13) and (2-3.23), we have that $S_i^T S_i U_i^{(n+1)} = y_i$. To determine $U_i^{(n+1)}$, we first solve

$$S_i^T z = y_i \tag{2-3.25}$$

for z by a simple forward substitution and then solve

$$S_i U_i^{(n+1)} = z \tag{2-3.26}$$

for $U_i^{(n+1)}$ by a simple back substitution.

Note that the factorization (2-3.23) is independent of the source vector y_i. Thus the coefficients of the factorization matrix S_i need be computed only once. In the case of the discrete analog of the generalized Dirichlet problem, the use of line iteration requires roughly the same amount of work per iteration as would the corresponding point method. Solution procedures for the subsystems (2-3.13) are discussed further in Sections 5.6 and 8.5.

The Gauss–Seidel Method

The Gauss–Seidel method applied to the partitioned linear system (2-3.9) is defined by

$$A_{i,i} U_i^{(n+1)} = -\sum_{j=1}^{i-1} A_{i,j} U_j^{(n+1)} - \sum_{j=i+1}^{q} A_{i,j} U_j^{(n)} + F_i, \quad i = 1, 2, \dots, q.$$

(2-3.27)

As in the case of the Jacobi method, each step of a Gauss–Seidel iteration requires the solution of a subsystem of the form (2-3.13).

Using the matrix notation of (2-3.11), we can write (2-3.27) as $(D - C_L)u^{(n+1)} = C_U u^{(n)} + b$. Thus the Gauss–Seidel method may be expressed in the matrix form

$$u^{(n+1)} = \mathcal{L} u^{(n)} + k,$$

(2-3.28)

where

$$\mathcal{L} \equiv (I - L)^{-1} U, \qquad k \equiv (I - L)^{-1} D^{-1} b$$

(2-3.29)

and where

$$L \equiv D^{-1} C_L, \qquad U \equiv D^{-1} C_U.$$

(2-3.30)

The matrix $\mathcal{L}$ is called the *Gauss–Seidel iteration matrix*.

The splitting matrix for the Gauss–Seidel method is $(D - C_L)$, which is not SPD. Moreover, the Gauss–Seidel method is not in general symmetrizable.

Since A and D are SPD, it can be shown that the Gauss–Seidel method always converges. In general, the eigenvalues of $\mathcal{L}$, though less than unity in modulus, may be complex, and the set of eigenvectors for $\mathcal{L}$ may not include a basis for the associated vector space. When this is the case, the extrapolation method of (2-2.13) is not applicable.

In certain cases, however, extrapolation can be used. For example, the eigenvalues of $\mathcal{L}$ are real, nonnegative, and less than unity for the problems of Sections 1.6 and 1.7 when the natural ordering or the red/black ordering of the unknowns is used. For these problems, an improvement in the asymptotic convergence rate would be achieved by using extrapolation. We remark that for problems with the natural ordering, the average rate of convergence of the extrapolated Gauss–Seidel method can be significantly less than its asymptotic rate. This is caused by the fact (see, e.g., Miles *et al.* [1964]) that the matrix $\mathcal{L}$ for such problems can have principal vectors of grade two or greater associated with zero eigenvalues of $\mathcal{L}$. (How the convergence of an iterative procedure is affected by the presence of principal vectors of grade greater than one is discussed later in Sections 6.8 and 9.4.) The above remark does not apply to problems with a red/black ordering since, for this case, the set of eigenvectors for $\mathscr{L}$ includes a basis for the associated vector space. See, e.g., Tee [1963].

For the model problem, it can be shown for either the natural ordering or the red/black ordering that

$$\mathbf{S}(\mathscr{L}) = M(\mathscr{L}) = M(B)^2 = \begin{cases} \cos^2 \pi h & \text{for point Gauss–Seidel,} \\ \left(\dfrac{\cos \pi h}{2 - \cos \pi h}\right)^2 & \text{for line Gauss–Seidel,} \end{cases}$$

$$(2\text{-}3.31)$$

and that $m(\mathscr{L}) = 0$. Thus from (2-3.31) we obtain

$$R_\infty(\mathscr{L}) \sim \begin{cases} \pi^2 h^2 & \text{for point Gauss–Seidel,} \\ 2\pi^2 h^2 & \text{for line Gauss–Seidel,} \end{cases} \quad h \to 0. \quad (2\text{-}3.32)$$

Letting $\mathscr{L}_{[\bar{\gamma}]}$ denote the optimum extrapolated iteration matrix (2-2.17) based on the Gauss–Seidel method, we have by (2-2.15) and (2-2.16) that $\bar{\gamma} = 2/(2 - M(B)^2)$ and that $\mathbf{S}(\mathscr{L}_{[\bar{\gamma}]}) = M(B)^2/(2 - M(B)^2)$. Combining this with (2-3.19), we can easily calculate that

$$R_\infty(\mathscr{L}_{[\bar{\gamma}]}) \sim \begin{cases} 2\pi^2 h^2 & \text{for point method,} \\ 4\pi^2 h^2 & \text{for line method,} \end{cases} \quad h \to 0. \quad (2\text{-}3.33)$$

The Successive Overrelaxation (SOR) Method

Relative to the partitioning (2-3.9), the SOR method is defined by

$$A_{i,i} U_i^{(n+1)} = \omega \left\{ - \sum_{j=1}^{i-1} A_{i,j} U_j^{(n+1)} - \sum_{j=i+1}^{q} A_{i,j} U_j^{(n)} + F_i \right\}$$

$$+ (1 - \omega) A_{i,i} U_i^{(n)}, \qquad i = 1, 2, \ldots, q. \qquad (2\text{-}3.34)$$

Here ω is a real number known as the relaxation factor. With $\omega = 1$, the SOR method reduces to the Gauss–Seidel method. If $\omega > 1$ or $\omega < 1$, we have *overrelaxation* or *underrelaxation*, respectively. We shall be concerned only with overrelaxation. As before, each step requires the solution of a subsystem of the form (2-3.13).

In the matrix notation of (2-3.11), the SOR method (2-3.34) becomes $Du^{(n+1)} = \omega(C_L u^{(n+1)} + C_U u^{(n)} + b) + (1 - \omega) Du^{(n)}$. This can be rewritten in the form

$$u^{(n+1)} = \mathscr{L}_\omega u^{(n)} + k_\omega^{(F)}, \qquad (2\text{-}3.35)$$

where

$$\mathcal{L}_\omega \equiv (I - \omega L)^{-1}(\omega U + (1 - \omega)I) \quad \text{and} \quad k_\omega^{(F)} \equiv (I - \omega L)^{-1}\omega D^{-1}b.$$

(2-3.36)

Here L and U are given by (2-3.30). The matrix $\mathcal{L}_\omega$ is called the *SOR iteration matrix*.

The splitting matrix for the SOR method is $(\omega^{-1}D - C_L)$, which, as in the case of the Gauss–Seidel method, is not SPD. For $\omega > 1$, the SOR method is not symmetrizable. Moreover, for the overrelaxation case, the matrix $\mathcal{L}_\omega$ normally has some eigenvalues that are complex. Thus extrapolation based on the SOR method is not applicable for $\omega > 1$.

Since A and D are SPD, it can be shown that the SOR method converges for any ω such that $0 < \omega < 2$. Moreover, it is often possible to choose ω so that the SOR method converges rapidly; much more rapidly than the Jacobi method or the Gauss–Seidel method, for example. Normally, such an "optimum" value for ω can be prescribed if the coefficient matrix A, relative to the partitioning imposed, has property $\mathcal{A}$ and is consistently ordered. These terms are defined later in Chapter 9, where a more detailed summary of the theoretical properties of the SOR method is given. Here we note only that any block tridiagonal matrix A or any matrix A partitioned into a red/black form has property $\mathcal{A}$ and is consistently ordered.

In the case of the model problem, the optimum value of ω, which we denote by ω_b, is given by

$$\omega_b = \frac{2}{1 + \sqrt{1 - M(B)^2}} = \begin{cases} \dfrac{2}{1 + \sin \pi h} & \text{for point SOR,} \\[2mm] \dfrac{2(1 + 2\sin^2 \pi h/2)}{(1 + \sqrt{2}\sin \pi h/2)^2} & \text{for line SOR.} \end{cases}$$

(2-3.37)

The corresponding spectral radius of $\mathcal{L}_{\omega_b}$ is equal to $(\omega_b - 1)$ and can be expressed in the form

$$\mathbf{S}(\mathcal{L}_{\omega_b}) = \frac{1 - \sqrt{1 - M(B)^2}}{1 + \sqrt{1 - M(B)^2}} = \begin{cases} \dfrac{1 - \sin \pi h}{1 + \sin \pi h} & \text{for point SOR,} \\[2mm] \left(\dfrac{1 - \sqrt{2}\sin \pi h/2}{1 + \sqrt{2}\sin \pi h/2}\right)^2 & \text{for line SOR.} \end{cases}$$

(2-3.38)

From (2-3.38), it can be shown that the asymptotic rates of convergence for the SOR method satisfy

$$R_\infty(\mathcal{L}_{\omega_b}) \sim \begin{cases} 2\pi h & \text{for point SOR,} \\ 2\pi h \sqrt{2} & \text{for line SOR,} \end{cases} \quad h \to 0. \quad (2\text{-}3.39)$$

The Symmetric SOR (SSOR) Method

Relative to the partitioning (2-3.9), the SSOR method is defined by

$$A_{i,i} U_i^{(n+1/2)} = \omega \left\{ -\sum_{j=1}^{i-1} A_{i,j} U_j^{(n+1/2)} - \sum_{j=i+1}^{q} A_{i,j} U_j^{(n)} + F_i \right\}$$
$$+ (1 - \omega) A_{i,i} U_i^{(n)}, \quad i = 1, 2, \ldots, q, \quad (2\text{-}3.40)$$

and

$$A_{i,i} U_i^{(n+1)} = \omega \left\{ -\sum_{j=1}^{i-1} A_{i,j} U_j^{(n+1/2)} - \sum_{j=i+1}^{q} A_{i,j} U_j^{(n+1)} + F_i \right\}$$
$$+ (1 - \omega) A_{i,i} U_i^{(n+1/2)}, \quad i = q, q - 1, \ldots, 1. \quad (2\text{-}3.41)$$

Here one first successively computes $U_1^{(n+1/2)}, U_2^{(n+1/2)}, \ldots, U_q^{(n+1/2)}$, using the SOR method (2-3.40). Then one successively computes $U_q^{(n+1)}, U_{q-1}^{(n+1)}, \ldots,$ $U_1^{(n+1)}$, using the backward SOR method (2-3.41). As in the case of the SOR method, each step requires solutions to subsystems of the form (2-3.13).

Using the notation of (2-3.11), we may express the SSOR method in the matrix form

$$Du^{(n+1/2)} = \omega(C_L u^{(n+1/2)} + C_U u^{(n)} + b) + (1 - \omega) Du^{(n)},$$
$$Du^{(n+1)} = \omega(C_L u^{(n+1/2)} + C_U u^{(n+1)} + b) + (1 - \omega) Du^{(n+1/2)} \quad (2\text{-}3.42)$$

or equivalently in the form

$$u^{(n+1/2)} = \mathcal{L}_\omega u^{(n)} + k_\omega^{(F)}, \quad (2\text{-}3.43)$$
$$u^{(n+1)} = \mathcal{U}_\omega u^{(n+1/2)} + k_\omega^{(B)}, \quad (2\text{-}3.44)$$

where $\mathcal{L}_\omega$ and $k_\omega^{(F)}$ are given by (2-3.36) and where

$$\mathcal{U}_\omega \equiv (I - \omega U)^{-1}(\omega L + (1 - \omega)I),$$
$$k_\omega^{(B)} \equiv (I - \omega U)^{-1} \omega D^{-1} b. \quad (2\text{-}3.45)$$

Combining (2-3.43) and (2-3.44), we have

$$u^{(n+1)} = \mathcal{S}_\omega u^{(n)} + k_\omega, \quad (2\text{-}3.46)$$

where

$$\mathcal{S}_\omega \equiv \mathcal{U}_\omega \mathcal{L}_\omega,$$

$$k_\omega \equiv \omega(2 - \omega)(I - \omega U)^{-1}(I - \omega L)^{-1}D^{-1}b. \qquad (2\text{-}3.47)$$

One can verify that the splitting matrix for the SSOR method is

$$Q = \frac{\omega}{2 - \omega} \left(\frac{1}{\omega}D - C_L\right)D^{-1}\left(\frac{1}{\omega}D - C_U\right). \qquad (2\text{-}3.48)$$

Moreover, since $(\omega^{-1}D - C_L)^1 = \omega^{-1}D - C_U$, it follows that Q is SPD and hence that the SSOR method is symmetrizable. Choices for the symmetrization matrix include $W = A^{1/2}$, $W = S^{-1}(\omega^{-1}D - C_U)$, where S is any matrix such that $S^TS = D$, or

$$W = D^{-1/2}(\omega^{-1}D - C_U). \qquad (2\text{-}3.49)$$

Since the SSOR method is symmetrizable, the extrapolation method (2-2.13) may be applied to the SSOR method.

Since A is SPD, it can be shown that the SSOR method converges for any ω in the interval $0 < \omega < 2$. Further, since

$$\mathcal{S}_\omega = \left(\frac{1}{\omega}D - C_U\right)^{-1}D\left(\frac{1}{\omega}D - C_L\right)^{-1}\left(\frac{\omega - 1}{\omega}D - C_L\right)D^{-1}$$

$$\times \left(\frac{\omega - 1}{\omega}D - C_U\right) \qquad (2\text{-}3.50)$$

is similar to the symmetric and nonnegative definite matrix

$$\left[D^{1/2}\left(\frac{1}{\omega}D - C_L\right)^{-1}\left(\frac{\omega - 1}{\omega}D - C_L\right)D^{-1/2}\right]$$

$$\times \left[D^{1/2}\left(\frac{1}{\omega}D - C_L\right)^{-1}\left(\frac{\omega - 1}{\omega}D - C_L\right)D^{-1/2}\right]^T,$$

it follows that the eigenvalues of $\mathcal{S}_\omega$, in addition to being real, are also nonnegative. Because of this, it can easily be shown that the optimum extrapolated SSOR method converges twice as fast as the ordinary SSOR method.

The rate of convergence of the SSOR method is relatively insensitive to the exact choice of ω so that a precise optimum value of ω is not crucial. If the spectral radius of the matrix LU satisfies

$$\mathbf{S}(LU) \leq \tfrac{1}{4}, \qquad (2\text{-}3.51)$$

then a good value of ω is given by

$$\omega = \frac{2}{1 + \sqrt{2(1 - M(B))}}. \qquad (2\text{-}3.52)$$

Here L and U are given by (2-3.30) and B is the Jacobi iteration matrix (2-3.15). With this choice of ω, the spectral radius of $\mathscr{S}_\omega$ satisfies

$$\mathbf{S}(\mathscr{S}_\omega) \le \left(1 - \sqrt{\frac{1 - M(B)}{2}}\right) \Bigg/ \left(1 + \sqrt{\frac{1 - M(B)}{2}}\right). \qquad (2\text{-}3.53)$$

In order for the SSOR method to be effective, $\mathbf{S}(LU)$ should either be less than $\frac{1}{4}$ or, at least, only slightly greater than $\frac{1}{4}$. This condition need not be satisfied for matrix problems resulting from discretizations of some boundary-value problems involving discontinuous coefficients or some problems involving nonuniform mesh subdivisions of the geometric domain. See, for example, Habetler and Wachspress [1961] and Benokraitis [1974]. More details concerning the SSOR method can be found in papers by Sheldon [1955], Ehrlich [1963, 1964], Young [1971, 1972, 1977], Hayes and Young [1977], and Axelsson [1972, 1974].

We now consider the use of the SSOR method in solving the model problem of Section 1.7. If the red/black ordering of the unknowns is used,† then the optimum value of ω, in the sense of minimizing $\mathbf{S}(\mathscr{S}_\omega)$, is unity. For this choice of relaxation factor, the SSOR method reduces to the forward and backward Gauss–Seidel method. Thus any advantage in using overrelaxation is lost. However, if the natural ordering is used, then the spectral radius of $\mathscr{S}_\omega$ with optimum ω is often substantially less than $\mathbf{S}(\mathscr{L}_1)$.

If the natural ordering is used, it can be shown that the "SSOR condition" (2-3.51) holds for the model problem. Using the ω given by (2-3.52), we have by (2-3.19) and (2-3.53) that

$$\mathbf{S}(\mathscr{S}_\omega) \le \begin{cases} (1 - \sin \tfrac{1}{2}\pi h)/(1 + \sin \tfrac{1}{2}\pi h) & \text{for point SSOR,} \\[2ex] \left(1 - \sqrt{\dfrac{1 - \cos \pi h}{2 - \cos \pi h}}\right)^2 \Bigg/ \left(\dfrac{1}{2 - \cos \pi h}\right) & \text{for line SSOR.} \end{cases}$$

$$(2\text{-}3.54)$$

From this it can be shown that the corresponding rates of convergence for the SSOR method satisfy

$$R_\infty(\mathscr{S}_\omega) \sim \begin{cases} \ge \pi h & \text{for point SSOR,} \\[2ex] \ge \sqrt{2}\,\pi h & \text{for line SSOR,} \end{cases} \qquad h \to 0. \quad (2\text{-}3.55)$$

The rates of convergence for the optimum extrapolated point and line SSOR methods would be twice that of the unextrapolated point and line methods.

---

† In this case, inequality (2-3.51) usually is not satisfied since here $\mathbf{S}(LU) = \mathbf{S}(\mathscr{L}_1)$, which is normally close to unity.

2.4 COMPARISON OF BASIC METHODS

We present in Table 2-4.1 the asymptotic convergence rates of the basic methods given in the previous section when applied to the model problem. The convergence rates given are approximations for small h. Unless noted otherwise, the values given are valid for either the natural ordering or the

TABLE 2-4.1

Approximate Asymptotic Convergence Rates for the Model Problem of Section 1.7

	Basic method				
	RF	Jacobi	Gauss–Seidel	SOR	SSOR
Unextrapolated					
point	—	$\frac{1}{2}\pi^2 h^2$	$\pi^2 h^2$	$2\pi h$	πh [a]
line	—	$\pi^2 h^2$	$2\pi^2 h^2$	$2\sqrt{2}\,\pi h$	$\sqrt{2}\,\pi h$ [a]
Optimum extrapolated					
point	$\frac{1}{2}\pi^2 h^2$	$\frac{1}{2}\pi^2 h^2$	$2\pi^2 h^2$ [b]	—	$2\pi h$ [a]
line	—	$\pi^2 h^2$	$4\pi^2 h^2$ [b]	—	$2\sqrt{2}\,\pi h$ [a]
Optimum Chebyshev accelerated					
point	πh	πh	$2\pi h$ [b]	—	$2\sqrt{\pi h}$ [a]
line	—	$\sqrt{2}\,\pi h$	$2\sqrt{2}\,\pi h$ [b]	—	$2^{5/4}\sqrt{\pi h}$ [a]
Conjugate gradient accelerated					
point	$\geq \pi h$	$\geq \pi h$	—	—	$\geq 2\sqrt{\pi h}$ [a]
line	—	$\geq \sqrt{2}\,\pi h$	—	—	$\geq 2^{5/4}\sqrt{\pi h}$ [a]
Cyclic Chebyshev accelerated					
point	—	$2\pi h$ [b]	—	—	—
line	—	$2\sqrt{2}\,\pi h$ [b]	—	—	—
Cyclic Conjugate gradient accelerated					
point	—	$\geq 2\pi h$ [b]	—	—	—
line	—	$\geq 2\sqrt{2}\,\pi h$ [b]	—	—	—

[a] Natural ordering required.

[b] Red/black ordering required.

red/black ordering of the unknowns. A dash indicates that a particular procedure is not applicable.

For comparison purposes, we include in Table 2-4.1 data for the Chebyshev and conjugate gradient polynomial acceleration methods given in subsequent chapters. Briefly, the Chebyshev and conjugate gradient methods may be used to accelerate the rate of convergence of any basic method that is symmetrizable. The Chebyshev method requires the use of iteration parameters, which must be properly chosen to obtain the greatest rate of convergence. The optimum Chebyshev parameters are functions of the extreme eigenvalues, $m(G)$ and $M(G)$, of the iteration matrix G for the related basic method. For the model problem data given in Table 2-4.1, the optimum Chebyshev parameters are based on the exact values for $m(G)$ and $M(G)$ given in the previous section. The conjugate gradient method also utilizes iteration parameters. However, these parameters are generated automatically during the iteration process and require no information concerning the eigenvalues of G. Formulas for the convergence rates of the conjugate gradient procedures are not known. However, when measured in a particular norm, the average convergence rate of the conjugate gradient method cannot be less than that of the Chebyshev method. The cyclic methods are special Chebyshev and conjugate gradient procedures that are applicable when the linear system is partitioned into the red/black form (1-5.3).

Model problem analysis is useful as a first step in the evaluation of any iterative solution procedure. However, some methods give extremely rapid convergence for certain model type problems but are less attractive, for various reasons, when applied to a more general class of problems. Thus for completeness, we augment the model problem results with a brief discussion concerning the essential requirements and the expected behavior of the procedures of Table 2-4.1 when applied to more general problems.

Before considering particular procedures, we make two general observations. First, the fundamental requirement of all methods considered here is that the coefficient matrix A be SPD. If A is not SPD, the procedures of Table 2-4.1 usually are not effective except under special conditions (see, e.g., the discussions given in Section 6.8 and in Chapters 11 and 12). Second, recall that the Jacobi, Gauss–Seidel, SOR, and SSOR methods are defined relative to a fixed partitioning imposed on the coefficient matrix A. Hence for a given matrix A, each of these methods is really a family of procedures; each member corresponds to a different partitioning and each member possesses, possibly, different convergence properties. The partitioning imposed on A often is a key factor in the iterative behavior of procedures based on the aforementioned methods.

For any partitioning of an arbitrary SPD coefficient matrix A, the Jacobi method is symmetrizable. Thus the J–OE method is always convergent, and moreover, the Chebyshev method or the conjugate gradient method can always be used to accelerate the basic Jacobi method. For Chebyshev acceleration, it is important that nearly optimum iteration parameters be used for intrinsically slowly convergent problems. For general applications, such parameters are not known a priori but can be determined by using the computational algorithms given in Chapters 5 and 6. Although Chebyshev acceleration and conjugate gradient acceleration of the Jacobi method are effective for any choice of partitioning, we show in Chapter 8 and Section 9.9 that the fastest rates of convergence usually are obtained for partitionings which are red/black.

The SOR method with $0 < \omega < 2$ is also convergent for any partitioning of a coefficient matrix A that is SPD. But the effectiveness of the SOR method depends strongly upon the availability of a prescription for selecting the iteration parameter ω. In Chapter 9 we give a precise formula for the ω that maximizes $R_\infty(\mathscr{L}_\omega)$; however, this formula for the "optimum" ω is valid only for a certain class of partitionings. When the partitioning of A is such that a precise formula for the optimum ω can be given, the SOR method is competitive with the best acceleration methods applied to the Jacobi method. For other partitionings, the SOR method normally should not be considered as an effective general solution method. We remark that it is important to use an ω near the optimum value for intrinsically slowly convergent problems. Algorithms to determine numerically nearly optimum vaules of ω are given in Chapter 9.

For any partitioning of a SPD matrix A, the SSOR method is symmetrizable. Therefore, the SSOR method with either Chebyshev or conjugate gradient acceleration can be used to solve general problems. For the model problem, the convergence rates of the best accelerated SSOR procedures are considerably larger than those of any of the other iterative procedures. Unfortunately, the SSOR condition (2-3.51) must be satisfied or nearly satisfied in order to realize this extremely rapid convergence. When condition (2-3.51) is not satisfied, as is the case for many practical applications, the SSOR procedures are much less effective. Because of this and because the computational effort for a SSOR iteration step is sometimes twice that required by other methods, the SSOR procedures are not frequently used in the solution of large general problems.

In this book, the Gauss–Seidel method and related procedures are used primarily in the discussion of other methods and are not considered as distinctive general solution procedures. The reason for this is that any Gauss–Seidel related procedure which may be of interest to us can be described and treated more easily in terms of other procedures.

The RF method also is used primarily in the discussion of other methods. The simple point RF method we described is sufficient for our use but should not be considered as a general solution procedure since block techniques cannot be accommodated. We note in passing, however, that a viable generalized form of the RF method can be defined (see, e.g., Young [1971]).

2.5   OTHER METHODS

In this section we discuss briefly several useful solution procedures which, because of limitations of space and our own experience, will not be considered elsewhere in this book. We shall describe only the general ideas involved. No details for any particular method will be given; however, references are cited for the reader who wishes more detailed information. The methods that we consider are: (i) approximate factorization methods, (ii) alternating direction implicit methods, and (iii) fast direct methods.

Basic Methods Based on Approximate Factorization

In Section 2.2, we assumed that any basic method can be defined uniquely in terms of a splitting matrix Q. Let such a matrix Q be defined by the splitting $A = Q - R$ for the coefficient matrix A. By (2-2.1), the basic iterative method (2-1.2) defined by this splitting can be written in the form

$$Qu^{(n+1)} = Ru^{(n)} + b, \qquad (2\text{-}5.1)$$

with the corresponding iteration matrix $G = I - Q^{-1}A = Q^{-1}R$. We now consider those methods for which the matrix Q has the form $Q = HK$. Let the nonsingular matrices H and K be chosen such that they are easy to obtain and such that the matrix Q is "easily invertible." Moreover, to maximize the rate of convergence, H and K should also be chosen such that the spectral radius of the iteration matrix G is as small as possible.

If A is symmetric and positive definite, one choice is to let $H = L^T$ and $K = L$, where L is the upper triangular matrix defined by the Cholesky decomposition $A = L^T L$ of A. In this case, $R = 0$, and the process would converge in one iteration. Of course, this is just a form of the Gaussian direct solution method, and for large sparse matrix problems, the matrix $H = L$ may not to be sparse and/or may not be easy to obtain.

One approach to avoid this problem of completely decomposing A is to pick H and K to be lower and upper triangular matrices, respectively, but such that the product HK only approximates A. Usually this is done by defining a particular sparsity pattern† for H and K and then determining the nonzero elements of H and K so that the product HK approximates A as closely as possible. This approach encompasses a family of iterative techniques that differ mainly in the choice for the matrices H and K. Some members of

---

† The imposed sparsity pattern for H and K is often related to the sparsity pattern of A.

this family are called "primitive iterative methods" by Varga [1960], "approximate factorization" by Dupont et al. [1968] and Bracha-Barak and Saylor [1973], "strongly implicit" by Stone [1968], and "incomplete Cholesky factorization" by Meijerink and van der Vorst [1974, 1977]. Also see Axelsson [1978], Chandra [1978], and Beauwens [1979].

If A is symmetric and positive definite, then an incomplete symmetric factorization $Q = HH^T$ is usually used. If Q is positive definite,[†] then the basic method (2-5.1) is symmetrizable and the Chebyshev and conjugate gradient procedures described later may be used to accelerate the rate of convergence. Considerable success has been reported when the conjugate gradient method is used to accelerate approximate factorization splitting methods. See, for example, Chandra [1978] and the references cited there.

Alternating Direction Implicit Methods

Douglas [1955], Peaceman and Rachford [1955], and Douglas and Rachford [1956] introduced a class of methods that are called *alternating direction implicit methods*. We now describe briefly one method of this class which is sometimes referred to as the *Peaceman–Rachford method*. The Peaceman–Rachford method is based on representing the matrix A of (2-1.1) as a sum

$$A = H + V, \tag{2-5.2}$$

where A, H, and V are assumed to be SPD. The method is defined by

$$(H + \rho_n I)u^{(n+1/2)} = b - (V - \rho_n I)u^{(n)},$$
$$(V + \rho'_n I)u^{(n+1)} = b - (H - \rho'_n I)u^{(n+1/2)}. \tag{2-5.3}$$

Here it is assumed that for any positive numbers ρ_n and ρ'_n, the first system can be solved easily for $u^{(n+1/2)}$, given $u^{(n)}$, and that the second can be solved easily for $u^{(n+1)}$, given $u^{(n+1/2)}$. In a typical case involving a linear system arising from an elliptic partial differential equation, H and V might be tridiagonal matrices or at least matrices with small bandwidths. For finite difference methods over rectangular mesh subdivisions, H is the matrix corresponding to horizontal differences and V is the matrix corresponding to vertical differences.

Normally, one would use particular values $\rho_1, \rho'_1, \rho_2, \rho'_2, \ldots, \rho_m, \rho'_m$ of the parameters in a cyclic order. The values of the ρ_i often depend[‡] on bounds of the eigenvalues of H and V. For the model problem of Section 1.7, it can be

---

[†] For some variants, some care must be taken to ensure that the resulting incomplete factorization $Q = HH^T$ is positive definite. See, for example, Kershaw [1978].

[‡] For algorithms to generate the acceleration parameters ρ_i and ρ'_i, see Wachspress [1963] and Kellogg and Spanier [1965].

shown (e.g., see Birkhoff *et al.* [1962]) that the parameters can be so chosen that the number of iterations needed for convergence varies as $\log h^{-1}$ as the mesh size h tends to zero. This is in contrast, for example, to h^{-1} for the SOR method (see (2-3.39)).

The basic theory for the Peaceman–Rachford method is valid only if the matrices H and V commute. For elliptic partial differential equations, this requirement implies that the differential equation is separable and that the region is a rectangle (see Birkhoff and Varga [1959] and Birkhoff *et al.* [1962]). Widlund [1966, 1969] extended the theory to nonseparable equations with rectangular regions. The method works well in many cases in which the existing theory does not rigorously apply. However, Price and Varga [1962] have constructed a case in which the theory fails to apply in an essential way. In spite of many attempts to analyze the Peaceman–Rachford method, a solid theoretical foundation still does not exist.

Fast Direct Methods

We now mention briefly a class of very fast methods that are often used to solve certain linear systems. Such systems frequently arise in the solution of linear elliptic partial differential equations with constant coefficients over rectangular regions. These methods depend on the fact that a closed-form solution to the discretized problem exists. This solution has a form similar to the Fourier series solution that can be obtained for the continuous problem in the "separable" case. For the discretized case, the solution is expressed in the form of a finite sum with special properties which make it possible to find the sum very rapidly. Sometimes these methods are referred to as *fast Fourier methods*.

Fast direct methods also can be considered for more general linear systems for which it is possible to apply certain transformations to successively reduce the order of the system by a factor of two. Thus we have methods known as *cyclic reduction, odd–even reduction*, etc. (Buzbee *et al.*, 1970; Sweet, 1974).

For problems in which fast direct methods cannot be applied directly, they can often be used in conjunction with other methods. Thus Concus and Golub [1973] and Concus *et al.* [1976] use fast direct methods combined with Chebyshev and conjugate gradient acceleration to solve certain partial differential equations. Proskurowski and Widlund [1976] describe the use of *capacitance matrices* to treat problems in which the region is not a rectangle. Other fast direct methods are the marching algorithms of Bank [1975]; see also Bank and Rose [1975] and Bank [1976].

For additional information on fast direct methods the reader is referred to the papers cited above and to Buzbee *et al.* [1971], Dorr [1970], Hockney [1965], Swartztrauber [1974], and Sweet [1973].

CHAPTER

3

Polynomial Acceleration

3.1 INTRODUCTION

In this chapter we describe a general procedure for accelerating the rates of convergence of basic iterative methods. This acceleration procedure, which we call *polynomial acceleration*, involves the formation of a new vector sequence from linear combinations of the iterates obtained from the basic method. As noted by Varga [1962], such a procedure is suggested by the theory of summability of sequences.

We define the general polynomial procedure, assuming only that the basic method is completely consistent (see Section 2.2). However, later when we consider Chebyshev and conjugate gradient polynomial methods, we generally assume that the basic method is also symmetrizable.

The polynomial procedure we present is but one of many approaches that may be used to accelerate the convergence of basic iterative methods. Some nonpolynomial acceleration methods are discussed briefly in Section 3.3.

3.2 POLYNOMIAL ACCELERATION OF BASIC ITERATIVE METHODS

Suppose the completely consistent basic method (2-1.2) is used to obtain approximations for the solution $\bar{u}$ of the nonsingular matrix problem $Au = b$.

Let the sequence of iterates generated by the basic method be given by $\{w^{(n)}\}$, i.e., given $w^{(0)}$, the sequence $\{w^{(n)}\}$ is formed by

$$w^{(n)} = Gw^{(n-1)} + k, \qquad n = 1, 2, \ldots. \tag{3-2.1}$$

From (2-2.6), the error vector $\tilde{\varepsilon}^{(n)} \equiv w^{(n)} - \bar{u}$ associated with the nth iterate of (3-2.1) satisfies

$$\tilde{\varepsilon}^{(n)} = G^n\tilde{\varepsilon}^{(0)}. \tag{3-2.2}$$

As a means to enhance the convergence of the $w^{(n)}$ iterates,[†] we consider a new vector sequence $\{u^{(n)}\}$ determined by the linear combination

$$u^{(n)} = \sum_{i=0}^{n} \alpha_{n,i} w^{(i)}, \qquad n = 0, 1, \ldots. \tag{3-2.3}$$

The only restriction we impose on the real numbers $\alpha_{n,i}$ is that

$$\sum_{i=0}^{n} \alpha_{n,i} = 1, \qquad n = 0, 1, \ldots. \tag{3-2.4}$$

This condition is imposed in order to ensure that $u^{(n)} = \bar{u}$ for all $n \geq 0$ whenever the initial guess vector $w^{(0)}$ is equal to the solution $\bar{u}$.

If we let $\varepsilon^{(n)} \equiv u^{(n)} - \bar{u}$ denote the error vector associated with the vectors $u^{(n)}$ of (3-2.3), we have from (3-2.3) and (3-2.4) that

$$\varepsilon^{(n)} = \sum_{i=0}^{n} \alpha_{n,i} w^{(i)} - \bar{u} = \sum_{i=0}^{n} \alpha_{n,i}(w^{(i)} - \bar{u}) = \sum_{i=0}^{n} \alpha_{n,i} \tilde{\varepsilon}^{(i)}.$$

Using (3-2.2), we then may express $\varepsilon^{(n)}$ in the form

$$\varepsilon^{(n)} = \left(\sum_{i=0}^{n} \alpha_{n,i} G^i \right) \tilde{\varepsilon}^{(0)}.$$

It follows from (3-2.3)–(3-2.4) that $\tilde{\varepsilon}^{(0)} = \varepsilon^{(0)}$. Thus, we may express $\varepsilon^{(n)}$ in the form

$$\varepsilon^{(n)} = Q_n(G)\varepsilon^{(0)}, \tag{3-2.5}$$

where $Q_n(G)$ is the *matrix polynomial* $Q_n(G) \equiv \alpha_{n,0} I + \alpha_{n,1} G + \cdots + \alpha_{n,n} G^n$. If $Q_n(x) \equiv \alpha_{n,0} + \alpha_{n,1} x + \cdots + \alpha_{n,n} x^n$ is the associated algebraic polynomial (see Section 1.3), then condition (3-2.4) requires that $Q_n(1) = 1$. This condition is the only restriction imposed thus far on $Q_n(x)$.

Because of the form (3-2.5) for the associated error vector, we call the combined procedure of (3-2.1) and (3-2.3) a *polynomial acceleration method*

---

† Since we have assumed only that the basic method (3-2.1) is completely consistent, convergence of the iterates $w^{(n)}$ to $\bar{u}$ is not guaranteed.

applied to the basic method (3-2.1). Varga [1962] calls this procedure a semi-iterative method with respect to the iterative method (3-2.1).

The high arithmetic cost and the large amount of storage required in using (3-2.3) to obtain $u^{(n)}$ make it necessary to seek alternative, less costly ways to compute $u^{(n)}$. We now show that a simpler computational form for $u^{(n)}$ is possible whenever the polynomials $Q_n(x)$ satisfy the recurrence relation

$$Q_0(x) = 1,$$

$$Q_1(x) = \gamma_1 x - \gamma_1 + 1,$$

$$Q_{n+1}(x) = \rho_{n+1}(\gamma_{n+1}x + 1 - \gamma_{n+1})Q_n(x) + (1 - \rho_{n+1})Q_{n-1}(x) \quad \text{for} \ n \geq 1,$$

$$(3\text{-}2.6)$$

where $\gamma_1, \rho_2, \gamma_2, \ldots$ are real numbers. Note that the $Q_n(x)$, defined by (3-2.6), satisfy $Q_n(1) = 1$ for all $n \geq 0$. We remark that the set of polynomial sequences $\{Q_n(x)\}$ satisfying (3-2.6) is large. For example, any properly normalized real orthogonal polynomial sequence is in such a set (see, e.g., Davis [1963]).

Theorem 3-2.1. Let the basic method (3-2.1) be completely consistent. If the polynomial sequence $\{Q_n(x)\}$ is given by (3-2.6), then the iterates $u^{(n)}$ of (3-2.3) may be obtained using the three-term relation

$$u^{(1)} = \gamma_1(Gu^{(0)} + k) + (1 - \gamma_1)u^{(0)},$$

$$u^{(n+1)} = \rho_{n+1}\{\gamma_{n+1}(Gu^{(n)} + k) + (1 - \gamma_{n+1})u^{(n)}\} + (1 - \rho_{n+1})u^{(n-1)}$$

$$\text{for} \ n \geq 1.$$

$$(3\text{-}2.7)$$

Conversely, any iterative procedure with iterates $u^{(n)}$ defined by (3-2.7) is equivalent to the polynomial procedures (3-2.1) and (3-2.3), with the polynomials $\{Q_n(x)\}$ given by (3-2.6).

Proof. Let the polynomials $\{Q_n(x)\}$ be given by (3-2.6) and let $\varepsilon^{(n)}$ be the error vector associated with the vector $u^{(n)}$ of (3-2.3). For $n \geq 1$, we have from (3-2.5) and (3-2.6) that

$$\varepsilon^{(n+1)} = \{\rho_{n+1}[\gamma_{n+1}G + (1 - \gamma_{n+1})I]Q_n(G) + (1 - \rho_{n+1})Q_{n-1}(G)\}\varepsilon^{(0)},$$

$$(3\text{-}2.8)$$

and thus, again using (3-2.5), that

$$\varepsilon^{(n+1)} = \rho_{n+1}[\gamma_{n+1}G + (1 - \gamma_{n+1})I]\varepsilon^{(n)} + (1 - \rho_{n+1})\varepsilon^{(n-1)}. \quad (3\text{-}2.9)$$

By adding $\bar{u}$ to both sides of (3-2.9), we then obtain

$$u^{(n+1)} = \rho_{n+1}[\gamma_{n+1}G + (1 - \gamma_{n+1})I]u^{(n)} + (1 - \rho_{n+1})u^{(n-1)}$$

$$- \rho_{n+1}\gamma_{n+1}(G - I)\bar{u}. \quad (3\text{-}2.10)$$

Now, using the fact that $\bar{u}$ also is a solution to the related system (2-2.2), we obtain the three-term form (3-2.7). The special case for $\varepsilon^{(1)}$ follows similarly. Conversely, let $\hat{\varepsilon}^{(n)} \equiv u^{(n)} - \bar{u}$ be the error vector associated with the vectors $u^{(n)}$ of (3-2.7). By reversing the above steps using $\hat{\varepsilon}^{(n+1)}$ in place of $\varepsilon^{(n+1)}$, we get $\hat{\varepsilon}^{(n+1)} = Q_{n+1}(G)\hat{\varepsilon}^{(0)}$ with $Q_{n+1}(x)$ defined by (3-2.6). From this, it follows that the iterative procedure (3-2.7) is equivalent to a polynomial acceleration method, with the polynomials given by (3-2.6). ■

As noted previously, there are many polynomial sequences $\{Q_n(x)\}$ that satisfy (3-2.6). In this book, we consider only those polynomial sequences that are associated with the Chebyshev and conjugate gradient acceleration methods. We discuss below the general basis on which the polynomials $\{Q_n(x)\}$ are chosen for these methods.

We now assume that the basic method (3-2.1) is also *symmetrizable* with a symmetrization matrix W. From (3-2.5), we have for any vector norm $\|\cdot\|_L$ that

$$\|\varepsilon^{(n)}\|_L = \|Q_n(G)\varepsilon^{(0)}\|_L. \tag{3-2.11}$$

For the conjugate gradient method, the polynomial sequence $\{Q_n(x)\}$ is chosen to minimize $\|Q_n(G)\varepsilon^{(0)}\|_L$, or equivalently $\|\varepsilon^{(n)}\|_L$ for a particular choice of the vector norm $\|\cdot\|_L$. The conjugate gradient method is discussed in Chapter 7.

Since WGW^{-1} is symmetric, it follows that the matrix $WQ_n(G)W^{-1}$ is also symmetric. Thus from (1-4.7) and (1-4.11), we have that

$$\|\varepsilon^{(n)}\|_W \leq \|Q_n(G)\|_W \|\varepsilon^{(0)}\|_W = \mathbf{S}(Q_n(G))\|\varepsilon^{(0)}\|_W. \tag{3-2.12}$$

For the Chebyshev acceleration procedure, the error norm $\|\varepsilon^{(n)}\|_W$ is made small by picking the polynomials $\{Q_n(x)\}$ such that the spectral radius $\mathbf{S}(Q_n(G))$ is small. More precisely, let $\{\mu_i\}_{i=1}^N$ be the set of eigenvalues for the $N \times N$ matrix G. Then $\{Q_n(\mu_i)\}_{i=1}^N$ is the set of eigenvalues for the matrix $Q_n(G)$. (See Section 1.3.) Thus we have

$$\mathbf{S}(Q_n(G)) = \max_{1 \leq i \leq N} |Q_n(\mu_i)|. \tag{3-2.13}$$

Since the complete eigenvalue spectrum of G is seldom known, it is more convenient to consider the *virtual spectral radius* of $Q_n(G)$ in place of $\mathbf{S}(Q_n(G))$. If $M(G)$ and $m(G)$ denote, respectively, the algebraically largest and smallest eigenvalues of G, then the virtual spectral radius of $Q_n(G)$ is defined by

$$\bar{\mathbf{S}}(Q_n(G)) \equiv \max_{m(G) \leq x \leq M(G)} |Q_n(x)|. \tag{3-2.14}$$

Since the set of eigenvalues $\{\mu_i\}_{i=1}^N$ is in the interval $[m(G), M(G)]$, we have that

$$\mathbf{S}(Q_n(G)) \leq \bar{\mathbf{S}}(Q_n(G)). \tag{3-2.15}$$

For the Chebyshev method, the polynomial sequence $\{Q_n(x)\}$ is chosen such that $\bar{\mathbf{S}}(Q_n(G))$ is minimized. The Chebyshev method is described in Chapter 4.

Analogous to the definitions given in Section 2.2, the *virtual average rate of convergence for a polynomial method* is defined by

$$\bar{R}_n(Q_n(G)) \equiv -(1/n) \log \bar{\mathbf{S}}(Q_n(G)), \tag{3-2.16}$$

and provided the limit exists, the *virtual asymptotic rate of convergence* is given by

$$\bar{R}_\infty(Q_n(G)) \equiv \lim_{n \to \infty} \bar{R}_n(Q_n(G)). \tag{3-2.17}$$

3.3 EXAMPLES OF NONPOLYNOMIAL ACCELERATION METHODS

In this section we describe briefly several alternatives to the polynomial approach for accelerating the convergence of basic iterative methods. The classes of general acceleration procedures we present are called acceleration by additive correction and acceleration by multiplicative correction.

Suppose the basic iterative process

$$u^{(n)} = Gu^{(n-1)} + k \tag{3-3.1}$$

is used to obtain approximations for the solution $\bar{u}$ of the matrix problem $Au = b$. It easily follows that the error vector $\varepsilon^{(n)} \equiv u^{(n)} - \bar{u}$ and the *residual vector* $r^{(n)} \equiv Au^{(n)} - b$ satisfy the residual equation

$$A\varepsilon^{(n)} = r^{(n)}. \tag{3-3.2}$$

If Eq. (3-3.2) can be solved for $\varepsilon^{(n)}$, then we immediately have the solution $\bar{u} = u^{(n)} - \varepsilon^{(n)}$. However, it is as difficult to solve (3-3.2) for $\varepsilon^{(n)}$ as it is to solve $Au = b$ for $\bar{u}$. On the other hand, it is not always necessary to determine $\varepsilon^{(n)}$ with great precision to improve the accuracy of $u^{(n)}$. Thus the basic method (3-3.1) can often be accelerated by using the following procedure:

(1) Do L iterations of the basic method (3-3.1), using $u^{(0)}$ as the initial guess.

(2) Compute $\tilde{\varepsilon}^{(L)}$, where $\tilde{\varepsilon}^{(L)}$ is some approximation to $\varepsilon^{(L)}$ of (3-3.2), and set $u^{(0)} = u^{(L)} - \tilde{\varepsilon}^{(L)}$. Then go to step (1) again.

There are many ways to obtain the approximation $\tilde{\varepsilon}^{(L)}$. For example, if A corresponds to the discretization with a mesh π_h of a continuous operator, then $\tilde{\varepsilon}^{(L)}$ may be taken to satisfy $\tilde{A}\tilde{\varepsilon}^{(L)} = \tilde{r}^{(L)}$, where $\tilde{A}$ corresponds to a discretization over a coarser mesh, say π_{2h}. Some methods that utilize this general approach are "the synthetic method" (e.g., see Kopp [1963] and Gelbard and Hageman [1969]), "multilevel methods" (e.g., see Brandt [1977]), and "multigrid methods" (e.g., see Nicolaides [1975, 1977] and Hackbusch [1977]). Another approach is to obtain the approximation $\tilde{\varepsilon}^{(L)}$ using the method of weighted residuals. To do this, let $\tilde{\varepsilon}^{(L)}$ be written as

$$\tilde{\varepsilon}^{(L)} = \sum_{i=1}^{M} c_i \alpha_i, \tag{3-3.3}$$

where the α_i are some known vectors.† The unknown constants c_i are determined from the M equations

$$w_i^T A \tilde{\varepsilon}^{(L)} = w_i^T r^{(L)}, \qquad i = 1, \ldots, M, \tag{3-3.4}$$

where the w_i are known weighting vectors. See, for example, de la Vallee Poussin [1968] and Setturi and Aziz [1973].

The methods discussed above are called *additive correction acceleration methods*. However, *multiplicative correction methods* have also been used. Multiplicative correction methods attempt to improve the accuracy of $u^{(L)}$ by multiplying $u^{(L)}$ by some matrix E instead of adding a vector $\tilde{\varepsilon}^{(L)}$ as in step (2) above. Usually, E is a diagonal matrix whose diagonal entries are determined by some weighted residual or variational method. Descriptions of methods based on the multiplicative correction approach are given, for example, by Kellogg and Noderer [1960] (scaled iterations), Nakamura [1974] (coarse mesh rebalancing techniques), and Wachspress [1966] (coarse mesh variational techniques). Wachspress [1977] considers an acceleration procedure based on combined additive and multiplicative correction.

---

† Often, the elements of α_i are chosen to be either 0 or 1.

CHAPTER

4

Chebyshev Acceleration

4.1   INTRODUCTION

In this chapter we consider Chebyshev polynomial acceleration applied to basic iterative methods of the form

$$u^{(n)} = Gu^{(n-1)} + k, \qquad n = 1, 2, \ldots. \tag{4-1.1}$$

We assume throughout this chapter that the iterative method (4-1.1) is symmetrizable.

Recall from Chapter 3 that the error vector associated with a general polynomial acceleration procedure applied to (4-1.1) can be expressed as

$$\varepsilon^{(n)} = Q_n(G)\varepsilon^{(0)}, \tag{4-1.2}$$

where $Q_n(G) \equiv \alpha_{n,0} I + \alpha_{n,1} G + \cdots + \alpha_{n,n} G^n$ is a matrix polynomial subject only to the condition that $\sum_{i=0}^{n} \alpha_{n,i} = 1$. Using the algebraic polynomial $Q_n(x)$ associated with $Q_n(G)$, we defined

$$\bar{S}(Q_n(G)) \equiv \max_{m(G) \le x \le M(G)} |Q_n(x)| \tag{4-1.3}$$

as the virtual spectral radius of the matrix $Q_n(G)$. As before, $M(G)$ and $m(G)$ denote, respectively, the algebraically largest and smallest eigenvalues of G. That particular polynomial method which is obtained by choosing the

45

polynomial sequence $\{Q_n(G)\}$ such that $\bar{S}(Q_n(G))$, $n = 1, 2, \ldots$, is minimized is called the Chebyshev polynomial acceleration method.

In Section 4.2, we show that, indeed, the matrix polynomial $Q_n(G)$ that minimizes $\bar{S}(Q_n(G))$ can be defined in terms of Chebyshev polynomials. The basic Chebyshev computational procedure is also derived in Section 4.2. It turns out that the proper application of Chebyshev acceleration requires the use of "iteration parameters" whose optimum values are functions of the extreme eigenvalues $M(G)$ and $m(G)$ of G. When optimum iteration parameters are used, we show that Chebyshev acceleration can significantly improve the convergence rate. For most practical applications, however, the optimum parameters will not be known a priori and must be approximated by some means. In Sections 4.3 and 4.4, we study the behavior of the Chebyshev method when iteration parameters which are not optimum are used. Computational algorithms which generate the necessary Chebyshev iteration parameters adaptively during the iteration process are presented in Chapters 5 and 6.

4.2   OPTIMAL CHEBYSHEV ACCELERATION

We first show that the matrix polynomial $Q_n(G)$ which minimizes $\bar{S}(Q_n(G))$ is unique and can be defined in terms of Chebyshev polynomials. For any nonnegative integer n, the Chebyshev polynomial† of degree n in w may be defined by the recurrence relation

$$T_0(w) = 1, \qquad T_1(w) = w,$$
$$T_{n+1}(w) = 2wT_n(w) - T_{n-1}(w), \qquad n \geq 1. \tag{4-2.1}$$

It can be shown by mathematical induction (see, e.g., Young [1971]) that the $T_n(w)$ may also be expressed by

$$
\begin{aligned}
T_n(w) &= \tfrac{1}{2}[(w + \sqrt{w^2 - 1})^n + (w + \sqrt{w^2 - 1})^{-n}] \\
&= \tfrac{1}{2}[(w - \sqrt{w^2 - 1})^n + (w - \sqrt{w^2 - 1})^{-n}] \\
&= \cosh(n \cosh^{-1} w) \qquad \text{when} \quad w > 1, \\
&= \cos(n \cos^{-1} w) \qquad \text{when} \quad -1 \leq w \leq 1. \tag{4-2.2}
\end{aligned}
$$

We note that $T_n(w)$ is an even function of w for n even and an odd function of w for n odd.

The fundamental properties of Chebyshev polynomials that we shall use are given in the following theorem.

---

† The Chebyshev polynomials utilized in this book are the so-called Chebyshev polynomials of the first kind.

Theorem 4-2.1. Let n be a fixed integer and let d be any fixed real number such that $d > 1$. If we let

$$H_n(w) = T_n(w)/T_n(d), \tag{4-2.3}$$

where $T_n(w)$ is the Chebyshev polynomial (4-2.1), then

$$H_n(d) = 1 \tag{4-2.4}$$

and

$$\max_{-1 \leq w \leq 1} |H_n(w)| = 1/T_n(d). \tag{4-2.5}$$

Moreover, if $Q(w)$ is any polynomial of degree n or less such that $Q(d) = 1$ and

$$\max_{-1 \leq w \leq 1} |Q(w)| \leq \max_{-1 \leq w \leq 1} |H_n(w)|,$$

then

$$Q(w) = H_n(w). \tag{4-2.6}$$

Proof. See, for example, Young [1971] or Flanders and Shortly [1950]. ∎

Returning now to the problem of minimizing $\bar{S}(Q_n(G))$, we seek that polynomial $P_n(x)$ such that $P_n(1) = 1$ and such that

$$\max_{m(G) \leq x \leq M(G)} |P_n(x)| \leq \max_{m(G) \leq x \leq M(G)} |Q_n(x)|, \tag{4-2.7}$$

where $Q_n(x)$ is any polynomial of degree n or less satisfying $Q_n(1) = 1$. The existence and definition of such a polynomial follows from Theorem 4-2.1. Specifically, let

$$w(x) \equiv (2x - M(G) - m(G))/(M(G) - m(G)) \tag{4-2.8}$$

be the linear transformation which maps the interval $m(G) \leq x \leq M(G)$ onto the interval $-1 \leq w \leq 1$ and let

$$H_n(w) \equiv T_n(w(x))/T_n(w(1)).$$

Since the basic method (4-1.1) is symmetrizable, it follows from Theorem 2-2.1 that $M(G) < 1$. Thus $w(1) > 1$. If we now define $P_n(x)$ as

$$P_n(x) \equiv T_n\left(\frac{2x - M(G) - m(G)}{M(G) - m(G)}\right) \bigg/ T_n\left(\frac{2 - M(G) - m(G)}{M(G) - m(G)}\right), \tag{4-2.9}$$

then obviously

$$\max_{m(G) \leq x \leq M(G)} |P_n(x)| = \max_{-1 \leq w \leq 1} |H_n(w)|. \tag{4-2.10}$$

It now follows from Theorem 4-2.1 that the $P_n(x)$ of (4-2.9) is the unique polynomial satisfying (4-2.7). Thus we have

Theorem 4-2.2. Let $\mathscr{S}_n$ be the set of polynomials $\{Q_n(x)\}$ of degree n or less satisfying $Q_n(1) = 1$. Then, the polynomial $P_n(x)$ of (4-2.9) is the unique polynomial in the set $\mathscr{S}_n$ which satisfies

$$\max_{m(G)\leq x\leq M(G)} |P_n(x)| \leq \max_{m(G)\leq x\leq M(G)} |Q_n(x)|$$

for any $Q_n(x) \in \mathscr{S}_n$.

We now consider computational and convergence aspects of polynomial acceleration using $P_n(x)$ when applied to the basic method (4-1.1). Using (4-2.1), it is easy to show that the polynomials $P_n(x)$ satisfy the recurrence relation

$$P_0(x) = 1, \qquad P_1(x) = \bar{\gamma}x - \bar{\gamma} + 1,$$

$$P_{n+1}(x) = \bar{\rho}_{n+1}(\bar{\gamma}x + 1 - \bar{\gamma})P_n(x) + (1 - \bar{\rho}_{n+1})P_{n-1}(x) \qquad \text{for} \quad n \geq 1,$$

$$(4\text{-}2.11)$$

where

$$\bar{\gamma} = 2/(2 - M(G) - m(G)), \tag{4-2.12}$$

$$\bar{\rho}_{n+1} = 2w(1)T_n(w(1))/T_{n+1}(w(1)), \tag{4-2.13}$$

and where $w(x)$ is defined by (4-2.8). It now follows from Theorem 3-2.1 that the iterates for the polynomial procedure based on $P_n(x)$ may be obtained by using the three-term recurrence relation

$$u^{(n+1)} = \bar{\rho}_{n+1}\{\bar{\gamma}(Gu^{(n)} + k) + (1 - \bar{\gamma})u^{(n)}\} + (1 - \bar{\rho}_{n+1})u^{(n-1)}. \quad (4\text{-}2.14)$$

We refer to the method defined by (4-2.14) with $\bar{\gamma}$ and $\bar{\rho}_{n+1}$ given by (4-2.12) and (4-2.13) as the *optimal Chebyshev acceleration procedure*. The term "optimal" is used to distinguish the method from other (nonoptimal) procedures in which estimates m_E and M_E are used for $m(G)$ and $M(G)$, respectively. Such procedures are described in Section 4.3.

Again making use of the Chebyshev polynomial recurrence relation (4-2.1), we can write the parameters $\bar{\rho}_{n+1}$ of (4-2.13) in the more computationally convenient form

$$\bar{\rho}_1 = 1, \qquad \bar{\rho}_2 = (1 - \tfrac{1}{2}\bar{\sigma}^2)^{-1},$$

$$\bar{\rho}_{n+1} = (1 - \tfrac{1}{4}\bar{\sigma}^2\bar{\rho}_n)^{-1}, \qquad n \geq 2, \tag{4-2.15}$$

where

$$\bar{\sigma} = 1/w(1) = (M(G) - m(G))/(2 - M(G) - m(G)). \tag{4-2.16}$$

It can also be shown (Varga, 1962) that

$$\lim_{n \to \infty} \bar{\rho}_n \equiv \bar{\rho}_\infty = 2/(1 + \sqrt{1 - \bar{\sigma}^2}). \tag{4-2.17}$$

We now examine the convergence rate of the optimal Chebyshev procedure. From (4-1.3), (4-2.5), and (4-2.10), the virtual spectral radius of $P_n(G)$ is

$$\bar{S}(P_n(G)) = [T_n(w(1))]^{-1} = [T_n(1/\bar{\sigma})]^{-1}. \tag{4-2.18}$$

Using (4-2.2), and after a small amount of algebra, we can write $T_n(1/\bar{\sigma})$ in the form

$$T_n\left(\frac{1}{\bar{\sigma}}\right) = \frac{1 + \bar{r}^n}{2\bar{r}^{n/2}},$$

where

$$\bar{r} \equiv (1 - \sqrt{1 - \bar{\sigma}^2})/(1 + \sqrt{1 - \bar{\sigma}^2}). \tag{4-2.19}$$

Thus we obtain

$$\bar{S}(P_n(G)) = 2\bar{r}^{n/2}/(1 + \bar{r}^n). \tag{4-2.20}$$

Thus from (3-2.16) and (4-2.20), the average virtual rate of convergence for the optimal Chebyshev method is

$$\bar{R}_n(P_n(G)) = -\frac{1}{2} \log \bar{r} - \frac{1}{n} \log\left(\frac{2}{1 + \bar{r}^n}\right). \tag{4-2.21}$$

It is easy to see from (4-2.21) that the asymptotic rate of convergence defined by (3-2.17) can be expressed in the form

$$\bar{R}_\infty(P_n(G)) = -\tfrac{1}{2} \log \bar{r}. \tag{4-2.22}$$

Note from (4-2.17) and (4-2.19) that $\bar{r} = \bar{\rho}_\infty - 1$. Thus we also have that

$$\lim_{n \to \infty} [\bar{S}(P_n(G))]^{1/n} = (\bar{\rho}_\infty - 1)^{1/2} \quad \text{and that} \quad \bar{R}_\infty(P_n(G)) = -\tfrac{1}{2}\log(\bar{\rho}_\infty - 1).$$

From (4-2.21) and (4-2.22), it easily follows that $\bar{R}_n(P_n(G)) < \bar{R}_\infty(P_n(G))$ for all finite n. In fact, it can be shown that $\bar{R}_n(P_n(G))$ is an increasing function of n. We omit the proof. However, the data given in Table 4-2.1 show that many iterations often are required before the asymptotic convergence is achieved. In Table 4-2.1, we tabulate the values of the ratio

$$\frac{\bar{R}_n(P_n(G))}{\bar{R}_\infty(P_n(G))} = \frac{-\log(2\bar{r}^{n/2}/(1 + \bar{r}^n))}{-\log \bar{r}^{n/2}} = 1 + \frac{\log(2/(1 + \bar{r}^n))}{\log \bar{r}^{n/2}} \tag{4-2.23}$$

as a function of $\bar{r}$. Thus we see that if, after n iterations, $\bar{r}^n \doteq 0.1$, then the average virtual convergence rate for these n iterations is only about one-half

TABLE 4-2.1

Values of $\bar{R}_n(P_n(G))/\bar{R}_\infty(P_n(G))$ as a Function of $\bar{r}^n$

$\bar{r}^n$	$\bar{R}_n(P_n(G))/\bar{R}_\infty(P_n(G))$	$\bar{r}^n$	$\bar{R}_n(P_n(G))/\bar{R}_\infty(P_n(G))$
10^{-1}	0.481	10^{-6}	0.900
10^{-2}	0.703	10^{-12}	0.948
10^{-3}	0.800	10^{-24}	0.975
10^{-4}	0.850	0	1.000
10^{-5}	0.880		

of its value for later iterations when the asymptotic convergence rate is achieved.

We now compare the optimum Chebyshev acceleration procedure with the optimum extrapolated procedure defined in Section 3.2. For the optimum extrapolated method applied to the basic method (4-1.1), we have by (2-2.16) and (4-2.16) that $\mathbf{S}(G_{[\bar{\gamma}]}) = \bar{\sigma}$ and hence that

$$R_\infty(G_{[\bar{\gamma}]}) = -\log \bar{\sigma}. \tag{4-2.24}$$

For the optimal Chebyshev procedure applied to (4-1.1), we have from (4-2.22) that $\bar{R}_\infty(P_n(G)) = -\frac{1}{2}\log \bar{r}$, where $\bar{r}$ is given by (4-2.19). It is easy to show that

$$-\log \bar{\sigma} \sim 1 - \bar{\sigma}, \qquad \bar{\sigma} \to 1-$$

and that

$$-\tfrac{1}{2}\log \bar{r} \sim \sqrt{1 - \bar{\sigma}^2} \sim \sqrt{2}\sqrt{1 - \bar{\sigma}}, \qquad \bar{\sigma} \to 1-.$$

Combining this with (4-2.22) and (4-2.24), we have

$$\bar{R}_\infty(P_n(G)) \sim \sqrt{2}\sqrt{R_\infty(G_{[\bar{\gamma}]})}, \qquad \bar{\sigma} \to 1-. \tag{4-2.25}$$

Thus for $\bar{\sigma}$ close to unity, the optimum Chebyshev procedure is an order of magnitude faster than the optimum extrapolated procedure.

As an example, consider the case in which $M(G) = -m(G) = 0.99$. In this case we have $\bar{\sigma} = 0.99$ and $\bar{r} \doteq 0.753$. Thus each iteration of the optimum extrapolated method reduces the error by approximately a factor of 0.99, while each iteration of the optimum Chebyshev method reduces the error by approximately a factor of 0.868. The number of iterations needed to reduce the norm of the error vector by a factor of 10^{-6}, as compared with the norm of the initial error vector, would be approximately 1375 for the extrapolated method and 98 for the optimum Chebyshev method. The factor of improvement would be greater for a larger value of $\bar{\sigma}$.

4.3   CHEBYSHEV ACCELERATION WITH ESTIMATED EIGENVALUE BOUNDS

We now study the behavior of the Chebyshev acceleration process when estimates m_E and M_E are used for $m(G)$ and $M(G)$, respectively. When these estimates are used, the normalized Chebyshev polynomial of (4-2.9) is written as

$$P_{n,E}(x) \equiv T_n\left(\frac{2x - M_E - m_E}{M_E - m_E}\right) \Big/ T_n\left(\frac{2 - M_E - m_E}{M_E - m_E}\right) = \frac{T_n(w_E(x))}{T_n(w_E(1))}, \quad (4\text{-}3.1)$$

where $w_E(x) \equiv (2x - M_E - m_E)/(M_E - m_E)$. If we let

$$\gamma \equiv 2/(2 - M_E - m_E), \quad (4\text{-}3.2)$$

$$\sigma_E \equiv (M_E - m_E)/(2 - M_E - m_E) = 1/w_E(1), \quad (4\text{-}3.3)$$

and

$$\rho_{n+1} \equiv 2w_E(1)T_n(w_E(1))/T_{n+1}(w_E(1)), \quad (4\text{-}3.4)$$

then it follows as before that the $P_{n,E}(x)$ satisfy the recurrence relation (4-2.11) with γ and ρ_{n+1} replacing the $\bar{\gamma}$ and $\bar{\rho}_{n+1}$, respectively. Thus from Theorem 3-2.1, the iterates for the polynomial acceleration procedure based on $P_{n,E}(x)$ may be given by

$$u^{(n+1)} = \rho_{n+1}\{\gamma(Gu^{(n)} + k) + (1 - \gamma)u^{(n)}\} + (1 - \rho_{n+1})u^{(n-1)}. \quad (4\text{-}3.5)$$

We remark that the error vector $\varepsilon^{(n)} \equiv u^{(n)} - \bar{u}$ associated with (4-3.5) now satisfies

$$\varepsilon^{(n)} = P_{n,E}(G)\varepsilon^{(0)}. \quad (4\text{-}3.6)$$

Relations analogous to (4-2.15) and (4-2.16) also are valid for ρ_{n+1}; i.e.,

$$\rho_1 = 1, \qquad \rho_2 = (1 - \tfrac{1}{2}\sigma_E^2)^{-1},$$
$$\rho_{n+1} = (1 - \tfrac{1}{4}\sigma_E^2\rho_n)^{-1}, \qquad n \geq 2, \quad (4\text{-}3.7)$$

and

$$\lim_{n \to \infty} \rho_n \equiv \rho_\infty = \frac{2}{1 + \sqrt{1 - \sigma_E^2}}. \quad (4\text{-}3.8)$$

In the above discussion, we have tacitly assumed that $M_E \neq m_E$. If $M_E = m_E$, then $\gamma = 1/(1 - M_E), \rho_1 = \rho_2 = \cdots = 1$ and $\sigma_E = 0$. For this case, the Chebyshev acceleration procedure (4-3.5) reduces to the (nonoptimum) extrapolation procedure (2-2.13). We shall not consider this case separately in the balance of this section. The correct formulas can be obtained by a limiting process.

It was shown in the previous section that the $P_n(x)$, defined by (4-2.9), is the *unique* polynomial satisfying the inequality (4-2.7). Thus $\bar{S}(P_{n,E}(G))$ is minimized only when $P_{n,E}(x) \equiv P_n(x)$; i.e., when $m_E = m(G)$ and $M_E = M(G)$. We note that $P_{n,E}(x)$ is that polynomial satisfying $P_{n,E}(1) = 1$ and which has the minimum "maximum" magnitude over the interval $m_E \leq x \leq M_E$. Because of this, equations for $\bar{S}(P_{n,E}(G))$ and $\bar{R}_\infty(P_{n,E}(G))$ analogous to (4-2.18) and (4-2.22) need not be valid when $P_{n,E}(x)$ is used. In the remainder of this section, we shall study the behavior of $\bar{S}(P_{n,E}(G))$ as a function of m_E and M_E in more detail.

In order to simplify the discussion while at the same time retaining adequate generality, we will make one of the following two sets of assumptions:†

Case I

$$m_E \leq m(G), \tag{4-3.9}$$

$$M(G) < 1, \tag{4-3.10}$$

$$m_E < M_E < 1. \tag{4-3.11}$$

Case II

$$m_E = -M_E, \tag{4-3.12}$$

$$|m(G)| \leq M(G) < 1, \tag{4-3.13}$$

$$0 \leq M_E < 1. \tag{4-3.14}$$

In later chapters, we shall strengthen the assumptions of each case by assuming additionally that $M_E < M(G)$.

We now graphically illustrate the behavior of $P_{n,E}(x)$ for two cases, each involving the assumptions of Case I. Figure 4-3.1 shows the behavior of $P_{10,E}(x)$ when $M_E > M(G)$. Here r is defined in (4-3.21). For this case, it is easy to see that

$$\max_{m(G) \leq x \leq M(G)} |P_{10,E}(x)| \leq P_{10,E}(M_E). \tag{4-3.15}$$

The equality holds for the value of $M(G)$ shown in the figure. However, if the value of $M(G)$ were equal to the M^* shown, then strict inequality would hold. Figure 4-3.2 shows the behavior of $P_{10,E}(x)$ when $M_E < M(G)$. Here it is clear that

$$\max_{m(G) \leq x \leq M(G)} |P_{10,E}(x)| = P_{10,E}(M(G)). \tag{4-3.16}$$

† The practicalities of these assumptions are discussed in Section 5.3.

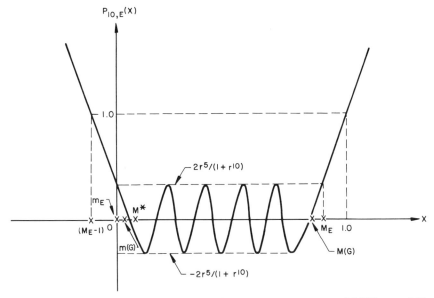

Fig. 4-3.1. Behavior of $P_{10, E}(x)$ for Case I when $M_E > M(G)$. $M_E = 0.94737$, $m_E = 0.00$, $\sigma_E = 0.90$, $r = 0.39$.

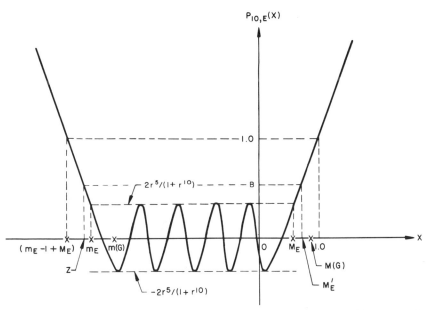

Fig. 4-3.2. Behavior of $P_{10, E}(x)$ for Case I when $M_E < M(G)$. $M_E = 0.789$, $m_E = -3.0$, $\sigma_E = 0.90$, $r = 0.39$.

The symbols B, z, and M'_E shown in Fig. 4-3.2 are used later in Chapter 5. From the above discussion we have, for either Case I or Case II,† that

$$\bar{\mathbf{S}}(P_{n,\,E}(G)) = P_{n,\,E}(M(G)) \qquad \text{if} \quad M_E \leq M(G),$$
$$\leq P_{n,\,E}(M_E) \qquad \text{if} \quad M_E \geq M(G). \tag{4-3.17}$$

It then follows from (4-3.1) that

$$\bar{\mathbf{S}}(P_{n,\,E}(G)) = T_n\!\left(\frac{\sigma^*}{\sigma_E}\right)\!\Big/T_n\!\left(\frac{1}{\sigma_E}\right) \qquad \text{if} \quad M_E \leq M(G),$$

$$\leq \left[T_n\!\left(\frac{1}{\sigma_E}\right)\right]^{-1} \qquad \text{if} \quad M_E \geq M(G), \tag{4-3.18}$$

where σ_E is given by (4-3.3) and

$$\sigma^* \equiv (2M(G) - M_E - m_E)/(2 - M_E - m_E). \tag{4-3.19}$$

Note that $w_E(M(G)) = \sigma^*/\sigma_E$ and that $\sigma^*/\sigma_E = 1$ when $M_E = M(G)$. We use σ^*/σ_E instead of $w_E(M(G))$ merely for notation purposes in discussing $\bar{\mathbf{S}}(P_{n,\,E}(G))$ and $\bar{R}_\infty(P_{n,\,E}(G))$. Using (4-2.2), we can write the relations (4-3.18) equivalently as

$$\bar{\mathbf{S}}(P_{n,\,E}(G)) = \frac{2r^{n/2}}{1+r^n}\,\Big/\,\frac{2\hat{r}^{n/2}}{1+\hat{r}^n} \qquad \text{if} \quad M_E \leq M(G),$$

$$\leq \frac{2r^{n/2}}{1+r^n} \qquad \text{if} \quad M_E \geq M(G), \tag{4-3.20}$$

where

$$r \equiv \frac{1-\sqrt{1-\sigma_E^2}}{1+\sqrt{1-\sigma_E^2}}, \qquad \hat{r} \equiv \frac{1-\sqrt{1-(\sigma_E/\sigma^*)^2}}{1+\sqrt{1-(\sigma_E/\sigma^*)^2}}. \tag{4-3.21}$$

From (4-3.20), it follows that for either Case I or Case II we have

$$\bar{\mathbf{S}}(P_{n,\,E}(G)) < 1. \tag{4-3.22}$$

From (3-2.17) and (4-3.20), the virtual asymptotic rate of convergence can be given by

$$\bar{R}_\infty(P_{n,\,E}(G)) = -\log\sqrt{r/\hat{r}} \qquad \text{if} \quad M_E \leq M(G),$$

$$\geq -\log\sqrt{r} \qquad \text{if} \quad M_E \geq M(G). \tag{4-3.23}$$

We now show, under the assumptions of Case II, that the above bound on $\bar{R}_\infty(P_{n,\,E}(G))$ is an *increasing* function of M_E for $M_E \leq M(G)$ and a *decreasing*

---

† Note that $\bar{\mathbf{S}}(P_{n,\,E}(G)) = \mathbf{S}(P_{n,\,E}(G))$ whenever $M_E \leq M(G)$.

function of M_E for $M_E \geq M(G)$. Indeed, by (4-3.12), (4-3.3), (4-3.19), and (4-3.21) we have $\sigma_E = M_E$, $\sigma^* = M(G)$,

$$r = \frac{1 - \sqrt{1 - M_E^2}}{1 + \sqrt{1 - M_E^2}}, \qquad (4\text{-}3.24)$$

and

$$\hat{r} = \frac{1 - \sqrt{1 - (M_E/M(G))^2}}{1 + \sqrt{1 - (M_E/M(G))^2}}. \qquad (4\text{-}3.25)$$

The fact that $\bar{R}_\infty(P_{n,\,E}(G))$ is a decreasing function of M_E for $M_E \geq M(G)$ follows from (4-3.23) and (4-3.24). It also follows from (4-3.23) that $\bar{R}_\infty(P_{n,\,E}(G))$ is an increasing function of M_E for $M_E \leq M(G)$ if we can show that $-\log(r/\hat{r})$ is an increasing function of M_E. But by (4-3.24) and (4-3.25) we have

$$\frac{d}{dM_E}\left[-\log\frac{r}{\hat{r}}\right] = -\frac{2}{M_E}\left\{\frac{1}{\sqrt{1 - M_E^2}} - \frac{1}{\sqrt{1 - (M_E/M(G))^2}}\right\}, \quad (4\text{-}3.26)$$

which is positive since $M(G) < 1$. Hence the desired result follows.

We remark that the above result can also be shown to be true for $\bar{R}_n(P_{n,\,E}(G))$ as well as for $\bar{R}_\infty(P_{n,\,E}(G))$, and for Case I as well as for Case II.

4.4   SENSITIVITY OF THE RATE OF CONVERGENCE TO THE ESTIMATED EIGENVALUES

In this section, we give quantitative results to illustrate the sensitivity of the asymptotic virtual rate of convergence $\bar{R}_\infty(P_{n,\,E}(G))$ to the estimates M_E of $M(G)$ and m_E of $m(G)$.

We first consider the behavior of $\bar{R}_\infty(P_{n,\,E}(G))$ as a function of M_E when $M(G)$ is close to one. For convenience of exposition, we shall assume here that the assumptions of Case II hold with $m(G) = -M(G)$. With these assumptions, we shall show that if $(1 - M(G))$ and $(1 - M_E)$ are both small, then approximately

$$\frac{[\bar{R}_\infty(P_{n,\,E}(G))]^{-1}}{[\bar{R}_\infty(P_n(G))]^{-1}} \doteq \begin{cases} \sqrt{\theta} + \sqrt{\theta - 1} & \text{if } M_E \leq M(G), \\ \leq 1/\sqrt{\theta} & \text{if } M_E \geq M(G), \end{cases} \quad (4\text{-}4.1)$$

where

$$\theta = (1 - M_E)/(1 - M(G)). \qquad (4\text{-}4.2)$$

For the case $M_E \le M(G)$, we have by (4-3.23) that

$$\bar{R}_\infty(P_{n,E}(G)) = -\log \sqrt{r} - (-\log \sqrt{\hat{r}}). \tag{4-4.3}$$

But for small $(1 - M_E)$ and small $(1 - M(G))$, by (4-3.24) and (4-3.25) we have, approximately, that

$$r \doteq 1 - 2\sqrt{2}\sqrt{\theta}\sqrt{1 - M(G)}, \qquad \hat{r} \doteq 1 - 2\sqrt{2}\sqrt{\theta - 1}\sqrt{1 - M(G)}. \tag{4-4.4}$$

Thus both r and $\hat{r}$ are close to unity so that $\bar{R}_\infty(P_{n,E}(G))$ of (4-4.3) may be approximated by $\bar{R}_\infty(P_{n,E}(G)) \doteq \frac{1}{2}[(1 - r) - (1 - \hat{r})]$. From this we obtain, using (4-4.4),

$$\bar{R}_\infty(P_{n,E}(G)) \doteq \frac{\sqrt{2}\sqrt{(1 - M(G))}}{\sqrt{\theta} + \sqrt{\theta - 1}}. \tag{4-4.5}$$

Moreover, from (4-2.22), $\bar{R}_\infty(P_n(G)) = -\log \bar{r}^{1/2}$, where $\bar{r} = [1 - (1 - \bar{\sigma}^2)^{1/2}]/[1 + (1 - \bar{\sigma}^2)^{1/2}]$ and where $\bar{\sigma}$ is given by (4-2.16). But $\bar{\sigma} = M(G)$ here since we have assumed that $m(G) = -M(G)$. Combining these facts, we obtain

$$\bar{R}_\infty(P_n(G)) \doteq \frac{1}{2}[1 - \bar{r}] \doteq \sqrt{2}\sqrt{1 - M(G)}. \tag{4-4.6}$$

Thus the first part of (4-4.1) follows from (4-4.5) and (4-4.6). A similar argument can be used to show the second part of (4-4.1).

To illustrate these results, we consider the following examples. First, consider the case $M(G) = 0.99$. If $\theta = 1.1$, we have $M_E = 0.989$, which at first sight would seem to be a very close estimate. However, we have $\hat{r} = 0.91400$, $r = 0.74229$, and $\bar{r} = 0.75274$, so that

$$[\bar{R}_\infty(P_{n,E}(G))]^{-1} = 9.61120$$

as compared with

$$[\bar{R}_\infty(P_n(G))]^{-1} = 7.04138.$$

Thus the actual ratio of convergence rates in (4-4.1) is 1.36496. This implies that the expected number of iterations when using $M_E = 0.989$ is 36% more than if $M_E = M(G)$ were used. We note that

$$\sqrt{\theta} + \sqrt{\theta - 1} = 1.3650.$$

Thus the approximation (4-4.1) is quite accurate.

Suppose, on the other hand, that $\theta = 0.9$. In this case, $M_E = 0.991$. Here $r = 0.76388$ and

$$[\bar{R}_\infty(P_{n,E}(G))]^{-1} = (-\tfrac{1}{2}\log r)^{-1} = 7.42544.$$

The left ratio in (4-4.1) is 1.05455. Thus the expected number of iterations, using $M_E = 0.991$, is only about 5.5% more than if $M_E = M(G)$ were used. We note that

$$(\sqrt{\theta})^{-1} = 1.054.$$

Again, the approximation (4-4.1) is quite accurate.

Let us now consider the case $\theta = 2$, in which again $M(G) = 0.99$. In this case, $M_E = 0.98$, $r = 0.66806$, $\hat{r} = 0.75167$, and $[\bar{R}_\infty(P_{n,E}(G))]^{-1} = 16.961$. Thus the left ratio in (4-4.1) is 2.408. This should be compared with the approximation (4-4.1); i.e.,

$$\sqrt{\theta} + \sqrt{\theta - 1} = 2.414.$$

If $1 - M(G)$ is very small, the approximation (4-4.1) is accurate even for quite large θ. Suppose $M(G) = 0.9999$. If $\theta = 100$, then $M_E = 0.99$, $r = 0.75274$, and $\hat{r} = 0.75383$. Moreover, $[\bar{R}_\infty(P_{n,E}(G))]^{-1} = 1382.17$ as compared with $[\bar{R}_\infty(P_n(G))]^{-1} = (-\frac{1}{2} \log \bar{r})^{-1} = 70.706$. (Note that $\bar{r} = 0.97211$.) Thus the left ratio in (4-4.1) is 19.548. This should be compared with the approximation

$$\sqrt{\theta} + \sqrt{\theta - 1} = 19.950.$$

We remark that the behavior of $\bar{R}_\infty(P_{n,E}(G))$ as a function of M_E for Case I conditions is similar to that given above for Case II.

From the above discussion, it is clear that if we underestimate $M(G)$, we increase the expected number of iterations much more than if we overestimate $M(G)$ by an equivalent amount. However, upper bounds for $M(G)$ which are nontrivial in the sense that they are less than unity and yet close to $M(G)$ are very difficult to obtain. Moreover, it is very difficult to improve an overestimated value even if one is available. On the other hand, if M_E is an underestimate for $M(G)$, then improved estimates for $M(G)$ can be obtained by using the adaptive procedures given in Chapters 5 and 6. In addition, as we shall see later, very accurate estimates of the iteration error may be obtained when $M_E < M(G)$. For these reasons, the adaptive procedures described in subsequent chapters are designed so that the estimates M_E converge to $M(G)$ from below.

Turning our attention now to the estimate m_E, we first note that $\bar{R}_\infty(P_{n,E}(G))$ does not depend on m_E when Case II conditions hold. For Case I conditions, we shall show that $\bar{R}_\infty(P_{n,E}(G))$ is relatively insensitive to the choice of m_E as long as $m_E \leq m(G)$.† Specifically, if $(1 - M(G))$ is small, if

† The case in which m_E is greater than $m(G)$ is considered later in Chapter 6.

$M_E = M(G)$, and if $m_E \leq m(G) < 0$, we show that, approximately,

$$\frac{[\bar{R}_\infty(P_{n,E}(G))]^{-1}}{[\bar{R}_\infty(P_n(G))]^{-1}} \doteq \left[\frac{1 + \delta - m_E}{1 + \delta - m(G)}\right]^{1/2} \leq \left[\frac{1 - m_E}{1 - m(G)}\right]^{1/2}$$

$$\leq \left[1 + \frac{\lambda}{1 - m(G)}\right]^{1/2}, \qquad (4\text{-}4.7)$$

where

$$\delta = 1 - M(G), \qquad \lambda = m(G) - m_E. \qquad (4\text{-}4.8)$$

Since $M_E = M(G)$, we have from (4-3.23) that $\bar{R}_\infty(P_{n,E}(G)) = -\log r^{1/2}$, where $r = [1 - (1 - \sigma_E^2)^{1/2}]/[1 + (1 - \sigma_E^2)^{1/2}]$ and where σ_E is given by (4-3.3). Using (4-4.8) with $M_E = M(G)$, we obtain

$$(1 - \sigma_E) = 2\delta(1 + \delta - m(G))^{-1}.$$

Since δ is small, it follows that r may be approximated by

$$r \doteq 1 - 2\sqrt{2}\sqrt{1 - \sigma_E} = 1 - 2\sqrt{2}\sqrt{2\delta/(1 + \delta - m_E)}. \qquad (4\text{-}4.9)$$

Thus we have approximately that

$$\bar{R}_\infty(P_{n,E}(G)) \doteq \tfrac{1}{2}(1 - r) \doteq \sqrt{2}\sqrt{2\delta/(1 + \delta - m_E)}. \qquad (4\text{-}4.10)$$

From (4-2.22) together with (4-2.16) and (4-2.19), it follows similarly that $\bar{R}_\infty(P_n(G))$ may be approximated by

$$\bar{R}_\infty(P_n(G)) \doteq \sqrt{2}\sqrt{2\delta/(1 + \delta - m(G))}. \qquad (4\text{-}4.11)$$

Thus

$$\frac{[\bar{R}_\infty(P_{n,E}(G))]^{-1}}{[\bar{R}_\infty(P_n(G))]^{-1}} \doteq \sqrt{\frac{1 + \delta - m_E}{1 + \delta - m(G)}} \qquad (4\text{-}4.12)$$

and (4-4.7) follows.

Thus it is clear from (4-4.7) that Chebyshev acceleration is relatively insensitive to the estimate m_E as long as $m_E \leq m(G)$. For example, if $\lambda(=m(G) - m_E)$ is equal to 0.1, then the expected number of iterations using m_E is only about 4% more than if $m_E = m(G)$ were used. Further, if $m(G) \leq -1.0$, then m_E need satisfy only

$$(m(G) - m_E)/|m(G)| \leq 0.1 \qquad (4\text{-}4.13)$$

in order that the increase in the number of iterations using m_E be less than 4%.

5

An Adaptive Chebyshev Procedure Using Special Norms

5.1   INTRODUCTION

In this chapter we develop parameter estimation and stopping procedures for the Chebyshev acceleration method applied to the basic iterative process

$$u^{(n+1)} = Gu^{(n)} + k. \tag{5-1.1}$$

As in Chapter 4, we assume that the basic method (5-1.1) is symmetrizable (Definition 2-2.1). Thus there exists a symmetrization matrix W such that

$$W(I - G)W^{-1} \tag{5-1.2}$$

is symmetric and positive definite. The adaptive procedures given in this chapter are based on certain inequalities satisfied by the W-vector norm. Examples of symmetrization matrices W for various basic methods were given in Chapter 2. In this chapter, we assume that the matrix W, or at least $W^{T}W$, is conveniently available for computational purposes. Adaptive procedures not utilizing the W-norm are given in Chapter 6.

It follows from Theorem 2-2.1 that the eigenvalues $\{\mu_i\}_{i=1}^{i=N}$ of the $N \times N$ matrix G are real and may be ordered as

$$(m(G) \equiv) \mu_N \leq \mu_{N-1} \leq \cdots \leq \mu_2 \leq \mu_1 (\equiv M(G)) < 1. \qquad (5\text{-}1.3)$$

If m_E and M_E are estimates for $m(G)$ and $M(G)$, respectively, the Chebyshev procedure is defined by (4-3.5) and has the form

$$u^{(n+1)} = \rho_{n+1}[\gamma(Gu^{(n)} + k) + (1 - \gamma)u^{(n)}] + (1 - \rho_{n+1})u^{(n-1)}, \qquad (5\text{-}1.4)$$

where

$$\gamma = 2/(2 - M_E - m_E), \qquad (5\text{-}1.5)$$

$$\rho_{n+1} = \begin{cases} 1 & \text{if} \quad n = 0, \\ (1 - \tfrac{1}{2}\sigma_E^2)^{-1} & \text{if} \quad n = 1, \\ (1 - \tfrac{1}{4}\sigma_E^2\rho_n)^{-1} & \text{if} \quad n \geq 2, \end{cases} \qquad (5\text{-}1.6)$$

and

$$\sigma_E = (M_E - m_E)/(2 - M_E - m_E). \qquad (5\text{-}1.7)$$

In the adaptive procedure given here, a test is made during each iteration to determine whether or not the acceleration parameters are satisfactory. If the acceleration parameters are judged unsatisfactory, the adaptive procedure then gives new estimates for the required parameters. The decisions made by the adaptive procedure are based on a comparison of the observed error reduction of the iterates (5-1.4) with the expected error reduction when the estimated parameters are optimal.

To obtain the optimum Chebyshev parameters, both $m(G)$ and $M(G)$ must be known. However, by making certain assumptions, we simplify the problem in this chapter to that of determining only $M(G)$. We state and discuss these assumptions in Section 5.3.

In Section 5.4, the basic relationships used in the adaptive parameter and stopping procedures are developed. The overall adaptive procedure is then presented in Section 5.5 and some computational aspects are discussed in Section 5.6. The effectiveness of the adaptive process is illustrated in Section 5.7, using a simulated iteration process.

We remark that the adaptive procedure given in Chapter 6 is applicable to more general problems than is the procedure given here. However, the fundamental ideas discussed here provide the basis for the procedure of Chapter 6.

Before starting the development of the adaptive procedure, we introduce in Section 5.2 the pseudoresidual vector, which plays an important role in our adaptive procedure.

5.2   THE PSEUDORESIDUAL VECTOR $\delta^{(n)}$

For the iterate $u^{(n)}$ of (5-1.4), the *pseudoresidual vector* $\delta^{(n)}$ is defined by

$$\delta^{(n)} \equiv Gu^{(n)} + k - u^{(n)}. \tag{5-2.1}$$

With the exact solution to $(I - G)u = k$ denoted by $\bar{u}$, the error vector $\varepsilon^{(n)} \equiv u^{(n)} - \bar{u}$ and the residual vector $\delta^{(n)}$ are related by

$$\delta^{(n)} = (G - I)\varepsilon^{(n)}. \tag{5-2.2}$$

Indeed, since $k = (I - G)\bar{u}$, (5-2.2) is obtained directly from (5-2.1). Moreover, by (5-1.2), $(I - G)$ is nonsingular so that

$$\varepsilon^{(n)} = (G - I)^{-1}\delta^{(n)}. \tag{5-2.3}$$

For any vector norm $\|\cdot\|$ we have

$$\|\varepsilon^{(n)}\| \leq \|(G - I)^{-1}\| \, \|\delta^{(n)}\| \tag{5-2.4}$$

and in particular from (1-4.11)

$$\|\varepsilon^{(n)}\|_W \leq (1 - M(G))^{-1} \|\delta^{(n)}\|_W. \tag{5-2.5}$$

Thus an upper bound for $\|\varepsilon^{(n)}\|_W$ can be obtained from $\|\delta^{(n)}\|_W$ provided that $M(G)$ is known. *The relationships (5-2.3) and (5-2.5) are fundamental to the stopping procedures used in this book.* We now show that the behavior of the $\delta^{(n)}$ vectors is also related to the normalized Chebyshev polynomial associated with (5-1.4).

By (4-3.6), the error vector associated with (5-1.4) satisfies

$$\varepsilon^{(n)} = P_{n, \mathrm{E}}(G)\varepsilon^{(0)}, \tag{5-2.6}$$

where

$$P_{n, \mathrm{E}}(x) \equiv T_n\!\left(\frac{2x - M_\mathrm{E} - m_\mathrm{E}}{M_\mathrm{E} - m_\mathrm{E}}\right) \bigg/ T_n\!\left(\frac{2 - M_\mathrm{E} - m_\mathrm{E}}{M_\mathrm{E} - m_\mathrm{E}}\right) \tag{5-2.7}$$

and $T_n(w)$ is the Chebyshev polynomial defined by (4-2.2). Using (5-2.3), we may express the relation (5-2.6) in terms of $\delta^{(n)}$ as

$$\delta^{(n)} = P_{n, \mathrm{E}}(G)\delta^{(0)}. \tag{5-2.8}$$

Thus from (1-4.11) and (3-2.15), the W-norm of $\delta^{(n)}$ satisfies

$$\|\delta^{(n)}\|_W \leq \mathbf{S}(P_{n, \mathrm{E}}(G))\|\delta^{(0)}\|_W \leq \bar{\mathbf{S}}(P_{n, \mathrm{E}}(G))\|\delta^{(0)}\|_W, \tag{5-2.9}$$

where $\mathbf{S}(P_{n, \mathrm{E}}(G))$ and $\bar{\mathbf{S}}(P_{n, \mathrm{E}}(G))$ denote, respectively, the spectral radius and the virtual spectral radius of the matrix $P_{n, \mathrm{E}}(G)$. *The relationships (5-2.8) and (5-2.9) are fundamental to the adaptive parameter estimation procedures given in this book.*

We remark that it is easy to show that $\varepsilon^{(n+1)}$ and $\delta^{(n+1)}$ also satisfy the same three-term recurrence relation. By (3-2.9), $\varepsilon^{(n+1)}$ may be expressed as

$$\varepsilon^{(n+1)} = \rho_{n+1}[\gamma G\varepsilon^{(n)} + (1 - \gamma)\varepsilon^{(n)}] + (1 - \rho_{n+1})\varepsilon^{(n-1)}, \quad (5\text{-}2.10)$$

which, because of (5-2.2), may also be written in terms of $\delta^{(n)}$ as

$$\delta^{(n+1)} = \rho_{n+1}[\gamma G\delta^{(n)} + (1 - \gamma)\delta^{(n)}] + (1 - \rho_{n+1})\delta^{(n-1)}. \quad (5\text{-}2.11)$$

5.3   BASIC ASSUMPTIONS

To estimate the optimum Chebyshev parameters γ and ρ_n, both $m(G)$ and $M(G)$ must be estimated. However, as shown in Section 4.4, the estimate M_E for $M(G)$ is by far the most important. This fact is used in the following alternative sets of assumptions on the estimates m_E and M_E and on the eigenvalue bounds $m(G)$ and $M(G)$. By making either set of assumptions, we simplify our parameter estimation problem to that of determining only $M(G)$. The two sets of assumptions are referred to as Case I* and Case II*. We have

Case I*: $m_E < M_E < M(G) < 1$ and $m_E \le m(G)$.   (5-3.1)

Case II*: $0 < -m_E = M_E < M(G) < 1$ and $|m(G)| \le M(G)$. (5-3.2)

We remark that Case I* and Case II* are the same as Case I and Case II, respectively, discussed in Chapter 4, except that here we require that $M_E < M(G)$. The assumption that $M_E \ne m_E$ is imposed for convenience and will be relaxed later. The adaptive procedures given later in this chapter are based only on Case I* conditions. For reasons given below, Case II* is of considerably less practical interest than Case I*.

For Case I* conditions, the assumption that $M_E < M(G)$ can be satisfied easily. As we shall see later, all that is required is that the initial guess $M_E^{(0)}$ for $M(G)$ satisfy $M_E^{(0)} < M(G)$; if no better guess is available, this can be done, for example, by picking $M_E^{(0)}$ close to m_E. The inequality $M(G) < 1$ follows from the fact that the basic method is symmetrizable.

Discussion of Case I*

The basic assumption for Case I* is that a lower bound $\underline{m}$ for $m(G)$ is available. The estimate m_E then can be chosen to be $\underline{m}$. For Case I* conditions, the adaptive procedure given later assumes $m_E = \underline{m}$ to be fixed throughout the calculations. Thus only $M(G)$ must be determined. Note that for this case, we do not attempt to obtain the truly optimum Chebyshev parameters

$\bar{\gamma}$ and $\bar{\rho}_{n+1}$ defined by (4-2.12) and (4-2.13). Instead we seek to obtain only the "pseudo"-optimum parameters $\hat{\gamma}$ and $\hat{\rho}_{n+1}$, which are defined by (4-2.12) and (4-2.13), except that the lower bound $\underline{m}$ is used for $m(G)$.

As shown in Section 4.4, the convergence rate of the Chebyshev process is relatively insensitive to the estimate m_E for $m(G)$. Thus if $\underline{m}$ is a reasonable lower bound for $m(G)$, the use of pseudoparameters rather than true optimum parameters will not significantly affect the effectiveness of the Chebyshev process. If $M(G)$ is close to unity and if $m(G)$ and $\underline{m}$ satisfy

$$0 \leq m(G) - \underline{m} \leq 0.1 \qquad \text{if} \quad -1.0 \leq m(G) \leq 0,$$

$$0 \leq \frac{m(G) - \underline{m}}{|m(G)|} \leq 0.1 \qquad \text{if} \quad m(G) < -1.0, \tag{5-3.3}$$

then (see Section 4.4) the use of "pseudo"-optimum parameters will cause roughly a 5% increase in the number of iterations required for convergence. Any $\underline{m}$ satisfying (5-3.3) may be considered reasonable.

We remark that the Chebyshev convergence rate may be significantly decreased if $\underline{m}$ is too loose a bound for $m(G)$. For example, it can be shown that the use of "pseudo"-optimum parameters can increase the number of iterations by more than 40% whenever $\underline{m} < 2m(G) - 1.0$. Thus some care should be taken in the choice of $\underline{m}$. *We warn the reader* that iterative divergence may result if $\underline{m}_E$ does not satisfy $\underline{m}_E \leq m(G)$. An adaptive procedure for estimating $m(G)$ and for detecting if possibly $m_E > m(G)$ is given in Chapter 6.

Case I* is important since a reasonable lower bound for $m(G)$ is readily available for many problems. Some examples of this are the following:

(a) if the basic method (5-1.1) is the SSOR method or the "double method" discussed below, then $m(G) \geq 0$;

(b) if the basic method (5-1.1) is a convergent process, then $\mathbf{S}(G) < 1$ so that $m(G) > -1$; and

(c) if $\|G\|_\infty$ is conveniently calculable, then $\mathbf{S}(G) \leq \|G\|_\infty$ so that $m(G) \geq -\|G\|_\infty$.

It is also sometimes possible to utilize special properties of the coefficient matrix A in (2-1.1) to obtain a priori bounds for $m(G)$. Examples of this are given in Chapters 8 and 10. Also, as we shall presently show, any Case II* problem may be transformed into a problem satisfying Case I* conditions.

Discussion of Case II*

An adaptive procedure for Case II* conditions is of considerably less practical interest than Case I*. There are two reasons for this. First, problems that satisfy (5-3.2) often satisfy the so-called property $\mathscr{A}$ condition (defined in Chapter 9). For these problems, the special methods discussed in Chapters 8 and 9 should be used since they are more efficient than the Chebyshev acceleration method described in this chapter.

The second reason is that any problem satisfying (5-3.2) may also be solved efficiently as a Case I* problem. This can be done in two ways. First, since (5-3.2) implies $m(G) > -1$, any problem satisfying (5-3.2) can be solved by using procedures for Case I* with $\underline{m} = -1$. Moreover, if $M(G)$ is close to unity, the use of $\underline{m} = -1$ causes only a minimal loss of efficiency. This follows from the discussion given in Section 4.4. Second, any basic method (5-1.1) satisfying (5-3.2) may be reformulated as a so-called "double" method, for which Case I* conditions hold with $\underline{m} = 0$. The double method associated with (5-1.1) is

$$u^{(n+1)} = G(Gu^{(n)} + k) + k = G^2 u^{(n)} + (I + G)k. \qquad (5-3.4)$$

If G satisfies the Case II* conditions, then $M(G^2) < 1$ and $m(G^2) \geq 0$. Thus Case I* conditions are satisfied for the basic method (5-3.4) with $\underline{m} = 0$. No loss of efficiency occurs when the double method is used (see, e.g., Hageman et al. [1977]).

In order to avoid the complication of developing two similar but different adaptive procedures, henceforth, we shall be concerned only with problems satisfying the more important Case I* conditions.

5.4   BASIC ADAPTIVE PARAMETER AND STOPPING RELATIONS

In this section we develop stopping and adaptive parameter estimation procedures for the Chebyshev acceleration process (5-1.4) when the estimates m_E and M_E satisfy the Case I* conditions of (5-3.1).

The iteration stopping procedure we give is based on inequality (5-2.5). The adaptive parameter estimation procedure we give consists of two principal subprocedures, both of which are based on inequality (5-2.9). One subprocedure is designed to determine whether or not the parameters currently being used are satisfactory. The other subprocedure is designed to give new improved estimates for the optimum acceleration parameters when the present parameters are judged to be unsatisfactory.† The algorithm given in the next section is divided into analogous subprocedures. It may be instructive to the reader to look ahead at the algorithm while reading the material given below.

---

† As we shall see, these improved parameter estimates will also be utilized in the iteration stopping procedure.

For the procedures presented here, we assume that the estimate M_E for $M(G)$ will be changed several times during the iteration process. To indicate more clearly that M_E is changed and, hence, that different polynomial sequences are generated during the iteration process, we introduce the following notation:

$P_{p,E_s}(x)$ denotes the normalized Chebyshev polynomial of degree p using the most recent estimate M_{E_s} for $M(G)$; here, the subscript s indicates that this is the sth estimate used for $M(G)$

q denotes the last iteration step at which the previous estimate $M_{E_{s-1}}$ was used; for the initial estimate in which $s = 1$, we let $q = 0$

n denotes the current iteration step

Note that M_{E_s} is used first at iteration step $q + 1$ and that $\varepsilon^{(q)} = u^{(q)} - \bar{u}$ is the error vector prior to the first use of M_{E_s}. From (5-2.6), the error vector after step $q + 1$ is $\varepsilon^{(q+1)} = P_{1,E_s}(G)\varepsilon^{(q)}$, and, in general, for step $n = q + p$, the error vector $\varepsilon^{(n)}$ may be expressed as

$$\varepsilon^{(n)} = \varepsilon^{(q+p)} = P_{p,E_s}(G)\varepsilon^{(q)}. \tag{5-4.1}$$

Substituting for $\varepsilon^{(n)}$ and $\varepsilon^{(q)}$ in (5-4.1) and using (5-2.3), we obtain a similar equation for $\delta^{(n)}$, namely,

$$\delta^{(n)} = \delta^{(q+p)} = P_{p,E_s}(G)\delta^{(q)}. \tag{5-4.2}$$

Thus using the above notation, we may express the fundamental inequality (5-2.9) as

$$\|\delta^{(n)}\|_W \leq \mathbf{S}(P_{p,E_s}(G))\|\delta^{(q)}\|_W \leq \bar{\mathbf{S}}(P_{p,E_s}(G))\|\delta^{(q)}\|_W. \tag{5-4.3}$$

For notational convenience, we drop the subscript s on M_{E_s} and $P_{p,E_s}(G)$ when the meaning is clear.

Evaluation of Current Parameters

If the current estimate M_E for $M(G)$ satisfies $M_E \geq M(G)$, it follows from (5-4.3) and (4-3.20) that

$$\|\delta^{(n)}\|_W / \|\delta^{(q)}\|_W \leq 2r^{p/2}/(1 + r^p), \tag{5-4.4}$$

where $p = n - q$ and r is given by (4-3.21). Thus if inequality (5-4.4) is not satisfied, i.e., if

$$\|\delta^{(n)}\|_W / \|\delta^{(q)}\|_W > 2r^{p/2}/(1 + r^p),$$

then we have conclusive evidence that $M_E < M(G)$.

Even though at a given stage we have definite proof that $M(G)$ is greater than M_E, it may nevertheless not be efficient to change M_E. Any time we change M_E, we must start the Chebyshev process over again. In the limiting case in which M_E is changed after every iteration, we would always have $\rho_n = 1$. Thus for this case, we would be carrying out the basic method with extrapolation but without Chebyshev acceleration. Moreover, as shown in Section 4.2, the average virtual rate of convergence for the Chebyshev process increases to an asymptotic value and many iterations are often required before the asymptotic state is reached. Thus if M_E is changed too frequently, the optimum asymptotic convergence rate will never be achieved.

We use a "damping factor" to prevent M_E from being changed too often in the following test: The *current acceleration parameters are deemed to be inadequate if*

$$\frac{\|\delta^{(n)}\|_W}{\|\delta^{(q)}\|_W} > \left(\frac{2r^{p/2}}{1 + r^p}\right)^F. \tag{5-4.5}$$

Here F is a strategy parameter† which is chosen in the interval $[0, 1]$. We shall refer to F as the *damping factor*. The purpose of F is to introduce a kind of "damping" into the adaptive process. As we have already stated, choosing $F = 1$ may result in changing parameters very frequently. On the other hand, with $F = 0$ one would never change parameters no matter what value of $M_E < 1$ is chosen since we always have $\|\delta^{(n)}\|_W < \|\delta^{(q)}\|_W$. Our numerical studies indicate that some value of F in the range 0.65–0.85 is appropriate but that the effectiveness of the adaptive procedure is relatively insensitive to F. By choosing $F < 1$, we show in Section 5.7 that we are in effect resigning ourselves to an average convergence rate which may be only F times the optimum attainable. On the other hand, too large a value of F could result in the parameters being changed too frequently. Further discussion on the choice of F is given in Section 5.7.

The Estimation of New Acceleration Parameters

Once the present acceleration parameters are judged unsatisfactory, new estimates must be obtained. For Case I* conditions, this requires only a new estimate M'_E for $M(G)$. To obtain this new estimate M'_E, we utilize again inequality (5-4.3). First, since $M_E < M(G)$, we have from (4-3.17) that $\bar{\mathbf{S}}(P_{p,\,E}(G)) = P_{p,\,E}(M(G))$. Thus inequality (5-4.3) may be expressed in the form

$$\|\delta^{(n)}\|_W/\|\delta^{(q)}\|_W \leq \bar{\mathbf{S}}(P_{p,\,E}(G)) = P_{p,\,E}(M(G)) = T_p(w_E(M(G)))/T_p(w_E(1)), \tag{5-4.6}$$

---

† By strategy parameter, we mean that no mathematical basis exists for choosing this parameter and that the optimum parameter value is likely to be problem dependent. Usually, the effectiveness of the process is relatively insensitive to the value chosen for a strategy parameter.

where, as in Section 4.3, $w_E(x)$ is defined by

$$w_E(x) \equiv (2x - M_E - m_E)/(M_E - m_E). \quad (5\text{-}4.7)$$

We take as the new estimate M_E' for $M(G)$, the largest real x which satisfies the *Chebyshev equation*

$$T_p(w_E(x))/T_p(w_E(1)) = \|\delta^{(n)}\|_W / \|\delta^{(q)}\|_W. \quad (5\text{-}4.8)$$

The situation can be illustrated, using Fig. 4-3.2. Let $B = \|\delta^{(n)}\|_W / \|\delta^{(q)}\|_W$ and assume that (5-4.5) is satisfied. Thus we have

$$B > 2r^{p/2}/(1 + r^p) = 1/T_p(w_E(1)). \quad (5\text{-}4.9)$$

The solutions to the Chebyshev equation (5-4.8) can be obtained by determining the points at which the horizontal line through $(0, B)$ intersect the curve $P_{p, E}(x)$. For the case of Fig. 4-3.2, the horizontal line through $(0, B)$ intersects $P_{10, E}(x)$ at the points (M_E', B) and (z, B). Note that both z and M_E' lie outside the interval $[m_E, M_E]$, but only M_E' lies in the interval $[m(G), M(G)]$. Thus since the solution z is not relevant here, we take M_E' to be the new estimate for $M(G)$. We note that the horizontal line through $(0, B)$ and the curve $P_{p, E}(x)$ intersect only once if p is odd.

The largest real solution to the Chebyshev equation (5-4.8) may be easily obtained in closed form.

Theorem 5-4.1. Let

$$B \equiv \|\delta^{(n)}\|_W / \|\delta^{(q)}\|_W \quad (5\text{-}4.10)$$

and

$$Q \equiv 1/T_p(w_E(1)) = 2r^{p/2}/(1 + r^p), \quad (5\text{-}4.11)$$

where r is given by (4-3.21) and $w_E(x)$ by (5-4.7). If Case I* conditions (5-3.1) are satisfied and if $B > Q$, then the largest real solution M_E' to the Chebyshev equation (5-4.8) is given by

$$M_E' = \tfrac{1}{2}[M_E + m_E + \tfrac{1}{2}(M_E - m_E)((Y^2 + 1)/Y)], \quad (5\text{-}4.12)$$

where Y is the positive real number

$$Y = [(B/Q) + \sqrt{(B/Q)^2 - 1}]^{1/p}. \quad (5\text{-}4.13)$$

Moreover, M_E' satisfies

$$M_E < M_E' \leq M(G). \quad (5\text{-}4.14)$$

Proof. From (5-4.10) and (5-4.11), the Chebyshev equation (5-4.8) may be written as

$$T_p(w_E(x)) = B/Q, \quad (5\text{-}4.15)$$

where $(B/Q) > 1$ by assumption. Thus by (4-2.2), a solution M'_E to (5-4.15) satisfies

$$w_E(M'_E) = \cosh\left(\frac{1}{p}\cosh^{-1}\left(\frac{B}{Q}\right)\right) = \frac{1}{2}\left(\frac{Y^2 + 1}{Y}\right), \qquad (5\text{-}4.16)$$

where Y is given by (5-4.13). Note that $w_E(M'_E)$ is positive. Then solving (5-4.16) for M'_E we obtain (5-4.12). To show that M'_E is the largest real solution, suppose that $x_1 > M'_E$ is also a solution to (5-4.15). Since this implies by (5-4.7) that $w_E(x_1) > w_E(M'_E) > 0$, we have by (5-4.16) that

$$0 < w_E(x_1) - w_E(M'_E) = w_E(x_1) - \cosh((1/p)\cosh^{-1}(B/Q)); \quad (5\text{-}4.17)$$

but x_1 also satisfies (5-4.8) so that $T_p(w_E(x_1)) = (B/Q) > 1$. Thus we get from (5-4.17)

$$0 < \{w_E(x_1) - \cosh((1/p)\cosh^{-1}T_p(w_E(x_1)))\} = w_E(x_1) - w_E(x_1) = 0,$$

which is a contradiction.

To prove (5-4.14), we use (5-4.6) and the fact that $(B/Q) > 1$ to obtain

$$1 < B/Q \le T_p(w_E(M(G))). \qquad (5\text{-}4.18)$$

Since $\cosh((1/p)\cosh^{-1}x)$ is an increasing function of x for $x \ge 1$, we have that

$$\cosh\left(\frac{1}{p}\cosh^{-1}1\right) < \cosh\left(\frac{1}{p}\cosh^{-1}\frac{B}{Q}\right) \le \cosh\left(\frac{1}{p}\cosh^{-1}T_p(w_E(M(G)))\right)$$

$$(5\text{-}4.19)$$

or equivalently from (5-4.16) and (4-2.2) that

$$1 < w_E(M'_E) \le w_E(M(G)). \qquad (5\text{-}4.20)$$

Inequality (5-4.14) now follows from the definition of $w_E(x)$ and the proof of Theorem 5-4.1 is complete. ∎

Thus when the test (5-4.5) to change parameters is satisfied, the new estimate M'_E for $M(G)$, given by (5-4.12), can be used to obtain improved Chebyshev acceleration parameters γ and ρ_n, given by (5-1.5) and (5-1.6). We note that the required condition $B/Q > 1$ of Theorem 5-4.1 is satisfied whenever inequality (5-4.5) is satisfied. Result (5-4.14) ensures that any new estimate M'_E for $M(G)$ will satisfy the Case I* condition that $M_E \le M(G)$.

Remark. For the algorithm given in the next section, we permit an initial estimate M_E equal to m_E. This is done primarily for convenience when no other reasonable choice for M_E is available. When $M_E = m_E$, the Chebyshev polynomial method (5-1.4) reduces to the extrapolation method (2-2.13) with $\gamma = 1/(1 - m_E)$. For this case, we have

$$\delta^{(n)} = (G_{[\gamma]})^n \delta^{(0)}, \tag{5-4.21}$$

where $G_{[\gamma]} = \gamma G + (1 - \gamma)I$. With $m_E \leq m(G)$, it follows that $\mathbf{S}(G_{[\gamma]}) = (M(G) - m_E)/(1 - m_E)$ and that

$$\|\delta^{(n)}\|_W \leq \left(\frac{M(G) - m_E}{1 - m_E}\right)^n \|\delta^{(0)}\|_W. \tag{5-4.22}$$

Thus when $m_E = M_E$, an estimate M'_E for $M(G)$ may be obtained using

$$M'_E = m_E + (1 - m_E)\left[\frac{\|\delta^{(n)}\|_W}{\|\delta^{(0)}\|_W}\right]^{1/n}. \tag{5-4.23}$$

Formula (5-4.12) for M'_E does not reduce to (5-4.23) when $M_E = m_E$. However, an alternative expression for the solution M'_E to the Chebyshev equation may be given which agrees with (5-4.23) in the limit as $M_E \to m_E$. This alternative expression may be obtained easily from (5-4.12). We initially assume, as in (5-4.12), that $M_E \neq m_E$. It can be shown, using (4-3.21) that $\sigma_E = 2\sqrt{r}/(1 + r)$. But by definition (4-3.3),

$$\sigma_E = (M_E - m_E)/(2 - M_E - m_E).$$

Thus $M_E - m_E = 2\sqrt{r}(2 - M_E - m_E)/(1 + r)$. Using this expression to eliminate $(M_E - m_E)$ in (5-4.12), we obtain

$$M'_E = \frac{1}{2}\left[M_E + m_E + (2 - M_E - m_E)\left(\frac{1}{1 + r}\right)(X + rX^{-1})\right], \tag{5-4.24}$$

where

$$X = [\tfrac{1}{2}(B + \sqrt{B^2 - Q^2})(1 + r^p)]^{1/p}.$$

Since $r \to 0$ and $Q \to 0$ as $M_E \to m_E$, it follows that (5-4.24) agrees with (5-4.23) in the limit. The use of (5-4.24) instead of (5-4.12) in the algorithm given in the next section enables us to treat the special case $M_E = m_E$ in a normal way.

Termination of the Iterative Process

As before, we define the iteration error vector by

$$\varepsilon^{(n)} \equiv u^{(n)} - \bar{u}, \tag{5-4.25}$$

where $\bar{u}$ is the unique solution to $(I - G)u = k$ and $u^{(n)}$ is the Chebyshev iterate of (5-1.4). The iterations are to be terminated whenever some measure of the error becomes sufficiently small; i.e., whenever

$$\|\varepsilon^{(n)}\|_\beta \leq \zeta, \tag{5-4.26}$$

where $\|\cdot\|_\beta$ denotes some vector norm and ζ is the desired accuracy. Since $\bar{u}$ is not known, $\|\varepsilon^{(n)}\|_\beta$ cannot be computed directly. We shall use the pseudo-residual vector $\delta^{(n)}$ to approximate $\varepsilon^{(n)}$.

By (5-2.5) we have $\|\varepsilon^{(n)}\|_W \leq [1/(1 - M(G))]\|\delta^{(n)}\|_W$. Thus if

$$\|\delta^{(n)}\|_W/(1 - M(G)) \leq \zeta, \tag{5-4.27}$$

then (5-4.26) is satisfied whenever $\beta = W$. There are two difficulties with the use of (5-4.27) to terminate the iterative process. First, $M(G)$ may not be known. This, however, causes no great difficulty since sufficiently accurate estimates for $M(G)$ usually are available from the adaptive parameter procedure or by some other means. The second difficulty is that the desired measure of the error vector may not be the W-norm. We note that the symmetrization matrix W depends on the basic iteration method being used as well as on the problem to be solved. The norm used to measure the error vector should be adjustable to the type of problem under consideration. For example, in solving heat conduction problems, the user is often interested in the maximum temperature. For this case, the maximum component norm $\|\cdot\|_\infty$ would be a meaningful measure for the error vector. For other types of problems, appropriate measures for the error vector include the energy norm,[†] the 2-norm, the W-norm, or some area weighted norm.

We now show that the error vector $\varepsilon^{(n)}$ can be measured approximately by

$$\|\varepsilon^{(n)}\|_\beta \doteq \|\delta^{(n)}\|_\beta/(1 - M'_E) \tag{5-4.28}$$

for any vector norm $\|\cdot\|_\beta$ provided that $M'_E \doteq M(G)$ and that n is sufficiently large. The following lemma follows directly from Theorems 6-2.3 and 6-2.4, given later in Section 6.2.

Lemma 5-4.1. If m_E and M_E satisfy the Case I* conditions of (5-3.1), then[‡] for any vector norm β

$$\lim_{n \to \infty} (\|\delta^{(n)}\|_\beta/\|\varepsilon^{(n)}\|_\beta) = 1 - M(G). \tag{5-4.29}$$

[†] The energy norm is defined by $\|\cdot\|_{A^{1/2}}$, where A is the symmetric and positive definite coefficient matrix of (2-1.1). For some problems, the energy norm is related to the energy of the system.

[‡] We also require that the expansion for $\varepsilon^{(0)}$ in terms of the eigenvectors of G not be void in the eigenvector corresponding to $M(G)$.

Thus provided that n is sufficiently large, we have for any vector norm β that

$$\|\varepsilon^{(n)}\|_\beta \doteq \|\delta^{(n)}\|_\beta/(1 - M(G)). \qquad (5\text{-}4.30)$$

With $M(G)$ in (5-4.30) replaced by the estimate M'_E, we get (5-4.28).

Often, a relative error measure is desired; i.e., to terminate the iterative process whenever

$$\|\varepsilon^{(n)}\|_\beta/\|\bar{u}\|_\eta \leq \zeta. \qquad (5\text{-}4.31)$$

Using (5-4.28) and the approximation $\|\bar{u}\|_\eta \doteq \|u^{(n+1)}\|_\eta$, we obtain the termination test

$$\frac{\|\varepsilon^{(n)}\|_\beta}{\|\bar{u}\|_\eta} \doteq \frac{1}{1 - M'_E} \frac{\|\delta^{(n)}\|_\beta}{\|u^{(n+1)}\|_\eta} \leq \zeta. \qquad (5\text{-}4.32)$$

Although any norm η may be used in the normalization, some caution should be taken in the choice of η. Some vector norms are functions of the vector dimension and others are not. For example, if $\bar{u}$ were nearly constant, then $\|\bar{u}\|_2 \doteq \sqrt{N}\,\|\bar{u}\|_\infty$, where N is the dimension of $\bar{u}$. For this case, the choice of $\beta = \infty$ and $\eta = 2$ would result in an equivalent error measurement of $\|\varepsilon^{(n)}\|_\infty/[\sqrt{N}\,\|\bar{u}\|_\infty]$, which can be unfortunate since it depends on the vector dimension N. Usually the normalizing norm $\|\cdot\|_\eta$ should be the maximum component norm $\|\cdot\|_\infty$ or the same β-norm used to measure $\varepsilon^{(n)}$.

Perhaps a better, but sometimes more costly, way to obtain a relative error measure is to use the relative vector norm. This is defined as follows: Let z be any vector with components $z_i \neq 0$ and let $F = (f_{ii})$ be the nonsingular diagonal matrix with elements $f_{ii} = |z_i|$. If β is any vector norm, the β, z-relative norm of a vector ε is then defined as

$$\|\varepsilon\|_{\beta,z} \equiv \|F^{-1}\varepsilon\|_\beta. \qquad (5\text{-}4.33)$$

Obviously, $\|\cdot\|_{\beta,z}$ is a vector norm. The additional computational cost results from the division required to obtain $F^{-1}\varepsilon$.

To terminate the iterative process based on the β, $\bar{u}$-relative norm, we use the obvious approximation

$$\|\varepsilon^{(n)}\|_{\beta,\bar{u}} \doteq \frac{\|\delta^{(n)}\|_{\beta,u^{(n+1)}}}{1 - M'_E} \leq \zeta. \qquad (5\text{-}4.34)$$

If any component $u_i^{(n+1)}$ of $u^{(n+1)}$ is zero or very small, $f_{ii} = 1$ should be used. If β is the ∞-norm, then (5-4.34) becomes

$$\|\varepsilon^{(n)}\|_{\infty,\bar{u}} \doteq \frac{1}{1 - M'_E} \max_i \frac{|\delta_i^{(n)}|}{|v_i|} \leq \zeta, \qquad (5\text{-}4.35)$$

where $|v_i| = 1$ if $u_i^{(n+1)}$ is small and $|v_i| = |u_i^{(n+1)}|$ otherwise.

In Section 5.7 and in later chapters, numerical results are given which show that the iteration error vector can be accurately measured using the pseudo-residual vector $\delta^{(n)}$, as described above.

5.5   AN OVERALL COMPUTATIONAL ALGORITHM

We now describe a computational procedure based on the above discussion. The overall algorithm is given below as an informal program. The notation used is similar to that given previously. In addition, we use the underbar, as $\underline{u}$, to indicate more clearly which variables are vectors.

The input required is summarized below.

ζ the stopping criterion number ζ used in (5-4.32)

m_E the lower bound for $m(G)$, the smallest eigenvalue of G. For Algorithm 5-5.1, m_E *must satisfy* $m_E \leq m(G)$†.

M_E the initial estimate for $M(G)$, the largest eigenvalue of G. For Algorithm 5-5.1, M_E *must satisfy* $m_E \leq M_E < 1$. If $m_E < 0$ and it is known that $M(G) > 0$, then $M_E = 0.0$ is appropriate. If no better choice is available, set $M_E = m_E$.

F the damping factor F used in the parameter change test (5-4.5). The choice of F is discussed in Section 5.7. Typically, F should satisfy $0.65 \leq F \leq 0.85$.

$\underline{u}_r$ the initial guess vector.

The counter n is the current iteration step number, while the counter p is the degree of the Chebyshev polynomial generated (see Section 5.4). DELNE is the norm of δ used in the measurement of the error vector ε, while DELNP is the norm of δ used in the adaptive parameter calculations. Comments are enclosed by the symbols $\langle \ \rangle$.

If an initial estimate $M_E \geq M(G)$ is used, it follows from (5-4.4) that the parameter change test (5-4.5) will never be satisfied. Thus such an estimate will be used for all iterations until convergence. We also note that the initial estimate M_E will be used for all iterations if $F = 0$.

Algorithm 5-5.1. An adaptive procedure for Chebyshev acceleration using the W-norm.

Input: $(\zeta, M_E, m_E, F, \underline{u}_r)$

---

† If an appropriate lower bound for $m(G)$ is not available, we give in Section 6.5 a numerical procedure for estimating $m(G)$.

Initialize:

$n := 0; p := -1; M'_E := M_E; \underline{u} := \underline{0};$

Next Iteration:

$n := n + 1; p := p + 1;$

If $p = 0$ *then* ⟨Initialize for start of new polynomial⟩
    Begin
    $M_E := M'_E; \rho := 1.0; \gamma := 2/(2 - M_E - m_E);$
    $\sigma_E := (M_E - m_E)/(2 - M_E - m_E);$
    $r := (1 - \sqrt{1 - \sigma_E^2})/(1 + \sqrt{1 - \sigma_E^2});$
    End
    else ⟨Continue polynomial generation⟩
    Begin
    If $p = 1$ *then* $\rho := 1/(1 - \frac{1}{2}\sigma_E^2);$ *else* $\rho := 1/(1 - \frac{1}{4}\sigma_E^2\rho);$
    End

Calculate New Iterate:

$\underline{u}_\phi := \underline{u}; \underline{u} := \underline{u}_\Gamma$

$\underline{\delta} := G\underline{u} + \underline{k} - \underline{u};$ DELNP $:= \|\underline{\delta}\|_W;$ DELNE $:= \|\underline{\delta}\|_\beta;$

$\underline{u}_\Gamma := \rho(\gamma\underline{\delta} + \underline{u}) + (1 - \rho)\underline{u}_\phi;$ YUN $:= \|\underline{u}_\Gamma\|_n.$

Calculate New Estimate M'_E

If $p = 0$ *then*
    Begin
    DELNPI $:=$ DELNP;
    Go to **Next Iteration**
    End
    else
    Begin
    $Q := 2r^{p/2}/(1 + r^p); B := \text{DELNP}/\text{DELNPI};$
    If $B > Q$ *then*
        Begin
        $X := [(\frac{1}{2}(1 + r^p))(B + \sqrt{B^2 - Q^2})]^{1/p};$
        $M'_E := \frac{1}{2}\left[M_E + m_E + \left(\frac{2 - M_E - m_E}{1 + r}\right)\left(\frac{X^2 + r}{X}\right)\right];$
        End
        else $M'_E := M_E;$
    End

Convergence Test

If $\dfrac{DELNE}{YUN} \leq \zeta(1 - M'_E)$ *then* print final output and STOP (converged);

$$else \text{ continue};$$

Parameter Change Test

If $B > (Q)^F$ *then* $p := -1$; *else* $p := p + 1$;

Go to **Next Iteration**

It would appear that storage is needed for each of the vectors $\underline{u}$, $\underline{u}_\phi$, and $\underline{u}_\Gamma$ in Algorithm 5-5.1. However, upon closer inspection it is easy to see that storage for only two $\underline{u}$-type vectors is required and that the elements of these vectors need not be moved each iteration. To show how this can be done, let $\underline{u}^{[1]}$ and $\underline{u}^{[2]}$ denote two vector storage arrays. Then replace **Input, Initialize,** and **Calculate New Iterate** in Algorithm 5-5.1 by the following:

Modification 5-5.1.   A modification for Algorithm 5-5.1.

Input: $(\zeta, M_E, m_E, F, u^{[1]})$

Initialize

$n := 0$; $p := -1$; $M'_E := M_E$; $\underline{u}^{[2]} := \underline{0}$; $a := 1$; $b := 2$;

Calculate New Iterate

$\underline{\delta} := G\underline{u}^{[a]} + \underline{k} - \underline{u}^{[a]}$; $DELNP := \|\underline{\delta}\|_W$; $DELNE := \|\underline{\delta}\|_\beta$;

$\underline{u}^{[b]} := \rho(\gamma\delta + \underline{u}^{[a]}) + (1 - \rho)\underline{u}^{[b]}$; $YUN := \|\underline{u}^{[b]}\|_\eta$;

$c := a$; $a := b$; $b := c$; $\langle$Relabeling to interchange $\underline{u}^{[a]}$ and $\underline{u}^{[b]}\rangle$

5.6   TREATMENT OF THE W-NORM

Almost all of the computer cost in carrying out Algorithm 5-5.1 lies in the calculation of the vector δ and in the calculation of the vector norms $\|\delta\|_W$ and $\|\delta\|_\beta$. Our object in this section is to discuss the choice of W and the overhead cost involved in the calculation of $\|\delta\|_W$. We show that the extra work in the computation of $\|\delta\|_W$ can be minimized in some cases. In Chapter 6 we give an adaptive parameter estimation procedure that uses only the two-norm and thus avoids any need for the more general W-norm.

We first discuss the role that the W-norm plays in the overall computational scheme. Recall from Section 5.4 that while the symmetrization matrix

W is used to show that the Chebyshev acceleration process is effective, the matrix W is not used directly in the iterative process. For example, the "optimal" Chebyshev acceleration procedure of (4-2.12)–(4-2.14) does not involve the matrix W. In Algorithm 5-5.1, the matrix W is used in the computation of $\|\delta\|_W$, which is used only in the adaptive parameter estimation procedure. Thus the matrix W affects the Chebyshev acceleration process only through the γ and ρ acceleration parameters.

As shown in Chapter 2, more than one choice of the matrix W is often possible. We do not consider here the interesting problem as to which matrix W is best. However, in the following theorem, we show that a given symmetrization matrix W may be replaced under certain conditions by another matrix W_1 without any effect on the Chebyshev process.

Theorem 5-6.1.   Let W be a symmetrization matrix for G. If the non-singular matrix W_1 satisfies

$$W_1^T W_1 = W^T W, \tag{5-6.1}$$

then W_1 is also a symmetrization matrix for G. Moreover, W_1 and W are norm equivalent in the sense that for any matrix H and any vector v we have

$$\|H\|_{W_1} = \|H\|_W \tag{5-6.2}$$

and

$$\|v\|_{W_1} = \|v\|_W. \tag{5-6.3}$$

Proof.   It is easy to show that if $W(I - G)W^{-1}$ is SPD, then so is $W_1(I - G)W_1^{-1}$. This follows since

$$W^T[W(I - G)W^{-1}]W = W^T W(I - G) = W_1^T W_1(I - G)$$

is SPD, and hence $(W_1^T)^{-1}[W_1^T W_1(I - G)]W_1^{-1} = W_1(I - G)W_1^{-1}$ is also. The relation (5-6.2) follows from

$$
\begin{aligned}
\|H\|_W^2 &= \|WHW^{-1}\|^2 = \mathbf{S}((WHW^{-1})(WHW^{-1})^T) \\
&= \mathbf{S}(WH(W^T W)^{-1} H^T W^T) \\
&= \mathbf{S}((W^T W)H(W^T W)^{-1} H^T) = \mathbf{S}((W_1^T W_1)H(W_1^T W_1)^{-1} H^T) \\
&= \|H\|_{W_1}^2 \tag{5-6.4}
\end{aligned}
$$

Equality (5-6.3) follows from (5-6.1) and the definition of $\|v\|_W$.   ∎

Thus if W_1 and W satisfy (5-6.1), it follows from the above theorem that the Chebyshev acceleration process is independent of the choice W or W_1 for the symmetrization matrix.

For the discussion which follows, we assume that the basic iteration matrix G is expressed, as in (2-2.1), in terms of a splitting matrix Q as

$$G = I - Q^{-1}A = Q^{-1}(Q - A). \tag{5-6.5}$$

Here A is the coefficient matrix (2-1.1) of the linear system to be solved. We assume that both A and Q are symmetric and positive definite. For this case, candidates for a symmetrization matrix W include the symmetric matrices $Q^{1/2}$ and $A^{1/2}$. Since the computation of $\|\delta\|_{A^{1/2}}^2 = (A^{1/2}\delta, A^{1/2}\delta) = (\delta, A\delta)$ is usually more costly than that of $\|\delta\|_{Q^{1/2}}^2 = (Q^{1/2}\delta, Q^{1/2}\delta) = (\delta, Q\delta)$, we assume $W = Q^{1/2}$ is the symmetrization matrix. The matrix $Q^{1/2}$ is not convenient, even if available, for computational use. Thus $\|\delta\|_{Q^{1/2}}^2$ is computed by $(\delta, Q\delta)$.

If carried out in a straightforward manner, the computation of $\|\delta\|_{Q^{1/2}}^2$ requires the matrix–vector product $Q\delta$ plus the inner product $(\delta, Q\delta)$. Compared with the cost to compute $u^{(n+1)}$ from $u^{(n)}$ (i.e., to carry out (5-1.4)), the overhead cost required to compute $\|\delta\|_{Q^{1/2}}$ can be significant. The exact cost is a strong function of the problem being solved and the basic iteration method used. One way to reduce these overhead costs is simply not to compute $\|\delta\|_{Q^{1/2}}$ (and $\|\delta\|_{\beta}$) at every step. Another way to avoid much of the overhead cost is described in the following procedure, which is applicable when Q is expressed in factored form as $Q = S^T S$.

An Alternative Computational Procedure for $\|\delta\|_W$

With G given by (5-6.5), the pseudoresidual vector $\delta^{(n)}$ is given by

$$\delta^{(n)} = Q^{-1}[Ru^{(n)} + b] - u^{(n)}, \tag{5-6.6}$$

where $b = Qk$ is defined by (2-2.1) and where

$$R = Q - A. \tag{5-6.7}$$

For many basic methods, Q is expressed in the factored form (e.g., see Section 2.3)

$$Q = S^T S, \tag{5-6.8}$$

where S is an upper triangular matrix. When Q is of the form (5-6.8), the vector $\delta^{(n)}$ can be obtained easily by successively solving the triangular systems

$$S^T z = Ru^{(n)} + b, \qquad Sy = z \tag{5-6.9}$$

and then obtaining $\delta^{(n)}$ from

$$\delta^{(n)} = y - u^{(n)}. \tag{5-6.10}$$

From (5-6.8), we have that $S^T S = Q = Q^{1/2}Q^{1/2} = (Q^{1/2})^T Q^{1/2}$. Thus from Theorem 5-6.1, the adaptive parameter estimation procedure is independent of whether $\|\delta\|_{Q^{1/2}}$ or $\|\delta\|_S$ is used. However, there are several reasons for using $\|\delta\|_S$ instead of $\|\delta\|_{Q^{1/2}}$. First, it is usually more efficient computationally† to compute $\|\delta\|_S^2 = (S\delta, S\delta)$ rather than $\|\delta\|_{Q^{1/2}}^2 = (\delta, Q\delta)$. Second,

---

† This is true whenever S has fewer nonzero elements than Q.

the additional matrix–vector product $S\delta$ needed for the computation of $\|\delta\|_S$ often can be avoided. As we now show, this can be done by working with the vectors $(Su^{(n)})$ and $(S\delta^{(n)})$ instead of $u^{(n)}$ and $\delta^{(n)}$.

To formulate the problem in terms of the Su and $S\delta$ vectors, we multiply (5-1.4) by S to obtain

$$Su^{(n+1)} = \rho_{n+1}[\gamma S\delta^{(n)} + Su^{(n)}] + (1 - \rho_{n+1})Su^{(n-1)}. \qquad (5\text{-}6.11)$$

Also, from (5-6.6) and (5-6.8), $S\delta^{(n)}$ may be expressed as

$$S\delta^{(n)} = z - Su^{(n)}, \qquad \text{where} \quad z = (S^T)^{-1}[Ru^{(n)} + b]. \qquad (5\text{-}6.12)$$

We now use (5-6.11) and (5-6.12) to reduce the computations required in Modification 5-5.1.

Let $\underline{Su}^{[1]}$, $\underline{Su}^{[2]}$, $\underline{S\delta}$, and $\underline{u}$ denote four storage arrays. (The notation $\underline{S\delta}$, for example, implies the storage of the vector $(S\delta)$.) Now replace **Input**, **Initialize**, and **Calculate New Iterate** in Algorithm 5.1 by the following:

Modification 5-6.1. A Modification for Algorithm 5.1.

Input: $(\zeta, M_E, m_E, F, \underline{u})$

Initialize:

$n := 0$; $p := -1$; $M'_E := M_E$; $\underline{Su}^{[1]} := \underline{Su}$; $\underline{Su}^{[2]} := \underline{0}$; $a := 1$; $b := 2$.

Calculate New Iterate:

$\underline{S\delta} := (S^T)^{-1}[R\underline{u} + \underline{b}]$
$\underline{S\delta} := \underline{S\delta} - \underline{Su}^{[a]}$; DELNP $:= \|\underline{S\delta}\|_2$; DELNE $:=$ DELNP;
$\underline{Su}^{[b]} := \rho(\gamma\underline{S\delta} + \underline{Su}^{[a]}) + (1 - \rho)\underline{Su}^{[b]}$;
$\underline{u} := S^{-1}(\underline{Su}^{[b]})$; YUN $:= \|\underline{u}\|_n$
$c := a$; $a := b$; $b := c$; $\langle$Relabeling to interchange $\underline{Su}^{[a]}$ and $\underline{Su}^{[b]}\rangle$

The procedure given in Modification 5-6.1 requires only the matrix–vector product $R\underline{u}$ and the solution of the two triangular systems, $S^T\underline{z} = R\underline{u} + \underline{b}$ and $S\underline{u} = \underline{Su}^{[b]}$. But to carry out the optimum Chebyshev process (4-2.14) without the acceleration parameter estimation procedure requires these same matrix–vector operations. (In fact, the basic method (5-1.1) also requires these same operations.) Thus, using the above procedure, the only overhead cost for the adaptive parameter estimation procedure is the computation of the inner product $(\underline{S\delta}, \underline{S\delta})$. However, relative to Modification 5-5.1, the above procedure requires that one additional vector be stored.

The above procedure uses the S-norm in the measurement of the error vector. If the maximum component or some measure other than the S-norm is required for the stopping test, then the vector $\underline{\delta}$ is needed. If $\underline{\delta} = S^{-1}(\underline{S\delta})$ is calculated, the above procedure has no advantage over that of Modification 5-5.1. However, $\underline{\delta}$ need not be calculated every iteration. Also, one may require that the $\|\underline{\delta}\|_S$ measure of the error vector be sufficiently small before any estimates using $\|\underline{\delta}\|_\beta$ are computed.

For the balance of this section, we briefly discuss the use of the adaptive procedure, utilizing the W-norm when the basic method is either the Jacobi method or the SSOR method.

The Jacobi Method

For the Jacobi method, the splitting matrix Q is taken to be the block diagonal matrix D of (2-3.11). Thus

$$Q = D = \begin{bmatrix} A_{1,1} & & & 0 \\ & A_{2,2} & & \\ & & \ddots & \\ 0 & & & A_{q,q} \end{bmatrix}. \tag{5-6.13}$$

We still assume that the coefficient matrix A is symmetric and positive definite. Hence each of the diagonal blocks A_{ii} of D is also symmetric and positive definite. The factorization of D can be done by blocks and can be expressed as

$$D = S^T S = \begin{bmatrix} S_1^T S_1 & & & 0 \\ & S_2^T S_2 & & \\ & & \ddots & \\ 0 & & & S_q^T S_q \end{bmatrix}. \tag{5-6.14}$$

The fact that D is a block diagonal matrix has some advantages in the solution of large matrix problems for which disk bulk storage is required. To illustrate this, let the vectors $\underline{S\delta}$, $\underline{Su}^{[b]}$, etc. of Modification 5-6.1 be partitioned in a subvector form which is consistent with the partitioning of D; i.e., for $i = 1, \ldots, q$, we have $\underline{S\delta} = [(\underline{S\delta})_i]$, etc. From the diagonal block form (5-6.14) of $S^T S$, it is clear that the subvectors $(\underline{S\delta})_i$, $(\underline{Su}^{[b]})_i$, and $(\underline{u})_i$ may be computed for each i without reference to the other subvectors. Thus the submatrix S_i can be moved from disk storage, used to calculate $(\underline{S\delta})_i$ and $(\underline{u})_i$, and then discarded until the next iteration. Note that the subvector $(\underline{S\delta})_i$ is not used after $(\underline{Su}^{[b]})_i$ is calculated. Thus for this case, the vector $\underline{S\delta}$ does not have to be stored; only temporary storage for $(\underline{S\delta})_i$ is required.

If S were an arbitrary upper triangular matrix, $\underline{u}$ could not be computed until all components of $\underline{Su}^{[b]}$ were calculated. Thus if S could not be stored in fast memory, the calculations $\underline{S\delta} := (S^T)^{-1}[R\underline{u} + \underline{b}]$ and $\underline{u} := S^{-1}(\underline{Su}^{[b]})$

probably would require that the matrix S (or equivalently S^T) be moved from bulk storage to fast memory twice every iteration. This can increase the computer time required because data transfer from disk storage is slow.

The SSOR Method

From Section 2.3, the SSOR method can be written as

$$u^{(n+1)} = \mathcal{L}_\omega u^{(n)} + k_\omega = \mathcal{U}_\omega(\mathcal{L}_\omega u^{(n)} + k_\omega^{(F)}) + k_\omega^{(B)}. \qquad (5\text{-}6.15)$$

Here $\mathcal{L}_\omega$ is the iteration matrix corresponding to the "forward" SOR method and $\mathcal{U}_\omega$ is the iteration matrix corresponding to the "backward" SOR method. From (2-3.49), symmetrization matrices for the SSOR method include $W = A^{1/2}$ and

$$W = D^{-1/2}[\omega^{-1}D - C_U] = D^{1/2}[\omega^{-1}I - U], \qquad (5\text{-}6.16)$$

where D, C_U, and U are defined by (2-3.11) and (2-3.30).

The pseudoresidual vector $\delta^{(n)}$ for the SSOR method is defined as $\delta^{(n)} \equiv \mathcal{L}_\omega u^{(n)} + k_\omega - u^{(n)}$. From (5-6.15), $\delta^{(n)}$ may also be expressed as

$$\delta^{(n)} = \mathcal{U}_\omega(\Delta^{(n)} + u^{(n)}) + k_\omega^{(B)} - u^{(n)}, \qquad (5\text{-}6.17)$$

where $\Delta^{(n)}$ is the difference vector for the "forward" SOR method, i.e.,

$$\Delta^{(n)} = \mathcal{L}_\omega u^{(n)} + k_\omega^{(F)} - u^{(n)}. \qquad (5\text{-}6.18)$$

When W is given by (5-6.16), Hayes and Young [1977] show that $W\delta^{(n)}$ can be expressed in terms of $\Delta^{(n)}$ as

$$W\delta^{(n)} = \frac{2 - \omega}{\omega} D^{1/2}\Delta^{(n)}. \qquad (5\text{-}6.19)$$

Thus in the adaptive parameter estimation procedure of Algorithm 5-5.1, $\|\delta^{(n)}\|_W$ for the SSOR method may be computed, using

$$\|\delta^{(n)}\|_W = \frac{2 - \omega}{\omega} \|\Delta^{(n)}\|_{D^{1/2}} = \frac{2 - \omega}{\omega} (\Delta^{(n)}, D\Delta^{(n)})^{1/2}. \qquad (5\text{-}6.20)$$

5.7 NUMERICAL RESULTS

In this section we describe results of numerical experiments that illustrate the effectiveness and behavior of the adaptive procedure given in Algorithm 5-5.1. The numerical experiments were carried out using a "simulation" iteration procedure, which we now describe.

The "Simulation" Iteration Procedure†

From (5-1.4), the Chebyshev acceleration procedure applied to the basic method (5-1.1) can be expressed as

$$u^{(n+1)} = \rho_{n+1}[\gamma(Gu^{(n)} + k) + (1 - \gamma)u^{(n)}] + (1 - \rho_{n+1})u^{(n-1)}. \quad (5\text{-}7.1)$$

We assume the basic method is symmetrizable (Definition 2-2.1). From (5-2.3), (5-2.6), and (5-2.8), the error vectors and pseudoresidual vectors associated with (5-7.1) satisfy

$$\varepsilon^{(n)} = (G - I)^{-1}\delta^{(n)} = P_{n,E}(G)\varepsilon^{(0)},$$
$$\delta^{(n)} = P_{n,E}(G)\delta^{(0)}. \qquad (5\text{-}7.2)$$

Let the W-norm be used to measure the error vector. Once $\varepsilon^{(0)}$ is given, the behavior of the Chebyshev procedure of Algorithm 5-5.1 is completely determined by $\|\delta^{(n)}\|_W$. If the eigenvectors of G are known, $\|\delta^{(n)}\|_W$ may be computed by a less costly procedure than that given in the algorithm.

To show this, let $v(1), \ldots, v(N)$ be a set of orthonormal eigenvectors of WGW^{-1} with corresponding eigenvalues $\mu_1, \ldots, \mu_N$. We order the eigenvalues as

$$m(G) \equiv \mu_N \le \mu_{N-1} \le \cdots \le \mu_2 \le \mu_1 \equiv M(G) < 1. \quad (5\text{-}7.3)$$

It follows that the vectors $\{W^{-1}v(i)\}_{i=1}^{i=N}$ are eigenvectors of G and that $\varepsilon^{(0)}$ can be written as

$$\varepsilon^{(0)} = \sum_{i=1}^{N} c_i(W^{-1}v(i)),$$
$$\delta^{(0)} = \sum_{i=1}^{N} c_i(\mu_i - 1)(W^{-1}v(i)), \qquad (5\text{-}7.4)$$

where $c_1, \ldots, c_N$ are suitable constants. From (5-7.2), we have

$$\|\varepsilon^{(n)}\|_W \le \|\delta^{(n)}\|_W/(1 - M(G)),$$
$$\|\varepsilon^{(n)}\|_W^2 = \sum_{i=1}^{N} [c_i P_{n,E}(\mu_i)]^2, \qquad (5\text{-}7.5)$$
$$\|\delta^{(n)}\|_W^2 = \sum_{i=1}^{N} [c_i(1 - \mu_i)P_{n,E}(\mu_i)]^2.$$

† The "simulation" iteration procedure was developed independently by M. M. Sussman at Westinghouse and by D. M. Young and K. C. Jea at the University of Texas (Center for Numerical Analysis Technical Memorandum TM 77.1).

Thus, given the coefficients c_i of (5-7.4) and the eigenvalues μ_i of G, $\|\delta^{(n)}\|_W$ may be computed using (5-7.5). By the "simulated" Chebyshev iteration procedure, we mean the adaptive procedure of Algorithm 5-5.1 with the calculation of $\|\delta^{(n)}\|_W$ carried out by (5-7.5). In addition to being less costly computationally, the simulation procedure permits different problem conditions (i.e., the choice of the c_i and μ_i) to be imposed easily.

The input required by the simulation procedure is the same as that for Algorithm 5-5.1 except for the initial guess vector $u^{(0)}$. Instead of $u^{(0)}$, the simulation procedure requires input values for $\{c_i\}_{i=1}^{i=N}$ and $\{\mu_i\}_{i=1}^{i=N}$. We assume that the initial guess vector $u^{(0)}$ is the null vector. Thus the unique solution $\bar{u}$ to $(I - G)u = k$ is simply $(-\varepsilon^{(0)})$. This follows since then

$$\bar{u} = -[u^{(0)} - \bar{u}] = -\varepsilon^{(0)}. \tag{5-7.6}$$

We always let E_T denote the *true error measure*

$$E_T \equiv \frac{\|\varepsilon^{(n)}\|_W}{\|\bar{u}\|_W} = \frac{\|\varepsilon^{(n)}\|_W}{\|\varepsilon^{(0)}\|_W} \tag{5-7.7}$$

and let E_A denote the approximation (5-4.28) to E_T; i.e.,

$$E_A \equiv \frac{1}{(1 - M'_E)} \frac{\|\delta^{(n)}\|_W}{\|\bar{u}\|_W}. \tag{5-7.8}$$

If ζ is the input stopping number, we let $n(E_T)$ and $n(E_A)$ denote, respectively, the number of iterations required for the inequalities $E_T \leq \zeta$ and $E_A \leq \zeta$ to be first satisfied.

For the "optimal" nonadaptive Chebyshev procedure (4-2.14), by (3-2.12) and (4-2.20) we have that

$$E_T \leq 2\bar{r}^{n/2}/(1 + \bar{r}^n), \tag{5-7.9}$$

where

$$\bar{r} = [1 - \sqrt{1 - \bar{\sigma}^2}]/[1 + \sqrt{1 - \bar{\sigma}^2}] \quad \text{and}$$
$$\bar{\sigma} = [M(G) - m(G)]/[2 - M(G) - m(G)]. \tag{5-7.10}$$

The "theoretical" number of iterations, $n(\text{TNA})$, is defined to be the smallest integer n satisfying $[2\bar{r}^{n/2}/(1 + \bar{r}^n)] \leq \zeta$. It is clear that $n(\text{TNA})$ may be closely approximated by

$$n(\text{TNA}) \doteq [-\log(\zeta/2)/-\log\sqrt{\bar{r}}]. \tag{5-7.11}$$

As a measure of the effectiveness of the adaptive Chebyshev process, we use the *effectiveness ratio* $\mathscr{E}$, where

$$\mathscr{E} \equiv n(E_A)/n(\text{TNA}). \tag{5-7.12}$$

We often use a graph of $-\log E_{\mathrm{T}}$ versus n to indicate the average virtual rate of convergence for a particular iteration process. From (5-7.7), (3-2.12), (3-2.15), and (3-2.16), we have

$$-\log E_{\mathrm{T}} \geq n\left(-\frac{1}{n}\log \bar{\mathbf{S}}(P_{n,\,\mathrm{E}}(G))\right) = n[\bar{R}_n(P_{n,\,\mathrm{E}}(G))]. \quad (5\text{-}7.13)$$

Thus the slope of the curve $-\log E_{\mathrm{T}}(n)$ as a function of n approximates the average virtual rate of convergence.

The numerical results given here are designed to illustrate the following:

the choice of the damping factor F,
the overall effectiveness of the Chebyshev adaptive process,
the effect of the initial error distribution,
the effectiveness of the stopping test,
the sensitivity to the initial estimate M_{E} for $M(G)$, and
the sensitivity to the choice of $m_{\mathrm{E}} = \underline{m}$.

The Choice of Damping Factor F

We first show that if an estimate M_{E} is considered adequate by the test (5-4.5), then (roughly) the average convergence rate obtained with the estimate M_{E} is at least F times the optimum average convergence rate.

From (4-2.21), the optimum average virtual convergence rate for the Chebyshev method is

$$\bar{R}_p(P_p(G)) = -\frac{1}{p}\log \frac{2\bar{r}^{p/2}}{1 + \bar{r}^p}, \quad (5\text{-}7.14)$$

where $\bar{r}$ is given by (5-7.10).

Now suppose that the current estimate $M_{\mathrm{E}} \leq M(G)$ is first used on the $(q + 1)$st iteration to obtain $u^{(q+1)}$. We assume, for reasons of simplicity, that $m_{\mathrm{E}} = m(G)$ is used. From (3-2.16) and (4-3.20), the average (virtual) convergence rate with the estimate M_{E} is given by

$$\bar{R}_p(P_{p,\,\mathrm{E}}(G)) = -\frac{1}{p}\log \bar{\mathbf{S}}(P_{p,\,\mathrm{E}}(G)) = \frac{1}{p}\log\left\{\left[\frac{2r^{p/2}}{1 + r^p}\right]\Big/\left[\frac{2\hat{r}^{p/2}}{1 + \hat{r}^p}\right]\right\}, \quad (5\text{-}7.15)$$

where

$$r = [1 - \sqrt{1 - \sigma_{\mathrm{E}}^2}]/[1 + \sqrt{1 - \sigma_{\mathrm{E}}^2}],$$

$$\sigma_{\mathrm{E}} = [M_{\mathrm{E}} - m(G)]/[2 - M_{\mathrm{E}} - m(G)],$$

$$\hat{r} = [1 - \sqrt{1 - (\sigma_{\mathrm{E}}/\sigma^*)^2}]/[1 + \sqrt{1 - (\sigma_{\mathrm{E}}/\sigma^*)^2}], \quad (5\text{-}7.16)$$

$$\sigma^* = [2M(G) - M_{\mathrm{E}} - m(G)]/[2 - M_{\mathrm{E}} - m(G)].$$

From (5-4.3), it is reasonable to use $\|\delta^{(n)}\|/\|\delta^{(q)}\|$ as an approximation for $\bar{S}(P_{p,\,E}(G))$. Thus we have

$$\frac{\bar{R}_p(P_{p,\,E}(G))}{\bar{R}_p(P_p(G))} \doteq \frac{-\log[\|\delta^{(n)}\|_W/\|\delta^{(q)}\|_W]}{-\log[2\bar{r}^{p/2}/(1+\bar{r}^p)]}. \tag{5-7.17}$$

Now M_E is deemed satisfactory, using (5-4.5), only if

$$\frac{\|\delta^{(n)}\|_W}{\|\delta^{(q)}\|_W} \le \left(\frac{2r^{p/2}}{1+r^p}\right)^F. \tag{5-7.18}$$

Since $r \le \bar{r}$, it follows from (5-7.17) and (5-7.18) that, approximately, we have

$$\frac{\bar{R}_p(P_{p,\,E}(G))}{\bar{R}_p(P_p(G))} \gtrdot -\log\left(\frac{2r^{p/2}}{1+r^p}\right)^F \Bigg/ -\log\left(\frac{2\bar{r}^{p/2}}{1+\bar{r}^p}\right) \ge F. \tag{5-7.19}$$

Thus the estimate M_E is deemed satisfactory only if the ratio of actual to optimum convergence rates is greater than F.

We use the simulation procedure to illustrate the above discussion. Consider the simulative problem given by

$$N = 1000,$$

$$\mu_i = ((1000 - i)/999)M(G), \qquad i = 1, \ldots, 1000, \tag{5-7.20}$$

$$c_i = 1/(1 - \mu_i), \qquad i = 1, \ldots, 1000.$$

Both optimum nonadaptive Chebyshev acceleration and adaptive Chebyshev acceleration were used to solve (5-7.20) with $M(G) = 0.95$. For the adaptive procedure, we used $F = 0.75$, $m_E = 0$, and $M_E = 0.01$ as the initial guess for M_E.

A graph of $-\log_{10} E_T$ versus n is given in Fig. 5-7.1 for both the adaptive and optimal Chebyshev procedures. Numbers in circles (e.g., ①, ②) indicate the iterations for which the estimate M_E was changed. Note that each time M_E is changed the slope of the curve (which approximates $\bar{R}_p(P_{p,\,E}(G))$) becomes small and then increases. If one did not change parameters, the slope would approach an asymptotic value of $-\log_{10}\sqrt{r/\hat{r}}$, where r and $\hat{r}$ are given by (5-7.16). For the optimum nonadaptive procedure, $\bar{R}_8(P_8(G)) = 0.19756$. For the adaptive procedure, the average rate of convergence on iteration 46 (i.e., $\bar{R}_8(P_{8,\,E}(G))$ with $M_E = 0.9479$) was measured to be 0.154. Thus for this iteration, the ratio of actual to optimum convergence rates is $(\bar{R}_8(P_{8,\,E}(G))/\bar{R}_8(P_8(G))) > 0.78$, which indeed satisfies (5-7.19).

We now illustrate the behavior of the adaptive procedure as a function of F. In Fig. 5-7.2, graphs of $-\log_{10} E_T$ versus n are given for different values of F in solving the simulated problem (5-7.20) with $M(G) = 0.99$. We again used $m_E = 0$ and $M_E = 0.01$ as the initial guess for M_E. At the top of each

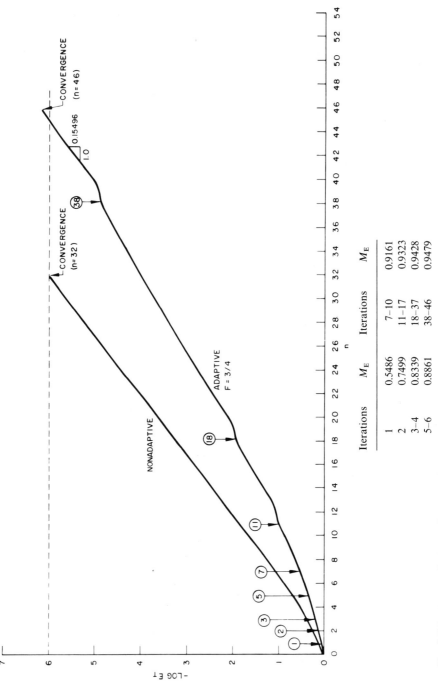

Fig. 5-7.1. Graphs of $-\log E_T$ for the adaptive and nonadaptive Chebyshev procedures. $m_E = 0$, $M(G) = 0.95$, $c_i = 1/(1 - \mu_i)$, $\mu_i = ((1000 - i)/999)$ (0.95), and $i = 1, 2, \ldots, 1000$.

Iterations	M_E	Iterations	M_E
1	0.5486	7–10	0.9161
2	0.7499	11–17	0.9323
3–4	0.8339	18–37	0.9428
5–6	0.8861	38–46	0.9479

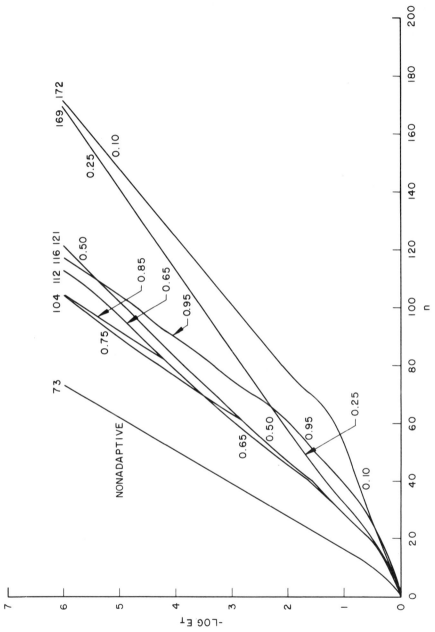

Fig. 5-7.2. Graphs of $-\log E_T$ versus n for various values of the damping factor F. $m_E = 0$, $M(G) = 0.99$.

curve we give the number of iterations required to satisfy $E_T \leq 10^{-6}$. For this problem (and others), the value $F = 0.75$ was found to be as good as any considered.

The problem with $M(G) = 0.99$ was also solved, using the uniform coefficient distribution $c_i = 1.0$. The number of iterations $n(E_T)$ required to satisfy $E_T \leq 10^{-6}$ for different values of F are summarized below.

F	1	0.95	0.9	0.8	0.75	0.7	0.6	0.5	0.25	0.1	Optimal nonadaptive
$n(E_T)$	276	115	107	100	103	102	107	118	151	202	70

For this problem the value of $n(\text{TNA})$ is 73. The actual number of iterations required, using the optimal nonadaptive process, was 70.

Overall Effectiveness of Adaptive Chebyshev Acceleration

In Table 5-7.1 we summarize the iteration data obtained in solving problem (5-7.20) with $M(G) = 0.95, 0.99, 0.999,$ and 0.9999. Results using a

TABLE 5-7.1

Overall Effectiveness[a]

		Initial error distribution			
		$c_i = 1$		$c_i = (1 - \mu_i)^{-1}$	
$M(G)$	Chebyshev procedure	n	$\mathcal{E}$	n	$\mathcal{E}$
0.95	TNA	32		32	
($\bar{r} = 0.40260$)	NA	32		32	
	A	44	1.375	46	1.435
0.99	TNA	73		73	
($\bar{r} = 0.66942$)	NA	71		71	
	A	98	1.342	104	1.425
0.999	TNA	230		230	
($\bar{r} = 0.88114$)	NA	224		227	
	A	284	1.235	312	1.355
0.9999	TNA	726		726	
($\bar{r} = 0.96079$)	NA	708		726	
	A	847	1.165	855	1.176

[a] TNA: Chebyshev acceleration, optimal nonadaptive (theoretical); NA: Chebyshev acceleration, optimal nonadaptive; A: Chebyshev acceleration, adaptive ($F = \frac{3}{4}$); $\mathcal{E} = n(A)/n(\text{TNA})$; n = number of iterations.

uniform coefficient distribution $c_i = 1.0$ are also given. The Chebyshev procedures used are defined in the table footnote. The values of n for the TNA procedure refer to the theoretical number of iterations $n(\text{TNA})$ given by (5-7.11) with $\zeta = 10^{-6}$. The values of n for the optimum nonadaptive procedure NA and the adaptive procedure A are the actual number of iterations required for E_T to satisfy $E_T \leq 10^{-6}$. For the adaptive procedure, $F = 0.75$, $m_E = 0$, and the initial estimate $M_E = 0.01$ were used. It can be seen that the effectiveness ratio never exceeds 1.5. In fact, $\mathscr{E}$ decreases as $M(G)$ increases and is only about 1.17 for the case $M = 0.9999$.

The Effect of the Initial Error Distribution $\{c_i\}$

We now consider the effect of the initial error distribution $\{c_i\}$ in solving problem (5-7.20). It was found that the use of the distribution $c_i = (1 - \mu_i)^{-1}$ leads to about as many iterations as any distribution and that one can expect an effectiveness ratio of not more than 1.50 no matter what distribution is used.

We now illustrate that the estimates M_E obtained by the adaptive process reflect the distribution $\{c_i\}$ used (or equivalently the initial guess vector used). Consider the simulative problem

$$N = 100,$$

$$\mu_i = \alpha_i \frac{\cos(i\pi/101)}{2 - \cos(\pi/101)}, \quad \text{where } \alpha_i = \begin{cases} 1 & \text{for } i = 1, \dots, 50, \\ 3 & \text{for } i = 51, \dots, 100, \end{cases} \tag{5-7.21}$$

and the three $\{c_i\}$ distributions

$$c_i = \begin{cases} R_i, & i = 1, \dots, 50, \\ 10^{-4} R_i, & i = 51, \dots, 100, \end{cases} \tag{5-7.21a}$$

$$\begin{aligned} c_1 &= c_2 = 0 \\ c_i &= \text{same as (5-7.21a)} \quad \text{for } i = 3, \dots, 50, \end{aligned} \tag{5-7.21b}$$

and

$$\begin{aligned} c_1 &= 0.01 R_1, \\ c_2 &= 0.01 R_2, \\ c_i &= \text{same as (5-7.21a)} \quad \text{for } i = 3, \dots, 50. \end{aligned} \tag{5-7.21c}$$

Here $\{R_i\}_{i=1}^{N}$ is a set of random numbers between 0 and 1.

In Fig. 5-7.3, graphs are given which indicate the estimate M_E obtained by the adaptive procedure in solving the three problems (5-7.21a)–(5-7.21c). A jump in the curve indicates that a new estimate M_E for $M(G)$ was obtained

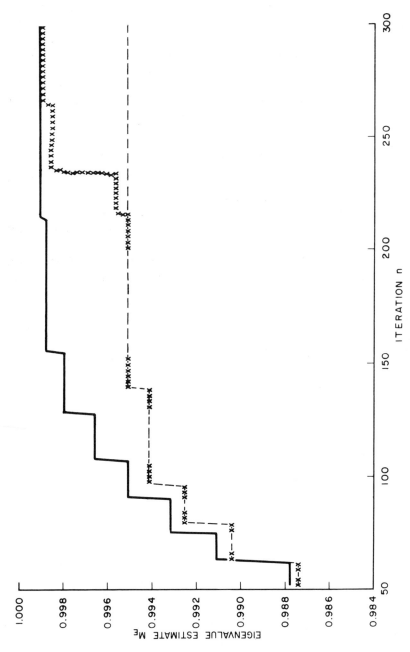

Fig. 5-7.3. Graphs of eigenvalue estimates M_E for different initial guess vectors. Guess vectors: —— (5-7.21a), --- (5-7.21b), × × × (5-7.21c).

at that point. For problem (5-7.21b), the M_E estimates converged to $\mu_3 = 0.995168$ instead of $\mu_1 = 0.999033$. For problem (5-7.21c) with c_1 and c_2 only about $\frac{1}{100}$ times the value of c_3, the estimates M_E behaved initially as in problem (5-7.21b) but then increased and converged to μ_1. Although not shown in Fig. 5-7.4, the M_E estimates prior to iteration 50 were almost the same for all three problems. For all problems, $m_E = -3.0$ $F - 0.75$, and initially $M_F = 0.2$.

The Effectiveness of the Stopping Tests

We now describe some experiments designed to study the effectiveness of the stopping procedure

$$E_A \equiv \frac{1}{(1 - M'_E)} \frac{\|\delta^{(n)}\|_W}{\|\bar{u}\|_W} \le \zeta. \tag{5-7.22}$$

The effectiveness of the stopping procedure (5-4.31), using a general norm β, will be illustrated later in Chapters 8 and 9.

Problem (5-7.20) with $M(G) = 0.99$, but with coefficients $c_i = (1 - \mu_i)^{-\alpha}$, was used for the numerical experiments. In Table 5-7.2, we give the number

TABLE 5-7.2

Iterations Required by the Adaptive Chebyshev Procedure, Using Various Stopping Tests

						α									
	0.2	0.4	0.6	0.8	1.0	1.25	1.5	1.75	2.0	2.25	2.50	2.75	3.0	3.5	5.0	10
$n(E_T)$	104	108	107	106	107	106	104	102	102	97	96	98	93	97	91	82
$n(E_A)$	103	108	106	106	106	104	104	101	101	97	96	98	93	97	91	82
$n(E'_A)$	104	108	107	106	107	106	104	102	102	97	96	98	93	97	91	82

of iterations $n(E)$ required to satisfy $E \le 10^{-6}$ for various values of α and for $E = E_T, E_A$, and E'_A. Here E_A and E_T are given by (5-7.7) and (5-7.22) and E'_A is the same as E_A, except that M'_E is replaced by $M(G)$. The adaptive Chebyshev procedure was used with the input values $m_E = 0$, $F = 0.75$, and $M_E = 0.01$. For this problem, $n(TNA) = 73$. The value of $n(E_A)$ was slightly less than $n(E_A)$ in some cases because M'_E was used instead of $M(G)$. However, $n(E'_A)$ agreed exactly with $n(E_T)$ in each case.

It can be seen that as α becomes large the values of $n(E_T)$ decrease. Eventually, for large α, $n(E_T)$ would come very close to the corresponding value for the optimum nonadaptive process. This follows since $(c_i/c_1) \sim 0$ as $\alpha \to \infty$ for all $i \ne 1$; thus for large α, $\varepsilon^{(n)}$ and $\delta^{(n)}$ for all n closely approximate the eigenvector $v(1)$ corresponding to $M(G)$. Moreover, in this case, the estimate M'_E obtained by the adaptive process would closely approximate $M(G)$ after one iteration.

The stopping test (5-7.22) based on E_A is somewhat less effective for the optimum nonadaptive Chebyshev acceleration procedure for which $M_E = M(G)$. This is to be expected since the assumption of Lemma 5-4.1 that $M_E < M(G)$ is not valid for this case. As we shall see in the next chapter, the basic cause of this loss of accuracy in the approximation E_A is that the $\varepsilon^{(n)}$ and $\delta^{(n)}$ vectors do not necessarily converge to an eigenvector of G associated with $M(G)$. In Table 5-7.3, we give the values of $n(E_T)$ and $n(E_A)$ $(=n(E'_A))$

TABLE 5-7.3

Iterations Required by the Optimum Chebyshev Nonadaptive Procedure

	α							
	0.4	0.6	0.8	1.0	1.5	5.0	10.0	50.0
$n(E_T)$	71	71	71	71	71	72	72	73
$n(E_A)$	89	87	85	82	76	72	72	73

when the optimum nonadaptive Chebyshev process was used to solve the problems of Table 5-7.2. Note that the stopping test based on E_A is more accurate for the larger values of α. This is due to the fact, as noted previously, that the vectors $\varepsilon^{(n)}$ and $\delta^{(n)}$ for all n closely approximate the eigenvector $v(1)$ when α is large.

When the optimum nonadaptive Chebyshev procedure is used, it is often possible to obtain a good a priori upper bound on the number of iterations required for convergence. Assume it is desired to terminate the iterative process when a certain error reduction has been achieved, i.e., when

$$\|\varepsilon^{(n)}\|_W / \|\varepsilon^{(0)}\|_W \leq \zeta. \tag{5-7.23}$$

For the optimum Chebyshev process, we have from (3-2.12) and (4-2.20) that

$$\|\varepsilon^{(n)}\|_W / \|\varepsilon^{(0)}\|_W \leq \bar{\mathbf{S}}(P_n(G)) = 2\bar{r}^{n/2}/(1 + \bar{r}^n), \tag{5-7.24}$$

where $\bar{r}$ is given by (5-7.10). Thus (5-7.23) is satisfied for all $n \geq n(\text{TNA})$, where $n(\text{TNA})$ is given by (5-7.11). For the problems of Table 5-7.3, $n(\text{TNA}) = 73$, which indeed is a good upper bound for $n(E_T)$. We remark that $n(\text{TNA})$ is a rigorous upper bound on the iterations required for convergence only when the error measure (5-7.23) is desired. However, the error measure

$$\|\varepsilon^{(n)}\|_W / \|\bar{u}\|_W \leq \zeta$$

is equivalent to (5-7.23) when the initial guess $u^{(0)}$ is the null vector.

Sensitivity to the Initial Estimate $M_E^{(0)}$ for $M(G)$

In Table 5-7.4, we give the number of iterations $n(E_T)$ required by the adaptive Chebyshev procedure to satisfy $E_T \leq 10^{-6}$ for different values of the initial estimate $M_E^{(0)}$ for $M(G)$. We also give in Table 5-7.4 the number of iterations required by the nonadaptive (nonoptimal) Chebyshev procedure, using $M_E = M_E^{(0)}$ fixed for all iterations. The problem solved was (5-7.20) with $M(G) = 0.99$. Again, $F = 0.75$ and $m_E = 0$ was used.

TABLE 5-7.4

Sensitivity of the Adaptive and Nonadaptive Procedures to the Initial Choice of M_E

	$n(E_T)$			$n(E_T)$	
$M_E^{(0)}$	Adaptive Chebyshev procedure	Nonadaptive (nonoptimal) Chebyshev procedure	$M_E^{(0)}$	Adaptive Chebyshev procedure	Nonadaptive (nonoptimal) Chebyshev procedure
0.0	107		0.95	105	268
0.2	106		0.96	102	236
0.4	105		0.97	102	199
0.6	107	795	0.98	100	153
0.8	106	559	0.985	97	122
0.85	105		0.99	71	71
0.90	105	390			

It can be seen that the behavior of the adaptive procedure is relatively insensitive to the initial choice of M_E unless the initial choice is very close to $M(G)$. On the other hand, the number of iterations required, using the nonadaptive procedure, increases very rapidly as the fixed choice of M_E is decreased below $M(G)$. For this problem $n(TNA) = 73$.

Sensitivity to the Choice of m_E

We now discuss the sensitivity of the Chebyshev procedure to the choice of m_E, assuming only that $m_E \leq m(G)$. The numerical results obtained verified the theoretical result of (4-4.12). Rather than present the numerical results, we use the theoretical result to illustrate the sensitivity of the Chebyshev procedure to the choice of m_E.

Let $M_E = M(G)$, where $M(G)$ is close to unity. Let $\bar{n}$ be the number of iterations required by the *optimum* nonadaptive Chebyshev procedure and let n' be the number of iterations required by the nonadaptive procedure, using the estimate m_E for $m(G)$. From (4-4.12), we have approximately that

$$n'/\bar{n} \doteq \sqrt{(1 - m_E)/(1 - m(G))}. \tag{5-7.25}$$

We first illustrate the effect of a large underestimation of $m(G)$. Suppose $m(G) = -3$ but $m_E = -10$. From (5-7.25), we have

$$n'/\bar{n} \doteq \sqrt{\tfrac{11}{4}} \doteq 1.658. \tag{5-7.26}$$

Hence a gross underestimation of $m(G)$ results in only a moderate, but significant, increase in the number of iterations.

Next, consider the example in which $m(G) = 0$ but $m_E = -1$. In this case

$$n'/\bar{n} \doteq \sqrt{2} \doteq 1.414. \tag{5-7.27}$$

We remark that this result indicates that problems whose eigenvalues lie in the interval $[-M(G), M(G)]$ should require about $\sqrt{2}$ times as many iterations as problems whose eigenvalues lie in $[0, M(G)]$.

As the final example, consider the case in which $m(G) = -0.99$ but $m_E = -1.0$. Here we have

$$n'/\bar{n} \doteq 1.0025, \tag{5-7.28}$$

which is insignificant.

We again caution the reader that using an estimate $m_E > m(G)$ may cause iterative divergence.

6

Adaptive Chebyshev Acceleration

6.1   INTRODUCTION

In Chapter 5, procedures utilizing the W-norm were used to estimate the required iteration parameters for Chebyshev acceleration applied to the basic method

$$u^{(n+1)} = Gu^{(n)} + k. \tag{6-1.1}$$

Properties of the W-norm were used to obtain inequality (5-2.9)

$$\|\delta^{(n)}\|_W / \|\delta^{(0)}\|_W \leq \bar{\mathbf{S}}(P_{n,\mathrm{E}}(G)), \tag{6-1.2}$$

which is the fundamental relation for the adaptive parameter procedure, and inequality (5-2.5)

$$\|\varepsilon^{(n)}\|_W \leq (1 - M(G))^{-1} \|\delta^{(n)}\|_W, \tag{6-1.3}$$

which is the fundamental relation for the stopping procedure.

The W-norm procedures require the existence of a "conveniently available" matrix W, or at least $W^{\mathrm{T}}W$, such that $W(I - G)W^{-1}$ is SPD. However, there are problems when neither W nor $W^{\mathrm{T}}W$ is conveniently available, or else it is not computationally convenient or efficient to work with W-norms. In this chapter, we consider adaptive procedures that involve the use of the 2-norm rather than the W-norm of the pseudoresidual vectors. The procedures given are based on the convergence of the error and pseudoresidual vectors to an eigenvector of the basic iteration matrix G.

As in Chapter 5, we assume the basic method (6-1.1) is symmetrizable (Definition 2-2.1); i.e., there exists a nonsingular matrix W such that $W(I - G)W^{-1}$ is SPD. In this chapter the matrix W will not be used in the computational procedure. Here, the existence of such a matrix W is used only to ensure that the set of eigenvalues $\{\mu_i\}_{i=1}^{i=N}$ for G is real, and to ensure that the set of eigenvectors $\{v(i)\}$ for G includes a basis for the associated vector space. We note that it is often possible to use Chebyshev acceleration, even though the basic method is not symmetrizable. Singular and eigenvector deficient nonsymmetrizable problems are discussed in Section 6.8. There we show that the Chebyshev algorithms given in this chapter and in Chapter 5 may be used without modification to solve certain singular problems but that these algorithms generally should not be used when the basic iteration matrix G has an eigenvector deficiency. The use of Chebyshev acceleration when some of the eigenvalues of G are complex is discussed in Chapters 11 and 12.

The basic adaptive procedures given in this chapter are similar to those given in Chapter 5. For example, if inequality (6-1.2) is satisfied in the 2-norm; i.e., if

$$\|\delta^{(n)}\|_2/\|\delta^{(0)}\|_2 \leq \bar{\mathbf{S}}(P_{n,\mathrm{E}}(G)), \qquad (6\text{-}1.4)$$

then using arguments similar to those given in Chapter 5, the same parameter change test (5-4.5) and the same Chebyshev equation (5-4.8) can be obtained as before except for the norm used. The key question here is that concerning the validity of (6-1.4).

The arguments presented concerning the validity of (6-1.4) are based on the convergence of the pseudoresidual vector $\delta^{(n)}$ to an eigenvector of G. Thus in this chapter, two iterative processes are of interest. The *main iterative process* concerns the convergence of the Chebyshev iterates $u^{(n+1)}$ of (5-1.4) to the solution $\bar{u}$ of $(I - G)u = k$. The *secondary iterative process* concerns convergence of the pseudoresidual iterates $\delta^{(n)}$ of (5-2.8) to an eigenvector of G. The requirements for convergence and the rate of convergence for the secondary iterative process differ from those of the main iterative process. In Section 6.2, convergence theorems are given which are applicable to the secondary iterative process. More complete and rigorous arguments, together with error bounds for the stopping test (5-4.32), are also given in Section 6.2.

The basic relationships for the adaptive parameter estimation procedure utilizing the 2-norm are developed in Section 6.3. An overall computational algorithm is then given in Section 6.4. The algorithm given is based on the assumption that the estimates M_E and m_E satisfy the Case I* conditions of (5-3.1); i.e.,

$$\text{Case } I^*: \quad m_E < M_E < M(G) < 1 \quad \text{and} \quad m_E \leq m(G). \tag{6-1.5}$$

Here $M(G)$ and $m(G)$ are the algebraically largest and smallest eigenvalues of G, respectively. Since the assumption $M_E < M(G)$ need not always be valid when the 2-norm is used, the possibility that some M_E may be larger than $M(G)$ is considered.

In Section 6.5, the behavior of the Chebyshev iterations when $m_E > m(G)$ is discussed. A procedure is given for detecting when this occurs and for obtaining a new estimate for $m(G)$. The procedure described may also be used to obtain an initial estimate for $m(G)$. In Section 6.6, results of numerical experiments are given. Finally, the possibility of accelerating the Chebyshev iterations by intentionally overestimating $M(G)$ is discussed in Section 6.7.

6.2   EIGENVECTOR CONVERGENCE THEOREMS

In this section, convergence theorems for the secondary iterative process are given. Theorem proofs are included for completeness and may be skimmed if desired. Let the vector sequence $\{\delta^{(n)}\}$ be generated by

$$\delta^{(n)} = Q_n(G)\delta^{(0)}, \quad n = 0, 1, \ldots, \tag{6-2.1}$$

where $Q_n(G)$ is a matrix polynomial of degree n (see Section 1.3) with $Q_0(G) = I$ and G is an $N \times N$ matrix. We assume that G is similar to a symmetric matrix. Thus the eigenvalues $\{\mu_i\}_{i=1}^{i=N}$ of G are real and, assuming that $\mu_1 \neq \mu_j$ for some j, may be ordered as

$$m(G) \equiv \mu_N \leq \mu_{N-1} \leq \cdots \leq \mu_t < \mu_{t-1} = \cdots = \mu_1 \equiv M(G). \tag{6-2.2}$$

Note that the largest eigenvalue μ_1 is assumed to be repeated $t - 1$ times. We let $v(i)$ denote an eigenvector of G associated with μ_i; i.e., $Gv(i) = \mu_i v(i)$. We are concerned with the convergence of the vector sequence $\{\delta^{(n)}\}$ to the eigenvector $v(1)$. We show that a sufficient condition for convergence is that the polynomial sequence $\{Q_n(x)\}$ be G-uniformly convergent.

Definition 6-2.1. Let G be an $N \times N$ matrix with the real eigenvalues (6-2.2). The polynomial sequence $\{Q_n(x)\}$ is said to be G-uniformly convergent if

$$Q_n(\mu_1) \neq 0 \quad \text{for all} \quad n \geq 0 \tag{6-2.3a}$$

and

$$\lim_{n \to \infty} |Q_n(x)/Q_n(\mu_1)| = 0 \quad \text{for all} \quad x \in [\mu_N, \mu_t]. \tag{6-2.3b}$$

At the end of this section, we show that the sequence of normalized Chebyshev polynomials $\{P_{n,E}(x)\}$ given by (5-2.7) is G-uniformly convergent when the Case I* conditions of (6-1.5) are satisfied.

Before giving the convergence results, we develop some notation. Since G is similar to a symmetric matrix, there exists a set of eigenvectors $\{v(i)\}_{i=1}^{N}$ for G that is a basis for the associated vector space. Hence there exist constants $d_1, \ldots, d_N$ such that $\delta^{(0)}$ may be expressed as

$$\delta^{(0)} = d_1 v(1) + \cdots + d_{t-1} v(t-1) + \sum_{i=t}^{N} d_i v(i). \qquad (6\text{-}2.4)$$

If some $d_i \neq 0$ for $i = 1, \ldots, t-1$, then $\tilde{v}(1) \equiv d_1 v(1) + \cdots + d_{t-1} v(t-1)$ is also an eigenvector of G with eigenvalue μ_1. Thus no generality is lost by expressing the eigenvector expansion of $\delta^{(0)}$ as

$$\delta^{(0)} = d_1 v(1) + \sum_{i=t}^{N} d_i v(i). \qquad (6\text{-}2.5)$$

In what follows, *we always assume that* $d_1 \neq 0$. As a measure of how closely $\delta^{(0)}$ approximates the eigenvector $v(1)$, we use the *contamination factor* K,

$$K \equiv \left(\sum_{i=t}^{N} \|d_i v(i)\|_{\psi} \right) \Big/ \|d_1 v(1)\|_{\psi}, \qquad (6\text{-}2.6)$$

where the norm ψ is determined by the context in which K is used. Note that if $K = 0$, then $\delta^{(0)}$ is an eigenvector of G associated with the eigenvalue μ_1.

If $Q_n(\mu_1) \neq 0$, it follows from the expansion (6-2.5) that the vectors $\delta^{(n)}$ of (6-2.1) can be expressed as

$$\delta^{(n)} = Q_n(G)\delta^{(0)} = Q_n(\mu_1)\left\{ d_1 v(1) + \sum_{i=t}^{N} d_i \frac{Q_n(\mu_i)}{Q_n(\mu_1)} v(i) \right\}. \qquad (6\text{-}2.7)$$

Thus if the polynomial sequence $\{Q_n(x)\}$ is G-uniformly convergent, we have from (6-2.7) and (6-2.3b) that the sequence $\{\delta^{(n)}\}$ approaches (in direction) the eigenvector $v(1)$ at a rate governed by $\max |Q_n(\mu_i)/Q_n(\mu_1)|$, $t \leq i \leq N$. Since the μ_i for $t \leq i \leq N$ usually are not known, we use the factor

$$\alpha(n) \equiv \max_{x \in [\mu_N, \mu_t]} |Q_n(x)/Q_n(\mu_1)| \qquad (6\text{-}2.8)$$

to indicate the rate of convergence of $\delta^{(n)}$ to $v(1)$. If the polynomial sequence $\{Q_n(x)\}$ is G-uniformly convergent, we have from (6-2.3b) that

$$\lim_{n \to \infty} \alpha(n) = 0. \qquad (6\text{-}2.9)$$

We now have

Theorem 6-2.1. Let the matrix G be similar to a symmetric matrix and let the vector sequence $\{\delta^{(n)}\}$ be defined by (6-2.1), where $\delta^{(0)}$ is given by (6-2.5). If $\{Q_n(x)\}$ is a G-uniformly convergent polynomial sequence, then

$$\delta^{(n)}/Q_n(\mu_1) = d_1 v(1) + \alpha(n)\eta^{(n)}, \qquad (6\text{-}2.10a)$$

where

$$\eta^{(n)} = \sum_{i=t}^{N} \frac{d_i}{\alpha(n)} \frac{Q_n(\mu_i)}{Q_n(\mu_1)} v(i) \qquad \text{and} \qquad \|\eta^{(n)}\| \le K\|d_1 v(1)\|. \quad (6\text{-}2.10b)$$

Moreover,

$$\lim_{n \to \infty} (\delta^{(n)}/Q_n(\mu_1)) = d_1 v(1). \qquad (6\text{-}2.11)$$

Proof. From (6-2.8)

$$\left[\frac{1}{\alpha(n)} \left| \frac{Q_n(\mu_i)}{Q_n(\mu_1)} \right| \right] \le 1 \qquad \text{for} \quad i \ge t.$$

Thus (6-2.10) follows from (6-2.7) and the definition (6-2.6) of K. The limit (6-2.11) is an immediate consequence of (6-2.9) and (6-2.10). ∎

The next theorem relates the behavior of the vector norm sequence $\{\|\delta^{(n)}\|\}$ to the values of $|Q_n(\mu_1)|$.

Theorem 6-2.2. Let G and $\{Q_n(x)\}$ satisfy the conditions of Theorem 6-2.1. Then for any vector norm $\|\cdot\|$,

$$\frac{\|\delta^{(n)}\|}{\|\delta^{(0)}\|} \le |Q_n(\mu_1)| \frac{1 + \alpha(n)K}{\|d_1 v(1) + \cdots + d_N v(n)\|/\|d_1 v(1)\|} \qquad (6\text{-}2.12)$$

and

$$\frac{\|\delta^{(n)}\|}{\|\delta^{(n-1)}\|} \le \left| \frac{Q_n(\mu_1)}{Q_{n-1}(\mu_1)} \right| \frac{1 + \alpha(n)K}{\|d_1 v(1) + \alpha(n-1)\eta^{(n-1)}\|/\|d_1 v(1)\|}, \qquad (6\text{-}2.13)$$

where $\|\eta^{(n)}\| \le K\|d_1 v(1)\|$. Moreover, for sufficiently large n

$$\frac{\|\delta^{(n)}\|}{\|\delta^{(n-1)}\|} \le \left| \frac{Q_n(\mu_1)}{Q_{n-1}(\mu_1)} \right| \frac{1 + \alpha(n)K}{1 - \alpha(n-1)K}. \qquad (6\text{-}2.14)$$

Proof. From (6-2.10), we have that

$$\|\delta^{(n)}\| \le Q_n(\mu_1)[1 + \alpha(n)K]\|d_1 v(1)\|. \qquad (6\text{-}2.15)$$

Inequalities (6-2.12) and (6-2.13) then follow from (6-2.5), (6-2.10), and (6-2.15). From (6-2.9), $\alpha(n - 1)K < 1$ for sufficiently large n. When this is the case, we have from (6-2.10) that

$$\|\delta^{(n-1)}\| \geq Q_{n-1}(\mu_1)[1 - \alpha(n - 1)K]\|d_1v(1)\| > 0, \qquad (6\text{-}2.16)$$

which together with (6-2.15) gives (6-2.14). ∎

In the next theorem, we compare two vector sequences $\{\delta^{(n)}\}$ and $\{\varepsilon^{(n)}\}$ that satisfy

$$\delta^{(n)} = Q_n(G)\delta^{(0)},$$

$$\varepsilon^{(n)} = Q_n(G)\varepsilon^{(0)}, \qquad (6\text{-}2.17)$$

$$\delta^{(n)} = (I - G)\varepsilon^{(n)}.$$

Theorem 6-2.3.   Let $(I - G)$ be similar to a SPD matrix, and let the vector sequences $\{\delta^{(n)}\}$ and $\{\varepsilon^{(n)}\}$ satisfy (6-2.17), where $\delta^{(0)}$ is given by (6-2.5). If $\{Q_n(x)\}$ is a G-uniformly convergent polynomial sequence, then for sufficiently large n and for any vector norm $\|\cdot\|$,

$$\|\varepsilon^{(n)}\| \leq \frac{\|\delta^{(n)}\|}{1 - \mu_1}\left\{1 + \alpha(n)\left[\frac{K}{1 - \alpha(n)K}\right]\right\}. \qquad (6\text{-}2.18)$$

In addition,

$$\lim_{n \to \infty} (\varepsilon^{(n)}/Q_n(\mu_1)) = d_1v(1)/(1 - \mu_1) \qquad (6\text{-}2.19)$$

and

$$\lim_{n \to \infty} \frac{\|\varepsilon^{(n)}\|}{\|\delta^{(n)}\|} = \frac{1}{1 - \mu_1}. \qquad (6\text{-}2.20)$$

Proof.   Since $(I - G)$ is similar to a positive definite matrix, the eigenvalues μ_i of G satisfy $\mu_i < 1$ and hence the matrix $(I - G)$ is nonsingular. Thus, using the eigenvector expansion (6-2.7) for $\delta^{(n)}$, we may write $\varepsilon^{(n)}$ as

$$\varepsilon^{(n)} = (I - G)^{-1}\delta^{(n)} = Q_n(\mu_1)\left\{\frac{d_1v(1)}{1 - \mu_1} + \sum_{i=t}^{N} \frac{d_i}{1 - \mu_i}\frac{Q_n(\mu_i)}{Q_n(\mu_1)}v(i)\right\}. \qquad (6\text{-}2.21)$$

The limit (6-2.19) now follows from (6-2.3b). To show (6-2.18), we first substitute for $Q_n(\mu_1)d_1v(1)$ in (6-2.21), using (6-2.7) to obtain

$$\varepsilon^{(n)} = \frac{\delta^{(n)}}{1 - \mu_1} - \sum_{i=t}^{N}\left(\frac{\mu_1 - \mu_i}{1 - \mu_i}\right)\frac{d_i}{1 - \mu_1}Q_n(\mu_i)v(i). \qquad (6\text{-}2.22)$$

Since $|(\mu_1 - \mu_i)/(1 - \mu_1)| < 1$, we have that

$$\|\varepsilon^{(n)}\| \leq \frac{1}{1 - \mu_1}\left[\|\delta^{(n)}\| + \alpha(n)|Q_n(\mu_1)|\sum_{i=t}^{N}\frac{1}{\alpha(n)}\left|\frac{Q_n(\mu_i)}{Q_n(\mu_1)}\right|\|d_iv(i)\|\right]. \qquad (6\text{-}2.23)$$

From (6-2.8),

$$\left[\frac{1}{\alpha(n)}\left|\frac{Q_n(\mu_i)}{Q_n(\mu_1)}\right|\right] \le 1 \qquad \text{for} \quad i \ge t$$

so that

$$\|\varepsilon^{(n)}\| \le \frac{1}{1 - \mu_1} \left[\|\delta^{(n)}\| + K\alpha(n)Q_n(\mu_1)\|d_1 v(1)\|\right], \qquad (6\text{-}2.24)$$

where K is defined by (6-2.6). The result (6-2.18) then follows since, from (6-2.16), $Q_n(\mu_1)\|d_1 v(1)\| \le \|\delta^{(n)}\|/[1 - \alpha(n)K]$ for sufficiently large n. To prove (6-2.20), from (6-2.9) and (6-2.18) we have that $\lim_{n \to \infty}[\|\varepsilon^{(n)}\|/\|\delta^{(n)}\|] \le 1/(1 - \mu_1)$. But for sufficiently large n, it also can be shown that

$$\|\varepsilon^{(n)}\| \ge \left\{1 - \alpha(n)\left[\frac{K}{1 - \alpha(n)K}\right]\right\}\frac{\|\delta^{(n)}\|}{1 - \mu_1}. \qquad (6\text{-}2.25)$$

Thus we also have that $\lim_{n \to \infty}[\|\varepsilon^{(n)}\|/\|\delta^{(n)}\|] \ge 1/(1 - \mu_1)$ and (6-2.20) follows. ∎

We now give a sufficient condition for the sequence of normalized Chebyshev polynomials $\{P_{n, E}(x)\}$ defined by (5-2.7) to be G-uniformly convergent.

Theorem 6-2.4. Let $P_{n, E}(x) = T_n(w(x))/T_n(w(1))$, where $T_n(w)$ is the Chebyshev polynomial (4-2.1) and where

$$w(x) \equiv (2x - M_E - m_E)/(M_E - m_E). \qquad (6\text{-}2.26)$$

If m_E and M_E satisfy the Case I* conditions of (6-1.5), then the polynomial sequence $\{P_{n, E}(x)\}$ is G-uniformly convergent. Moreover, the associated convergence factors $\alpha(n)$ of (6-2.8) satisfy

$$\alpha(n) \le T_n(w^*)/T_n(w(\mu_1)), \qquad (6\text{-}2.27)$$

where $w^* = w(M_E) = 1$ for $\mu_t < M_E$ and $w^* = w(\mu_t)$ for $\mu_t \ge M_E$.

Proof. We first show (6-2.27). By definition, we have

$$\alpha(n) = \max_{x \in [\mu_N, \mu_t]}\left|\frac{P_{n, E}(x)}{P_{n, 1}(\mu_1)}\right| = \frac{1}{|T_n(w(\mu_1))|}\max_{x \in [\mu_N, \mu_t]}|T_n(w(x))|. \qquad (6\text{-}2.28)$$

Since m_E and M_E satisfy (6-1.5), it follows from (6-2.26) that

$$|w(x)| \le w(M_E) = 1 \qquad \text{for} \quad x \in [m_E, M_E],$$

$$w(x) > 1 \qquad \text{and monotone increasing for} \quad x \in (M_E, 1).$$

$$(6\text{-}2.29)$$

Using (6-2.29) and the fact that $\max_{-1 \leq w \leq 1} |T_n(w)| = T_n(1) = 1$, we obtain

$$\max_{x \in [\mu_N, \mu_t]} |T_n(w(x))| = T_n(w(\mu_t)) \qquad \text{if} \quad \mu_t \geq M_E$$
$$\leq T_n(w(M_E)) \qquad \text{if} \quad \mu_t < M_E, \qquad (6\text{-}2.30)$$

which from (6-2.28) proves (6-2.27). From (4-2.2) and (6-2.29), it easily follows that $P_{n,E}(\mu_1) > 0$. Thus the sequence $\{P_{n,E}(x)\}$ is G-uniformly convergent if we can show that $\lim_{n \to \infty} \alpha(n) = 0$. To do this, let $h(w) \equiv w + \sqrt{w^2 - 1}$ so that by (4-2.2)

$$T_n(w) = \tfrac{1}{2}[h(w)^n + h(w)^{-n}]. \qquad (6\text{-}2.31)$$

Thus from (6-2.27), $\alpha(n)$ satisfies

$$\alpha(n) \leq \left\{ \left[\frac{h(w^*)}{h(\bar{w})} \right]^n + \frac{1}{[h(w^*)h(\bar{w})]^n} \right\} \bigg/ \left\{ 1 + \frac{1}{[h(\bar{w})]^{2n}} \right\}, \qquad (6\text{-}2.32)$$

where $\bar{w} \equiv w(\mu_1)$. Since $h(w) > 1$ for $w > 1$ and since $h(w^*)/h(\bar{w}) < 1$ for $M_E < \mu_1$, we have that $\lim_{n \to \infty} \alpha(n) = 0$, which completes the proof. ∎

From (6-2.10), the factors $\alpha(n)$ give a good indication of the rate at which $\delta^{(n)}$ (and also $\varepsilon^{(n)}$) converges to the eigenvector $v(1)$ of G. If M_E and m_E satisfy the conditions of Case I*, it is easy to show that $\alpha(n) < 1$ for all n. Moreover, for fixed n, it can be shown (after considerable manipulation) that $\alpha(n)$ is a decreasing function of M_E for $M_E \leq \mu_t$ and an increasing function for $\mu_t \leq M_E < \mu_1$. Thus $\alpha(n)$ is minimized when $M_E = \mu_t$. Note that since $\alpha(n) = 1$ for all n when $M_E = \mu_1$, the rate at which $\delta^{(n)}$ converges to $v(1)$ approaches zero as the estimate M_E approaches μ_1 from μ_t.

6.3 ADAPTIVE PARAMETER AND STOPPING PROCEDURES

In this section we develop stopping and adaptive parameter estimation procedures not utilizing the W-norms for the Chebyshev acceleration method applied to the basic method (6-1.1). The adaptive procedures we give are basically the same as those given in Chapter 5, except that the W-norm is replaced by the two-norm. We still assume that the basic method (6-1.1) is symmetrizable. Let the eigenvalues $\{\mu_i\}_{i=1}^{i=N}$ for the $N \times N$ matrix G be ordered as in (6-2.2), and let $v(i)$ be an eigenvector for G associated with the eigenvalue μ_i.

We assume in this section that the estimates M_E and m_E for μ_1 and μ_N, respectively, satisfy the Case I* conditions of (6-1.5) and that the estimate M_E will be changed several times during the iteration process. Whenever a new estimate M_E for μ_1 is used, the generation of a new Chebyshev polynomial sequence is started. Thus when adaptive parameter estimation procedures are used, the total iteration process involves the successive application of different Chebyshev polynomial sequences. We show later in this section that successive polynomial generation can enhance the adaptive procedures.

Successive Polynomial Generation

To indicate more clearly the generation of successive polynomials, we use the notation introduced in Section 5.4, i.e.,

$P_{p, E_s}(x)$ denotes the normalized Chebyshev polynomial of degree p using the most recent estimate M_{E_s} for μ_1; the subscript s indicates that this is the sth estimate used for μ_1

q denotes the last iteration on which the previous estimate $M_{E_{s-1}}$ was used; for the initial estimate where $s = 1$, we let $q = 0$

$P_{\bar{p}_m, E_m}(x)$ denotes the maximum degree polynomial generated, using the estimate M_{E_m}, where $m < s$

n denotes the current iteration step, note that $p = n - q$

With m_E and M_{E_s} as the current estimates for μ_N and μ_1, respectively, the Chebyshev acceleration procedure applied to (6-1.1) is defined by (5-1.4), which in the above notation takes the form

$$u^{(n+1)} = \rho_{n+1}[\gamma(Gu^{(n)} + k) + (1 - \gamma)u^{(n)}] + (1 - \rho_{n+1})u^{(n-1)}, \quad (6\text{-}3.1)$$

where $\gamma = 2/(2 - M_{E_s} - m_E)$ and

$$\rho_{n+1} = \begin{cases} 1 & \text{if } n = q, \\ (1 - \tfrac{1}{2}\sigma_E^2)^{-1} & \text{if } n = q + 1, \\ (1 - \tfrac{1}{4}\sigma_E^2\rho_n)^{-1} & \text{if } n \geq q + 2. \end{cases} \quad (6\text{-}3.2)$$

Here $\sigma_E = (M_{E_s} - m_E)/(2 - M_{E_s} - m_E)$. From Section 5.2, the error vector $\varepsilon^{(n)} \equiv u^{(n)} - \bar{u}$ and the pseudoresidual vector $\delta^{(n)} \equiv Gu^{(n)} + k - u^{(n)}$ associated with (6-3.1) satisfy

$$\delta^{(n)} = P_{p, E_s}(G)\delta^{(q)},$$
$$\varepsilon^{(n)} = P_{p, E_s}(G)\varepsilon^{(q)}, \quad (6\text{-}3.3)$$
$$\delta^{(n)} = (G - I)\varepsilon^{(n)},$$

where $p = n - q$ and

$$P_{p, E_s}(x) = T_p\left(\frac{2x - M_{E_s} - m_E}{M_{E_s} - m_E}\right)\bigg/ T_p\left(\frac{2 - M_{E_s} - m_E}{M_{E_s} - m_E}\right) \quad (6\text{-}3.4)$$

is the normalized Chebyshev polynomial. Here $T_p(w)$ is the Chebyshev polynomial defined by (4-2.1).

Since $n = p + q$, we have $\delta^{(n)} = \delta^{(p+q)} = P_{p,E_s}\delta^{(q)}$. To indicate more clearly the dependence of $\delta^{(p+q)}$ on the current estimate M_{E_s}, we write $\delta^{(p+q)}(=\delta^{(n)})$ as $\delta^{(s,p)}$.† Thus we have (see Eq. (5-4.2))

$$\delta^{(s,p)} \equiv \delta^{(p+q)} = P_{p,E_s}(G)\delta^{(q)} \equiv P_{p,E_s}(G)\delta^{(s,0)}, \tag{6-3.5}$$

which clearly indicates the relationship between δ and the normalized Chebyshev polynomial currently being used. We note that $\delta^{(s,0)} = \delta^{(s-1,\bar{p}_{s-1})}$. Thus

$$\delta^{(s,p)} = P_{p,E_s}(G)\delta^{(s,0)} = P_{p,E_s}(G)\delta^{(s-1,\bar{p}_{s-1})} = P_{p,E_s}(G)P_{\bar{p}_{s-1},E_{s-1}}(G)\delta^{(s-1,0)}. \tag{6-3.6}$$

and in general

$$\delta^{(s,p)} = P_{p,E_s}(G)P_{\bar{p}_{s-1},E_{s-1}}(G)\cdots P_{\bar{p}_1,E_1}(G)\delta^{(1,0)}. \tag{6-3.7}$$

Similarly, we express the error vector $\varepsilon^{(n)}$ as $\varepsilon^{(s,p)}$. Relationships similar to (6-3.5)–(6-3.7) are also valid for the $\varepsilon^{(s,p)}$ vectors.

As in (6-2.5), let the expansion of $\delta^{(m,0)}$, $m = 1, \ldots, s$, in terms of the eigenvectors $\{v(i)\}$ of G be given by

$$\delta^{(m,0)} = d_{1,m}v(1) + d_{t,m}v(t) + \cdots + d_{N,m}v(N). \tag{6-3.8}$$

Then, analogous to (6-2.6), the *contamination factor* K_m for $\delta^{(m,0)}$ is defined by

$$K_m \equiv (\textstyle\sum_{i=t}^N \|d_{i,m}v(i)\|_\psi)/\|d_{1,m}v(1)\|_\psi. \tag{6-3.9}$$

We now show that $\{K_m\}$ is a decreasing sequence under certain conditions.

Theorem 6-3.1. If M_{E_m} and m_E satisfy the Case I* conditions of (6-1.5), then

$$K_{m+1} \leq \alpha(\bar{p}_m)K_m, \tag{6-3.10}$$

where $\alpha(\bar{p}_m)$ satisfies

$$0 < \alpha(\bar{p}_m) \leq T_{\bar{p}_m}(w^*)/T_{\bar{p}_m}(w(\mu_1)) < 1. \tag{6-3.11}$$

Here $w^* = \max[1, w(\mu_t)]$ and $w(x) = (2x - M_{E_m} - m_E)/(M_{E_m} - m_E)$.

Proof. From (6-3.6), $\delta^{(m+1,0)} = \delta^{(m,\bar{p}_m)} = P_{\bar{p}_m,E_m}(G)\delta^{(m,0)}$. Thus the coefficient $d_{i,m+1}$ in the expansion (6-3.8) for $\delta^{(m+1,0)}$ satisfies $d_{i,m+1} = P_{\bar{p}_m,E_m}(\mu_i)d_{i,m}$. Hence we have

$$K_{m+1} \leq \max_{x\in[\mu_N,\mu_t]} |P_{\bar{p}_m,E_m}(x)/P_{\bar{p}_m,E_m}(\mu_1)|K_m = \alpha(\bar{p}_m)K_m, \tag{6-3.12}$$

---

† This notation is more general than that used previously in Section 5.4. Note that $\delta^{(s,0)} = \delta^{(q)}$.

where $\alpha(\bar{p}_m)$ is the convergence factor (6-2.8) associated with the polynomial $P_{\bar{p}_m, E_m}(x)$. It now follows from Theorem 6-2.4 that $\alpha(\bar{p}_m)$ satisfies (6-3.11). ∎

Feasibility of the Basic Inequalities (6-1.2) and (6-1.3) Not Using the W-Norm

For our discussion here, we assume that M_{E_m} and m_E satisfy the Case I* conditions for $m = 1, \ldots, s$.

With $\delta^{(s, 0)}$ given by (6-3.8), we have from Theorem 6-2.2 that

$$\|\delta^{(s, p)}\|_2 / \|\delta^{(s, 0)}\|_2 \leq P_{p, E_s}(\mu_1)\beta(p, K_s), \qquad (6\text{-}3.13)$$

where

$$\beta(p, K_s) \equiv \frac{1 + \alpha(p)K_s}{\|d_{1, s}v(1) + \cdots + d_{N, s}v(N)\|_2 / \|d_{1, s}v(1)\|_2}. \qquad (6\text{-}3.14)$$

Moreover, if $\alpha(p)K_s < 1$, from Theorem 6-2.3 we have for any vector norm that

$$\|\varepsilon^{(s, p)}\| \leq \frac{\|\delta^{(s, p)}\|}{1 - \mu_1}\left\{1 + \frac{\alpha(p)K_s}{1 - \alpha(p)K_s}\right\}. \qquad (6\text{-}3.15)$$

From Theorem 6-2.4, the factors $\alpha(p)$ satisfy (6-2.27) and $\lim_{p \to \infty} \alpha(p) = 0$. Moreover, from Theorem 6-3.1, the contamination factor K_s satisfies

$$K_s \leq \alpha(\bar{p}_{s-1})K_{s-1} \leq \alpha(\bar{p}_{s-1}) \cdots \alpha(\bar{p}_1)K_1, \qquad (6\text{-}3.16)$$

where K_1 is the initial contamination factor associated with $\delta^{(1, 0)}$ $(= \delta^{(0)})$. Note that K_s, for $s > 1$, can be made arbitrarily small by requiring each polynomial generated to be of sufficiently large degree, i.e., by requiring each $\bar{p}_m$ for $m < s$ to be sufficiently large.

We use (6-3.15) to show that inequality (6-1.3) is feasible in any vector norm, and we use (6-3.13) to show that inequality (6-1.2) is feasible using the 2-norm. (Note that $\bar{S}(P_{p, E_s}(G)) = P_{p, E_s}(\mu_1)$ in (6-1.2) since M_{E_s} and m_E are assumed to satisfy Case I* conditions.)

For sufficiently large p, it follows from (6-2.20) that the error vector $\varepsilon^{(s, p)}$ can be approximated by

$$\|\varepsilon^{(s, p)}\| \doteq \frac{\|\delta^{(s, p)}\|}{1 - \mu_1}, \qquad (6\text{-}3.17)$$

where $\|\cdot\|$ denotes any vector norm. Moreover, from (6-3.15) and the discussion following (6-3.15), approximation (6-3.17) is enhanced when s is not small. Thus the fundamental stopping relation (6-1.3) is feasible for any vector norm provided p or s is not small.

Relative to the inequality (6-3.13), we make the following observations concerning $\beta(p, K_s)$:

(a) $\beta(p, K_s) \leq 1$ if the eigenvectors $v(i)$ are orthogonal.
(b) $\beta(p, K_s)$ decreases as p increases. Moreover, if

$$\|d_{1,s}v(1) + \cdots + d_{N,s}v(N)\|_2 > \|d_{1,s}v(1)\|_2,$$

then $\beta(p, K_s) \leq 1$ for sufficiently large p.

(c) Since $\beta(p, K_s) \leq (1 + \alpha(p)K_s)/(1 - K_s)$ if $K_s < 1$, the upper bound for β approaches unity as the contamination factor K_s approaches zero.

Hence if K_s is small or p is sufficiently large, it is likely that $\beta(p, K_s) \leq 1$. When this is the case, from (6-3.13), we have that

$$\|\delta^{(s, \, p)}\|_2/\|\delta^{(s, \, 0)}\|_2 \leq P_{p, \, E_s}(\mu_1) = \bar{\mathbf{S}}(P_{p, \, E_s}(G)). \tag{6-3.18}$$

Thus when p or s is not small, it is reasonable to assume (but with some caution) that the fundamental relation (6-1.2) for the adaptive parameter procedure is approximately valid in the 2-norm.

The validity of both (6-3.17) and (6-3.18) is enhanced if K_s is small and p is large. As noted above, K_s can be made small by requiring each polynomial generated to be of sufficiently large degree. However, if p is required to be too large, the primary Chebyshev iterative process (6-3.1) will be slowed down since many of the iterations will be carried out with a nonoptimum estimate for μ_1. For the calculational procedure given in the next section, we require each polynomial generated to be at least of degree p^*, but we try to pick p^* such that the convergence rate of the primary Chebyshev iterative process is not reduced significantly. With r defined by (4-3.21), the strategy we employ is to pick p^* to be the smallest positive integer greater than 5 that satisfies

$$r^{p^*} \leq d \tag{6-3.19}$$

or, equivalently, to pick p^* to be the smallest integer greater than 5 that satisfies

$$p^* \geq (\log d)/(\log r).$$

The constant d is a *strategy parameter* lying in the range 0–1. Numerical studies indicate that d in the range 0.15–0.03 is appropriate. In Table 6-3.1 we tabulate p^* for different values of r with $d = 0.1$.

We note that p^* could equivalently be defined as the smallest positive integer (> 5) such that the ratio of average to asymptotic convergence rates is greater than some constant, say $\hat{d}$. That is, p^* is the smallest integer greater than 5 satisfying

$$\bar{R}_{p^*}(P_{p, \, E_s}(G))/\bar{R}_\infty(P_{p, \, E_s}(G)) \geq \hat{d}, \tag{6-3.20}$$

TABLE 6-3.1

Values of p^* versus r When
$d = 0.1$

r	p^*	r	p^*
0.5	6	0.95	45
0.7	7	0.975	91
0.8	11	0.98	114
0.9	22		

where the ratio $\bar{R}_{p^*}/\bar{R}_\infty$ is given by (4-2.23). A value of $\hat{d}$ in the range 0.42–0.6 is equivalent to d of (6-3.19) in the range 0.15–0.03. If $\hat{d} = 0.481$, the p^* obtained from (6-3.20) is identical to the p^* obtained from (6-3.19) with $d = 0.1$. (See Table 4-2.1.)

The assumption that M_{E_m} and m_E satisfy the Case I* conditions is discussed in Section 6.4.

Parameter Estimation and Stopping Procedures

If (6-3.17) and (6-3.18) are valid, then using arguments similar to those given in Chapter 5, the same adaptive procedure is obtained as before except for the norm used.

For the parameter change test we use (5-4.5); i.e., the present estimate M_{E_s} is deemed inadequate if

$$\|\delta^{(s,\,p)}\|_2/\|\delta^{(s,\,0)}\|_2 > [2r^{p/2}/(1 + r^p)]^F, \tag{6-3.21}$$

where F is the damping factor and r is defined by (4-3.21). If the present estimate M_E is judged unsatisfactory, a new estimate M_E' is obtained by solving the Chebyshev equation (5-4.8); i.e., M_E' is the largest real x satisfying

$$T_p(w_E(x))/T_p(w_E(1)) = \|\delta^{(s,\,p)}\|_2/\|\delta^{(s,\,0)}\|_2, \tag{6-3.22}$$

where $w_E(x)$ is defined by (5-4.7). Moreover, if $B \equiv \|\delta^{(s,\,p)}\|_2/\|\delta^{(s,\,0)}\|_2$, the solution M_E' can be obtained in closed form by (5-4.12) or by (5-4.24).

For the stopping test, we use (5-4.32), i.e.,

$$[1/(1 - M_E')][\|\delta^{(s,\,p)}\|_\beta/\|u^{(n+1)}\|_\eta] \le \zeta, \tag{6-3.23}$$

where ζ is the stopping number, M_E' the best available estimate for μ_1, and $\|\cdot\|_\beta$ and $\|\cdot\|_\eta$ are any appropriate vector norms.

For a discussion on the choice of norms in (6-3.23), see the remarks given after Eq. (5-4.32) in Chapter 5.

6.4 AN OVERALL COMPUTATIONAL ALGORITHM USING THE 2-NORM

An adaptive procedure for Chebyshev acceleration based on the discussion given in Section 6.3 is given as an informal program in Algorithm 6-4.1. The adaptive parameter estimation procedure given utilizes only the 2-norm of the pseudoresidual vector δ. Relative to the W-norm procedure given in Algorithm 5-5.1, the 2-norm procedure given here requires fewer computations per iteration, is applicable to a wider class of problems, and often requires less storage. However, since the basic inequality (6-1.2) need not be valid for the 2-norm, the estimate M'_E for μ_1 obtained by solving the Chebyshev equation (6-3.22) may be greater than μ_1. Certain precautionary steps are taken in the overall procedure of Algorithm 6-4.1 in an attempt to ensure that all estimates M'_E used are less than μ_1. When used to solve the same problem, the W-norm procedure often requires fewer iterations for convergence than does the 2-norm procedure. The difference is usually less than 10% and is primarily caused by the additional precautionary steps mentioned above.

Except for the additional strategy parameter d, the input required for Algorithm 6-4.1 is the same as that for Algorithm 5-5.1. Again, we assume that the input estimate m_E for μ_N satisfies $m_E \leq \mu_N$.† If such an estimate m_E is not available, we give in Section 6.5 an algorithm for numerically estimating μ_N. For completeness, we summarize each input quantity.

ζ the stopping criterion number ζ used in (6-3.23)

m_E the lower bound for μ_N, the smallest eigenvalue of G; for Algorithm 6-4.1, m_E *must satisfy* $m_E \leq \mu_N$

M_E the initial estimate for μ_1, the largest eigenvalue of G; for Algorithm 6-4.1, M_E *must satisfy* $m_E \leq M_E < 1$. If $m_E < 0$ and it is known that $M(G) > 0$, then $M_E = 0$ is appropriate; if no better choice for M_E is available, set $M_E = m_E$

F the damping factor F used in the parameter change test (6-3.21); the choice of F is discussed in Section 5.7; typically, F should satisfy $0.65 \leq F \leq 0.8$

$\underline{u}_r$ the input guess vector

d the strategy parameter defined by (6-3.19) used to determine the minimum degree required for each Chebyshev polynomial generated; typically, d should satisfy $0.03 \leq d \leq 0.15$

---

† For discussions on possible methods for obtaining a priori bounds m_E such that $m_E \leq \mu_N$, see Section 5.3.

The control variables and counters used are

n	counter for the current iteration step
p	counter for the degree of the Chebyshev polynomial currently being used
s	counter for the number of different estimates M_E used for μ_1
τ_s	upper bound for the sth estimate M_E for μ_1; τ_s is given in (6-4.1)
p^*	each Chebyshev polynomial must be at least of degree p^* before the estimate M_E can be changed; p^* is defined by (6-3.19)
T	counter that has no effect on the procedure of Algorithm 6-4.1; the counter T will be used in the next section
IE	control variable that has no effect on the procedure of Algorithm 6-4.1; IE will be an input quantity in the next section

As in Algorithm 5-5.1, we use the underline mark, as $\underline{u}$, to indicate more clearly which variables are vectors.

Algorithm 6-4.1. An adaptive procedure for Chebyshev acceleration using the 2-norm.

Input: $(\zeta, M_E, m_E, F, \underline{u}_\Gamma, d)$

Initialize:

$n := 0$; $p := -1$; $M'_E := M_E$; $\underline{u}_\phi := \underline{0}$; $s := 0$; $IE := 1$; $R := 1.0$; $\text{DELNP} := 1.0$.

Next Iteration:

$n := n + 1$; $p := p + 1$; $\text{DELNØ} := \text{DELNP}$;
If $p = 0$, *then* ⟨Initialize for start of new polynomial⟩
> *Begin*
> $s := s + 1$; $T := 0$; *if* $M'_E > \tau_s$, *then* $M'_E = \tau_s$; *else* continue;
> $M_E := M'_E$; $\rho := 1.0$; $\gamma := 2/(2 - M_E - m_E)$;
> $\sigma_E := (M_E - m_E)/(2 - M_E - m_E)$;
> $r := (1 - \sqrt{1 - \sigma_E^2})/(1 + \sqrt{1 - \sigma_E^2})$;
> $p^* := [\log d/\log r]$;
> *If* $p^* < 6$, *then* $p^* = 6$; *else* continue;
> *If* $IE \geq 0$, *then* continue; *else* $p^* = 8$;
> *End*
> *else* ⟨Continue polynomial generation⟩
> *Begin*
> *If* $p = 1$, *then* $\rho := 1/(1 - \frac{1}{2}\sigma_E^2)$; *else* $\rho := 1/(1 - \frac{1}{4}\sigma_E^2\rho)$;
> *End*

Calculate New Iterate:

$\underline{\delta} := G\underline{u}_r + \underline{k} - \underline{u}_r$; $DELNP := \|\underline{\delta}\|_2$; $DELNE := \|\underline{\delta}\|_\beta$;
$\underline{u} := \rho(\gamma\underline{\delta} + \underline{u}_r) + (1 - \rho)\underline{u}_\phi$; $YUN := \|\underline{u}\|_\eta$.
$\underline{u}_\phi := \underline{u}_r$; $\underline{u}_r := \underline{u}$

Calculate New Estimate M'_E:

$R\varnothing := R$; $R := DELNP/DELN\varnothing$;
If $p \le 2$, *then*
  Begin
  If $p = 0$, *then* $DELNPI := DELNP$; *else* continue;
  Go to **Next Iteration**
  End
  else
  Begin
  $Q := 2r^{p/2}/(1 + r^p)$; $B := DELNP/DELNPI$;
  If $B \ge 1.0$, *then*
    Begin
    $T = T + 1$;
    If $IE > 0$, *then go to* **Next Iteration**;
      else go to **Parameter Change Test**;
    End
    else continue;
  If $B > Q$, *then*
    Begin

$$X := \left[\left(\frac{1 + r^p}{2} \right) (B + \sqrt{B^2 - Q^2}) \right]^{1/p};$$

$$M'_E := \frac{1}{2}\left[M_E + m_E + \left(\frac{2 - M_E - m_E}{1 + r} \right)\left(\frac{X^2 + r}{X} \right) \right];$$

    End
    else $M'_E = M_E$;
  End

Convergence Test:

If $\dfrac{DELNE}{YUN} \le \zeta(1 - M'_E)$, *then* print final output and STOP (converged);

    else continue;

Parameter Change Test:

If $p \geq p^*$, *then*
        Begin
        If $B > (Q)^F$, *then* $p := -1$; *else* continue;
        End
        else continue;

Go to **Next Iteration**

In an attempt to ensure that all estimates M_E are less than or equal to μ_1, two precautionary steps are taken in the procedure of Algorithm 6-4.1. First is the requirement discussed in the previous section that each Chebyshev polynomial generated be at least of degree p^*. The second precautionary step is that of imposing upper bounds on the M_E estimates. Specifically, if $s - 1$ Chebyshev polynomials have been generated, then for polynomial s, we require that $M_E \leq \tau_s$, where the strategy parameters τ_s are chosen to be

$$\tau_1 = 0.948, \qquad \tau_2 = 0.985, \qquad \tau_3 = 0.995, \qquad \tau_4 = 0.9975,$$
$$\tau_5 = 0.9990, \qquad \tau_6 = 0.9995, \qquad \tau_s = 0.99995 \text{ for } s \geq 7.$$

$$(6\text{-}4.1)$$

We have found that the upper bounds of (6-4.1) are imposed infrequently. Note that the iteration convergence test is made only if p is greater than 2. This is a minor precautionary step whose effect on the iterations required is of little significance.

If Modification 5-5.1 is used, storage for only two $\underline{u}$-type vectors is required. Also note that the vector $\underline{\delta}$ need not be stored in some situations. For a discussion on this, see the remarks given in Section 5.6 concerning the use of the Jacobi method.

As mentioned previously, it may happen when the 2-norm is used that the estimate M_E used to generate $P_{p,E}(G)$ is greater than μ_1. When this happens, we show in what follows that it is highly unlikely that the parameter change test (6-3.21) will ever be satisfied. Thus the iteration process will continue, using M_E until convergence occurs. Normally, when $M_E > \mu_1$, $\delta^{(s,p)}$ and $\varepsilon^{(s,p)}$ will already closely approximate an eigenvector of G and M_E also will be close to μ_1. Thus the validity of the stopping test and the convergence rate of the iterative process are usually not seriously hampered by the fact that $M_E > \mu_1$. In fact, as we shall see in Section 6.7, the convergence rate is often accelerated when $M_E > \mu_1$.

We now give an intuitive argument to show why the parameter change test (6-3.21) is not likely to be satisfied† and why the convergence rate is

---

† With $M_E > \mu_1$, it is clear from (5-4.4) that inequality (6-3.21) can never be satisfied if the W-norm is used instead of the 2-norm.

sometimes accelerated when $M_E > \mu_1$. Since the occurrence of $M_E > \mu_1$ is most likely for s not small, it is reasonable to assume that $\delta^{(s, 0)}$ closely approximates the eigenvector $v(1)$ of G. Thus, approximately, $\delta^{(s, 0)} \doteq d_{1,s} v(1)$ and†

$$\delta^{(s, p)} \doteq P_{p, E}(\mu_1) d_{1, s} v(1). \tag{6-4.2}$$

Let $w = \sigma_E(\mu_1)/\sigma_E(M_E)$, where $\sigma_E(x) = (2x - M_E - m_E)/(2 - M_E - m_E)$. Since $M_E > \mu_1$, w is less than unity. Thus from the definition (4-2.2) of $T_p(w)$, we can write $P_{p, E}(\mu_1)$ as

$$P_{p, E}(\mu_1) = \frac{T_p(w)}{T_p(1/\sigma_E(M_E))} = \frac{\cos p\theta}{T_p(1/\sigma_E(M_E))}, \tag{6-4.3}$$

where

$$\theta = \cos^{-1} w = \tan^{-1}[\sqrt{1 - w^2}/w]. \tag{6-4.4}$$

From (6-4.2), we then have that

$$\frac{\|\delta^{(s, p)}\|}{\|\delta^{(s, 0)}\|} \doteq \frac{|\cos p\theta|}{T_p(1/\sigma_E(M_E))} = |\cos p\theta| \left[\frac{2r^{p/2}}{1 + r^p}\right]. \tag{6-4.5}$$

It is clear for this case that the parameter change test will never be satisfied for any value of p.

Since the error vector $\varepsilon^{(s, p)}$ may be expressed as $\varepsilon^{(s, p)} = (G - I)^{-1}\delta^{(s, p)}$, we have from (6-4.2) and (6-4.3) that

$$\|\varepsilon^{(s, p)}\| \doteq |P_{p, E}(\mu_1)| \|\varepsilon^{(s, 0)}\| = \frac{|\cos p\theta|}{T_p(1/\sigma_E(M_E))} \|\varepsilon^{(s, 0)}\|. \tag{6-4.6}$$

Thus the error vector will become small as $p\theta$ approaches $\pi/2$. We shall discuss this occurrence again in Section 6.7.

Remark 1. The success of the adaptive procedures of Algorithm 6-4.1 is not guaranteed mathematically for every problem. However, the procedures are based on sound mathematical reasoning, complemented by numerical experimentation, and have worked well in practice. Moreover, by printing certain iteration data, an a posteriori measure of the effectiveness of the adaptive procedure can be obtained. The following two numbers can be useful in this appraisal.

The first number is

$$L^{(s, p)} = \left[\frac{\|\delta^{(s, p)}\|_2}{\|\delta^{(s, 0)}\|_2}\right] \bigg/ \left[\frac{2r^{p/2}}{1 + r^p}\right]. \tag{6-4.7}$$

† For notational convenience, we drop the subscript s on M_E and $P_{p, E}(x)$.

With $\sigma_E(x) = (2x - M_E - m_E)/(2 - M_E - m_E)$, we first show that if $M_E < \mu_1$, then

$$L^{(s,\,p)} \le T_p(\sigma_E(\mu_1)/\sigma_E(M_E))\beta(p, K_s), \qquad (6\text{-}4.8)$$

where $\beta(p, K_s)$ is defined by (6-3.14). Indeed, from (6-3.4) we have that

$$P_{p,\,E}(\mu_1) = \frac{T_p(\sigma_E(\mu_1)/\sigma_E(M_E))}{T_p(1/\sigma_E(M_E))} \quad \text{and} \quad P_{p,\,E}(M_E) = \frac{1}{T_p(1/\sigma_E(M_E))}$$

$$= \frac{2r^{p/2}}{1 + r^p}, \qquad (6\text{-}4.9)$$

which when combined with (6-3.13), gives (6-4.8). Now, $[\sigma_E(M_E)/\sigma_E(\mu_1)] < 1$ if $M_E < \mu_1$. Hence, using (4-2.2), we may express (6-4.8) in the form

$$L^{(s,\,p)} \le [(1 + \hat{r}^p)/2\hat{r}^{p/2}]\beta(p, K_s), \qquad (6\text{-}4.10)$$

where $\hat{r}$ is given by (4-3.21) and is

$$\hat{r} = \left[1 - \sqrt{1 - \left(\frac{\sigma_E(M_E)}{\sigma_E(\mu_1)}\right)^2}\right] \bigg/ \left[1 + \sqrt{1 - \left(\frac{\sigma_E(M_E)}{\sigma_E(\mu_1)}\right)^2}\right].$$

Thus if $L^{(s,\,p)}$ is greater than unity and is increasing with p, the implication from (6-4.10) is that the current estimate M_{E_s} is less than μ_1. If $L^{(s,\,p)}$ behaves as $|\cos p\theta|$, the implication from (6-4.5) and (6-4.7) is that $M_{E_s} > \mu_1$. We note that $L^{(s,\,p)} = 1$ under the ideal circumstances that $M_{E_s} = \mu_1$ and that $\delta^{(s,\,0)} = d_{1,\,s}\,v(1)$.

The second number suggested for printing is

$$C^{(s,\,p)} = \frac{\log[\|\delta^{(s,\,p)}\|_2/\|\delta^{(s,\,0)}\|_2]}{\log[2r^{p/2}/(1 + r^p)]}. \qquad (6\text{-}4.11)$$

We shall show that $C^{(s,\,p)}$ approximates the ratio of actual to optimum average convergence rates and is useful mainly as a quick indicator of the effectiveness of the adaptive process.† We define the optimum convergence rate $\hat{R}_p$ to be that obtained when $M_E = \mu_1$ or

$$\hat{R}_p(P_p(G)) = -\frac{1}{p}\log P_p(\mu_1) = -\frac{1}{p}\log\frac{1}{T_p(1/\sigma^*)}, \qquad (6\text{-}4.12)$$

where $\sigma^* = (\mu_1 - m_E)/(2 - \mu_1 - m_E)$. If $M_E < \mu_1$, the average convergence rate, by (3-2.16) and (4-3.17), is $\bar{R}_p(P_{p,\,E}(G)) = -(1/p)\log P_{p,\,E}(\mu_1)$. For sufficiently large p and s we have, from (6-3.13) and the discussion given there,

† Note that $C^{(s,\,p)} = 1 + \{(\log L^{(s,\,p)})/\log[2r^{p/2}/(1 + r^p)]\}$.

that $P_{p,E}(\mu_1) \doteq \|\delta^{(s,p)}\|_2/\|\delta^{(s,0)}\|_2$. Thus $\bar{R}_p(P_{p,E}(G))$ may be approximated by

$$\bar{R}_p(P_{p,E}(G)) \doteq -\frac{1}{p} \log \frac{\|\delta^{(s,p)}\|_2}{\|\delta^{(s,0)}\|_2}. \tag{6-4.13}$$

Moreover, since $M_E < \mu_1$, $\sigma_E(M_E) < \sigma^*$, and $T_p(1/\sigma_E(M_E)) > T_p(1/\sigma^*) > 1$. Combining this with (6-4.12) and using the fact that $T_p(1/\sigma_E(M_E)) = (1 + r^p)/(2r^{p/2})$, we have

$$\hat{R}_p(P_p(G)) \leq -\frac{1}{p} \log \frac{2r^{p/2}}{1 + r^p}. \tag{6-4.14}$$

Inequality (6-4.14) together with (6-4.13) then gives approximately that

$$\bar{R}_p(P_{p,E}(G))/\hat{R}_p(P_p(G)) \gtrdot C^{(s,p)}. \tag{6-4.15}$$

Thus when $M_E < \mu_1$, $C^{(s,p)}$ may be given the following interpretation: The average rate of convergence being obtained with the present estimate M_E for μ_1 is at worst only $C^{(s,p)}$ times the optimum convergence rate. When $M_E < \mu_1$, obviously $C^{(s,p)}$ should be less than unity. If $M_E > \mu_1$, $C^{(s,p)}$ will be greater than unity for sufficiently large p.

As in Section 5.7, it can be shown that the parameter change test (6-3.21) is satisfied only if $C^{(s,p)} < F$. Thus the estimate M_E will not be changed if $C^{(s,p)} \geq F$.

Remark 2. Since μ_1 may be known for some problems, an option should be added to the procedures of Algorithm 6-4.1 such that the input estimate M_E is never changed. This can be done easily by avoiding the test on τ_1 and setting p^* to a large number.

Remark 3. As indicated in Table 5-4.1, each time a new polynomial is started, its average rate of convergence starts at a low value and increases monotonically with p to its asymptotic value. If M_E or r is close to unity, p is quite large before the asymptotic convergence rate is approached. Thus if the stopping test is close to being satisfied, it may be more efficient to continue to use the present estimate M_E rather than to start a new polynomial with a new estimate. Such a decision can be based on estimates for the additional iterations required for convergence, using the present estimate M_E and using the new estimate M_E'.

6.5 THE ESTIMATION OF THE SMALLEST EIGENVALUE μ_N

The adaptive parameter estimation procedures given in Sections 5.5 and 6.4 are based on the assumption that the eigenvalue estimate m_E satisfies $m_E \leq \mu_N$. It was shown in Chapter 4 that the convergence rate of the Chebyshev method does not depend critically on the estimate m_E, provided $m_E \leq \mu_N$. However, iterative divergence may result if $m_E > \mu_N$. Since the a priori methods given in Section 5.3 for obtaining appropriate estimates for μ_N are not applicable for all problems, numerical methods for estimating μ_N and for detecting when $m_E > \mu_N$ must be considered.

In this section, the behavior of the Chebyshev iterations when $m_E > \mu_N$ is discussed and a procedure is given for detecting when this occurs and for obtaining a new estimate of μ_N. This procedure may also be used at the start of the regular Chebyshev iterations to obtain an initial estimate for μ_N.

The procedures given here do not attempt to estimate μ_N and μ_1 simultaneously; a lower bound for μ_N must be obtained before μ_1 can be estimated accurately. However, methods for estimating μ_1 and μ_N simultaneously have been developed. See, for example, Diamond [1971].

Behavior of the Chebyshev Iterations When $m_E > \mu_N$

From Eqs. (6-3.5) and (6-3.8), the pseudoresidual vector $\delta^{(s,p)}$ can be expressed as

$$\delta^{(s,p)} = P_{p,E_s}(\mu_1)d_{1,s}v(1) + \cdots + P_{p,E_s}(\mu_N)d_{N,s}v(N), \qquad (6\text{-}5.1)$$

where

$$P_{p,E_s}(x) = T_p(w_{E_s}(x))/T_p(w_{E_s}(1)) \qquad (6\text{-}5.2)$$

and where

$$w_{E_s}(x) = (2x - M_{E_s} - m_{E_s})/(M_{E_s} - m_{E_s}). \qquad (6\text{-}5.3)$$

In what follows, we assume† that $M_E < \mu_1$. With this assumption we have

$$\max_{1 \leq i \leq N} |P_{p,E}(\mu_i)| < \max[P_{p,E}(\mu_1), |P_{p,E}(\mu_N)|].$$

Thus the divergence or convergence of $\delta^{(s,p)}$ for large p is determined by the maximum absolute value of $P_{p,E}(x)$ at $x = \mu_1$ or μ_N. From (6-5.2), (6-5.3), and the definition (4-2.2) of $T_p(w)$, it follows that (see, for example, Figs. 6-5.1 and 4-3.2)

$$|P_{p,E}(\mu_N)| < P_{p,E}(\mu_1), \qquad \text{if} \quad (m_E - \mu_N) < (\mu_1 - M_E), \qquad (6\text{-}5.4)$$

$$P_{p,E}(\mu_1) < |P_{p,E}(\mu_N)| \leq 1, \qquad \text{if} \quad (\mu_1 - M_E) < (m_E - \mu_N) \leq (1 - M_E), \qquad (6\text{-}5.5)$$

$$1 < |P_{p,E}(\mu_N)|, \qquad \text{if} \quad (1 - M_E) < (m_E - \mu_N). \qquad (6\text{-}5.6)$$

---

† For notational convenience, we drop the subscript s on M_E, m_E, and $P_{p,E}(w)$ when the meaning is clear. We also drop the subscript E_s on w.

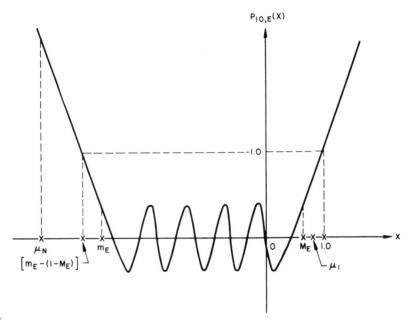

Fig. 6-5.1.   Graph of $P_{10,\mathrm{E}}(x)$ when $M_{\mathrm{E}} < \mu_1$ and $m_{\mathrm{E}} > \mu_N$, $M_{\mathrm{E}} = 0.789$, $m_{\mathrm{E}} = -3.0$, $\sigma_{\mathrm{E}} = 0.90$, $r = 0.39$.

From these inequalities, it follows easily from (6-5.1) that $\delta^{(s,\,p)}$ for sufficiently large p behaves in a distinctive way, depending on the value of m_{E}. The characteristics of $\delta^{(s,\,p)}$ as a function of m_{E} are summarized in Fig. 6-5.2. (We assume that the coefficients $d_{1,s}$ and $d_{N,s}$ in (6-5.1) are nonzero.) If m_{E} lies in the interval $(-\infty, A)$, where A is defined in Fig. 6-5.2, inequality (6-5.4) implies that the behavior of $\delta^{(s,\,p)}$ for large p is dictated by $P_{p,\mathrm{E}}(\mu_1)$. This is basically the assumption of Case I* in (6-1.5), and the previously given procedures are valid.

If m_{E} lies in the interval $[A, B)$ of Fig. 6-5.2, $\delta^{(s,\,p)}$ will converge to the null

Fig. 6-5.2.   Behavior of $\delta^{(s,\,p)}$ as a function of m_{E}. m_{E} on line segment $(-\infty, A)$; $\delta^{(s,\,p)}$ converges with rate governed by $P_{p,\mathrm{E}}(\mu_1)$. m_{E} on line segment $[A, B)$; $\delta^{(s,\,p)}$ converges with rate governed by $P_{p,\mathrm{E}}(\mu_N)$. m_{E} on line segment $[B, M_{\mathrm{E}})$; $\delta^{(s,\,p)}$ is nonconvergent.

vector, but its behavior is now dictated by $P_{p,\,E}(\mu_N)$. A special procedure† for detecting the occurrence will not be given since the likelihood of m_E lying in this interval of length $1 - \mu_1$ is small.

Inequality (6-5.6) implies that $\delta^{(s,\,p)}$ will not converge to the null vector if m_E lies in the range (B, M_E). We now give a procedure for detecting this occurrence and for obtaining a new estimate for μ_N.

A Procedure for Detecting When $m_E > \mu_N$

We assume m_E lies on the line segment (B, M_E) of Fig. 6-5.2; i.e., m_E satisfies

$$m_E > \mu_N + (1 - M_E). \qquad (6\text{-}5.7)$$

For notational convenience, we also assume that $\mu_{N-1} \neq \mu_N$. Inequality (6-5.7) then implies that $|w(\mu_N)| > |w(\mu_i)|$ for $i \neq N$ and that $-w(\mu_N) > w(1) > 1$. From the definition (4-2.2) of $T_p(w)$, it then follows that

$$|P_{p,\,E}(\mu_N)| > 1, \qquad -[P_{p,\,E}(\mu_N)/P_{p-1,\,E}(\mu_N)] > 1,$$
$$\lim_{p \to \infty} [P_{p,\,E}(\mu_i)/P_{p,\,E}(\mu_N)] = 0 \qquad \text{for} \quad i \neq N. \qquad (6\text{-}5.8)$$

Thus for sufficiently large p and provided that $d_{N,\,s} \neq 0$, $\delta^{(s,\,p)}$ of (6-5.1) may be approximated by

$$\delta^{(s,\,p)} \doteq P_{p,\,E}(\mu_N)d_{N,\,s}v(N). \qquad (6\text{-}5.9)$$

With this approximation, we conclude that for sufficiently large s

$$B^{(s,\,p)} \equiv \frac{\|\delta^{(s,\,p)}\|}{\|\delta^{(s,\,0)}\|} \doteq |P_{p,\,E}(\mu_N)| > 1, \qquad (6\text{-}5.10)$$

$$R^{(s,\,p)} \equiv \frac{\|\delta^{(s,\,p)}\|}{\|\delta^{(s,\,p-1)}\|} \doteq -\frac{P_{p,\,E}(\mu_N)}{P_{p-1,\,E}(\mu_N)} > 1. \qquad (6\text{-}5.11)$$

Thus $B^{(s,\,p)} > 1$ and $R^{(s,\,p)} > 1$ imply that condition (6-5.7) is satisfied and may be used as a signal that $m_E > \mu_N$. We remark that $R^{(s,\,p)}$ may be greater than unity under other conditions. For example, if $M_E \geq \mu_1$, it is shown in Section 6.7 that $R^{(s,\,p)}$ can sometimes become greater than unity. However, $B^{(s,\,p)} > 1$ usually occurs only when $m_E > \mu_N$.

If $B^{(s,\,p)}$ and $R^{(s,\,p)}$ are greater than unity, new estimates for μ_N may be obtained by solving Eqs. (6-5.10) and (6-5.11).

To obtain a new estimate m'_E for μ_N, using (6-5.10), we use a procedure similar to that used in Chapter 5 to obtain the estimate M'_E for μ_1. Analogous

---

† If $(\delta^{(s,\,p)})_k \neq 0$ is the kth component of $\delta^{(s,\,p)}$, then $-1 < ((\delta^{(s,\,p)})_k/(\delta^{(s,\,p-1)})_k) < 0$ would be indicative of this occurrence.

to (5-4.8), we take m_E' to be the algebraically smallest real x that satisfies

$$|T_p(w(x))|/T_p(w(1)) = B, \qquad (6\text{-}5.12)$$

where $B \equiv B^{(s, p)} = \|\delta^{(s, p)}\|_2/\|\delta^{(s, 0)}\|_2$. If $B > 1$, it follows from the definition of $T_p(w)$ that

$$w(m_E') < -1 \qquad \text{and} \qquad |T_p(w(m_E'))| = T_p(|w(m_E')|). \qquad (6\text{-}5.13)$$

Thus analogous to (5-4.16), if $B > 1$, a solution m_E' to (6-5.12) satisfies

$$-w(m_E') = |w(m_E')| = \cosh\left(\frac{1}{p}\cosh^{-1}\left(\frac{B}{Q}\right)\right) = \frac{1}{2}\left(\frac{Y^2 + 1}{Y}\right), \qquad (6\text{-}5.14)$$

where Q and Y are given by (5-4.11) and (5-4.13), respectively. Substituting (6-5.3) into (6-5.14) and solving for m_E', we obtain

$$m_E' = \frac{1}{2}\left[M_E + m_E - \left(\frac{M_E - m_E}{2}\right)\left(\frac{Y^2 + 1}{Y}\right)\right]. \qquad (6\text{-}5.15)$$

The fact that m_E' of (6-5.15) is the smallest real solution to (6-5.12) follows from arguments similar to those given in the proof of Theorem 5-4.1.

Another estimate m_E'' for μ_N may be obtained from the relation (6-5.11). Let m_E'' be the algebraically smallest real x which satisfies

$$\left|\frac{T_p(w(x))}{T_{p-1}(w(x))}\right| = U, \qquad (6\text{-}5.16)$$

where

$$U \equiv \frac{T_p(w(1))}{T_{p-1}(w(1))}R^{(s, p)} = \frac{T_p(w(1))}{T_{p-1}(w(1))}\frac{\|\delta^{(s, p)}\|_2}{\|\delta^{(s, p-1)}\|_2}.$$

Note that $[T_p(w(1))/T_{p-1}(w(1))] = [1/\sqrt{r}][(1 + r^p)/(1 + r^{p-1})]$, where r is defined by (4-3.21). If $U > 1$, then m_E'' satisfies (6-5.13); i.e., $w(m_E'') < -1$ and $|T_p(w(m_E''))| = T_p(|w(m_E'')|)$. Thus with

$$h \equiv |w(m_E'')| + \sqrt{|w(m_E'')|^2 - 1}, \qquad (6\text{-}5.17)$$

the solution m_E'' satisfies

$$(h + h^{-2p+1})/(1 + h^{-2p-2}) = U. \qquad (6\text{-}5.18)$$

Equation (6-5.18) is obtained by substituting the definition (4-2.2) for $T_p(|w(m_E'')|)$ into (6-5.16). But $h > 1$. Thus for sufficiently large p, $h \doteq U$ or, equivalently, $[|w(m_E'')| + \sqrt{|w(m_E'')|^2 - 1}] \doteq U$. Solving for $|w(m_E'')|$, we then obtain the approximation

$$|w(m_E'')| \doteq (U^2 + 1)/2U, \qquad (6\text{-}5.19)$$

which from (6-5.3) gives

$$m_E'' \doteq \frac{1}{2}\left[M_E + m_E - \left(\frac{M_E - m_E}{2}\right)\left(\frac{U^2 + 1}{U}\right)\right]. \qquad (6\text{-}5.20)$$

The solutions (6-5.15) and (6-5.20) to Eqs. (6-5.12) and (6-5.16), respectively, are reasonable estimates for μ_N only if $w(\mu_N) < -1$ or, equivalently, only if $m_E > \mu_N$. This necessary condition is implied whenever $B^{(s,\,p)} > 1$. However, as mentioned previously, $R^{(s,\,p)} > 1$ need not imply that $w(\mu_N) < -1$. Thus m_E'' should be used as an estimate for μ_N only if both $B^{(s,\,p)}$ and $R^{(s,\,p)}$ are greater than unity. Usually, m_E'' is a better estimate for μ_N than is m_E'. This is so because m_E' utilizes the ratio $\|\delta^{(s,\,p)}\|/\|\delta^{(s,\,0)}\|$, while m_E'' utilizes the ratio $\|\delta^{(s,\,p)}\|/\|\delta^{(s,\,p-1)}\|$, which reflects more recent data.

An informal program which utilizes the estimates m_E' and m_E'' is given in Algorithm 6-5.1. The algorithm includes procedures for detecting when an initial estimate m_E is greater than μ_N and for obtaining a new estimate for μ_N if needed. An option is provided whereby the initial estimate m_E for μ_N is obtained numerically. Except for the **Update** m_E calculations, Algorithm 6-5.1 is similar to Algorithm 6-4.1.

Algorithm 6-5.1.   An adaptive procedure for Chebyshev acceleration which includes a procedure for estimating μ_N.

Input: $(\zeta, M_E, m_E, F, \underline{u}_\Gamma, d, \text{IE}, \text{ILIM})$ $\langle M_E$ must satisfy $m_E < M_E < 1\rangle$

Initialize:

$n := 0; p := -1; M_E' = M_E; \underline{u}_\phi := \underline{0}; s := 0;$
$\tilde{m}_E := m_E; \text{DELNP} := 1.0; R := 1.0;$
If $\text{IE} \geq 0$, *then* continue
   else
   Begin
    $M_E' := \tau_1; m_E := -\tau_1;$
   End

Next Iteration: Same as Algorithm 6-4.1.

Calculate New Iterate: Same as Algorithm 6-4.1.

Calculate New Estimate M_E': Same as Algorithm 6-4.1.

Convergence Test: Same as Algorithm 6-4.1.

Parameter Change Test:

If p is even *then* continue; *else Go to* **Next Iteration**;
If $p \geq p^*$, *then*

Begin
If IE $\geq$ 0, *then* continue; *else Go to* **Update m_E**;
If $B > (Q)^F$, *then* continue; *else Go to* **Next Iteration**
If $T = 0$, *then*
        Begin
        $p := -1$; *Go to* **Next Iteration**
        End
        else continue;
If $R > 1$ and $T \geq 5$, *then Go to* **Update m_E**;
                  else Go to **Next Iteration**
End
else Go to **Next Iteration**

Update m_E:

$m'_E := 0.0$; $m''_E := 0.0$; $DR = 0$;
If $R > 1.0$, *then*
    Begin
    $U := [1/\sqrt{r}][(1 + r^p)/(1 + r^{p-1})]R$;
    $m''_E := \frac{1}{2}[M_E + m_E - \frac{1}{2}(M_E - m_E)((U^2 + 1)/U)]$;
    If $|R\emptyset - R| < 0.1$, *then* $DR = 1$; *else* continue
    End
    else continue
If $B > 1.0$, *then*
    Begin
    $Y := [(B/Q) + \sqrt{(B/Q)^2 - 1}]^{1/p}$
    $m'_E := \frac{1}{2}[M_E + m_E - \frac{1}{2}(M_E - m_E)((Y^2 + 1)/Y)]$;
    End
    else
    Begin
    If $IE = 0$, *then Go to* **Next Iteration**; *else* continue
    End
If $IE = 0$, *then*
    Begin
    If $DR = 0$, *then Go to* **Next Iteration**;
        else
        Begin
        $m_E := \min[1.1m'_E, 1.1m''_E, m_E]$;
        $M'_E := 0.1$; $p := -1$;
        End
    End
    else

Begin
If $DR = 0$ and $n <$ ILIM, then Go to **Next Iteration**;
      else
      Begin
        $m_E = \min[1.1m'_E, 1.1m''_E, \tilde{m}_E, -1.0]$;
        $M'_E := 0.1; \ IE := 0; \ p := -1; \ \jmath := 0,$
      End
  End
Go to **Next Iteration**

Two additional input parameters are required for Algorithm 6-5.1. They are

$$\text{IE} = \begin{cases} 0 & \text{implies that the initial estimate for } \mu_N \text{ is input;} \\ -1 & \text{implies that an initial estimate for } \mu_N \text{ is to be calculated by} \\ & \text{the adaptive procedure.} \end{cases}$$

$$\text{ILIM} = \begin{cases} \text{the limit on the number of preliminary iterations to be done in} \\ \text{obtaining an initial estimate for } \mu_N \text{ (ILIM must be at least 8).} \end{cases}$$

The formulas for m'_E and m''_E are not valid if $M_E = m_E$. Thus for Algorithm 6-5.1, we insist that $m_E < M_E$. The other input quantities have been discussed previously in Algorithm 6-4.1.

T is an additional control variable which counts the number of times the ratio $B^{(s, p)}$ of (6-5.10) is greater than unity.

First consider the case in which $IE = 0$ and $m_E > \mu_N + (1 - M_E)$. By (6-5.10), $B^{(s, p)}$ must become larger than unity for sufficiently large p provided, of course, that $d_{N, s} \neq 0$. When $B^{(s, p)}$ becomes larger† than unity, the T index is increased in **Calculate New Estimate** M'_E. No attempt is made, however, to calculate new estimates for μ_N until the index T becomes larger than four and the ratio $R^{(s, p)}$ of (6-5.11) is also greater than unity. When these conditions are satisfied, new estimates m'_E and m''_E are obtained, using (6-5.15) and (6-5.20). The new estimates m''_E and m'_E are considered to be sufficiently accurate if the change in $R^{(s, p)}$ is less‡ than 0.1; i.e., if

$$|R^{(s, p)} - R^{(s, p-1)}| < 0.1. \tag{6-5.21}$$

---

† Note that M_E may be updated or convergence may occur before $B^{(s, p)}$ becomes greater than unity.

‡ The estimates m''_E obtained from $R^{(s, p)}$ are usually considerably more accurate than those obtained from $B^{(s, p)}$. Thus the change in $R^{(s, p)}$ is used to measure the accuracy or convergence of the estimates for μ_N.

Note that the control word DR equals zero if (6-5.21) is not satisfied. If (6-5.21) is satisfied, the new estimate m_E for μ_N is taken to be†

$$m_E = \min[1.1m'_E, 1.1m''_E, m_E^{(old)}]. \qquad (6\text{-}5.22)$$

A new Chebyshev polynomial is then started, using m_E as the estimate for μ_N and $M_E = 0.1$ as the estimate for μ_1. The fact that a new estimate for μ_N was needed implies that the eigenvector modes with algebraically small eigenvalues have been magnified in the eigenvector expansion of the error vector. The Chebyshev polynomial method, using a small value for M_E, will reduce the magnitude of these eigenvectors quickly. It is for this reason that we use $M_E = 0.1$. *For some problems, some other value for M_E may be appropriate.*‡

The procedure used to obtain an initial estimate for μ_N, i.e., when $IE = -1$, is basically the same as that described above. The differences are

(a) $M_E = -m_E = \tau_1 = 0.948$ are used as the eigenvalue estimates for the initial Chebyshev polynomial generated;
(b)   no updating of M_E is allowed;
(c)   the eigenvalue estimates for the first Chebyshev polynomial after the special initial iterations are completed are

$$m_E = \min[1.1m'_E, 1.1m''_E, \tilde{m}_E, -1.0],$$
$$M_E = 0.1, \qquad (6\text{-}5.23)$$

where $\tilde{m}_E$ is the input estimate for μ_N.

The eigenvalue estimates used in the Chebyshev polynomial generated by the special initial iterations are based on the assumption that $\mu_N < -1.0$. If $\mu_N \geq -1.0$, these special initial iterations will not give a good estimate for μ_N. In the determination of m_E in (6-5.23), -1.0 is included to ensure that any estimate m_E obtained is less than -1.0.

In Appendix A, we give a Fortran listing of a subroutine, called CHEBY, which implements Algorithm 6-5.1 with the exception of the **Calculate New Iterate** portion. The CHEBY subroutine is designed for use as a software package to provide the required acceleration parameters and to provide an estimate of the iteration error for the Chebyshev polynomial method.

6.6   NUMERICAL RESULTS

In this section we describe results of numerical experiments that were designed to test the effectiveness of the Chebyshev adaptive procedures used

---

†The multiplication by 1.1 may be considered as a safety factor to ensure that m_E will satisfy $m_E \leq \mu_N$.

‡The choice $M_E = 0.1$ here is based on the assumption that $M(G)$ is close to unity. If $M(G)$ is close to zero, for example, then some other value should be used; perhaps $M_E = m_E/2$.

in Algorithm 6-5.1. We shall be concerned primarily with numerical experiments that illustrate the effectiveness and the behavior of the adaptive procedures for estimating μ_N. The "simulation" iteration procedure of Chapter 5 is used. That is, we assume that the set of eigenvectors $\{v(i)\}_{i=1}^{i=N}$ for the iteration matrix G is an orthonormal basis for the associated vector space.† Thus there exist constants c_i, $i = 1, \ldots, N$, such that the initial error vector $\varepsilon^{(0)} = u^{(0)} - \bar{u}$ may be written as

$$\varepsilon^{(0)} = c_1 v(1) + c_2 v(2) + \cdots + c_N v(N). \tag{6-6.1}$$

If μ_i is the eigenvalue of G corresponding to $v(i)$, then the initial pseudo-residual vector $\delta^{(0)} \equiv \delta^{(1, 0)}$ can be given by

$$\delta^{(1, 0)} = d_1 v(1) + d_2 v(2) + \cdots + d_N v(N), \tag{6-6.2}$$

where, since $\delta^{(1, 0)} = \delta^{(0)} = (G - I)\varepsilon^{(0)}$, d_i satisfies

$$d_i = (\mu_i - 1)c_i. \tag{6-6.3}$$

Moreover, from (6-3.7), we have that

$$\delta^{(s, p)} = \sum_{i=1}^{N} \{P_{p, E_s}(\mu_i) P_{\bar{p}_{s-1}, E_{s-1}}(\mu_i) \cdots P_{\bar{p}_1, E_1}(\mu_i) d_i v(i)\}, \tag{6-6.4}$$

and since $\varepsilon^{(s, p)} = (G - I)^{-1} \delta^{(s, p)}$,

$$\varepsilon^{(s, p)} = \sum_{i=1}^{N} \{P_{p, E_s}(\mu_i) P_{\bar{p}_{s-1}, E_{s-1}}(\mu_i) \cdots P_{\bar{p}_1, E_1}(\mu_i) c_i v(i)\}. \tag{6-6.5}$$

Since the vectors $v(i)$ are orthonormal, we also have

$$\|\delta^{(s, p)}\|_2^2 = \sum_{i=1}^{N} \{P_{p, E_s}(\mu_i) \cdots P_{\bar{p}_1, E_1}(\mu_i) d_i\}^2 \tag{6-6.6}$$

and

$$\|\varepsilon^{(s, p)}\|_2^2 = \sum_{i=1}^{N} \{P_{p, E_s}(\mu_i) \cdots P_{\bar{p}_1, E_1}(\mu_i) c_i\}^2. \tag{6-6.7}$$

The adaptive procedure of Algorithm 6-5.1 requires only $\|\delta^{(s, p)}\|_2$ from the iteration process. But from (6-6.6), $\|\delta^{(s, p)}\|_2$ can be calculated easily, given only the set of eigenvalues $\{\mu_i\}$ of G, the set of coefficients $\{c_i\}$ for the expansion (6-6.1), and data for the Chebyshev polynomials used. Since the Chebyshev polynomial data are determined from the adaptive procedure, the

---

† The orthogonality assumption on the eigenvectors of G is equivalent to assuming that the matrix G is symmetric or, equivalently, that the matrix W in (5-7.4) and (5-7.5) may be chosen to be the identity matrix.

iterative behavior of any problem is uniquely determined by the eigenvalues μ_i, the coefficients c_i, and the adaptive procedure.

If

$$\mu_i = \begin{cases} \dfrac{\cos(i\pi/(N+1))}{2 - \cos(\pi/N + 1)} & \text{if } \cos\dfrac{i\pi}{N+1} > 0, \\[4mm] \alpha\,\dfrac{\cos(i\pi/N + 1)}{2 - \cos(\pi/N + 1)} & \text{if } \cos\dfrac{i\pi}{N+1} < 0, \end{cases} \tag{6-6.8}$$

then the three sets of eigenvalues $\{\mu_i\}_{i=1}^{i=N}$ we consider for our test problems are given in Table 6-6.1.

TABLE 6-6.1

Specification of Eigenvalues for Test Problems

Problem	N	α
1	40	1
2	100	3.0
3	100	10.0

If $\{R_i\}_{i=1}^{i=N}$ is a set of random numbers between 0 and 1 and if

$$c_i = \begin{cases} R_i & \text{if } \mu_i > 0, \\ \beta R_i & \text{if } \mu_i < 0, \end{cases} \tag{6-6.9}$$

the different combinations of c_i we consider are defined by the β's given in Table 6-6.2.

TABLE 6-6.2

Specification of Coefficients c_i for Test Problems

Guess	β
A	1.0
B	10^{-4}
C	10^{-6}

Normally, the eigenvectors $v(i)$ for small i are smoother than those for large i. Since for most practical problems, the solution $\bar{u}$ and initial guess $u^{(0)}$ are also relatively smooth, the coefficients c_i for small i are usually larger than those for large i. We try to incorporate this property in our choices for c_i. We also note that the worst case for detecting if $m_E > \mu_N$ occurs when the c_i is small for large i.

As shown in Section 5.7, we may generate the same adaptive Chebyshev behavior using the simulated iteration process as that obtained from the adaptive Chebyshev procedure applied to the general problem (6-1.1) provided that the W-norm is used. Thus the simulated iterative process simulates W-norm Chebyshev procedures, which is what we are trying to avoid in this chapter. However, we chose to use the simulated iteration process here since, as discussed in Chapter 3, it is easier to generate different problem conditions with the use of this method. We note that the adaptive process does not use the fact that the W-norm, in essence, is being used. For the numerical results given in Chapter 8 on the cyclic Chebyshev method, $\delta^{(s, p)}$ is obtained from actual iterations. There it is shown that the behavior of the iteration process is basically independent of whether or not the two- or W-norm of $\delta^{(s, p)}$ is used. In all likelihood, the same conclusion will also be valid for the adaptive procedures of this chapter.

For the problems considered here, we wish to terminate the iterations when†

$$\|\varepsilon^{(s, p)}\|_2 \leq 10^{-6}. \tag{6-6.10}$$

As the estimate for $\|\varepsilon^{(s, p)}\|_2$, we use (5-4.28). Thus we consider the problem converged when

$$\|\delta^{(s, p)}\|_2/(1 - M'_E) \leq 10^{-6}, \tag{6-6.11}$$

where M'_E is the best approximation to μ_1 and is as defined in Algorithm 6-5.1.

Unless stated otherwise, we use the following input values for the adaptive procedure: $F = 0.75$, $d = 0.1$, and ILIM $= 25$.

Tables 6-6.3–6-6.5 contain summaries of the iterative behavior for the three problems of Table 6-6.1 and under different problem conditions. The $m_E^{(0)}$ and $M_E^{(0)}$ values in the "problem conditions" column are the input values for m_E and M_E. Recall from Section 6.5 that IE is the input parameter which signals whether or not the initial estimate for μ_N is to be obtained by the adaptive procedure. The column headed by "last est. used for μ," gives the value of M_E used in the generation of the Chebyshev polynomial when convergence was achieved. The $C^{(s, p)}$ in the fourth column is defined by (6-4.11). The true error at convergence was obtained using (6-6.7), while the estimated error was obtained from (6-6.11). When applicable, the columns under the heading "re-estimation of μ_N" give the iteration on which a new estimate for μ_N was first used and the value of this estimate. For all problems, a new estimate for μ_N was obtained at most one time.

---

† For reasons of simplicity, we ignore here the normalization quantity $\|u^{(n+1)}\|$ normally used in most stopping tests.

TABLE 6-6.3

Iteration Summary for Problem 1: $\mu_1 = 0.994149$ and $\mu_N = -0.994149$

Problem conditions	Iterations to converge	Last est. used for μ_1	$C^{(s,\,p)}$ at convergence	Error at convergence — True	Error at convergence — Estimated	Reestimation of μ_N — Value of new est. for μ_N	Reestimation of μ_N — Calculated on iteration
Guess = B $IE = 0$ $m_E^{(0)} = -1.0$ $M_E^{(0)} = 0.994149$ (fixed)	159	—	0.9943	1.13×10^{-6}	0.970×10^{-6}	—	—
Guess = B $IE = 0$ $m_E^{(0)} = -1.0$ $M_E^{(0)} = 0.2$	204	0.993778	0.787	0.961×10^{-6}	0.968×10^{-6}	—	—
Guess = B $IE = 0$ $m_E^{(0)} = 0.0$ $M_E^{(0)} = 0.2$	198	0.994135	0.9803	1.00×10^{-6}	0.939×10^{-6}	-1.074	18
Guess = B $IE = -1$ $m_E^{(0)} = 0.0$ $M_E^{(0)} = 0.2$	197	0.993856	0.814	0.945×10^{-6}	0.939×10^{-6}	-1.000	25
Guess = C $IE = 0$ $m_E^{(0)} = 0.0$ $M_E^{(0)} = 0.2$	210	0.994146	0.9979	0.843×10^{-6}	0.937×10^{-6}	-1.079	23

TABLE 6-6.4

Iteration Summary for Problem 2: $\mu_1 = 0.999033$ and $\mu_N = -2.99710$

Problem conditions	Iterations to converge	Last est. used for μ_1	$C^{(s, p)}$ at convergence	Error at convergence		Reestimation of μ_N	
				True	Estimated	Value of new est. for μ_N	Calculated on iteration
Guess = B $IE = 0$ $m_E^{(0)} = -3.0$ $M_E^{(0)} = 0.999033$ (fixed)	548	—	1.018	0.298×10^{-6}	0.986×10^{-6}	—	—
Guess = B $IE = 0$ $m_E^{(0)} = -3.0$ $M_E^{(0)} = 0.2$	647	0.999023	0.9403	0.953×10^{-6}	0.981×10^{-6}	—	—
Guess = B $IE = -1$ $m_E^{(0)} = 0.0$ $M_E^{(0)} = 0.2$	683	0.999006	0.8753	0.960×10^{-6}	0.985×10^{-6}	-3.208	9
Guess = C $IE = 0$ $m_E^{(0)} = -2.0$ $M_E^{(0)} = 0.2$	727	0.998982	0.8104	0.975×10^{-6}	0.998×10^{-6}	-3.276	30
Guess = A $IE = 0$ $m_E^{(0)} = -2.0$ $M_E^{(0)} = 0.2$	707	0.998998	0.8496	0.959×10^{-6}	0.983×10^{-6}	-3.250	11

TABLE 6-6.5

Iteration Summary for Problem 3: $\mu_1 = 0.999033$ *and* $\mu_N = -9.99033$

Problem conditions	Iterations to converge	Last est. used for μ_1	$C^{(s,p)}$ at convergence	Error at convergence		Reestimation of μ_N	
				True	Estimated	Value of new est. for μ_N	Calculated on iteration
Guess = B $IE = 0$ $m_E^{(0)} = -10.0$ $M_E^{(0)} = 0.999033$ (fixed)	908	—	1.019	0.302×10^{-6}	0.984×10^{-6}	—	—
Guess = B $IE = 0$ $m_E^{(0)} = -10.0$ $M_E^{(0)} = 0.2$	1174	0.998965	0.7845	0.985×10^{-6}	0.986×10^{-6}	—	—
Guess = B $IE = -1$ $m_E^{(0)} = 0.0$ $M_E^{(0)} = 0.2$	1179	0.998989	0.8457	0.981×10^{-6}	0.991×10^{-6}	-10.688	9
Guess = C $IE = 0$ $m_E^{(0)} = -1.0$ $M_E^{(0)} = 0.2$	1107	0.999029	0.9821	0.976×10^{-6}	0.988×10^{-6}	-10.752	11
Guess = C $IE = 0$ $m_E^{(0)} = -9.0$ $M_E^{(0)} = 0.2$	1140	0.999014	0.9176	0.984×10^{-6}	0.992×10^{-6}	-10.937	39

The first two rows in each table give the iteration data when the input m_E satisfies the Case I* condition of (6-1.5), i.e., $m_E^{(0)} \leq \mu_N$. The first row gives the iterations required for convergence if μ_1 were known, i.e., the nonadaptive case. The second row gives the iteration results when μ_1 was adaptively estimated. The last three rows in each table give iteration results for various problem conditions in which either the initial estimate for μ_N was obtained during the initial iterations or else an input value $m_E^{(0)} > \mu_N$ was used.

The following observations are made concerning the iterative behavior of these problems:

1. With $m_E^{(0)} \leq \mu_N$, the ratio of iterations required using the adaptive procedure to the iterations required using $M_E = \mu_1$ fixed (nonadaptive) varied from 1.19 to 1.33.

2. For all problem conditions considered, the procedure to reestimate μ_N worked well. Moreover, the iterations required for convergence did not increase significantly when μ_N was estimated or reestimated by the adaptive procedure.

3. Accurate estimates for μ_1 were obtained for all problems.

4. The true error was accurately estimated whenever the adaptive procedure was used. For the nonadaptive problems with $M_E^{(0)} = \mu_1$ fixed, the true error was one-third to one-ninth that of the estimated error.

We remark that using an input $m_E^{(0)}$ which is considerably less than μ_N may significantly increase the number of iterations required for convergence. For example, Problem 2 with $M_E = 0.999033$ fixed was rerun with $m_E^{(0)} = -10.0$. This problem required 908 iterations to converge as compared to the 548 iterations given in Table 6-6.4 when $m_E^{(0)} = -3.0$ was used.†

To give some indication of the behavior of the adaptive procedure for different problem conditions, graphs‡ of $R^{(n)}$ versus n and the eigenvalue estimates used are given in Figs. 6-6.1–6-6.3 for three different conditions of Problem 3 given in Table 6-6.5. No iteration data were printed for any Chebyshev polynomial of degree three or less. Thus a break in the curve for $R^{(n)}$ indicates that a new Chebyshev polynomial was started.

In Fig. 6-6.1, data are given for Case I* problem conditions (i.e., $m_E^{(0)} \leq \mu_N$). Figure 6-6.2 gives the behavior of $R^{(n)}$ when the initial estimate for μ_N is obtained by the adaptive procedure. Note that after the generation of the second Chebyshev polynomial, the behavior of $R^{(n)}$ is the same as that in Fig. 6-6.1. If μ_N is significantly less than minus one, it is probably better to

---

† Note that the iteration ratio (908/548 = 1.657) obtained here agrees very closely with the analytical bound given in Chapter 4. The bound given by (4-4.12) indicates that the ratio of iterations using $m_E = -10.0$ to that using $m_E = -3.0$ should be less than $\sqrt{11/4} = 1.658$.

‡ Recall from (6-5.11) that $R^{(n)} = \|\delta^{(n)}\|_2 / \|\delta^{(n-1)}\|_2$.

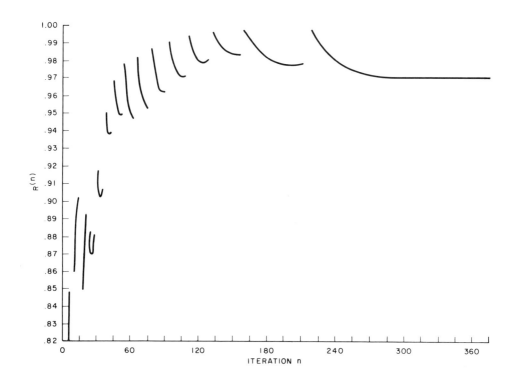

Iterations	M_E	m_E	Iterations	M_E	m_E
1–7	0.200000	−3.0	63–75	0.991036	−3.0
8–14	0.478640	−3.0	76–90	0.993164	−3.0
15–21	0.837447	−3.0	91–107	0.995085	−3.0
22–28	0.912525	−3.0	108–128	0.996680	−3.0
29–35	0.942898	−3.0	129–155	0.997983	−3.0
36–42	0.963442	−3.0	156–214	0.998798	−3.0
43–51	0.979501	−3.0	215–647	0.999023	−3.0
52–62	0.987788	−3.0			

Fig. 6-6.1.   Graph of $R^{(n)}$ versus n for problem 3 with $IE = 0$ and $m_E^{(0)} = -3.0$.

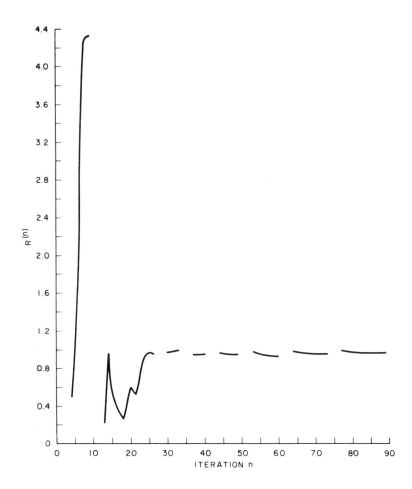

Iterations	M_E	m_E	Iterations	M_E	m_E
1–9	0.948000	−0.948	74–78	0.993005	−3.209
10–26	0.200000	−3.209	89–105	0.994908	−3.209
27–33	0.264741	−3.209	106–126	0.996509	−3.209
34–40	0.960730	−3.209	127–153	0.997845	−3.209
41–49	0.979351	−3.209	154–194	0.998741	−3.209
50–60	0.988040	−3.209	195–683	0.999006	−3.209
61–73	0.990902	−3.209			

Fig. 6-6.2. Graph of $R^{(n)}$ versus n for problem 3 with the initial estimate for μ_N obtained by the adaptive procedure.

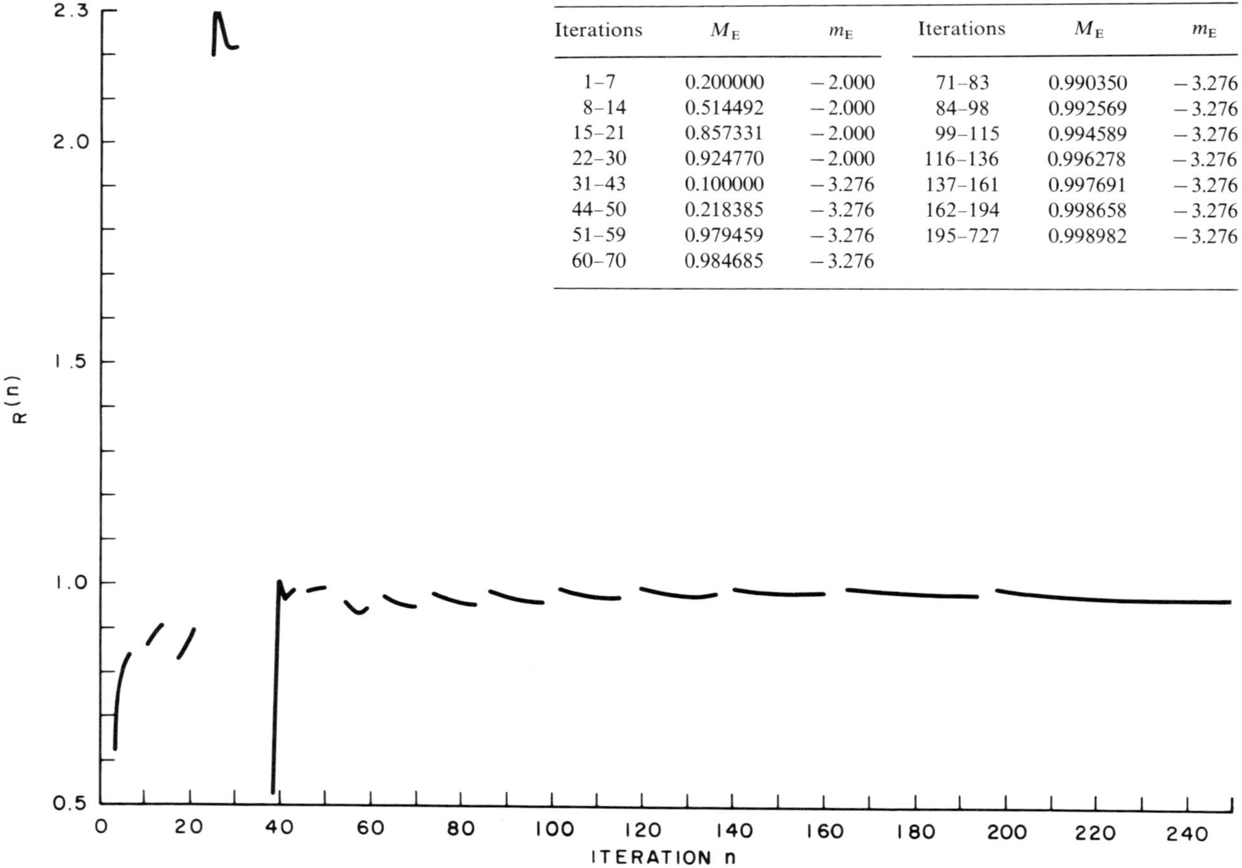

Iterations	M_E	m_E		Iterations	M_E	m_E
1–7	0.200000	–2.000		71–83	0.990350	–3.276
8–14	0.514492	–2.000		84–98	0.992569	–3.276
15–21	0.857331	–2.000		99–115	0.994589	–3.276
22–30	0.924770	–2.000		116–136	0.996278	–3.276
31–43	0.100000	–3.276		137–161	0.997691	–3.276
44–50	0.218385	–3.276		162–194	0.998658	–3.276
51–59	0.979459	–3.276		195–727	0.998982	–3.276
60–70	0.984685	–3.276				

Fig. 6-6.3. Graph of $R^{(n)}$ versus n for problem 3 with $IE = 0$ and $M_E^{(0)} = -2.0$.

use a new guess for $u^{(n)}$ in the first iteration after obtaining the initial estimate for μ_N. In Fig. 6-6.3, data are given for the condition in which the input estimate $m_E^{(0)}$ is greater than μ_N. Note that this condition is not detected until the fourth Chebyshev polynomial generated, and then it is indicated strongly. The increase in the total number of iterations for this case is due more to the last estimate used for μ_1 than to the fact that an initial $m_E^{(0)} > \mu_N$ was used.

We remark that the convergence rate of the Chebyshev acceleration method depends strongly on the values of μ_1 and μ_N but only weakly on the size of the matrix problem. Thus although the number of unknowns is small for the problems considered in this section, similar behavior would be expected for much larger matrix problems with similar maximum and minimum eigenvalues.

6.7 ITERATIVE BEHAVIOR WHEN $M_E > \mu_1$

For the following discussion, we assume that $M_{E_s} > \mu_1$ and that $m_E < \mu_N$. Moreover, the occurrence of $M_{E_s} > \mu_1$ is most likely for s not small. Thus as in (6-4.2), it is reasonable to assume that $\delta^{(s,0)}$ closely approximates the eigenvector $v(1)$ of G. When this is the case, we have approximately that

$$B^{(s,p)} = \frac{\|\delta^{(s,p)}\|_2}{\|\delta^{(s,0)}\|_2} \doteq P_{p,E_s}(\mu_1) = \frac{T_p(w_{E_s}(\mu_1))}{T_p(w_{E_s}(1))}, \qquad (6\text{-}7.1)$$

where $w_E(x)$ is defined by (6-5.3).† Since $M_E > \mu_1$, $w(\mu_1)$ is less than unity. Thus, as in (6-4.5), we have the approximation

$$B^{(s,p)} \doteq \frac{|\cos p\theta|}{T_p(w(1))} = \left[\frac{2r^{p/2}}{1+r^p}\right]|\cos p\theta|, \qquad (6\text{-}7.2)$$

where

$$\theta = \cos^{-1} w(\mu_1) = \tan^{-1}[\sqrt{1 - w(\mu_1)^2}/w(\mu_1)]. \qquad (6\text{-}7.3)$$

From (6-4.6), the error vector $\varepsilon^{(s,p)}$ also satisfies the same relationship, i.e.,

$$\frac{\|\varepsilon^{(s,p)}\|_2}{\|\varepsilon^{(s,0)}\|_2} \doteq \left[\frac{2r^{p/2}}{1+r^p}\right]|\cos p\theta|. \qquad (6\text{-}7.4)$$

Thus $\|\varepsilon^{(s,p)}\|_2$ achieves a local minimum whenever $(p\theta)$ is closest to $(i + \frac{1}{2})\pi$, $i = 0, \dots$. To illustrate this behavior, Problem 2, with $IE = -1$, $m_E^{(0)} = 0.0$, $M_E^{(0)} = 0.2$, and using Guess B, was resolved using estimates $M_E > \mu_1$ for the last Chebyshev polynomial generated. For the result given in Table 6-6.4,

---

† Again for notational simplicity, we shall drop the subscript s on w and M_E.

the last Chebyshev polynomial generated was started on iteration 195, using the estimates $m_E = -3.208551$ and $M_E = 0.9990061$. The M_E estimate was less than $\mu_1 = 0.999033$. This problem was rerun twice, using the same value for m_E, but using $M_E = 0.99905$ and $M_E = 0.9991$ as the estimate for μ_1 in the generation of the last Chebyshev polynomial. Graphs of $\|\varepsilon^{(n)}\|_2$ versus n for these three cases are given in Fig. 6-7.1. For the case in which $M_E = 0.99910$, θ is about 0.008 rad so that, from (6-7.4), a local minimum should occur for p that is approximately equal to 196. From Fig. 6-7.1, a local minimum was achieved on iteration 393 for which $p = 198$. For the case in which $M_E = 0.99905$, θ is about 0.004 rad, which implies that a local mini- mum should be achieved at about iteration 597 when p is 392. Indeed, no local minimum was detected for this case before convergence was achieved on iteration 561.

The iterations required for convergence were 683, 646, and 561, respectively, for the three cases $M_E = 0.9990061$, $M_E = 0.9991$, and $M_E = 0.99905$. Thus the overestimation of μ_1 can accelerate convergence under certain situations.

If $M_E > \mu_1$, the easily calculable quantities $L^{(s, p)}$ defined by (6-4.7), $C^{(s, p)}$ defined by (6-4.11), and $R^{(s, p)}$ defined by (6-5.11) usually behave in a particular way and may even be used to indicate the iteration on which the local minimum occurs. From (6-7.2), when $M_E > \mu_1$ and approximation (6-7.1) is valid, we have

$$L^{(s, p)} \equiv \frac{B^{(s, p)}}{[2r^{p/2}/(1 + r^p)]} \doteq |\cos p\theta| \qquad (6\text{-}7.5)$$

and

$$C^{(s, p)} = \frac{\log B^{(s, p)}}{\log[2r^{p/2}/(1 + r^p)]} \doteq 1 + \frac{\log|\cos p\theta|}{\log[2r^{p/2}/(1 + r^p)]}. \qquad (6\text{-}7.6)$$

Thus when $M_E > \mu_1$, $L^{(s, p)}$ oscillates between 0 and 1 and achieves a local minimum when $(p\theta)$ is closest to $(i + \frac{1}{2})\pi$, $i = 0, 1, \ldots$. Thus the occurrence of a local minimum for $L^{(s, p)}$ and a local maximum for $C^{(s, p)}$ may be taken to indicate that a local minimum for $\|\varepsilon^{(s, p)}\|_2$ was achieved on the same iteration. With the same assumptions, it is easy to show that $R^{(s, p)}$ approximately satisfies

$$R^{(s, p)} \equiv \frac{\|\delta^{(s, p)}\|_2}{\|\delta^{(s, p-1)}\|_2} \doteq \frac{\rho_p}{2w(1)} \left| \frac{\cos p\theta}{\cos(p - 1)\theta} \right|, \qquad (6\text{-}7.7)$$

where ρ_p is the acceleration parameter (6-3.2). Hence, when $M_E > \mu_1$, $R^{(s, p)}$ will oscillate about unity and will go through unity from below when $(p\theta)$ is close to $(i + \frac{1}{2})\pi$, $i = 0, 1, \ldots$. For the case $M_E = 0.9991$ of Fig. 6-7.1, $C^{(s, p)}$ achieved a local maximum on iteration 392, $L^{(s, p)}$ a local minimum on iteration 393, and $R^{(s, p)}$ first became greater than unity on iteration 394.

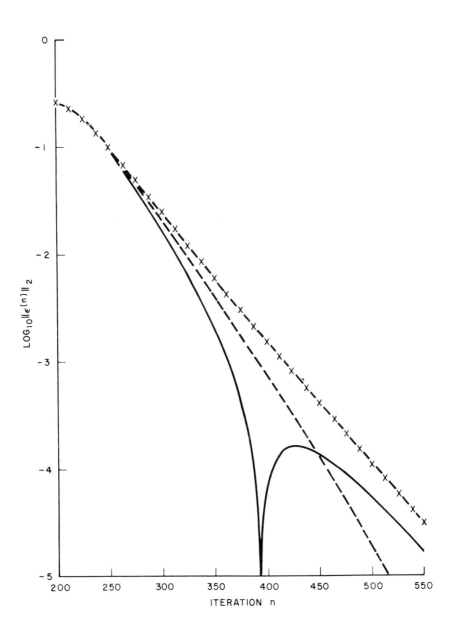

Fig. 6-7.1. Graph of $\log_{10} \|\varepsilon^{(n)}\|_2$ versus n for problem 2 with different estimates M_E in the generation of the last Chebyshev polynomial. $- \times - \times - \times -$, $M_E = 0.9990061 < \mu_1$; $---$, $M_E = 0.99905 > \mu_1$; ———, $M_E = 0.99910 > \mu_1$.

From the above discussion, it is clear that the Chebyshev method may be accelerated by intentionally overestimating μ_1 once the M_E estimates have converged to μ_1. The use of such a procedure would probably be beneficial only for slowly converging problems.

6.8 SINGULAR AND EIGENVECTOR DEFICIENT PROBLEMS

In this chapter as well as in the previous chapter, we have always assumed that the basic method (6-1.1) is symmetrizable. Because of this assumption, the iteration matrix G satisfied the important properties (a)–(c) given in Theorem 2-2.1. Now we shall discuss briefly some difficulties introduced when property (b) or property (c) is not valid. Problems for which property (a) is not valid (i.e., G has complex eigenvalues) will not be considered here since this case is discussed later in Chapters 11 and 12. Specifically, in this section we show that the algorithms given in this chapter and in Chapter 5 may be used without modification to solve certain singular matrix problems but that these algorithms generally should not be used when the basic iteration matrix G has an eigenvector deficiency.

For our discussion, we let A be an $N \times N$ matrix and let b be any vector for which the system

$$Au = b \qquad (6\text{-}8.1)$$

has at least one solution. We assume that the system (6-8.1) is to be solved by the polynomial acceleration method (3-2.7). Let $\bar{u}$ be any solution of (6-8.1) and let $\varepsilon^{(n)} \equiv u^{(n)} - \bar{u}$, $n = 0, 1, \ldots$, denote the error vector associated with the iterates $u^{(n)}$ of (3-2.7). From Theorem 3-2.1, $\varepsilon^{(n)}$ can be expressed in the form

$$\varepsilon^{(n)} = Q_n(G)\varepsilon^{(0)}. \qquad (6\text{-}8.2)$$

Here G is the iteration matrix associated with the basic method (6-1.1), and $Q_n(G)$ is the matrix polynomial defined by (3-2.6). We assume that G is given by (2-2.1), i.e.,

$$G = I - Q^{-1}A. \qquad (6\text{-}8.3)$$

We first discuss the important case in which the system (6-8.1) is singular.

A is Singular, Symmetric, and Positive Semidefinite

We remark that for this case the source vector b must be orthogonal to the null space of A in order for the system (6-8.1) to have a solution. (See, e.g.,

Noble and Daniel [1977].) Moreover, if the singular system (6-8.1) has a solution, then it has infinitely many solutions. Also, since the singular matrix A has an eigenvalue equal to zero, it follows from (6-8.3) that the iteration matrix G has an eigenvalue equal to unity and that any eigenvector of G with eigenvalue equal to unity is in the null space of A.

When the system (6-8.1) is singular and has a solution, we now show that under certain conditions a modified Chebyshev acceleration procedure applied to (6-1.1) leads to solution of (6-8.1) for any choice of $u^{(0)}$. However, the solution obtained will depend on $u^{(0)}$.

Theorem 6-8.1.   Let the $N \times N$ matrix A of (6-8.1) be singular, symmetric, and positive semidefinite, and let the system (6-8.1) have a solution. Let the iteration matrix G of the basic method (6-1.1) be given by (6-8.3), where Q is SPD, and let $\tilde{M}(G)$ denote the largest eigenvalue of G that is less than unity. If the polynomials $\{Q_n(x)\}$ of (3-2.6) are defined by

$$Q_n(x) \equiv \tilde{P}_n(x) \equiv T_n(\tilde{w}(x))/T_n(\tilde{w}(1)), \tag{6-8.4}$$

where $T_n(w)$ is the Chebyshev polynomial (4-2.2) and

$$\tilde{w}(x) \equiv (2x - \tilde{M}(G) - m(G))/(\tilde{M}(G) - m(G)), \tag{6-8.5}$$

then the polynomial acceleration procedure defined by (3-2.7) converges to a solution of (6-8.1).

Proof.   Since $Q^{1/2}(I - G)Q^{-1/2}$ is symmetric and positive semidefinite, it follows from the results of Section 1.3 that properties (a)–(c) given in Theorem 2-2.1 are valid for G except that the largest eigenvalue(s) of G is (are) equal to unity. Let the eigenvalues $\{\mu_i\}_{i=1}^{i=N}$ of G be ordered as

$$m(G) \equiv \mu_N \leq \mu_{N-1} \leq \cdots \leq \mu_t < \mu_{t-1} = \cdots = \mu_1 = 1, \tag{6-8.6}$$

and let the set of eigenvectors $\{v(i)\}_{i=1}^{i=N}$ be a basis for the associated vector space. We assume that $Gv(i) = \mu_i v(i)$.

Let $\bar{u}$ be any solution of (6-8.1), and let $\varepsilon^{(0)} = u^{(0)} - \bar{u}$. As in Section 6.2, the eigenvector expansion of $\varepsilon^{(0)}$ may be expressed as

$$\varepsilon^{(0)} = c_1 v(1) + \sum_{i=t}^{N} c_i v(i). \tag{6-8.7}$$

Also, by (6-8.2) we have that

$$\varepsilon^{(n)} = u^{(n)} - \bar{u} = \tilde{P}_n(G)\varepsilon^{(0)}. \tag{6-8.8}$$

Using the fact that $\tilde{P}_n(1) = 1$, we can express $\varepsilon^{(n)}$ in the form

$$\varepsilon^{(n)} = c_1 v(1) + \sum_{i=t}^{N} c_i \tilde{P}(\mu_i)v(i). \tag{6-8.9}$$

Note that $\tilde{P}_n(x)$ is the same as the normalized Chebyshev polynomial $P_n(x)$ of (4-2.9) except that $M(G)$ is replaced by $\tilde{M}(G)$. It then follows from (4-2.10) and (4-2.5) that

$$\max_{\mu_1 \leq x \leq \mu_t} |\tilde{P}_n(x)| = [T_n(\tilde{w}(1))]^{-1}. \tag{6-8.10}$$

Since $\tilde{w}(1) > 1$, it follows from (4-2.2) that $\lim_{n \to \infty} [T_n(\tilde{w}(1))]^{-1} = 0$. Hence, using (6-8.10) in (6-8.9), we have

$$\lim_{n \to \infty} (\varepsilon^{(n)} - c_1 v(1)) = 0. \tag{6-8.11}$$

Therefore,

$$\lim_{n \to \infty} u^{(n)} = \bar{u} + c_1 v(1). \tag{6-8.12}$$

Moreover, since $v(1)$ is in the null space of A, we have that

$$A(\bar{u} + c_1 v(1)) = A\bar{u} = b, \tag{6-8.13}$$

from which it follows that the limit (6-8.12) is a solution of (6-8.1). ∎

Computationally, the singular problem can be solved by using the Chebyshev procedures given previously for the nonsingular case. When estimates m_E and $\tilde{M}_E$ are used for $m(G)$ and $\tilde{M}(G)$, respectively, it is easy to show that the Chebyshev method for singular problems can be carried out, using (5-1.4)–(5-1.7) with M_E replaced by $\tilde{M}_E$. Moreover, the previously given parameter and error estimation procedures may also be used. To show why this is so, let $\tilde{P}_{n, E}(x)$ denote the normalized Chebyshev polynomial (6-8.4) with $\tilde{M}(G)$ and $m(G)$ replaced by the estimates m_E and $\tilde{M}_E$. From (5-2.2), (6-8.9), and (6-8.6), we then have

$$\delta^{(n)} = \tilde{P}_{n, E}(G)\delta^{(0)} = \sum_{i=t}^{N} (\mu_i - 1)\tilde{P}_{n, E}(\mu_i)c_i v(i). \tag{6-8.14}$$

Note that the expansion for $\delta^{(n)}$ is void of any eigenvectors of G with eigenvalue unity. Because of this, previously given results for $\delta^{(n)}$ with $\mu_1 \, (= M(G))$ replaced by $\mu_t \, (= \tilde{M}(G))$ are valid for the singular problem. Thus estimates for $\tilde{M}(G)$ can be obtained by using the procedures described previously. Note that the error estimation procedure measures the vector $\varepsilon^{(n)} - c_1 v(1)$ since now

$$\lim_{n \to \infty} \frac{\|\varepsilon^{(n)} - c_1 v(1)\|}{\|\delta^{(n)}\|} = (1 - \mu_t)^{-1}.$$

For singular problems, the algorithms given in this chapter and in Chapter 5 require only the additional assumption that the source vector b is orthogonal to the null space of A.†

---

† In the case in which the null space of $I - G$ is one dimensional and a vector w in the null space is known, the system $Au = b - ((b, w)/(w, w))w$ has a solution for any given vector. (See, for example, O'Carroll [1973].)

The Set of Eigenvectors for G Does Not Include a Basis

We now assume that the set of eigenvalues $\{\mu_i\}_{i=1}^{i=N}$ of A are real but that the iteration matrix G has an eigenvector deficiency. To fix ideas, we further assume that the set of eigenvectors for G, supplemented by one principal vector of grade 2 (see Section 1.3), includes a basis for the associated vector space.† Let μ_1 be the repeated eigenvalue associated with the principal vector. Thus the set of basis vectors can be given by $\{v(1), p(1), v(3), \ldots, v(N)\}$, where $v(i)$ is an eigenvector of G satisfying $Gv(i) = \mu_i v(i)$, $i = 1, 3, \ldots, N$, and $p(1)$ is a principal vector of grade 2 which, from (1-3.15), satisfies $Gp(1) = \mu_1 p(1) + v(1)$.

With $\varepsilon^{(0)}$ given by $\varepsilon^{(0)} = c_1 v(1) + c_2 p(1) + \sum_{i=3}^{N} c_i v(i)$, the error vector $\varepsilon^{(n)}$ of (6-8.2) may be expressed as

$$\varepsilon^{(n)} = c_1 Q_n(\mu_1)v(1) + c_2 Q_n(G)p(1) + \sum_{i=3}^{N} c_i Q_n(\mu_i)v(i). \qquad (6\text{-}8.15)$$

Since, by (1-3.15), $G^l p(1) = \mu_1^l p(1) + l(\mu_1)^{l-1}v(1)$, it follows that $Q_n(G)p(1) = Q_n(\mu_1)p(1) + dQ_n(x)/dx|_{\mu_1}v(1)$. Thus relation (6-8.15) takes the form

$$\varepsilon^{(n)} = c_1 Q_n(\mu_1)v(1) + c_2 \left.\frac{dQ_n(x)}{dx}\right|_{\mu_1}$$

$$+ c_2 Q_n(\mu_1)p(1) + \sum_{i=3}^{N} c_i Q_n(\mu_i)v(i). \qquad (6\text{-}8.16)$$

Comparing (6-8.16) with (6-2.21), we see that the presence of a principal vector of grade 2 introduces a derivative of $Q_n(x)$ into the expansion for $\varepsilon^{(n)}$. In general, principal vectors of grade f introduce derivatives of $Q_n(x)$ of orders up to $f - 1$ into the expansion for $\varepsilon^{(n)}$. When $Q_n(x)$ is the normalized Chebyshev polynomial (4-2.9), Manteuffel [1975] has shown that convergence still takes place, but usually at a much reduced rate.

Because of the introduction of the additional derivative term(s), neither polynomial acceleration nor the algorithms given in this book are recommended when the set of eigenvectors for the iteration matrix G does not include a basis.

---

† This assumption corresponds to assuming that $k = N - 1$ in the Jordan canonical form J of G given in Theorem 1-3.2.

7

Conjugate Gradient Acceleration

7.1 INTRODUCTION

In the early 1950s Hestenes and Stiefel [1952] presented a new iterative method for solving systems of linear algebraic equations. This new method was known as the "conjugate gradient method"; we shall refer to it as the *CG method*. The CG method, though an iterative method, converges to the true solution of the linear system in a finite number of iterations in the absence of rounding errors. Because of this and many other interesting properties, the CG method attracted considerable attention in the numerical analysis community when it was first presented. However, for various reasons the method was not widely used, and little was heard about it for many years. As noted by Concus *et al.* [1976b], there was hardly any mention of the CG method in the proceedings of a *Conference on Sparse Matrices and Their Applications* held in 1971 (see Rose and Willoughby [1972]).

Beginning in the mid-1960s there was a strong resurgence of interest in the CG method. A number of papers appeared, including those by Daniel [1965, 1967], J. K. Reid [1971, 1972], Bartels and Daniel [1974], Axelsson [1974], O'Leary [1975], Chandra *et al.* [1977], Concus *et al.* [1976b], and many others.

Although not generally recognized until fairly recently, the conjugate gradient method is not just one method, but a whole family of methods (Hestenes [1956]). Each such method can be regarded as an acceleration process for a particular (basic) linear stationary iterative method of first degree. (The classical CG method can be regarded as an acceleration procedure based on the RF method of Section 2.3). Moreover, as shown by Engeli *et al.* [1959], the CG method can be represented in a three-term form which resembles Chebyshev acceleration applied to the RF method. One can also develop a similar three-term form for conjugate gradient acceleration applied to more general basic methods which resembles Chebyshev acceleration applied to those methods. It can also be shown that conjugate gradient acceleration of a given iterative method converges with respect to a certain error measurement procedure at least as fast as the corresponding Chebyshev procedure. Furthermore, no parameter estimates are required in the implementation of conjugate gradient acceleration. Because of these and other attractive properties, conjugate gradient acceleration has been used extensively in recent years.

We describe the classical CG method in Section 7.2. In Section 7.3 we describe the equivalent three-term form. In Section 7.4 we describe conjugate gradient acceleration of a class of basic iterative methods. In Section 7.5 we describe procedures for deciding when to terminate the iterative process. Computational procedures are given in Section 7.6. In Section 7.7 numerical results based on simulation experiments are given.

7.2   THE CONJUGATE GRADIENT METHOD

We now describe the classical conjugate gradient method (CG method) of Hestenes and Stiefel [1952] as applied to the linear system $Au = b$ given by (2-1.1). We assume that the $N \times N$ matrix A is SPD.

The CG method can be regarded as a modification of the method of steepest descent. To derive the method of steepest descent we consider the quadratic form

$$F(u) = \tfrac{1}{2}(u, Au) - (b, u). \qquad (7\text{-}2.1)$$

Since

$$F(u) = F(\bar{u}) + \tfrac{1}{2}((u - \bar{u}), A(u - \bar{u})), \qquad (7\text{-}2.2)$$

where $\bar{u} = A^{-1}b$ is the solution of (2-1.1), and since A is SPD, it follows that the problem of solving $Au = b$ is equivalent to the problem of minimizing $F(u)$. Moreover, the gradient of $F(u)$ is given by

$$\text{grad } F(u) = b - Au. \qquad (7\text{-}2.3)$$

The direction of the vector grad $F(u)$ is the direction for which the functional $F(u)$ at the point u has the greatest instantaneous rate of change. If $u^{(n)}$ is some approximation to $\bar{u}$, then in the method of steepest descent we obtain an improved approximation $u^{(n+1)}$ by moving in the direction of grad $F(u^{(n)})$ to a point where $F(u^{(n+1)})$ is minimal, i.e., $u^{(n+1)} = u^{(n)} + \lambda_n$ grad $F(u^{(n)})$, where λ_n is chosen to minimize $F(u^{(n+1)})$. Using (7-2.1), we easily calculate that $\lambda_n = (r^{(n)}, r^{(n)})/(r^{(n)}, Ar^{(n)})$, where $r^{(n)} \equiv b - Au^{(n)}$. Since, from (7-2.3), grad $F(u^{(n)}) = r^{(n)}$, we can express the method of steepest descent in the form

$$u^{(0)} \text{ is arbitrary,}$$

$$u^{(n+1)} = u^{(n)} + \lambda_n r^{(n)} \qquad \text{for} \quad n = 0, 1, \ldots,$$

$$r^{(n)} = b - Au^{(n)} \tag{7-2.4}$$

$$\lambda_n = \frac{(r^{(n)}, r^{(n)})}{(r^{(n)}, Ar^{(n)})}.$$

For ill-conditioned matrices A, the convergence rate of the method of steepest descent can be very slow (see, e.g., Luenberger [1973]). However, by choosing our direction vectors differently, we obtain the CG method, which, as we shall see shortly, gives the solution in at most N iterations in the absence of rounding errors.

Let $u^{(0)}$ be arbitrary and let successive approximations to the solution $\bar{u}$ be given by $u^{(n+1)} = u^{(n)} + \lambda_n p^{(n)}$, where $p^{(n)}$ is a "direction vector." For the CG method, we let $p^{(0)} = r^{(0)}$ and $p^{(n)} = r^{(n)} + \alpha_n p^{(n-1)}$ for $n \geq 1$, where α_n is chosen so that $p^{(n)}$ is A-conjugate to $p^{(n-1)}$, i.e., $(p^{(n)}, Ap^{(n-1)}) = 0$. Evidently, $\alpha_n = -(r^{(n)}, Ap^{(n-1)})/(p^{(n-1)}, Ap^{(n-1)})$. As before, choosing λ_n to minimize $F(u^{(n+1)})$, we obtain $\lambda_n = (p^{(n)}, r^{(n)})/(p^{(n)}, Ap^{(n)})$. The formulas for the CG method are given by

$$u^{(0)} \text{ is arbitrary,}$$

$$u^{(n+1)} = u^{(n)} + \lambda_n p^{(n)}, \qquad\qquad n = 0, 1, \ldots,$$

$$p^{(n)} = \begin{cases} r^{(n)}, & \text{if} \quad n = 0, \\ r^{(n)} + \alpha_n p^{(n-1)}, & n = 1, 2, \ldots, \end{cases}$$

$$\alpha_n = -\frac{(r^{(n)}, Ap^{(n-1)})}{(p^{(n-1)}, Ap^{(n-1)})}, \qquad n = 1, 2, \ldots, \tag{7-2.5}$$

$$r^{(n)} = b - Au^{(n)}, \qquad\qquad n = 0, 1, \ldots,$$

$$\lambda_n = \frac{(p^{(n)}, r^{(n)})}{(p^{(n)}, Ap^{(n)})}, \qquad\qquad n = 0, 1, \ldots.$$

It can be shown that α_n, λ_n, and $r^{(n)}$ can be given equivalently by

$$\alpha_n = \frac{(r^{(n)}, r^{(n)})}{(r^{(n-1)}, r^{(n-1)})}, \qquad n = 1, 2, \ldots,$$

$$\lambda_n = \frac{(r^{(n)}, r^{(n)})}{(p^{(n)}, Ap^{(n)})}, \qquad n = 0, 1, \ldots, \qquad (7\text{-}2.6)$$

$$r^{(n)} = r^{(n-1)} - \lambda_{n-1} Ap^{(n-1)}, \qquad n = 1, 2, \ldots.$$

Hestenes and Stiefel [1952] show that the residuals $r^{(0)}$, $r^{(1)}$, ... and the direction vectors $p^{(0)}$, $p^{(1)}$, ... generated by (7-2.5) satisfy the relations

$$(r^{(i)}, r^{(j)}) = 0 \qquad \text{for} \quad i \neq j,$$

$$(p^{(i)}, Ap^{(j)}) = 0 \qquad \text{for} \quad i \neq j, \qquad (7\text{-}2.7)$$

$$(r^{(i)}, Ap^{(j)}) = 0 \qquad \text{for} \quad i \neq j \quad \text{and} \quad i \neq j + 1.$$

Thus the residual vectors $r^{(0)}$, $r^{(1)}$, ... are mutually orthogonal and the direction vectors $p^{(0)}$, $p^{(1)}$, ... are mutually A-conjugate. From the first relation in (7-2.7), it follows that $r^{(s)} = 0$ for some $s \leq N$. Thus the method (7-2.5) converges, in the absence of rounding errors, in at most N iterations. Hestenes and Stiefel [1952] also show that the error vector $\varepsilon^{(n)} \equiv u^{(n)} - \bar{u}$ associated with the CG method satisfies

$$\|\varepsilon^{(n+1)}\|_2 < \|\varepsilon^{(n)}\|_2 \qquad (7\text{-}2.8)$$

whenever $\varepsilon^{(n)} \neq 0$. In the next section, we shall have more to say concerning the average rate of convergence for the CG method.

It can be shown (see, e.g., Beckman [1960]) that the direction vector $p^{(n)}$ is a scalar multiple of the projection of the gradient vector $r^{(n)} = \text{grad } F(u^{(n)})$ in the linear space spanned by $p^{(n)}$, $p^{(n+1)}$, ..., $p^{(N-1)}$. This fact coupled with the fact that the direction vectors $p^{(0)}$, $p^{(1)}$, ... are mutually A-conjugate accounts for the name "conjugate gradient method."

The CG method is a special case of the more general conjugate direction (CD) method. In the CD method, the vectors $p^{(0)}$, $p^{(1)}$, ..., $p^{(N-1)}$ are selected to be nonzero and mutually A-conjugate but have no further restrictions. To describe the basic idea involved, suppose that the set $\{p^{(n)}\}_{n=0}^{n=N-1}$ of nonzero mutually A-conjugate vectors is given. Since A is SPD, it is easy to show that the set $\{p^{(n)}\}_{n=0}^{n=N-1}$ is also linearly independent. Thus there exist constants $c_0, c_1, \ldots, c_{N-1}$ such that

$$\bar{u} = u^{(0)} + c_0 p^{(0)} + c_1 p^{(1)} + \cdots + c_{N-1} p^{(N-1)}, \qquad (7\text{-}2.9)$$

where $u^{(0)}$ is some initial approximation to $\bar{u}$. Multiplying (7-2.9) by A and taking inner products with $p^{(n)}$, we obtain

$$c_n = \frac{(p^{(n)}, b - Au^{(0)})}{(p^{(n)}, Ap^{(n)})}, \qquad n = 0, \ldots, N - 1. \qquad (7\text{-}2.10)$$

Note that the constants c_n are easily calculable.†

The CD method is given by the formulas

$$u^{(0)} \text{ is arbitrary,}$$

$$r^{(n)} = b - Au^{(n)}, \qquad n = 0, 1, \ldots,$$

$$\lambda_n = \frac{(p^{(n)}, r^{(n)})}{(p^{(n)}, Ap^{(n)})}, \qquad n = 0, 1, \ldots, \qquad (7\text{-}2.11)$$

$$u^{(n+1)} = u^{(n)} + \lambda_n p^{(n)}, \qquad n = 0, 1, \ldots.$$

It can be shown (see, e.g., Luenberger [1973]) that the λ_n in (7-2.11) are equal to the c_n from (7-2.10) and that the iterates $u^{(n+1)}$ in (7-2.11) can be expressed in the form

$$u^{(n+1)} = u^{(0)} + c_0 p^{(0)} + c_1 p^{(1)} + \cdots + c_n p^{(n)}, \qquad (7\text{-}2.12)$$

where the c_i, $i = 0, \ldots, n$, are those given in (7-2.10). From (7-2.12) and (7-2.9), it follows that $u^{(n)} = \bar{u}$ for some $n \leq N$. Thus the CD method also enjoys the property that convergence is achieved, in the absence of rounding errors, in at most N iterations.

The CD method is not well defined in that no prescription is given for the computation of the direction vectors $p^{(0)}, p^{(1)}, \ldots$. Various formulas can be given, with each leading to a special method. We can generate an A-conjugate set of vectors from any set $\{v^{(n)}\}_{n=0}^{n=N-1}$ of linearly independent vectors, using the Gram–Schmidt orthogonalization procedure. Hestenes and Stiefel [1952] show that the CD method is equivalent to the Gauss elimination method when the set $\{v^{(n)}\}$ is chosen to be the unit basis vectors, i.e., when $v^{(0)} = [1, 0, \ldots, 0]^T$, $v^{(1)} = [0, 1, 0, \ldots, 0]^T$, etc. For the CG method, the set $\{v^{(n)}\}$ is chosen to be the residual vectors, i.e., $v^{(n)} = r^{(n)}$. The residual and direction vectors for the CG method are not defined beforehand but are determined sequentially in the order $r^{(0)}, p^{(0)}, r^{(1)}, p^{(1)}, \ldots$ as the iterations progress.

---

† If the set $\{p^{(n)}\}_{n=0}^{n=N-1}$ were orthogonal and not A-conjugate, then $c_n = (p^{(n)}, (\bar{u} - u^{(0)}))/ (p^{(n)}, p^{(n)})$. This expression for the c_n is of little help since $\bar{u}$ is not known.

7.3   THE THREE-TERM FORM OF THE CONJUGATE GRADIENT METHOD

Engeli *et al.* [1959] considered the following three-term form of the CG method:

$$u^{(n+1)} = \rho_{n+1}\{\gamma_{n+1}r^{(n)} + u^{(n)}\} + (1 - \rho_{n+1})u^{(n-1)}, \qquad (7-3.1)$$

where

$$\gamma_{n+1} = \frac{(r^{(n)}, r^{(n)})}{(r^{(n)}, Ar^{(n)})} \qquad (7-3.2)$$

and

$$\rho_{n+1} = \left[1 - \frac{\gamma_{n+1}}{\gamma_n} \frac{(r^{(n)}, r^{(n)})}{(r^{(n-1)}, r^{(n-1)})} \frac{1}{\rho_n} \right]^{-1}, \qquad \text{if} \quad n \geq 1 \quad (\rho_1 = 1). \quad (7-3.3)$$

The above formulas can be obtained from (7-2.5) by eliminating $p^{(n)}$ and $p^{(n-1)}$ from the pair of equations $u^{(n+1)} = u^{(n)} + \lambda_n p^{(n)}$ and $u^{(n)} = u^{(n-1)} + \lambda_{n-1}p^{(n-1)}$. Thus we obtain (7-3.1) with

$$\rho_{n+1} = 1 + \alpha_n \lambda_n / \lambda_{n-1}, \qquad \text{if} \quad n \geq 1 \quad (\rho_1 = 1) \qquad (7-3.4)$$

and

$$\gamma_{n+1} = \lambda_n / \rho_{n+1}. \qquad (7-3.5)$$

Formulas (7-3.2) and (7-3.3) can then be derived directly from (7-2.5) (see, e.g., Reid [1971]).

An alternative derivation of (7-3.2) and (7-3.3) can be given as follows (see Concus *et al.* [1976b]). By (7-2.5) and (7-3.1), we have

$$r^{(n+1)} = \rho_{n+1}\{-\gamma_{n+1}Ar^{(n)} + r^{(n)}\} + (1 - \rho_{n+1})r^{(n-1)}. \qquad (7-3.6)$$

We now use the fact that the residuals are mutually orthogonal. If we require that $(r^{(n+1)}, r^{(n)}) = 0$, we get (7-3.2), provided that $\rho_{n+1} \neq 0$. But by (7-3.4) and (7-2.6) it follows that

$$\rho_{n+1} \geq 1, \qquad n = 0, 1, \ldots. \qquad (7-3.7)$$

If we require that $(r^{(n+1)}, r^{(n-1)}) = 0$ we get, assuming that $(r^{(n)}, r^{(n-1)}) = 0$,

$$0 = (r^{(n+1)}, r^{(n-1)})$$

$$= \rho_{n+1}\{-\gamma_{n+1}(Ar^{(n)}, r^{(n-1)})\} + (1 - \rho_{n+1})(r^{(n-1)}, r^{(n-1)}). \qquad (7-3.8)$$

Replacing n by $n - 1$ in (7-3.6) and taking the inner product of both sides with $r^{(n)}$, we get

$$(r^{(n)}, r^{(n)}) = \rho_n(-\gamma_n(Ar^{(n-1)}, r^{(n)})) \tag{7-3.9}$$

or

$$(Ar^{(n-1)}, r^{(n)}) = (r^{(n-1)}, Ar^{(n)}) = (Ar^{(n)}, r^{(n-1)})$$

$$= -(r^{(n)}, r^{(n)})/\rho_n\gamma_n. \tag{7-3.10}$$

Substituting (7-3.10) into (7-3.8), we get (7-3.3).

Replacing $r^{(n)}$ by $b - Au^{(n)}$ in (7-3.1), we can express $u^{(n+1)}$ in the alternative form

$$u^{(n+1)} = \rho_{n+1}\{\gamma_{n+1}(I - A)u^{(n)} + b) + (1 - \gamma_{n+1})u^{(n)}\} + (1 - \rho_{n+1})u^{(n-1)}. \tag{7-3.11}$$

Since $(I - A)$ is the iteration matrix for the RF method (see Section 2.3), it follows from Theorem 3-2.1 that the iterates (7-3.11) correspond to a polynomial acceleration procedure applied to the RF method. Thus from (3-2.5), there exists a matrix polynomial $Q_n(G) = \alpha_{n,0}I + \alpha_{n,1}G + \cdots + \alpha_{n,n}G^n$ such that the error vector $\varepsilon^{(n)} \equiv u^{(n)} - \bar{u}$ associated with (7-3.11) can be expressed in the form

$$\varepsilon^{(n)} = Q_n(G)\varepsilon^{(0)} = Q_n(I - A)\varepsilon^{(0)}.$$

It can be shown (see, e.g., Young et al. [1980]) that the property that the residual vectors are mutually orthogonal characterizes the CG method among all polynomial acceleration procedures applied to the RF method.

It can also be shown (see, e.g., Young et al. [1980]) that the CG method minimizes the $A^{1/2}$-norm of the error vector among all polynomial acceleration methods applied to the RF method. Thus if $\varepsilon^{(n)}$ is the error vector associated with (7-3.11) and if $\tilde{\varepsilon}^{(n)}$, where $\tilde{\varepsilon}^{(0)} = \varepsilon^{(0)}$, is the error vector associated with any other polynomial method applied to the RF method, then $\|\varepsilon^{(n)}\|_{A^{1/2}} \le \|\tilde{\varepsilon}^{(n)}\|_{A^{1/2}}$. In particular, if $\tilde{\varepsilon}^{(n)}$ corresponds to the Chebyshev acceleration method, it then follows from (3-2.12) and (4-2.20), since $A^{1/2}$ is a symmetrization matrix for the RF method, that

$$\|\varepsilon^{(n)}\|_{A^{1/2}} \le \|\tilde{\varepsilon}^{(n)}\|_{A^{1/2}} \le \frac{2\bar{r}^{n/2}}{1 + \bar{r}^n}\|\varepsilon^{(0)}\|_{A^{1/2}}, \tag{7-3.12}$$

where

$$\bar{r} = (1 - \sqrt{1 - \bar{\sigma}^2})/(1 + \sqrt{1 - \bar{\sigma}^2}) \tag{7-3.13}$$

and

$$\bar{\sigma} = [M(A) - m(A)]/[M(A) + m(A)] = [\kappa(A) - 1]/[\kappa(A) + 1]. \tag{7-3.14}$$

Here $\kappa(A)$ is the spectral condition number of A defined by (1-4.18).

7.4 CONJUGATE GRADIENT ACCELERATION

In the previous section we have shown how the CG method may be regarded as a polynomial acceleration procedure based on the RF method. We now show how the CG method may be modified so as to correspond to a polynomial acceleration procedure applied to more general basic iteration methods.

Let us consider the basic iterative method defined by

$$u^{(n+1)} = Gu^{(n)} + k, \tag{7-4.1}$$

where for some nonsingular splitting matrix Q we have

$$G = I - Q^{-1}A, \qquad k = Q^{-1}b. \tag{7-4.2}$$

We assume that the method is *symmetrizable* in the sense of Definition 2-2.1. Thus there exists a nonsingular symmetrization matrix W such that $W(I - G)W^{-1}$ is SPD. If A is SPD, then our discussion includes the RF, Jacobi, and SSOR methods, as well as any method in which the splitting matrix Q is SPD. (In the latter case we can let W be any matrix such that $W^T W = Q$.)

To derive a CG acceleration procedure based on the iterative method (7-4.1), we first construct a new linear system which has the same solution as the original system (2-1.1). The coefficient matrix of the new system will be SPD and generally will have a much smaller spectral condition number than that of the matrix A of the original linear system.

To derive the new system, we first consider the related linear system (2-2.2):

$$(I - G)u = k, \tag{7-4.3}$$

which, by complete consistency,† has the same solution as (2-1.1). We next multiply both sides of (7-4.3) by a symmetrization matrix W, obtaining

$$W(I - G)u = Wk. \tag{7-4.4}$$

The matrix $W(I - G)$ is not in general symmetric. However, by introducing a new vector $\hat{u} = Wu$, we can write (7-4.4) in the form

$$\hat{A}\hat{u} = \hat{b}, \tag{7-4.5}$$

where

$$\hat{A} = W(I - G)W^{-1}, \qquad \hat{u} = Wu, \qquad \hat{b} = Wk. \tag{7-4.6}$$

† See Section 2.2.

The system (7-4.5) is often called the preconditioned system† since, in general, the condition number of $\hat{A}$ is much less than that of A.

We remark that one can also obtain the preconditioned system from (2-1.1) as follows. Let Q be the splitting matrix corresponding to (7-4.1). We first multiply both sides of (2-1.1) by Q^{-1}, obtaining

$$Q^{-1}Au = Q^{-1}b,$$

which is the same as (7-4.3) by (7-4.2). Then we multiply both sides by W and replace u by $W\hat{u}$. We then obtain (7-4.5), where

$$\hat{A} = WQ^{-1}AW^{-1}, \qquad \hat{u} = Wu, \qquad \hat{b} = WQ^{-1}b.$$

If we apply the CG method of Section 7.2 to the preconditioned system (7-4.5), we obtain, using (7-2.5) and (7-2.6), that

$u^{(0)}$ is arbitrary,

$$u^{(n+1)} = u^{(n)} + \lambda_n p^{(n)}, \qquad n = 0, 1, \ldots,$$

$$p^{(n)} = \begin{cases} \delta^{(0)}, & n = 0, \\ \delta^{(n)} + \alpha_n p^{(n-1)}, & n = 1, 2, \ldots, \end{cases}$$

$$\alpha_n = -\frac{(W\delta^{(n)}, W(I - G)p^{(n-1)})}{(Wp^{(n-1)}, W(I - G)p^{(n-1)})} \tag{7-4.7}$$

$$= \frac{(W\delta^{(n)}, W\delta^{(n)})}{(W\delta^{(n-1)}, W\delta^{(n-1)})}, \qquad n = 1, 2, \ldots,$$

$$\lambda_n = \frac{(Wp^{(n)}, W\delta^{(n)})}{(Wp^{(n)}, W(I - G)p^{(n)})}$$

$$= \frac{(W\delta^{(n)}, W\delta^{(n)})}{(Wp^{(n)}, W(I - G)p^{(n)})}, \qquad n = 0, 1, 2, \ldots.$$

Here $\delta^{(n)}$ is the pseudoresidual vector

$$\delta^{(n)} = Gu^{(n)} + k - u^{(n)}. \tag{7-4.8}$$

This extension of the CG method is equivalent to that given by Hestenes [1956]. (See also Daniel [1965], [1967].)

From the first relation given in (7-2.7), it follows that the pseudoresidual vectors $\delta^{(0)}, \delta^{(1)}, \ldots$, defined in (7-4.8), are mutually W-orthogonal in the sense that

$$(W\delta^{(i)}, W\delta^{(j)}) = 0, \qquad i \neq j. \tag{7-4.9}$$

---

† Preconditioning was used by Evans [1967], Axelsson [1974], and others. The preconditioning used here is slightly more general than that of Evans and Axelsson but reduces to theirs if Q is SPD and if $W = Q^{1/2}$ or if W is any matrix such that $W^T W = Q$.

The second relation in (7-2.7) implies that the direction vectors $p^{(0)}, p^{(1)}, \ldots$ of (7-4.7) are $W^T W(I - G)$-conjugate, i.e., that

$$(p^{(i)}, W^T W(I - G)p^{(j)}) = 0, \qquad i \neq j. \tag{7-4.10}$$

Analogous to (7-3.1)–(7-3.3) we can obtain the following three-term form of the CG acceleration procedure (7-4.7):

$$u^{(n+1)} = \rho_{n+1}\{\gamma_{n+1}\delta^{(n)} + u^{(n)}\} + (1 - \rho_{n+1})u^{(n-1)}, \tag{7-4.11}$$

where

$$\gamma_{n+1} = \frac{(W\delta^{(n)}, W\delta^{(n)})}{(W\delta^{(n)}, W(I - G)\delta^{(n)})} = \left[1 - \frac{(W\delta^{(n)}, WG\delta^{(n)})}{(W\delta^{(n)}, W\delta^{(n)})}\right]^{-1}, \tag{7-4.12}$$

$$\rho_1 = 1, \quad \rho_{n+1} = \left[1 - \frac{\gamma_{n+1}}{\gamma_n}\frac{(W\delta^{(n)}, W\delta^{(n)})}{(W\delta^{(n-1)}, W\delta^{(n-1)})}\frac{1}{\rho_n}\right]^{-1}, \qquad \text{if } n \geq 1. \tag{7-4.13}$$

The above method represents a slight extension of the "generalized conjugate gradient procedure" presented by Concus *et al.* [1976]; see also Axelsson [1974]. It is equivalent to their method if Q is SPD and $W = Q^{1/2}$ or W is any matrix such that $W^T W = Q$.

We refer to the methods obtained by applying CG acceleration to the RF, Jacobi, and SSOR methods as the RF–CG, J–CG, and SSOR–CG methods, respectively. Thus in our terminology, the classical CG method presented previously in Sections 7.2 and 7.3 is the same as the RF–CG method.

Concerning the choice between the two-term form (7-4.7) and the three-term form (7-4.11)–(7-4.13) of CG acceleration, results of Reid [1972] indicate that the two-term form is somewhat more efficient. On the other hand, the three-term form is the same as that used for Chebyshev acceleration. In any case, the difference between the two- and three-term forms does not appear to be very significant.

It can easily be shown that CG acceleration of the method (7-4.1) minimizes the $[W^T W(I - G)]^{1/2}$-norm† of the error as compared with any polynomial acceleration procedure based on (7-4.1). This follows from (7-4.5) and (7-4.6). Indeed, we have

$$(\hat{\varepsilon}^{(n)}, \hat{A}\hat{\varepsilon}^{(n)}) = (W\varepsilon^{(n)}, W(I - G)\varepsilon^{(n)}) = (\varepsilon^{(n)}, W^T W(I - G)\varepsilon^{(n)}), \tag{7-4.14}$$

where we let $\hat{\varepsilon}^{(n)} = W\varepsilon^{(n)}$. Thus the CG acceleration procedure applied to (7-4.3) minimizes $(\hat{\varepsilon}^{(n)}, \hat{A}\hat{\varepsilon}^{(n)})$, which is equal to the square of the $[W^T W(I - G)]^{1/2}$-norm of the error $\varepsilon^{(n)}$. If A and Q are SPD and if $W^T W = Q$, then we minimize the $A^{1/2}$-norm of the error as in the CG method.

---

† In order for the square root to be well defined, $W^T W(I - G)$ must be SPD; but this follows from the facts that $W^T W(I - G) = W^T W(I - G)W^{-1}W$ and that $W(I - G)W^{-1}$ is SPD.

As in (7-3.12), we can obtain a bound for the error vector associated with the CG procedure (7-4.7), using known results for the Chebyshev procedure. We first note that $[W^TW(I - G)]^{1/2}$ is a symmetrization matrix for the basic method (7-4.1). To show this, since $W^TW(I - G)$ is SPD, we can write

$$[W^TW(I - G)]^{1/2}(I - G)[W^TW(I - G)]^{-1/2}$$
$$= [W^TW(I - G)]^{1/2}(W^TW)^{-1}[W^TW(I - G)]^{1/2}$$
$$= \{[W^TW(I - G)]^{1/2}W^{-1}\}\{[W^TW(I - G)]^{1/2}W^{-1}\}^T, \quad (7\text{-}4.15)$$

from which the desired result follows. Now let $\varepsilon^{(n)}$ and $\tilde{\varepsilon}^{(n)}$, respectively, be the error vectors associated with the CG and Chebyshev acceleration methods applied to (7-4.1). We assume $\tilde{\varepsilon}^{(0)} = \varepsilon^{(0)}$. Using the error minimization property of CG acceleration and using (3-2.12) and (4-2.20) for Chebyshev acceleration, we obtain

$$\|\varepsilon^{(n)}\|_{[W^TW(I-G)]^{1/2}} \leq \|\tilde{\varepsilon}^{(n)}\|_{[W^TW(I-G)]^{1/2}} \leq \frac{2\bar{r}^{n/2}}{1 + \bar{r}^n} \|\varepsilon^{(0)}\|_{[W^TW(I-G)]^{1/2}}, \quad (7\text{-}4.16)$$

where $\bar{r}$ is given by (4-2.19). From (7-4.16) and (2-2.8), it follows that the average rate of convergence for the CG acceleration method, when measured in the $[W^TW(I - G)]^{1/2}$ norm, is at least as large as that for the corresponding Chebyshev procedure.

7.5 STOPPING PROCEDURES

We now describe a procedure for deciding when the CG iterative procedure of Section 7.4 should be terminated. Ideally, we should like to stop the iterative process and accept $u^{(n)}$ as a satisfactory approximation to the true solution $\bar{u}$ whenever $u^{(n)}$ satisfies the inequality

$$E_T \equiv \|u^{(n)} - \bar{u}\|_W / \|\bar{u}\|_W \leq \zeta, \quad (7\text{-}5.1)$$

where ζ is the stopping criterion number. As in previous chapters, we shall express the unknown quantity $u^{(n)} - \bar{u}$ in terms of $\delta^{(n)}$.

By (5-2.5), it follows that $\|u^{(n)} - \bar{u}\|_W \leq (1 - M(G))^{-1}\|\delta^{(n)}\|_W$, and hence we have

$$\frac{\|u^{(n)} - \bar{u}\|_W}{\|\bar{u}\|_W} \leq \frac{1}{1 - M(G)} \frac{\|\delta^{(n)}\|_W}{\|\bar{u}\|_W}. \quad (7\text{-}5.2)$$

Based on (7-5.2), we propose the stopping test

$$\frac{1}{1 - M_E^{(n)}} \frac{\|\delta^{(n)}\|_W}{\|u^{(n)}\|_W} \leq \zeta, \qquad (7\text{-}5.3)$$

where $M_E^{(n)}$ is an estimate for $M(G)$, whose determination we discuss below.

For the adaptive Chebyshev acceleration procedures described in Chapter 5, we used the latest estimate for $M(G)$, which was obtained from the adaptive process, for the stopping test. For the case of CG acceleration, we do not require estimates for $M(G)$ in order to carry out the iteration process. However, we can easily obtain estimates for $M(G)$ using

$$M_E^{(n)} = M(T_n), \qquad (7\text{-}5.4)$$

where T_n is the tridiagonal matrix given by

$$T_n = \begin{bmatrix} 1 - \gamma_1^{-1} & \dfrac{\rho_2 - 1}{\rho_2 \gamma_2} & & & 0 \\[2ex] \gamma_1^{-1} & 1 - \gamma_2^{-1} & \dfrac{\rho_3 - 1}{\rho_3 \gamma_3} & & \\[2ex] & (\rho_2 \gamma_2)^{-1} & 1 - \gamma_3^{-1} & \ddots & \\[2ex] & & \ddots & \ddots & \dfrac{\rho_n - 1}{\rho_n \gamma_n} \\[2ex] 0 & & & (\rho_{n-1} \gamma_{n-1})^{-1} & 1 - \gamma_n^{-1} \end{bmatrix}. \qquad (7\text{-}5.5)$$

Before discussing how to compute $M(T_n)$, we give a justification for the procedure. From (7-4.8) and (7-4.11), we have

$$\delta^{(n+1)} = \rho_{n+1}\{\gamma_{n+1} G\delta^{(n)} + (1 - \gamma_{n+1})\delta^{(n)}\} + (1 - \rho_{n+1})\delta^{(n-1)} \quad (7\text{-}5.6)$$

or

$$G\delta^{(n)} = \frac{1}{\rho_{n+1}\gamma_{n+1}} \delta^{(n+1)} - \frac{1 - \gamma_{n+1}}{\gamma_{n+1}} \delta^{(n)} - \frac{(1 - \rho_{n+1})}{\rho_{n+1}\gamma_{n+1}} \delta^{(n-1)}. \quad (7\text{-}5.7)$$

Let s be the smallest number such that $\delta^{(s+1)} = 0$. (We note that $s \leq N - 1$ in the absence of rounding errors.) We then have

$$G(\delta^{(0)}\delta^{(1)} \cdots \delta^{(s)}) = (\delta^{(0)}\delta^{(1)} \cdots \delta^{(s)})T_s. \qquad (7\text{-}5.8)$$

We note that since $\rho_n \geq 1$ by (7-3.7), the matrix T_n is similar to the symmetric matrix $\hat{T}_n$ given by

$$
\hat{T}_n = \begin{bmatrix}
1 - \gamma_1^{-1} & \sqrt{\dfrac{\rho_2 - 1}{\gamma_1 \rho_2 \gamma_2}} & & & & \\[2ex]
\sqrt{\dfrac{\rho_2 - 1}{\gamma_1 \rho_2 \gamma_2}} & 1 - \gamma_2^{-1} & \sqrt{\dfrac{\rho_3 - 1}{\gamma_2 \rho_2 \gamma_3 \rho_3}} & & 0 & \\[2ex]
& \sqrt{\dfrac{\rho_3 - 1}{\gamma_2 \rho_2 \gamma_3 \rho_3}} & 1 - \gamma_3^{-1} & \ddots & & \\[2ex]
& & \ddots & \ddots & & \\[2ex]
& 0 & & & & \sqrt{\dfrac{\rho_n - 1}{\gamma_{n-1} \rho_{n-1} \gamma_n \rho_n}} \\[2ex]
& & & & \sqrt{\dfrac{\rho_n - 1}{\gamma_{n-1} \rho_{n-1} \gamma_n \rho_n}} & 1 - \gamma_n^{-1}
\end{bmatrix}.
$$

$$(7\text{-}5.9)$$

It is well known (see, for instance, Barnard and Child [1952]) that the eigenvalues of $\hat{T}_{n-1}$ interlace those of $\hat{T}_n$. Thus, in particular, we have

$$M(\hat{T}_1) \leq M(\hat{T}_2) \leq \cdots \leq M(\hat{T}_n) \leq \cdots \leq M(G).$$

The fact that $M(\hat{T}_n) \leq M(G)$ for all n follows from (7-5.8). To show this, let λ be an eigenvalue of T_s with corresponding eigenvector v. Thus $T_s v = \lambda v$, and from (7-5.8) we have that

$$G(\delta^{(0)} \cdots \delta^{(s)})v = \lambda(\delta^{(0)} \cdots \delta^{(s)})v. \qquad (7\text{-}5.10)$$

From this it follows that

$$w = (\delta^{(0)} \cdots \delta^{(s)})v \qquad (7\text{-}5.11)$$

is an eigenvector of G and λ is the corresponding eigenvalue. It thus follows that every eigenvalue of T_s is an eigenvalue of G. Hence $M(T_s) \leq M(G)$. To summarize, we have

$$M(T_1) \leq M(T_2) \leq \cdots \leq M(T_s) \leq M(G) < 1. \qquad (7\text{-}5.12)$$

As noted by Concus et al. [1976a], the above procedure for finding $M(G)$ is essentially the Lanczos [1950] algorithm. It has been shown by Kaniel [1966] and Paige [1971] that good estimates of the extreme eigenvalues of G can often be obtained by computing the eigenvalues of T_n, where n is considerably less than s.

We now describe a numerical procedure for finding $M(T_n)$. We use the method of bisection (see, e.g., Young and Gregory [1973, Chap. 4]) as applied to the determinant function $\Delta_n(\lambda)$, defined by

$$\Delta_n(\lambda) \equiv \det(T_n - \lambda I). \tag{7-5.13}$$

The evaluation of $\Delta_n(\lambda)$ for a given λ is easily carried out, using the recurrence relation

$$\Delta_{n+1}(\lambda) = \left[-\frac{1 - \gamma_{n+1}}{\gamma_{n+1}} - \lambda \right] \Delta_n(\lambda) + \frac{1 - \rho_{n+1}}{\rho_{n+1}\gamma_{n+1}\rho_n\gamma_n} \Delta_{n-1}(\lambda). \tag{7-5.14}$$

Since the zeros of $\Delta_{n-1}(\lambda)$ interlace those of $\Delta_n(\lambda)$ and since all of the zeros of $\Delta_n(\lambda)$ are less than unity, by (7-5.12) and (7-5.13), it follows that there is precisely one zero of $\Delta_n(\lambda)$ between $M(T_{n-1})$ and 1. Thus we must have

$$\Delta_n(M(T_{n-1})) \Delta_n(1) < 0. \tag{7-5.15}$$

This condition is sufficient to guarantee the convergence of the method of bisection.

In practice, however, it may happen that since the computed value of $M(T_{n-1})$ is not exact, we may have $\Delta_n(M(T_{n-1})) \Delta_n(1) > 0$. In that case, we let $M(T_n) = M(T_{n-1})$. This should be sufficiently accurate for use with the stopping procedure.

The stopping test used for the adaptive Chebyshev procedure is extremely accurate. This is caused by the fact that as convergence is approached, the error vector and the pseudoresidual vector are each nearly proportional to the eigenvector corresponding to the largest eigenvalue of G. On the other hand, with the nonadaptive Chebyshev procedure the test is somewhat less accurate since little can be said about the relative sizes of the components of the expansion of the pseudoresidual vector in terms of the eigenvectors of G. The same is true of the CG method. However, we can assert that

$$\frac{1}{1 - m(G)} \frac{\|\delta^{(n)}\|_W}{\|\bar{u}\|_W} \leq \frac{\|u^{(n)} - \bar{u}\|_W}{\|\bar{u}\|_W} \leq \frac{1}{1 - M(G)} \frac{\|\delta^{(n)}\|_W}{\|\bar{u}\|_W}. \tag{7-5.16}$$

Thus the predicted value of $E_T = \|u^{(n)} - \bar{u}\|_W / \|\bar{u}\|_W$ cannot be too large by a factor of more than $(1 - m(G))(1 - M(G))^{-1}$. Thus if $m(G) = -1$ and $M(G) = 0.99$, then the predicted value of E_T may be as much as 200 times the actual value. If $\zeta = 10^{-6}$, this may mean that about 30% more iterations may have to be performed than necessary.

7.6 COMPUTATIONAL PROCEDURES

In this section we describe computational algorithms for carrying out the CG acceleration procedure (7-4.11)–(7-4.13). We assume that we are given the symmetrizable basic iterative method

$$u^{(n+1)} = Gu^{(n)} + k \qquad (7\text{-}6.1)$$

and a symmetrization matrix W such that $W(I - G)W^{-1}$ is SPD.

If we were to apply formulas (7-4.11)–(7-4.13) and (7-4.8) directly, there would be two matrix–vector multiplications involving G (to get $Gu^{(n)}$ and $G\delta^{(n)}$) and two involving W (to get $W\delta^{(n)}$ and $WG\delta^{(n)}$). (There would also be a matrix–vector multiplication to get $Wu^{(n)}$ for the stopping test (7-5.3); however, it is not actually necessary to recompute $\|u^{(n)}\|_W$ every iteration.) We can reduce the number of multiplications involving G from two to one by using the following formula (see (7-5.6)):

$$\delta^{(n+1)} = \rho_{n+1}(\gamma_{n+1}G\delta^{(n)} + (1 - \gamma_{n+1})\delta^{(n)}) + (1 - \rho_{n+1})\delta^{(n-1)}. \qquad (7\text{-}6.2)$$

Thus, assuming that $\rho_n, \gamma_n, \delta^{(n)}, u^{(n)}$, and $u^{(n-1)}$ are available from the previous iteration, we can compute $W\delta^{(n)}, WG\delta^{(n)}$, and then γ_{n+1} from (7-4.12) and ρ_{n+1} from (7-4.13). We then can get $u^{(n+1)}$ from (7-4.11) and $\delta^{(n+1)}$ from (7-6.2). Thus we have only one matrix–vector multiplication by G and two by W, not counting $Wu^{(n)}$, which is used in the stopping test. Later in this section we shall show how in some cases we can eliminate the matrix–vector multiplications by W.

We give below (Algorithm 7-6.1) an informal program that describes the overall algorithm. As in Section 5.5, we use the underline, as $\underline{u}$, to denote vectors. The input required is

ζ the stopping number used in (5.3)

$\underline{u}$ the initial guess vector $u^{(0)}$

In general, during an iteration, $\underline{u}$ represents $u^{(n)}$, $\underline{u}_\phi$ represents $u^{(n-1)}$, and $\underline{u}_N$ represents $u^{(n+1)}$; similarly, for γ and $\gamma_\phi, \underline{\delta}, \underline{\delta}_\phi$, and $\underline{\delta}_N$. Also, $\underline{\theta}$ represents $G\delta^{(n)}$, $\underline{\psi}$ represents $WG\delta^{(n)}$, and $\underline{\tau}$ represents $W\delta^{(n)}$. The rest of the notation is self-explanatory.

Details of the computation of $M(T_n)$, which is used for the stopping test, are not given. The reader is referred to Section 7.5. We note that we allow the η-norm to be used in the computation of the norm of $u^{(n+1)}$. This should cause no problem since $\|u^{(n+1)}\|_\eta$ is basically a normalization constant. We also allow for the computation of $\|\delta^{(n+1)}\|_\beta$ for some β-norm. If the test is satisfied, it is expected that the condition

$$\|\varepsilon^{(n)}\|_\beta / \|u^{(n)}\|_\eta \le \zeta \qquad (7\text{-}6.3)$$

will be satisfied approximately. However, this cannot be guaranteed unless $\beta = W$. If $\beta = W$, we replace DELNE by (DELNP)$^{1/2}$ in the stopping test.

Algorithm 7-6.1. Conjugate gradient acceleration.

Input: $(\zeta, \underline{u})$

Initialize:

$n := 0; \gamma := 1.0; \underline{u}_\phi := \underline{0}; \underline{\delta}_\phi := \underline{0};$
$\underline{\delta} := G\underline{u} + \underline{k} - \underline{u};$
$\underline{\tau} := W\underline{\delta}; \text{DELNP} := (\underline{\tau}, \underline{\tau});$

Next Iteration:

$n := n + 1; \gamma_\phi := \gamma;$

Calculate New Iterate:

$\underline{\theta} := G\underline{\delta}$

$\underline{\psi} := W\underline{\theta}; \text{DEN} := (\underline{\tau}, \underline{\psi}); \gamma := \left[1 - \dfrac{\text{DEN}}{\text{DELNP}} \right]^{-1};$

If $n = 1$, then $\rho := 1;$ else $\rho := \left[1 - \left(\dfrac{\gamma}{\gamma_\phi} \right)\left(\dfrac{1}{\rho} \right)\left(\dfrac{\text{DELNP}}{\text{DELNØ}} \right) \right]^{-1};$

$\underline{u}_N := \rho(\gamma\underline{\delta} + \underline{u}) + (1 - \rho)\underline{u}_\phi; \text{YUN} := \|\underline{u}_N\|_\eta;$
$\text{DELNØ} := \text{DELNP};$
$\underline{\delta}_N := \rho[\gamma\underline{\theta} + (1 - \gamma)\underline{\delta}] + (1 - \rho)\underline{\delta}_\phi; \text{DELNE} := \|\underline{\delta}_N\|_\beta$
$\underline{\tau} := W\underline{\delta}_N; \text{DELNP} := (\underline{\tau}, \underline{\tau});$
$\underline{u}_\phi := \underline{u}; \underline{\delta}_\phi := \underline{\delta};$
$\underline{u} := \underline{u}_N; \underline{\delta} := \underline{\delta}_N;$

Stopping Test:

Compute M_E (if needed) from the T_n matrix using bisection (see Section 7.5).
If $\dfrac{\text{DELNE}}{\text{YUN}} \leq \zeta(1 - M_E)$, then print final output and STOP (converged);
else continue;

Go to **Next Iteration**

We now describe how in some cases Algorithm 7-6.1 can be modified to eliminate the extra matrix–vector multiplications involving the symmetrization matrix W. We now assume that the matrix G can be written in the form

$$G = Z^{-1}Y = W^{-1}W^{-T}Y, \tag{7-6.4}$$

where

$$Z = W^T W. \tag{7-6.5}$$

We also assume that for any vector v, less work is required to compute WGv than to compute Gv. Thus we assume it is efficient to compute WGv by

$$WGv = W^{-T}Yv, \qquad (7\text{-}6.6)$$

whereas to compute Gv we would use

$$Gv = W^{-1}(WGv). \qquad (7\text{-}6.6')$$

An example of a situation for which the above assumptions apply is the (block) Jacobi method, where the splitting matrix Q is the block diagonal matrix D. Here we can take $W = S$, where S is an upper triangular matrix with $S^TS = D$. (Note from Section 3.2 that if D is tridiagonal, S will be bi-diagonal.) In this case, we have $Z = Q$ and $Y = Q - A$. Moreover, the elements of Y corresponding to nonzero elements of Q vanish. The computation of WGv for any vector v involves computing Yv and then $W^{-T}v$. Thus we have

$$WGv = W^{-T}(Yv). \qquad (7\text{-}6.7)$$

Our strategy is to work primarily with $Wu^{(n)}$ and $W\delta^{(n)}$ rather than with $u^{(n)}$ and $\delta^{(n)}$. We rewrite (7-4.11) and (7-6.2) in the form

$$Wu^{(n+1)} = \rho_{n+1}(Wu^{(n)} + \gamma_{n+1}W\delta^{(n)}) + (1 - \rho_{n+1})Wu^{(n-1)}, \qquad (7\text{-}6.8)$$

$$W\delta^{(n+1)} = \rho_{n+1}(\gamma_{n+1}WG\delta^{(n)} + (1 - \gamma_{n+1})W\delta^{(n)}) + (1 - \rho_{n+1})W\delta^{(n-1)}. \qquad (7\text{-}6.9)$$

Thus, assuming that ρ_n, γ_n, $W\delta^{(n)}$, $W\delta^{(n-1)}$, $Wu^{(n)}$, and $Wu^{(n-1)}$ are available from the previous iteration, we can compute $\delta^{(n)}$ and $WG\delta^{(n)}$ by

$$\delta^{(n)} = W^{-1}(W\delta^{(n)}), \qquad (7\text{-}6.10)$$

$$WG\delta^{(n)} = W^{-T}(Y\delta^{(n)}). \qquad (7\text{-}6.11)$$

(Actually, to get $y = W^{-T}(Y\delta^{(n)})$, we solve the system $W^Ty = Y\delta^{(n)}$ for y.) We then can get γ_{n+1} by (7-4.12) and ρ_{n+1} by (7-4.13). Then we can get $Wu^{(n+1)}$ by (7-6.8) and $W\delta^{(n+1)}$ by (7-6.9). Thus for each iteration, we carry out matrix–vector multiplications by W^{-T}, Y, and W^{-1}. This, by our assumption, is equivalent to a single matrix–vector multiplication by G.

An informal program that uses the above ideas is given in Algorithm 7-6.2. We let $\underline{Wu_\phi}$, $\underline{Wu}$, and $\underline{Wu_N}$ represent $Wu^{(n-1)}$, $Wu^{(n)}$, and $Wu^{(n+1)}$, respectively. Similarly, $\underline{W\delta_\phi}$, $\underline{W\delta}$, and $\underline{W\delta_N}$ represent $W\delta^{(n-1)}$, $W\delta^{(n)}$, and $W\delta^{(n+1)}$, respectively. We also let $\gamma_\phi = \gamma_n$ and $\gamma = \gamma_{n+1}$.

Algorithm 7-6.2. Special algorithm for conjugate gradient acceleration when $G = Z^{-1}Y$, Z is factored as $Z = W^TW$, and W is the symmetrization matrix.

Input: $(\zeta, \underline{u})$

Initialize:

$n := 0$; $\gamma := 1.0$; $\underline{Wu}_\phi := \underline{0}$; $\underline{W\delta}_\phi := \underline{0}$;
$\underline{W\delta} := (WG)\underline{u} + W\underline{k}$; $\underline{Wu} := W\underline{u}$;
$\underline{W\delta} := \underline{W\delta} - \underline{Wu}$; DELNP $:= (\underline{W\delta}, \underline{W\delta})$;

Next Iteration:

$n := n + 1$; $\gamma_\phi := \gamma$

Calculate New Iterate:

$\underline{\delta} := W^{-1}(\underline{W\delta})$;

$\underline{W\theta} := (WG)\underline{\delta}$; DEN $:= (\underline{W\delta}, \underline{W\theta})$; $\gamma := \left[1 - \dfrac{\text{DEN}}{\text{DELNP}}\right]^{-1}$;

$If\ n = 1,\ then\ \rho := 1;\ else\ \rho := \left[1 - \left(\dfrac{\gamma}{\gamma_\phi}\right)\left(\dfrac{1}{\rho}\right)\dfrac{\text{DELNP}}{\text{DELN}\varnothing}\right]^{-1}$;

$\underline{Wu}_N := \rho(\underline{Wu} + \gamma\underline{W\delta}) + (1 - \rho)\underline{Wu}_\phi$; YUN $:= (\underline{Wu}_N, \underline{Wu}_N)$;
DELN$\varnothing$:= DELNP
$\underline{W\delta}_N := \rho[\gamma\underline{W\theta} + (1 - \gamma)\underline{W\delta}] + (1 - \rho)\underline{W\delta}_\phi$; DELNP $:= (\underline{W\delta}_N, \underline{W\delta}_N)$;
$\underline{Wu}_\phi := \underline{Wu}$; $\underline{W\delta}_\phi := \underline{W\delta}$
$\underline{Wu} := \underline{Wu}_N$; $\underline{W\delta} := \underline{W\delta}_N$

Stopping Test:

DELNE $:= [\text{DELNP}]^{1/2}$; YUN $:= [\text{YUN}]^{1/2}$;
Compute M_E (if needed) from the T_n matrix, using bisection (see Section 7.5).

$If\ \dfrac{\text{DELNE}}{\text{YUN}} \le \zeta(1 - M_E),\ then\ \underline{u} := W^{-1}(\underline{Wu})$, print final output and STOP;

$else$ continue;

Go to **Next Iteration**

Remark. The error vector for the stopping test given here is measured in the W-norm. If some other measure β is desired, then DELNE $:= \|\underline{\delta}\|_\beta$ can be computed when $\underline{\delta}$ is obtained at the beginning of **Calculate New Iterate**. The norm used in the calculation of YUN may need to be modified also.

We remark that for both Algorithms 7-6.1 and 7-6.2 the storage required for the $\underline{u}$- and $\underline{\delta}$-type vectors can be reduced by using a procedure similar to that given in Chapter 5. (See Modification 5-5.1.)

It is also possible to reduce the amount of work caused by matrix–vector multiplications by W for the SSOR method. This can be done, using the relations (5-6.19) and (5-6.20) between $W\delta^{(n)}$ for the SSOR method and the difference vector $\Delta^{(n)}$ for the "forward" SOR method. Details are given in Hayes and Young [1977] and Grimes *et al.* [1978].

7.7 NUMERICAL RESULTS

In this section we describe results of numerical experiments that were designed to test the effectiveness of conjugate gradient acceleration as compared with Chebyshev acceleration. The experiments given here were carried out using the "simulated" procedure described in Section 5.7. In Chapter 8 we give results of other experiments which compare the effectiveness of CG and Chebyshev acceleration for solving linear systems arising in the numerical solution of elliptic partial differential equations.

Consider the CG procedure (7-4.11)–(7-4.13) applied to the basic iterative method $u^{(n+1)} = Gu^{(n)} + k$, where G is an $N \times N$ matrix. We assume that this basic method is symmetrizable with the symmetrization matrix W. The symmetrization property (see Definition 2-2.1) implies that WGW^{-1} is symmetric. Thus there exists a set of orthonormal eigenvectors $\{v(i)\}_{i=1}^{i=N}$ for WGW^{-1} with corresponding real eigenvalues $\{\mu_i\}_{i=1}^{i=N}$. We order the eigenvalues of G as

$$m(G) \equiv \mu_N \leq \mu_{N-1} \leq \cdots \leq \mu_1 \equiv M(G) < 1. \qquad (7\text{-}7.1)$$

Note that the vectors $\{W^{-1}v(i)\}_{i=1}^{i=N}$ are linearly independent and are eigenvectors of G with corresponding eigenvalues $\{\mu_i\}_{i=1}^{i=N}$. (Recall from Section 5.7 that the "simulated" procedure is based on the assumption that the eigenvectors $\{v(i)\}_{i=1}^{i=N}$ are known.)

Expanding the initial error vector $\varepsilon^{(0)}$ in terms of the eigenvectors of G, we have

$$\varepsilon^{(0)} = \sum_{i=1}^{N} c_i(W^{-1}v(i)), \qquad (7\text{-}7.2)$$

where the $c_1, c_2, \ldots, c_N$ are suitable constants. Since from (5-2.3) $\delta^{(0)} = (G - I)\varepsilon^{(0)}$, we can express $\delta^{(0)}$ in the form

$$\delta^{(0)} = \sum_{i=1}^{N} c_i(\mu_i - 1)W^{-1}v(i). \qquad (7\text{-}7.3)$$

From (7-4.12), we obtain using (7-7.3) that

$$\gamma_1 = \left[1 - \frac{\sum_{i=1}^{N}\mu_i[c_i(\mu_i - 1)]^2}{\sum_{i=1}^{N}[c_i(\mu_i - 1)]^2} \right]^{-1}. \qquad (7\text{-}7.4)$$

Now from (3-2.9), $\varepsilon^{(1)} = [\gamma_1 G + (1 - \gamma_1)]\varepsilon^{(0)}$. Thus $\varepsilon^{(1)}$ can be expressed in the form

$$\varepsilon^{(1)} = \sum_{i=1}^{N} c_{i,1}(W^{-1}v(i)), \tag{7-7.5}$$

where $c_{i,1} = [\gamma_1\mu_i + (1 - \gamma_1)]c_i$. With $\varepsilon^{(1)}$ given by (7-7.5), the above procedure may now be used to calculate $\delta^{(1)}$, γ_2, ρ_2, and $\varepsilon^{(2)}$. Continuing in this way, we "simulate" the CG procedure (7-4.11)–(7-4.13). Note that only the coefficients c_i of (7-7.2) and the eigenvalues μ_i of G are needed to carry out the simulation procedure.

We shall describe here only a few of the many cases which were run. As in Section 5.7, we assume that the initial guess vector $u^{(0)}$ is the null vector. Thus the unique solution $\bar{u} = (I - G)^{-1}k$ is simply $(-\varepsilon^{(0)})$.

For one set of experiments the iteration matrix G was chosen to have the 100 eigenvalues

$$\mu_i = \begin{cases} \dfrac{\cos(i\pi/101)}{2 - \cos(\pi/101)} & \text{if } \cos(i\pi/101) > 0, \\[3mm] \alpha \dfrac{\cos(i\pi/101)}{2 - \cos(\pi/101)} & \text{if } \cos(i\pi/101) < 0, \end{cases}$$

where $i = 1, 2, \ldots, 100$. For the c_i we choose a set of random numbers R_i uniformly distributed in $[0, 1]$. The results shown in Table 7-7.2 were obtained.

The number of iterations required using CG acceleration was substantially less than the number required using the optimal nonadaptive Chebyshev procedure. This was because the number N was relatively small. With a larger value of N but with $M(G)$ and $m(G)$ unchanged, the number of Chebyshev iterations would have been essentially unchanged while the number of conjugate gradient iterations would have increased. However, we would expect that, for any N, the number of CG accelerations would always be less than, or at worst, only very slightly greater than the number of Chebyshev iterations.

By comparing columns a and b in Table 7-7.1, we see that the approximation (7-5.3) for E_T is very accurate for the adaptive Chebyshev acceleration procedure and much less accurate for the CG and optimal nonadaptive Chebyshev procedures. As described in Chapter 6, the approximation (7-5.3) is very accurate for adaptive Chebyshev acceleration since in the late stages of the iterative process the pseudoresidual vector $\delta^{(n)}$ is approximately an eigenvector of G corresponding to the eigenvalue $M(G)$. Thus for any norm $\|\cdot\|_\beta$ we have $\|\varepsilon^{(n)}\|_\beta \sim (1 - M(G))^{-1}\|\delta^{(n)}\|_\beta$ as $n \to \infty$. For the case of optimal nonadaptive Chebyshev acceleration and CG acceleration, however, even in the late stages of the iteration process, $\delta^{(n)}$ will not, in general, be close to an

TABLE 7-7.1

Number of Iterations† Required by the Chebyshev and CG Methods

| Method | $\alpha = 3$ | | | | | | $\alpha = 10$ | | | | | |
| | $\zeta = 0.1$ | | $\zeta = 0.01$ | | $\zeta = 10^{-6}$ | | $\zeta = 0.1$ | | $\zeta = 0.01$ | | $\zeta = 10^{-6}$ | |
	a	b	a	b	a	b	a	b	a	b	a	b
Conjugate gradient	63	106	96	126	127	134	65	139	158	177	178	188
Chebyshev nonadaptive	94	132	166	201	463	501	154	218	274	333	772	822
Chebyshev adaptive	174	172	284	284	598	598	279	271	453	453	—	—

† (a) Iterations determined by the stopping test: $\|\varepsilon^{(n)}\|_W/\|\bar{u}\|_W \leq \zeta$.
  (b) Iterations determined by the stopping test: $(1/(1 - M(G)))\|\delta^{(n)}\|_W/\|\bar{u}\|_W \leq \zeta$.

eigenvector of G corresponding to $M(G)$. If W is a symmetrization matrix, we know that $\|\varepsilon^{(n)}\|_W \leq (1 - M(G))^{-1}\|\delta^{(n)}\|_W$, by (5-2.5). However, in general, $\|\varepsilon^{(n)}\|_W$ may be considerably less than $(1 - M(G))^{-1}\|\delta^{(n)}\|_W$. (In an extreme case we could have $\|\varepsilon^{(n)}\|_W \sim (1 - m(G))^{-1}\|\delta^{(n)}\|_W$.) Thus the number of iterations required for convergence, based on the stopping procedure (7-5.3), may be considerably larger than necessary for the CG and the optimal nonadaptive Chebyshev procedures.

We remark that for the conjugate gradient results given in this section, we used the true value of $M(G)$ rather than the largest eigenvalue of the appropriate tridiagonal matrices as described in Section 7.5. This is relatively unimportant since the largest eigenvalues of the tridiagonal matrices T_n converge rapidly to $M(G)$ in most cases.

In another set of experiments we assumed that the iteration matrix G has N eigenvalues $\mu_1, \mu_2, \ldots, \mu_N$ uniformly distributed in the interval $[-0.99, 0.99]$ so that

$$0.99 = \mu_1 > \mu_2 > \cdots > \mu_N = -0.99. \qquad (7\text{-}7.6)$$

The c_i were chosen by the formulas

$$c_i = \frac{1}{(1 - \mu_i)(1 - (\mu_i/M(G))^2)^{1/4}}, \qquad i \neq 1, N, \quad c_1 = c_2, \quad c_N = c_{N-1}.$$

The results shown in Table 7-7.2 were obtained. Here $n(E_T)$ denotes the number of iterations required for inequality (7-5.1) to be first satisfied. The quantity $n(\text{TNA})$ is defined in Section 5.7.

TABLE 7-7.2

N	200	400	800	1600	3200	5000	7500
Optimal nonadaptive Chebyshev							
$n(E_T)$	100	101	101	100	101	101	101
$n(\text{TNA})$	102	102	102	102	102	102	102
Conjugate gradient							
$n(E_T)$	65	80	93	99	100	101	101

It can be seen that the number of iterations required for the optimal non-adaptive Chebyshev process is nearly independent of the number of eigenvalues N. For CG acceleration, the number of iterations is much smaller for small N but increases until both procedures require the same number of iterations for large N. We remark that the distribution $\{c_i\}$ was chosen so that for large N the polynomials corresponding to the CG process are nearly the same as the normalized Chebyshev polynomials corresponding to the optimal nonadaptive Chebyshev process.

Another experiment involved a choice of the $\{c_i\}$ distribution, which was constructed so that for N very large the optimal nonadaptive Chebyshev procedure would be slightly better than CG acceleration. We used the distribution

$$c_i = \frac{1}{\sqrt{1 - \mu_i(1 - (\mu_i/M(G))^2)^{1/4}}}, \qquad i \neq 1, N, \quad c_1 = c_2, \quad c_N = c_{N-1}.$$

For this distribution and with the eigenvalues of G given by (7-7.6), we obtained the following results for the case $N = 10,000$:

optimal nonadaptive Chebyshev $\qquad n(E_T) = 100,$
$\qquad\qquad\qquad\qquad\qquad\qquad\qquad n(\text{TNA}) = 102,$
$\qquad\qquad$ conjugate gradient $\qquad\qquad n(E_T) = 102.$

Thus optimal nonadaptive Chebyshev acceleration was indeed slightly better than CG acceleration. However, with a smaller value of N we would expect that CG acceleration would be much superior.

For large $M(G)$, we would expect that the CG acceleration procedure would be significantly better than nonadaptive Chebyshev except for extremely large N. Thus with the distribution $c_i = (1 - \mu_i)^{-1/2}$ and with $M(G) = -m(G) = 0.9999$, we obtained

optimal nonadaptive Chebyshev $\qquad n(E_T) = 1020,$
$\qquad\qquad\qquad\qquad\qquad\qquad\qquad n(\text{TNA}) = 1025,$
$\qquad\qquad$ conjugate gradient $\qquad\qquad n(E_T) = 230.$

Here 1600 uniformly distributed eigenvalues were used. With the distribution $c_i = 1$, CG acceleration required only 368 iterations for $N = 5000$. For Chebyshev acceleration $n(\text{TNA})$ would still be 1025. For values of $M(G)$ very close to unity, it appears that N must be very large before $n(E_T)$ for the CG procedure approaches its limiting value (which would be close to the number of iterations required by Chebyshev acceleration).

The numerical results based on the simulation experiments given in this section, as well as the numerical results given in Chapter 8, support the following conclusions.

(a)   The rapidity of convergence of the CG acceleration procedure depends not only on $M(G)$ and $m(G)$, but also on the number and distribution of the eigenvalues of G as well as on the initial error vector $\varepsilon^{(0)}$. For optimal nonadaptive Chebyshev acceleration, on the other hand, the rapidity of convergence depends almost entirely on $M(G)$ and $m(G)$. The convergence rate of the CG procedure, when measured in a certain norm, is at least as fast as that of any polynomial acceleration procedure including optimal non-adaptive Chebyshev acceleration. As we have seen, while cases can be constructed in which CG acceleration takes as many iterations as optimal nonadaptive Chebyshev acceleration, in most cases CG acceleration requires substantially fewer iterations. The CG acceleration process can take advantage of certain properties of the distribution of the eigenvalues of G. Thus if there are relatively few eigenvalues or if there are some isolated eigenvalues, the convergence of CG acceleration can be substantially better than if there are a large number of eigenvalues of G densely distributed over the interval $[m(G), M(G)]$.

(b)   Unless good estimates are available for $m(G)$ and $M(G)$, one cannot use optimal nonadaptive Chebyshev acceleration. Instead, one would probably use adaptive Chebyshev acceleration. This will increase the advantage in using CG acceleration.

(c)   The approximation (7-5.3) for the iteration error is considerably more accurate for the adaptive Chebyshev procedure than for the CG and the optimal nonadaptive Chebyshev procedures.

In spite of the apparent advantages of CG acceleration over Chebyshev acceleration, there are situations in which the use of the latter might be preferable. If good estimates for $\dot{m}(G)$ and $M(G)$ are available, one might prefer to use optimal nonadaptive Chebyshev acceleration because of its computational simplicity. For example, the computation of inner products is not required except for the stopping tests. Even for adaptive Chebyshev acceleration one could test for parameter changes only on certain iterations. The use of the symmetrization matrix W can be avoided with Chebyshev acceleration, whereas W must be used with CG acceleration.

Another situation in which CG acceleration would appear to be less advantageous than Chebyshev acceleration occurs when few iterations are required. Such a situation might arise, for instance, when ζ is large or when a very accurate initial vector $u^{(0)}$ is available, as might be the case if one were solving a two-dimensional time-dependent problem by an implicit method. This phenomenon was observed by Wang [1977].

Additional comparisons of CG and Chebyshev accelerations are given in Chapter 8. Also included in Chapter 8 is a discussion of hybrid methods involving the use of both Chebyshev and CG acceleration. Numerical experiments that compare the effectiveness of CG and Chebyshev acceleration procedures are also given by Eisenstat *et al.* [1979b].

CHAPTER

8

Special Methods
for Red/Black
Partitionings

8.1 INTRODUCTION

We again seek to solve the matrix equation $Au = b$, where A is a given $N \times N$ symmetric and positive definite matrix (SPD matrix). In this chapter we describe several variants of the Chebyshev and conjugate gradient acceleration procedures which are applicable when the coefficient matrix is partitioned according to the red/black partitioning (1-5.3); i.e., when A is partitioned into the form

$$A = \begin{bmatrix} D_R & H \\ H^T & D_B \end{bmatrix} \tag{8-1.1}$$

If we partition the vectors u and b in a form consistent with (8-1.1), we can write the matrix equation $Au = b$ as

$$\begin{bmatrix} D_R & H \\ H^T & D_B \end{bmatrix} \begin{bmatrix} u_R \\ u_B \end{bmatrix} = \begin{bmatrix} b_R \\ b_B \end{bmatrix}. \tag{8-1.2}$$

The procedures developed in this chapter exploit the structure of the red/ black partition to obtain increased convergence rates for the Chebyshev and conjugate gradient acceleration procedures applied to the Jacobi basic method associated with the partitioning (8-1.2).

For a given matrix problem, many red/black partitionings exist; however not all of them will be computationally feasible. For the methods we give in this chapter, each iteration step requires that the subsystems $D_R u_R = y_R$ and $D_B u_B = y_B$ be solved for u_R and u_B, given y_R and y_B. Thus it is important that the red/black partitioning be chosen so that the D_R and D_B matrices are "easily invertible" (see Section 1.5). A common occurrence is that the D_R and D_B are block diagonal matrices whose diagonal blocks are "easily invertible." For this circumstance, the D_R and D_B matrices are also "easily invertible." The formulation of red/black partitionings, where the D_R and D_B matrices are block diagonal is discussed and illustrated in Sections 1.7 and 9.2. It may be instructive for the reader to refer to these sections before proceeding with the material of this chapter. We note that the iteration procedures we describe here are not restricted to situations where the D_R and D_B matrices are block diagonal or to situations where the subsystems involving D_R and D_B are solved by direct methods. For some matrix problems, especially those resulting from the discretization of coupled differential equations, the nature of the D_R and D_B matrices can be such that the solutions to the subsystems $D_R u_R = y_R$ and $D_B u_B = y_B$, for computational reasons, are best computed approximately by a *sub* or *inner iteration* process. Iterative procedures which utilize inner iterations are discussed later in Chapter 11.

Letting $F_R \equiv -D_R^{-1}H$, $F_B \equiv -D_B^{-1}H^T$, $c_R \equiv D_R^{-1}b_R$ and $c_B \equiv D_B^{-1}b_B$, we may express the matrix equation (8-1.2) in the form

$$\begin{bmatrix} I & -F_R \\ -F_B & I \end{bmatrix} \begin{bmatrix} u_R \\ u_B \end{bmatrix} = \begin{bmatrix} c_R \\ c_B \end{bmatrix}. \tag{8-1.3}$$

The Jacobi method (see Section 2.3) associated with (8-1.3) is

$$u^{(n+1)} = Bu^{(n)} + c, \tag{8-1.4}$$

where $c = (c_R^T, c_B^T)^T$ and B is the Jacobi iteration matrix

$$B = \begin{bmatrix} 0 & F_R \\ F_B & 0 \end{bmatrix}. \tag{8-1.5}$$

Because the matrix A is partitioned into a red/black form and because A is SPD, it follows (see, e.g., Young [1971] or Varga [1962]) that the eigenvalues of B are real, occur in $\pm$ pairs, and are less than unity. That is, the eigenvalues $\{\mu_i\}$ of B satisfy

$$-\mu_1 = \cdots = -\mu_{t-1} < -\mu_t \le \cdots \le 0 \le \cdots \le \mu_t < \mu_{t-1} = \cdots = \mu_1 < 1. \tag{8-1.6}$$

Thus the algebraically smallest, $m(B)$, and largest, $M(B)$, eigenvalues of B satisfy

$$-m(B) = M(B) < 1. \tag{8-1.7}$$

With $W = \text{diag}(D_R^{1/2}, D_B^{1/2})$, or more generally with $W = S$ where S is any matrix such that $S^T S = \text{diag}(D_R, D_B)$, it is clear that the Jacobi method (8-1.4) is symmetrizable (Definition 2-2.1). Thus the Chebyshev and conjugate gradient acceleration methods discussed previously may be applied to the basic Jacobi method (8-1.4). We denote the *Chebyshev and conjugate gradient acceleration methods applied to the Jacobi method by J–SI and J–CG*, respectively. From (5-1.4) and (7-4.11), both of these acceleration methods may be expressed as

$$u^{(n+1)} = \rho_{n+1}\{\gamma_{n+1}(Bu^{(n)} + c - u^{(n)}) + u^{(n)}\} + (1 - \rho_{n+1})u^{(n-1)}, \tag{8-1.8}$$

where ρ_{n+1} and γ_{n+1} are given by (5-1.5)–(5-1.6) and (7-4.12)–(7-4.13) for the J–SI and J–CG methods, respectively. In this chapter, we describe two Chebyshev (and two conjugate gradient) acceleration procedures which utilize the special red/black form (8-1.2) and which converge exactly twice as fast as the J–SI (J–CG) acceleration procedure (8-1.8).

Golub and Varga [1961] observed that if the acceleration procedure (8-1.8) is such that†

$$\gamma_n = 1, \qquad n \geq 1, \tag{8-1.9}$$

then the procedure (8-1.8) for the red/black partitioned problem can be carried out by computing $u_R^{(n)}$ only for n odd and $u_B^{(n)}$ only for n even. Thus if we let

$$U_R^{(n)} \equiv u_R^{(2n-1)} \qquad \text{and} \qquad U_B^{(n)} \equiv u_B^{(2n)}, \tag{8-1.10}$$

the process (8-1.8) can be carried out by‡

$$\begin{aligned}
U_R^{(n)} &= \rho_R^{(n)}(F_R U_B^{(n-1)} + c_R - U_R^{(n-1)}) + U_R^{(n-1)}, \\
U_B^{(n)} &= \rho_B^{(n)}(F_B U_R^{(n)} + c_B - U_B^{(n-1)}) + U_B^{(n-1)}
\end{aligned} \tag{8-1.11}$$

for $n \geq 1$, where

$$\rho_R^{(n)} = \rho_{2n-1} \qquad \text{and} \qquad \rho_B^{(n)} = \rho_{2n}. \tag{8-1.12}$$

We refer to (8-1.11) as the *cyclic accelerated method*. Note from (8-1.10) and (8-1.11) that the cyclic accelerated method converges exactly twice as fast as the original accelerated method (8-1.8). The procedure (8-1.11) for Chebyshev

---

† We show in Section 8.3 that (8-1.9) is satisfied by both the J–SI and J–CG methods when B has the special form (8-1.5).

‡ The derivation of (8-1.11) is given in Section 8.3.

acceleration was first presented by Golub and Varga [1961] and was called the *cyclic Chebyshev semi-iterative* (CCSI) *method* by them. Reid [1972] was the first to observe that a similar reduction in computational work could also be obtained for the conjugate gradient method. When conjugate gradient acceleration is used, we call the cyclic method (8-1.11) the *cyclic conjugate gradient* (CCG) *method*. The special CCSI and CCG procedures are discussed in Section 8.3.

To obtain the second special procedure, we use the fact that the u_R (or u_B) unknowns in (8-1.3) may easily be eliminated. Indeed, multiplying (8-1.3) by the nonsingular matrix $(I + B)$, we obtain $(I - B^2)u = (I + B)c$ or

$$\begin{bmatrix} I - F_R F_B & 0 \\ 0 & I - F_B F_R \end{bmatrix} \begin{bmatrix} u_R \\ u_B \end{bmatrix} = \begin{bmatrix} c_R + F_R c_B \\ c_B + F_B c_R \end{bmatrix}. \tag{8-1.13}$$

Equation (8-1.13) represents two uncoupled systems so that the solution to the original red/black problem (8-1.3) may be obtained by solving, for example,† the lower-order system

$$(I - F_B F_R)u_B = c_B + F_B c_R \tag{8-1.14}$$

for u_B, and then obtaining the u_R unknowns explicitly from $u_R = F_R u_B + c_R$. We refer to (8-1.14) as the *reduced system*. We show later that the matrix $(I - F_B F_R)$ is similar to a SPD matrix. Thus many of the iterative methods discussed in earlier chapters may be used to solve the reduced system (8-1.14). In the solution of the two-dimensional elliptic differential equation using a 5-point discretization formula, Hageman and Varga [1964] showed that a 4-line CCSI solution method applied to the reduced system (8-1.14) required only half the computer time as that required by the 2-line CCSI method applied to the original system (8-1.3). However, the 4-line CCSI method (and most iterative methods) applied to the reduced system (8-1.14) requires that the matrix $F_B F_R$ be determined explicitly. This is feasible only when D_R or D_B is a diagonal matrix or when certain other rather restrictive conditions hold. Thus for the iterative solution of the reduced system (8-1.14), we restrict our attention to the RF method which, as we shall see, does not require that the matrix $F_B F_R$ be computed explicitly.

The RF method (see Section 2.3) for the reduced system is given by

$$u_B^{(n+1)} = F_B F_R u_B^{(n)} + F_B c_R + c_B. \tag{8-1.15}$$

---

† One could solve equally well the companion reduced system $(I - F_R F_B)u_R = c_R + F_R c_B$ for u_R and then obtain u_B from $u_B = F_B u_R + c_B$. Since the nonzero eigenvalues of $F_B F_R$ are the same as those of $F_R F_B$, the primary difference between the reduced systems is one of size, i.e., the order of the system.

Note that the explicit calculation of $F_B F_R$ can be avoided by using the procedure

$$u_R^{(n+1)} = F_R u_B^{(n)} + c_R$$
$$u_B^{(n+1)} = F_B u_R^{(n+1)} + c_B. \tag{8-1.16}$$

Since $(I - F_B F_R)$ is similar to a SPD matrix, either Chebyshev or conjugate gradient acceleration may be applied to the basic RF method (8-1.15). We refer to the *Chebyshev and the conjugate gradient acceleration of* (8-1.15) *as the* RS–SI *and* RS–CG *methods,* respectively. The reduced system methods are discussed in Section 8.2.

For either Chebyshev or conjugate gradient acceleration, Hageman *et al.* [1980] have shown† that the cyclic accelerated method (8-1.11) and the corresponding acceleration method applied to (8-1.15) each converge exactly twice as fast as the acceleration method (8-1.8). That is, the CCSI and the RS–SI methods converge at the same rate, and each of these methods converges exactly twice as fast as the J–SI method. Similarly, the CCG and the RS–CG methods converge at the same rate, and each of these methods converges exactly twice as fast as the J–CG method.

The storage and computational requirements for the RS–SI method are slightly greater than that for the CCSI method. However, the difference is slight, and the choice of method often is based on other factors such as computer architecture, type of problem to be solved, and programming convenience. Similar remarks also are valid for the RS–CG and CCG methods. The development of the cyclic accelerated procedures is considerably more complicated than the development of the reduced system acceleration procedures. However, the algorithms presented are self-contained, and their utilization does not require a detailed reading of the developmental material.

8.2 THE RS–SI AND RS–CG METHODS

As given previously, the RF method for solving the reduced system (8-1.14) is

$$u_B^{(n+1)} = F_B F_R u_B^{(n)} + F_B c_R + c_B, \tag{8-2.1}$$

where F_B, F_R, c_R, and c_B are defined as in (8-1.3). In this section, we discuss Chebyshev and conjugate gradient acceleration applied to the basic method (8-2.1). We first show that the RF method possesses properties sufficient for

---

† See also Chandra [1978].

the application of these acceleration methods. We then describe the Chebyshev method (RS–SI) in Section 8.2(A) and the conjugate gradient method (RS–CG) in Section 8.2(B).

From Chapters 5–7, Chebyshev and conjugate gradient acceleration applied to (8-2.1) will be effective if the basic RF method (8-2.1) is symmetrizable. To show that this is the case, we use the fact that the coefficient matrix A of (8-1.1) is SPD. First, it is easy to show that $F_B F_R$ is similar to a symmetric, positive semidefinite matrix. Indeed, since D_B and D_R are positive definite, these matrices may be factored using a Cholesky decomposition (e.g., see Sections 2.3 and 5.6) as

$$D_B = S_B^T S_B \quad \text{and} \quad D_R = S_R^T S_R, \tag{8-2.2}$$

where S_B and S_R are upper triangular matrices. Thus

$$\begin{aligned} S_B(F_B F_R)S_B^{-1} &= S_B(D_B^{-1}H^T D_R^{-1}H)S_B^{-1} \\ &= [(S_B^T)^{-1}H^T S_R^{-1}][(S_B^T)^{-1}H^T S_R^{-1}]^T, \end{aligned} \tag{8-2.3}$$

from which the desired conclusion follows.

This similarity property, of course, implies that the eigenvalues of $F_B F_R$ are real and nonnegative. Moreover, since $B^2 = \text{diag}(F_R F_B,\ F_B F_R)$, the spectral radii of the Jacobi matrix B and the matrix $F_B F_R$ satisfy

$$\mathbf{S}(F_B F_R) = \mathbf{S}(B^2) = [\mathbf{S}(B)]^2 < 1, \tag{8-2.4}$$

where the right inequality follows from (8-1.7). Thus the algebraically smallest, $m(F_B F_R)$, and largest, $M(F_B F_R)$, eigenvalues of $F_B F_R$ satisfy

$$0 \le m(F_B F_R) \le M(F_B F_R) = [M(B)]^2 < 1. \tag{8-2.5}$$

From (8-2.3) and (8-2.5), it follows that $S_B(I - F_R F_B)S_B^{-1}$ is SPD. Thus the RF method of (8-2.1) is symmetrizable with $W = S_B$, which is the desired result.

A.   The RS–SI Procedure

The Chebyshev acceleration method applied to the RF method (8-2.1) is given by (5-1.4)–(5-1.7) with $G = F_B F_R$ and may be carried out using either Algorithm 5-5.1 or Algorithm 6-4.1. Since $m(F_B F_R) \ge 0$, $m_E = 0$ *should be used*. For both algorithms, the matrix–vector product $F_B F_R u_B$ required in the calculation of δ_B best carried out using a two-step procedure analogous to (8-1.16). For example the **Calculate New Iterate** procedure of Algorithm 5-5.1 can be carried out by

Calculate New Iterate

$$\underline{u}_\phi := \underline{u}; \ \underline{u} := \underline{u}_\Gamma$$
$$\underline{u}_R := F_R \underline{u} + \underline{c}_R$$
$$\underline{\delta}_B := F_B \underline{u}_R + \underline{c}_B - \underline{u}; \ \text{DELNP} := \|\underline{\delta}_B\|_{S_B}; \ \text{DELINE} := \|\underline{\delta}_B\|_\beta$$
$$\underline{u}_\Gamma := \rho(\gamma \underline{\delta}_B + \underline{u}) + (1 - \rho)\underline{u}_\phi; \qquad \text{YUN} := \|\underline{u}_\Gamma\|_\eta$$

In the above procedure, all vectors are of the same order as $u_B^{(n+1)}$ in (8-2.1) except for $\underline{u}_R$ and $\underline{c}_R$. We note that only two of the three vectors $\{\underline{u}, \underline{u}_\phi, \underline{u}_\Gamma\}$ need to be stored (see Modification 5-5.1). Moreover, as discussed in Section 5.6, it is not necessary to store the $\underline{\delta}_B$ vector if D_B is a block diagonal matrix.

We remind the reader that the matrices F_B and F_R usually are not computed explicitly. For example, if D_R and D_B are factored as in (8-2.2), then $\underline{u}_R = F_R \underline{u} + \underline{c}_R$ is obtained by solving the equation

$$S_R^T S_R \underline{u}_R = -H\underline{u} + \underline{b}_R \tag{8-2.6}$$

and $\underline{\delta}_B$ is computed by $\underline{\delta}_B = \underline{z} - \underline{u}$, where $\underline{z}$ is obtained by solving the equation

$$S_B^T S_B \underline{z} = -H^T \underline{u}_R + \underline{b}_B. \tag{8-2.7}$$

The stopping test in the RS–SI procedure measures the iteration error only for the unknowns of the vector u_B. However, when measured in a particular norm, $u_R^{(n)}$ is as accurate as $u_B^{(n)}$. To see why this is so, let $\varepsilon_B^{(n)} \equiv u_B^{(n)} - \bar{u}_B$ and $\varepsilon_R^{(n)} \equiv u_R^{(n)} - \bar{u}_R$, where

$$u_R^{(n)} = F_R u_B^{(n)} + c_R. \tag{8-2.8}$$

Since (8-2.8) is also satisfied by the solution subvectors, $\bar{u}_R$ and $\bar{u}_B$, we have

$$\varepsilon_R^{(n)} = F_R \varepsilon_B^{(n)}. \tag{8-2.9}$$

By definition, $F_R = -(S_R^T S_R)^{-1} H$ and $F_B = -(S_B^T S_B)^{-1} H^T$. Hence from (8-2.9), we have

$$\|\varepsilon_R^{(n)}\|_{S_R} = \|S_R (S_R^T S_R)^{-1} H S_B^{-1} S_B \varepsilon_B^{(n)}\|_2 \le \sqrt{\mathbf{S}(F_B F_R)} \|\varepsilon_B^{(n)}\|_{S_B} < \|\varepsilon_B^{(n)}\|_{S_B}. \tag{8-2.10}$$

Thus $\|\varepsilon_R^{(n)}\|_{S_R}$ is less than $\|\varepsilon_B^{(n)}\|_{S_B}$. Note that this implies that the error vector, $\varepsilon \equiv (\varepsilon_R^T, \varepsilon_B^T)^T$, for the whole system satisfies

$$\|\varepsilon\|_S = [\|\varepsilon_R\|_{S_R}^2 + \|\varepsilon_B\|_{S_B}^2]^{1/2} < \sqrt{2}\|\varepsilon_B\|_{S_B}. \tag{8-2.11}$$

Here, $S \equiv \text{diag}(S_R, S_B)$.

For an arbitrary β-norm, it is not possible in general to show that

$\|\varepsilon_R^{(n)}\|_\beta / \|u_R^{(n)}\|_\eta \leq \|\varepsilon_B^{(n)}\|_\beta / \|u_B^{(n)}\|_\eta$. However, if desired, $\|\varepsilon_R^{(n)}\|_\beta$ may be approximated by

$$\|\varepsilon_R^{(n)}\|_\beta \doteq \frac{\|\Delta_R^{(n)}\|_\beta}{1 - R^{(n)}}, \tag{8-2.12}$$

where $\Delta_R^{(n)} \equiv u_R^{(n)} - u_R^{(n+1)}$ and $R^{(n)} \equiv \|\delta_B^{(n+1)}\| / \|\delta_B^{(n)}\|$. Analogous to the stopping test (6-3.23), we show that (8-2.12) is valid provided that the eigenvalue estimate M_E satisfies $M_E < M(F_B F_R)$ and that n is sufficiently large. Using the successive polynomial notation of Section 6.3, we have from (5-2.6) that $\varepsilon_B^{(n)} \equiv \varepsilon_B^{(s,p)} = P_{p,E}(F_B F_R)\varepsilon_B^{(s,0)}$. Multiplying this equation by F_R, we obtain by using (8-2.9) that

$$\varepsilon_R^{(n)} \equiv \varepsilon_R^{(s,p)} = P_{p,E}(F_R F_B)\varepsilon_R^{(s,0)}. \tag{8-2.13}$$

If $M_E < M(F_B F_R)$, it follows from Theorem 6-2.4 that the sequence $\{P_{p,E}(y)\}$ is "$F_R F_B$-uniformly convergent." Thus from (6-2.11), for sufficiently large p we have approximately that

$$\varepsilon_R^{(s,p)} \doteq P_{p,E}(\lambda_1)v_R(1), \tag{8-2.14}$$

where $\lambda_1 = M(F_R F_B)$ and where $v_R(1)$ is an eigenvector of $F_R F_B$ corresponding to the eigenvalue λ_1. Using the approximation (8-2.14) in the expression $\Delta_R^{(s,p)} \equiv \Delta_R^{(n)} = u_R^{(n)} - u_R^{(n+1)} = \varepsilon_R^{(s,p)} - \varepsilon_R^{(s,p+1)}$, we obtain the approximation

$$\Delta_R^{(s,p)} \doteq \left[1 - \frac{P_{p+1,E}(\lambda_1)}{P_{p,E}(\lambda_1)}\right]\varepsilon_R^{(s,p)}. \tag{8-2.15}$$

Now $\delta_B^{(s,p)} = P_{p,E}(F_B F_R)\delta_B^{(s,0)}$. Thus from (6-2.14), for sufficiently large p we have that

$$\frac{\|\delta_B^{(s,p+1)}\|}{\|\delta_B^{(s,p)}\|} \doteq \frac{P_{p+1,E}(\lambda_1)}{P_{p,E}(\lambda_1)} \tag{8-2.16}$$

and (8-2.12) follows. In the above discussion, we used the fact that $F_B F_R$ and $F_R F_B$ have the same nonzero eigenvalues.

B.   The RS–CG Procedure

The conjugate gradient acceleration method applied to the RF method (8-2.1) is given by (7-4.11)–(7-4.13) and may be carried out using Algorithm 7-6.1 with $G = F_B F_R$. If D_B and D_R are factored as in (8-2.2), then $W = S_B$ may be used as the symmetrization matrix. Moreover, in this case, the matrix–vector multiplication $\psi := W\theta$ can be eliminated. To show how this can be done, we first note that since $F_B F_R = D_B^{-1} H^T D_R^{-1} H$ we have

$$\text{DEN} = (\underline{\tau}, \underline{\psi}) = (W\underline{\delta}, WG\underline{\delta}) = (S_B\underline{\delta}, S_B F_B F_R\underline{\delta}) = ((S_R^T)^{-1}H\underline{\delta}, (S_R^T)^{-1}H\underline{\delta}). \tag{8-2.17}$$

Using (8-2.17), we may compute DEN during the calculation of $\underline{\theta}$ and thus avoid the need for $\underline{\psi}$. For example, DEN, γ, and $\underline{\theta}$ in Algorithm 7-6.1 can be computed by using the procedure

$$\underline{z}_R := (S_R^T)^{-1} H \underline{\delta}; \qquad \text{DEN} := (\underline{z}_R, \underline{z}_R);$$
$$\underline{\delta}_R := (S_R)^{-1} \underline{z}_R;$$
$$\underline{\theta} := (S_B)^{-1}(S_B^T)^{-1} H^T \underline{\delta}_R; \qquad \gamma := \left[1 - \frac{\text{DEN}}{\text{DENP}} \right]^{-1};$$

The additional matrix–vector multiplication $\underline{\tau} := W \underline{\delta}_N$ must still be carried out in Algorithm 7-6.1. To eliminate this multiplication, Algorithm 7-6.2 must be used with $Z = S_B^T S_B$ and $Y = H^T D_R^{-1} H$.

As for the RS–SI procedure, any matrix–vector multiplication involving the matrix $S_B F_B F_R$ is best carried out using a two-step procedure. For example, in **Initialize** of Algorithm 7-6.2, $\underline{W\delta} \, (= S_B \delta)$ can be calculated by

$$\underline{u}_R := F_R \underline{u} + \underline{c}_R;$$
$$\underline{S_B \delta} := (S_B F_B) \underline{u}_R + S_B \underline{c}_B; \quad \underline{S_B u} := S_B \underline{u};$$
$$\underline{S_B \delta} := \underline{S_B \delta} - \underline{S_B u}; \quad \text{DELNP} := (\underline{S_B \delta}, \underline{S_B \delta});$$

Here $\underline{u}$ is the input guess vector for the solution vector u_B of (8-1.14). Similarly, in **Calculate New Iterate**, the $\underline{\delta}$ and $\underline{W\theta} \, (= S_B \theta)$ vectors can be calculated by

$$\underline{\delta} := (S_B)^{-1}(\underline{S_B \delta})$$
$$\underline{\delta}_R := F_R \underline{\delta}$$
$$\underline{S_B \theta} := (S_B F_B) \underline{\delta}_R; \quad \text{DEN} := (\underline{S_B \delta}, \underline{S_B \theta}); \gamma := \left[1 - \frac{\text{DEN}}{\text{DELNP}} \right]^{-1};$$

In the above calculations, a procedure similar to that given by (8-2.6)–(8-2.7) should be used to obtain the matrix–vector products involving the F_R and $(S_B F_B)$ matrices.

We remark that the alternate expression (8-2.17) for DEN also is valid here. For programming reasons, it often is more convenient to compute DEN (using the alternate expression) concurrently with $\underline{\delta}_R$.

8.3   THE CCSI AND CCG PROCEDURES

In this section, we discuss the cyclic acceleration method (8-1.11) for both the Chebyshev and conjugate gradient cases. We first show that condition (8-1.9), which is necessary for the use of the cyclic method, is valid for both Chebyshev and conjugate gradient acceleration. We then derive the cyclic method from the original acceleration method (8-1.8). In Section 8.3(A), we develop and describe the cyclic Chebyshev (CCSI) procedure. In Section

8.3(B), we develop and describe the cyclic conjugate gradient (CCG) procedure. The development of these cyclic procedures is somewhat involved. However, the algorithms presented are self-contained and their utilization does not require a detailed understanding of the developmental material. On a first reading, some readers may wish to skim the developmental discussion.

We now show that $\gamma_{n+1} = \gamma = 1$ for both Chebyshev and conjugate gradient acceleration whenever B has the special form (8-1.5). First, from (8-1.7), $-m(B) = M(B)$. Thus for Chebyshev acceleration, the Case II* condition (5-3.2) that $-m_E = M_E$ may be used. From (5-1.5), this condition then implies that $\gamma_n = \gamma = 1$ for the Chebyshev method.

For conjugate gradient acceleration, it follows from (7-4.12) that $(W\delta^{(n)}, WB\delta^{(n)}) = 0, n = 0, 1, \ldots$ is a sufficient condition for γ_{n+1} to equal unity for all n. We show that this sufficient condition holds when the following reasonable choices are made for the symmetrization matrix W and for the initial guess vector:

(a) $W = \text{diag}(S_R, S_B)$, where S_R and S_B, respectively, are the Cholesky decomposition matrices for D_R and D_B given by (8-2.2).

(b) $u^{(0)}$ is given by

$$u_B^{(0)} \quad \text{arbitrary,}$$
$$u_R^{(0)} = F_R u_B^{(0)} + c_R. \tag{8-3.1}$$

With W given by (a) above and with B given by (8-1.5), we have

$$(W\delta^{(n)}, WB\delta^{(n)}) = (S_R \delta_R^{(n)}, S_R F_R \delta_B^{(n)}) + (S_B \delta_B^{(n)}, S_B F_B \delta_R^{(n)}). \tag{8-3.2}$$

Here, $\delta_R^{(n)}$ and $\delta_B^{(n)}$ are the subvectors of $\delta^{(n)}$ consistent with the partitioning (8-1.3). We show that $(W\delta^{(n)}, WB\delta^{(n)}) = 0$ for all n by showing that

$$\delta_R^{(0)} = \delta_B^{(1)} = \delta_R^{(2)} = \delta_B^{(3)} = \cdots = 0. \tag{8-3.3}$$

To do this, we first express the pseudoresidual vector $\delta^{(n)}$ defined by (7-4.8) in the form

$$\delta^{(n+1)} = \rho_{n+1}\{\gamma_{n+1}B\delta^{(n)} + (1 - \gamma_{n+1})\delta^{(n)}\} + (1 - \rho_{n+1})\delta^{(n-1)}. \tag{8-3.4}$$

The recurrence relation (8-3.4) is obtained by substituting $u^{(n)}$ from (7-4.11) into the expression $\delta^{(n+1)} = (B - I)u^{(n)} + c$. Now from (8-3.1), $\delta_R^{(0)} = F_B u_B^{(0)} + c_R - u_R^{(0)} = 0$ so that $\gamma_1 = 1$. By (8-3.4), we also have $\delta_B^{(1)} = F_B \delta_R^{(0)} = 0$ so that $\gamma_2 = 1$. Now assume $\delta_R^{(2n)} = \delta_B^{(2n+1)} = 0$, which also implies that $\gamma_{2n+1} = \gamma_{2n+2} = 1$. Thus again from (8-3.4), $\delta_R^{(2n+2)} = \rho_{2n+2}[F_R \delta_B^{(2n+1)}] + (1 - \rho_{2n+2})\delta_R^{(2n)} = 0$, which implies that $\gamma_{2n+3} = 1$. Similarly, $\delta_B^{(2n+3)} = 0$, from which we get that $\gamma_{2n+4} = 1$. Hence by induction, it follows that $\gamma_{n+1} = 1$ for all $n \geq 0$. Thus the condition (8-1.9) is satisfied by both Chebyshev and conjugate gradient acceleration.

That the cyclic accelerated method (8-1.11) is nothing more than the original accelerated method (8-1.8) with half of the calculations bypassed can now be easily shown. With B given by (8-1.5) and with $\gamma_{n+1} = 1$, the iterates (8-1.8) can be written as

$$u_R^{(n+1)} = \rho_{n+1}[F_R u_B^{(n)} + c_R] + (1 - \rho_{n+1})u_R^{(n-1)},$$
$$u_B^{(n+1)} = \rho_{n+1}[F_B u_R^{(n)} + c_B] + (1 - \rho_{n+1})u_B^{(n-1)}. \tag{8-3.5}$$

If these iterates are represented as

$$\begin{array}{ccccccc} u_R^{(0)} & \!u_R^{(1)}\! & u_R^{(2)} & \!u_R^{(3)}\! & u_R^{(4)} & \!u_R^{(5)}\! & u_R^{(6)} \\ \!u_B^{(0)}\! & u_B^{(1)} & \!u_B^{(2)}\! & u_B^{(3)} & \!u_B^{(4)}\! & u_B^{(5)} & \!u_B^{(6)}\! \end{array}, \tag{8-3.6}$$

it is easily seen from (8-3.5) that the circled subvector iterates can be computed without calculating the subvector iterates which are not circled. Thus for the red/black problem, the original acceleration method can be carried out by calculating only the subsequences $\{u_R^{(2n+1)}\}_{n=0}^{\infty}$ and $\{u_B^{(2n)}\}_{n=1}^{\infty}$, from which (8-1.11) follows.

A.   The CCSI Procedure

The CCSI procedure is given by (8-1.11) where, from (8-1.12) and (5-1.6), $\rho_R^{(1)} = 1$, $\rho_B^{(1)} = 2/(2 - M_E^2)$, and for $n \geq 1$

$$\rho_R^{(n+1)} = \frac{1}{1 - \frac{1}{4}M_E^2 \rho_B^{(n)}}, \qquad \rho_B^{(n+1)} = \frac{1}{1 - \frac{1}{4}M_E^2 \rho_R^{(n+1)}}. \tag{8-3.7}$$

Here M_E is the estimate for the eigenvalue $M(B)$. Recall that because of (8-1.7), we assume $-m_E = M_E$. The difficulty with implementing the CCSI method is that the adaptive procedures given previously in Chapters 5 and 6 cannot be used directly. The reason for this, as we shall see, is that the error and residual vectors normally associated with (8-1.11) do not satisfy the basic Chebyshev polynomial relations (5-2.6) and (5-2.8).

The partitioned error vector associated with the CCSI method is

$$\tilde{\varepsilon}^{(n)} = \begin{bmatrix} \tilde{\varepsilon}_R^{(n)} \equiv U_R^{(n)} - \bar{u}_R \\ \tilde{\varepsilon}_B^{(n)} \equiv U_B^{(n)} - \bar{u}_B \end{bmatrix}, \tag{8-3.8}$$

where $(\bar{u}_R^T, \bar{u}_B^T)^T$ is the partitioned solution vector for (8-1.2). If $\varepsilon_R^{(n)} \equiv u_R^{(n)} - \bar{u}_R$ and $\varepsilon_B^{(n)} \equiv u_B^{(n)} - \bar{u}_B$ are the corresponding error subvectors for the J–SI iterates (8-3.5), then from (8-1.10), we have

$$\tilde{\varepsilon}_R^{(n)} = \varepsilon_R^{(2n-1)}, \qquad \tilde{\varepsilon}_B^{(n)} = \varepsilon_B^{(2n)}. \tag{8-3.9}$$

From (5-2.6), the subvectors $\varepsilon_R^{(m)}$ and $\varepsilon_B^{(m)}$ satisfy

$$\begin{bmatrix} \varepsilon_R^{(m)} \\ \varepsilon_B^{(m)} \end{bmatrix} = P_{m,E}(B) \begin{bmatrix} \varepsilon_R^{(0)} \\ \varepsilon_B^{(0)} \end{bmatrix}, \qquad (8\text{-}3.10)$$

where $P_{m,E}(x)$ is defined by (5-2.7).

Since $P_{n,E}(x)$ is an odd polynomial for n odd and an even polynomial for n even, it is obvious that there exists (see Varga [1962]) polynomials $R_{n,E}(y)$ and $L_{n,E}(y)$ of degree n in y such that

$$P_{2n,E}(x) = R_{n,E}(x^2) \qquad \text{and} \qquad P_{2n+1,E}(x) = (x/M_E)L_{n,E}(x^2). \qquad (8\text{-}3.11)$$

Hence since

$$B^2 = \begin{bmatrix} F_R F_B & 0 \\ 0 & F_B F_R \end{bmatrix},$$

we have

$$P_{2n,E}(B) = R_{n,E}(B^2) = \begin{bmatrix} R_{n,E}(F_R F_B) & 0 \\ 0 & R_{n,E}(F_B F_R) \end{bmatrix},$$

$$P_{2n+1,E}(B) = \frac{B}{M_E} L_{n,E}(B^2) = \frac{1}{M_E} \begin{bmatrix} 0 & F_R L_{n,E}(F_B F_R) \\ F_B L_{n,E}(F_R F_B) & 0 \end{bmatrix}.$$

$$(8\text{-}3.12)$$

Thus from (8-3.9)–(8-3.10) and (8-3.12), we have

$$\tilde{\varepsilon}_R^{(n)} = \varepsilon_R^{(2n-1)} = \frac{F_R}{M_E} L_{n-1,E}(F_B F_R)\varepsilon_B^{(0)},$$

$$\tilde{\varepsilon}_B^{(n)} = \varepsilon_B^{(2n)} = R_{n,E}(F_B F_R)\varepsilon_B^{(0)}.$$

$$(8\text{-}3.13)$$

Note from (8-3.11) that $R_{0,E}(B^2) = I$ and $L_{0,E}(B^2) = M_E I$.

We now seek to investigate procedures for estimating the iteration error vector $\tilde{\varepsilon}_B^{(n)}$. In Chapters 5 and 6, the pseudoresidual vector $\delta^{(n)}$ corresponding to the basic method is used to estimate the error vector. The difficulty here is to find a "residual" vector which is suitable for this purpose.

We first consider the vector $\tilde{\delta}_B^{(n)}$ defined by

$$\tilde{\delta}_B^{(n)} \equiv F_B F_R U_B^{(n)} + F_B c_R + c_B - U_B^{(n)}. \qquad (8\text{-}3.14)$$

Since the solution $\bar{u}_B$ satisfies $\bar{u}_B = F_B F_R \bar{u}_B + F_B c_R + c_B$, it easily follows that $\tilde{\delta}_B^{(n)}$ satisfies $\tilde{\delta}_B^{(n)} = (F_B F_R - I)\tilde{\varepsilon}_B^{(n)}$ and, since $(I - F_B F_R)$ is nonsingular,

$$\tilde{\varepsilon}_B^{(n)} = (F_B F_R - I)^{-1}\tilde{\delta}_B^{(n)}. \qquad (8\text{-}3.15)$$

Moreover, $S_B(F_B F_R)S_B^{-1}$ is symmetric so that

$$\|\tilde{\varepsilon}_B^{(n)}\|_{S_B} \le \frac{1}{1 - \mathbf{S}(F_B F_R)} \|\tilde{\delta}_B^{(n)}\|_{S_B} = \frac{1}{1 - [\mathbf{S}(B)]^2} \|\tilde{\delta}_B^{(n)}\|_{S_B}. \quad (8\text{-}3.16)$$

Hence estimates for $\tilde{\varepsilon}_B^{(n)}$ may be obtained from $\tilde{\delta}_B^{(n)}$ using procedures similar to those given in Chapters 5 and 6.

The residual vector $\tilde{\delta}_B^{(n)}$ also may be used to obtain estimates for $M(B) = \mu_1$. Indeed, substituting for $\tilde{\varepsilon}_B^{(n)}$ in (8-3.13) by using (8-3.15), we obtain

$$\tilde{\delta}_B^{(n)} = R_{n,\,E}(F_B F_R)\tilde{\delta}_B^{(0)}, \quad (8\text{-}3.17)$$

from which it follows that

$$\|\tilde{\delta}_B^{(n)}\|_{S_B} \le \mathbf{S}(R_{n,\,E}(F_B F_R))\|\tilde{\delta}_B^{(0)}\|_{S_B} = \mathbf{S}(P_{2n,\,E}(B))\|\tilde{\delta}_B^{(0)}\|_{S_B}, \quad (8\text{-}3.18)$$

where the right equality follows from (8-3.11). Thus $\tilde{\delta}_B^{(n)}$ satisfies the basic inequality (5-2.9), and estimates for μ_1 may be obtained using the adaptive procedures given in Chapters 5 and 6.

The problem with implementing any adaptive procedure based on the $\tilde{\delta}_B^{(n)}$ vector is that $\tilde{\delta}_B^{(n)}$ does not appear naturally in the CCSI process and is difficult computationally to obtain. In what follows, we use the easily obtainable *difference vector*

$$\Delta_B^{(n)} \equiv U_B^{(n)} - U_B^{(n+1)} \quad (8\text{-}3.19)$$

either directly in the adaptive procedure or else to determine $\tilde{\delta}_B^{(n)}$.

From (8-1.11), $\Delta_B^{(n)}$ satisfies $\Delta_B^{(n)} = -\rho_B^{(n+1)}(F_B U_R^{(n+1)} + c_B - U_B^{(n)})$. Since $\bar{u}_B = F_B u_R + c_B$, it follows that $\Delta_B^{(n)}$ can be expressed in terms of the CCSI error vector as

$$\Delta_B^{(n)} = -\rho_B^{(n+1)}[F_B \tilde{\varepsilon}_R^{(n+1)} - \tilde{\varepsilon}_B^{(n)}] \quad (8\text{-}3.20)$$

or from (8-3.13)

$$\Delta_B^{(n)} = Q_{n+1,\,E}(F_B F_R)\varepsilon_B^{(0)}, \quad (8\text{-}3.21)$$

where $Q_{n+1,\,E}(y)$ is a polynomial of degree $n+1$ in y and is defined by

$$Q_{n+1,\,E}(y) \equiv \rho_B^{(n+1)}[R_{n,\,E}(y) - (y/M_E)L_{n,\,E}(y)]. \quad (8\text{-}3.22)$$

Note from (8-3.11) that we also have

$$Q_{n+1,\,E}(x^2) = \rho_B^{(n+1)}[P_{2n,\,E}(x) - xP_{2n+1,\,E}(x)]. \quad (8\text{-}3.23)$$

Adaptive Parameter and Stopping Procedures

For the discussion on the use of $\Delta_B^{(n)}$, we use the successive polynomial notation of Section 6.3; that is,

$P_{p, E_s}(x)$ denotes the normalized Chebyshev polynomial of degree p using the most recent estimate M_{E_s} for $M(B)$. The subscript s indicates this is the sth estimate used for $M(B)$.

q denotes the last iteration on which the previous estimate $M_{E_{s-1}}$ was used. For the initial estimate where $s = 1$, we let $q = 0$.

$P_{\bar{p}_m, E_m}(x)$ denotes the maximum degree polynomial generated using the estimate M_{E_m}, where $m < s$.

n denotes the current iteration count. Note that $p = n - q$.

Also, as in Chapter 6, to indicate more clearly the relationship between n, s, and p, we write $\tilde{\delta}_B^{(n)}$ $(\equiv \tilde{\delta}_B^{(p+q)})$ as $\tilde{\delta}_B^{(s, p)}$, $\tilde{\varepsilon}^{(n)}$ $(\equiv \tilde{\varepsilon}^{(p+q)})$ as $\tilde{\varepsilon}^{(s, p)}$, and $\Delta_B^{(n)}$ $(\equiv \Delta_B^{(p+q)})$ as $\Delta_B^{(s, p)}$. Thus (8-3.13) and (8-3.21) may be written as

$$\tilde{\varepsilon}_B^{(n)} \equiv \tilde{\varepsilon}_B^{(s, p)} = R_{p, E_s}(F_B F_R)\tilde{\varepsilon}_B^{(s, 0)},$$

$$\Delta_B^{(n)} \equiv \Delta_B^{(s, p)} = Q_{p+1, E_s}(F_B F_R)\tilde{\varepsilon}_B^{(s, 0)}. \tag{8-3.24}$$

As before, $\tilde{\varepsilon}_B^{(s, 0)} = \tilde{\varepsilon}_B^{(s-1, \bar{p}_{s-1})}$, so that

$$\tilde{\varepsilon}_B^{(s, p)} = R_{p, E_s}(F_B F_R)R_{\bar{p}_{s-1}, E_{s-1}}(F_B F_R) \cdots R_{\bar{p}_1, E_1}(F_B F_R)\tilde{\varepsilon}_B^{(1, 0)}. \tag{8-3.25}$$

However, a relationship analogous to (8-3.25) is not valid for the $\Delta_B^{(s, p)}$ vectors since now $\Delta_B^{(s, 0)} \neq \Delta_B^{(s-1, \bar{p}_{s-1})}$.

In the following, we assume the eigenvalues $\{\mu_i\}$ of B are given by (8-1.6). Thus from Section 8.2, the eigenvalues $\{\lambda_i\}$ of $F_B F_R$ satisfy $\lambda_i = \mu_i^2$ and can be ordered as

$$0 \leq \lambda_{\tilde{N}} \leq \cdots \leq \lambda_t < \lambda_{t-1} = \cdots = \lambda_1 = (M(B))^2 < 1, \tag{8-3.26}$$

where $\tilde{N}$ $(< N)$ is the order of $F_B F_R$. Moreover, we let $v_B(i)$ be the eigenvector of $F_B F_R$ associated with λ_i. Since $F_B F_R$ is similar to a symmetric matrix, we may assume the set $\{v_B(i)\}_{i=1}^{i=\tilde{N}}$ is a basis and write $\tilde{\varepsilon}_B^{(s, 0)}$ as

$$\tilde{\varepsilon}_B^{(s, 0)} = c_{1, s}v_B(1) + c_{2, s}v_B(2) + \cdots + c_{\tilde{N}, s}v_B(\tilde{N}). \tag{8-3.27}$$

If $c_{1, s} \neq 0$, the contamination factor† of the vector $\tilde{\varepsilon}_B^{(s, 0)}$ is defined to be

$$K_s = \left[\sum_{i=2}^{\tilde{N}} \|c_{i, s}v_B(i)\|_\psi\right]/\|c_{1, s}v_B(1)\|_\psi. \tag{8-3.28}$$

---

† Note that the contamination factor here is defined in terms of the error vector instead of the pseudoresidual vector as was done in Chapter 6.

where the ψ-norm is determined by the context in which K_s is used. Henceforth we drop the subscript s on M_{E_s} and $P_{p,E_s}(x)$, etc., when the meaning is clear. Moreover, we use μ_1 instead of $M(B)$ to denote the largest eigenvalue of B.

The proof of the following theorem is given in Section 8.7.

Theorem 8-3.1. If $0 < M_E < \mu_1$, then the polynomial sequences $\{R_{p,E}(y)\}$ and $\{Q_{p,E}(y)\}$ are $F_B F_R$-uniformly convergent.‡

Hence if $M_E < \mu_1$, it follows from (8-3.24) and Theorems 6-2.1 and 6-2.2 that for large p and/or large s we have approximately that

$$\tilde{\varepsilon}_B^{(s,p)}/R_{p,E}(\mu_1^2) \doteq c_{1,s}v_B(1), \tag{8-3.29}$$

$$\Delta_B^{(s,p)}/Q_{p+1,E}(\mu_1^2) \doteq c_{1,s}v_B(1), \tag{8-3.30}$$

$$\|\Delta_B^{(s,p)}\|/\|\Delta_B^{(s,p-1)}\| \doteq Q_{p+1,E}(\mu_1^2)/Q_{p,E}(\mu_1^2). \tag{8-3.31}$$

Moreover, if $M_{E_s} < \mu_1$ for all s, it can be shown, as in Chapter 6, that the contamination factors (8-3.28) decrease with s.

If

$$R^{(s,p)} \equiv \frac{2 - \rho_B^{(s,p)}}{2 - \rho_B^{(s,p+1)}} \frac{\|\Delta_B^{(s,p)}\|_2}{\|\Delta_B^{(s,p-1)}\|_2}, \tag{8-3.32}$$

we show in Section 8.7 by using (8-3.29)–(8-3.31) that approximately

$$R^{(s,p)} \doteq |P_{2p,E}(\mu_1)|/|P_{2p-2,E}(\mu_1)| \tag{8-3.33}$$

and that

$$\tilde{\varepsilon}_B^{(s,p)} \doteq \frac{\Delta_B^{(s,p)}}{1 - P_{2p+2,E}(\mu_1)/P_{2p,E}(\mu_1)} \doteq \frac{\Delta_B^{(s,p)}}{1 - \beta^{(p)}R^{(s,p)}}, \tag{8-3.34}$$

where $\beta^{(p)} \equiv [(\rho_B^{(p+1)} - 1)/(\rho_B^{(p)} - 1)]$. The relationships (8-3.33)–(8-3.34) are valid for sufficiently large p and/or sufficiently large s provided, of course, that $M_E < \mu_1$.

Thus from (8-3.34), *for the stopping test* we use

$$\frac{\|\tilde{\varepsilon}_B^{(s,p)}\|_\beta}{\|U_B^{(n)}\|_\eta} \doteq \frac{1}{1 - \beta^{(p)}H} \frac{\|\Delta_B^{(s,p)}\|_\beta}{\|U_B^{(n)}\|_\eta} \leq \zeta, \tag{8-3.35}$$

where

$$H = \max[R^{(s,p)}, |P_{2p,E}(M_E)/P_{2p-2,E}(M_E)|].$$

If $R^{(s,p)} \geq 1$, no test for convergence is made. The quantity H, instead of $R^{(s,p)}$, is used in (8-3.35) as a precautionary step in case $M_E \geq \mu_1$. If $M_E \geq \mu_1$, we show later in this section that the quantity $R^{(s,p)}$ probably will oscillate about $(P_{2p,E}(M_E)/P_{2p-2,E}(M_E))$. Thus on some iterations, $(1 - R^{(s,p)})$ may

‡ See Definition 6-2.1.

be considerably larger than its theoretical maximum value. The use of H instead of $R^{(s, p)}$ prevents any exceptionally large value for $(1 - R^{(s, p)})$ from being used in the stopping test.

From (8-3.33), *for the parameter change test*, the present estimate M_E for μ_1 is deemed to be adequate if

$$R^{(s, p)} < \left[\frac{P_{2p, E}(M_E)}{P_{2p-2, E}(M_E)} \right]^F \equiv \left[r \left(\frac{1 + r^{2p-2}}{1 + r^{2p}} \right) \right]^F \qquad (8\text{-}3.36)$$

where $F(<1)$ is the damping factor discussed previously in Chapter 5 and r is defined by (4-3.21) and is

$$r = (1 - \sqrt{1 - M_E^2})/(1 + \sqrt{1 - M_E^2}). \qquad (8\text{-}3.37)$$

As in Chapters 5 and 6, inequality (8-3.36) implies that the convergence rate obtained using M_E is at least F times the maximum possible convergence rate. This follows from the fact that when $M_E < \mu_1$, then

$$(P_{2p, E}(M_E)/P_{2p-2, E}(M_E)) < (P_{2p}(\mu_1)/P_{2p-2}(\mu_1)),$$

where $P_n(x)$ is the optimum Chebyshev polynomial (4-2.9). Thus inequality (8-3.36) implies

$$R^{(s, p)} < \left[\frac{|P_{2p, E}(M_E)|}{|P_{2p-2, E}(M_E)|} \right]^F < \left[\frac{P_{2p}(\mu_1)}{P_{2p-2}(\mu_1)} \right]^F,$$

from which the desired conclusion concerning the convergence rates follows.

The parameter change test used in Chapters 5 and 6 to determine the "goodness" of M_E is based on the comparison of average convergence rates for p iterations. The test (8-3.36) is based on the comparison of convergence rates just for iteration p and is likely to reflect the need to reestimate μ_1 more quickly than that of (6-3.21); i.e., for the same value of F, the inequality (8-3.36) would probably fail to be satisfied for a smaller value of p than that for (6-3.21). We use the test (8-3.36) for the CCSI method here since the pseudoresidual vector $\tilde{\delta}_B^{(s, p)}$, which is needed to measure average convergence rates in (6-3.21), is assumed not readily available.

We now describe a method which may be used to obtain *new estimates for* μ_1. If the current estimate M_E satisfies $M_E < \mu_1$, then $\mathbf{S}(P_{2n, E}(B)) = P_{2n, E}(\mu_1)$. From (8-3.18), we then have

$$\frac{\|\tilde{\delta}_B^{(s, p)}\|_{S_B}}{\|\tilde{\delta}_B^{(s, 0)}\|_{S_B}} \leq \mathbf{S}(P_{2p, E}(B)) = P_{2p, E}(\mu_1) = \frac{T_{2p}(w_E(\mu_1))}{T_{2p}(w_E(1))}, \qquad (8\text{-}3.38)$$

where $w_E(x)$ is defined by

$$w_E(x) \equiv x/M_E. \qquad (8\text{-}3.39)$$

Thus as in Section 5.4, we may take the new estimate M'_E for μ_1 to be the largest real x which satisfies the Chebyshev equation

$$T_{2p}(w_E(x))/T_{2p}(w_E(1)) = \|\tilde{\delta}_B^{(s,\,p)}\|_{S_B}/\|\tilde{\delta}_B^{(s,\,0)}\|_{S_B}. \tag{8-3.40}$$

If the ratio $\|\tilde{\delta}_B^{(s,\,p)}\|_{S_B}/\|\tilde{\delta}_B^{(s,\,0)}\|_{S_B}$ is known, the new estimate M'_E can be obtained in closed form using the results of Theorem 5-4.1. We now show how this ratio can be computed whenever a new estimate M'_E is needed.

Since $\rho_R^{(s,\,1)} = 1$, it is obvious from (8-1.11) and (8-3.14) that

$$\|\tilde{\delta}_B^{(s,\,0)}\|_{S_B} = [1/\rho_B^{(s,\,1)}]\,\|\Delta_B^{(s,\,0)}\|_{S_B}.$$

Thus $\|\tilde{\delta}_B^{(s,\,0)}\|_{S_B}$ is readily available, and the problem now is to obtain $\|\tilde{\delta}_B^{(s,\,p)}\|_{S_B}$.

Suppose the inequality (8-3.36) is not satisfied for $R^{(s,\,p-1)}$ and a new estimate is sought for μ_1. We assume $R^{(s,\,p-1)}$ was computed on iteration n. Thus $U_B^{(n)}$ and $\Delta_B^{(n-1)} \equiv \Delta_B^{(s,\,p-1)} = U_B^{(n-1)} - U_B^{(n)}$ have been computed. The residual vector $\tilde{\delta}_B^{(n)} = \tilde{\delta}_B^{(s,\,p)}$ needed to obtain an estimate for μ_1 by using (8-3.40) is by definition (8-3.14)

$$\tilde{\delta}_B^{(s,\,p)} = \tilde{\delta}_B^{(n)} = \hat{U}_B^{(n+1)} - U_B^{(n)}, \tag{8-3.41}$$

where

$$\hat{U}_B^{(n+1)} = F_B F_R U_B^{(n)} + F_B c_R + c_B. \tag{8-3.42}$$

The $\hat{U}_B^{(n+1)}$ in (8-3.42) corresponds to a Gauss–Seidel iteration and may be easily obtained from the CCSI process (8-1.11) by setting $\rho_R^{(n+1)} = \rho_B^{(n+1)} = 1.0$. Thus if it is decided that a new estimate M'_E for μ_1 is needed at the end of iteration n, a Gauss–Seidel iteration† is carried out on iteration $(n + 1)$, $\|\tilde{\delta}_B^{(s,\,p)}\|_{S_B} = \|\Delta_B^{(n)}\|_{S_B}$ is computed, and then M'_E is obtained from the Chebyshev equation (8-3.40). A new Chebyshev polynomial using M'_E is then started on iteration $(n + 2)$.

To obtain the solution M'_E to (8-3.40), let

$$\hat{B} \equiv \rho_B^{(s,\,1)}\,\frac{\|\Delta_B^{(n)}\|_{S_B}}{\|\Delta_B^{(s,\,0)}\|_{S_B}} \quad\text{and}\quad Q \equiv \frac{1}{T_{2p}(1/M_E)} = \frac{2r^p}{1 + r^{2p}}. \tag{8-3.43}$$

---

† We remark that this Gauss–Seidel iteration is not necessary if the $\tilde{\delta}_B^{(s,\,p)}$ vector calculated by (8-3.41) is stored. For in this case, the new estimate $M_{E_{s+1}}$ can be obtained after $\|\tilde{\delta}_B^{(s,\,p)}\|_{S_B}$ is computed. Then, using the stored $\tilde{\delta}_B^{(s,\,p)}$ and $U_B^{(n)}$ vectors, the generation of the new Chebyshev polynomial can be started immediately on iteration $(n + 1)$ by calculating

$$U_B^{(n+1)} = \rho_B^{(s+1,\,1)}\tilde{\delta}_B^{(s,\,p)} - U_B^{(n)},$$

where

$$\rho_B^{(s+1,\,1)} = [1 - \tfrac{1}{2}M_{E_{s+1}}^2]^{-1}.$$

We chose the one Gauss–Seidel iteration path here since it is not always convenient to store the $\tilde{\delta}_B^{(s,\,p)}$ vector.

If $\hat{B} > Q$, it then follows from Theorem 5-4.1 and Eq. (5-4.24) that the desired solution M'_E may be given by

$$M'_E = (1/(1 + r))(X + rX^{-1}), \tag{8-3.44}$$

where

$$X = [\tfrac{1}{2}(\hat{B} + \sqrt{\hat{B}^2 - Q^2})(1 + r^{2p})]^{1/2p} \tag{8-3.45}$$

and $r = [1 - \sqrt{1 - M_E^2}]/[1 + \sqrt{1 - M_E^2}]$. Moreover, again by Theorem 5-4.1, we have that

$$M_E < M'_E \leq \mu_1. \tag{8-3.46}$$

Note that the formula for M'_E used here corresponds to the alternate expression (5-4.24) and is valid when $M_E = 0$ ($=m_E$). Since the CCSI procedure reduces to the Gauss–Seidel method when $M_E = 0$, often it is convenient to use $M_E = 0$ as the initial guess for μ_1.

For reasons analogous to those given in Section 6.3 concerning the validity of (6-3.18), it is reasonable to assume, but with some caution, that inequality (8-3.38) is also valid in the 2-norm. Thus as before, the 2-norm may be used instead of the S_B-norm in the calculation of M'_E; i.e., use $\hat{B}$ given by

$$\hat{B} \equiv \rho_B^{(s, 1)} \|\Delta_B^{(n+1)}\|_2/\|\Delta_B^{(s, 0)}\|_2. \tag{8-3.47}$$

Since (8-3.38) is not always valid in the 2-norm, the right inequality in (8-3.46) is not always satisfied when (8-3.47) is used.

Remark. Other procedures also exist for estimating μ_1. For example, instead of the "one Gauss–Seidel iteration" approach used above, the vector $\tilde{\delta}_B^{(s, p)}$ could be determined on each iteration using the recurrence relation†

$$\tilde{\delta}_B^{(s, 0)} = \frac{-1}{\rho_B^{(s, 1)}} \Delta_B^{(s, 0)}$$

$$\tilde{\delta}_B^{(s, p)} = \frac{-1}{\rho_R^{(s, p+1)}\rho_B^{(s, p+1)}} \tag{8-3.48}$$

$$\times \left\{ \Delta_B^{(s, p)} - \frac{\rho_B^{(s, p+1)}}{\rho_B^{(s, p)}} (1 - \rho_R^{(s, p+1)})(1 - \rho_B^{(s, p)})\Delta_B^{(s, p-1)} \right\}.$$

† To derive (8-3.48), we first use (8-1.11) and (8-3.14) to write

$$\Delta_B^{(s, p)} = -\rho_B^{(s, p+1)}[F_B U_R^{(n+1)} + c_B - U_B^{(n)}]$$

or equivalently

$$\Delta_B^{(s, p)} = -\rho_B^{(s, p+1)}\{\rho_R^{(s, p+1)}\tilde{\delta}_B^{(s, p)} + (1 - \rho_B^{(s, p+1)})(F_B U_R^{(n)} + c_B - U_B^{(n)})\}.$$

Also, it is easy to show that $\Delta_B^{(s, p-1)} = -\rho_B^{(s, p)}(F_B U_R^{(n)} + c_B - U_B^{(n)}) + \rho_B^{(s, p)}\Delta_B^{(s, p-1)}$. Using the equation for $\Delta_B^{(s, p-1)}$ to eliminate $(F_B U_R^{(n)} + c_B - U_B^{(n)})$ in the expression for $\Delta_B^{(s, p)}$, we obtain (8-3.48). The authors wish to acknowledge helpful discussions with Dr. David R. Kincaid in deriving the recurrence relation (8-3.48).

Since $\tilde{\delta}_B^{(s,\,p)}$ is now known for all p, the adaptive stopping and parameter estimation procedures described in Chapters 5 and 6 may be used. The use of (8-3.48) requires that the Δ_B vector be stored and that one additional multiplication be performed.

An alternative method for estimating μ_1 is to solve a Chebyshev type equation defined in terms of the $\Delta_B^{(s,\,p)}$ vectors. Since $\rho_R^{(s,\,1)} = 1$ and $\tilde{\varepsilon}_R^{(s,\,1)} = F_R \tilde{\varepsilon}_B^{(s,\,0)}$, we have from (8-3.20),

$$\Delta_B^{(s,\,0)} = \rho_B^{(s,\,1)}[I - F_B F_R]\tilde{\varepsilon}_B^{(s,\,0)}. \tag{8-3.49}$$

Thus from (8-3.21)–(8-3.22),

$$\Delta_B^{(s,\,p)} = \frac{\rho_B^{(s,\,p+1)}}{\rho_B^{(s,\,1)}}\left[R_{p,\,E}(F_B F_R) - \frac{F_B F_R}{M_E} L_{p,\,E}(F_B F_R) \right][I - F_B F_R]^{-1}\Delta_B^{(s,\,0)}. \tag{8-3.50}$$

But $S_B(F_B F_R)S_B^{-1}$ is symmetric. Hence if $M_E < \mu_1$, we have by using (8-3.11) that

$$\frac{\|\Delta_B^{(s,\,p)}\|_{S_B}}{\|\Delta_B^{(s,\,0)}\|_{S_B}} \le \frac{1}{1 - \mu_1^2}\frac{\rho_B^{(s,\,p+1)}}{\rho_B^{(s,\,1)}}[P_{2p,\,E}(\mu_1) - \mu_1 P_{2p+1,\,E}(\mu_1)]. \tag{8-3.51}$$

As in Chapters 5 and 6, we then take the new estimate M_E' to be the largest real x which satisfies the Chebyshev Δ_B-equation†

$$\frac{\rho_B^{(s,\,1)}}{\rho_B^{(s,\,p+1)}}\frac{\|\Delta_B^{(s,\,p)}\|_{S_B}}{\|\Delta_B^{(s,\,0)}\|_{S_B}} = \frac{1}{1 - x^2}[P_{2p,\,E}(x) - xP_{2p,\,E}(x)]. \tag{8-3.52}$$

Contrary to the Chebyshev equations of Chapters 5 and 6, a closed form expression for the solution M_E' to (8-3.52) does not seem possible. However, some iterative solution procedure can be used. The bisection method has been used for this purpose and has worked well for the few sample problems tried. However, since the "one Gauss–Seidel" approach has worked well in practice and since the cost is relatively minor, we have not investigated in depth any iterative solution method for solving the Chebyshev Δ_B-equation.

An Overall Computational Algorithm

An overall algorithm based on the previous discussions is given below as an informal program. The input required is summarized below.

ζ the stopping criterion number ζ used in (8-3.35).

M_E the initial estimate for μ_1, the largest eigenvalue of B. M_E must satisfy $0 \le M_E < 1$. If nothing is known about μ_1, it is sufficient to let $M_E = 0.0$.

---

† For reasons analogous to those given in Chapter 6, the S_B-norm in (8-3.52) may be replaced by the 2-norm.

F the damping factor F used in the parameter change test (8-3.36). The choice of F is discussed in Section 5.7. Typically, F should satisfy $0.65 \leq F \leq 0.8$.

$\underline{U}_B$ the initial guess vector for $\bar{u}_B$.

d the strategy parameter defined by (6-3.19) used to determine the minimum degree, p^*, required for each Chebyshev polynomial generated. Typically, d should satisfy $0.03 \leq d \leq 0.15$.

The control variables and counters used are similar to those used in Chapters 5 and 6. (See, for example, Section 6.4.) Suggested values of τ_s, the imposed upper bound for the sth estimate M_{E_s}, are given in (6-4.1). The additional control variables S and T are used to indicate when the ratio $R^{(s, p)}$ of (8-3.32) lies outside a particular interval and are discussed below.

As in the previous algorithms, we use the underline, $\underline{u}$, to indicate more clearly which variables are vectors. The iterates $\underline{U}_R^{(n)}$ and $\underline{U}_B^{(n)}$ are obtained by using (8-1.11). For remarks concerning the calculation of the matrix–vector products $F_R \underline{U}_B$ and $F_B \underline{U}_R$, see Eqs. (8-2.6) and (8-2.7). If D_B is a block diagonal matrix, it is not necessary to store the vector $\underline{\Delta}_B$. See the discussion given in Section 5.6 for the Jacobi method.

No checking of any type is done until $p = 3$; i.e., until a Chebyshev polynomial of degree 8 has been generated. Note that a new estimate M'_E for μ_1 is computed only if $\rho_B = 1.0$, and this occurs only if a Gauss–Seidel iteration is being carried out. Also, note that the initial guess $M_E = 0$ is treated as a special case in **Calculate New Estimate** M'_E. If this case were not treated in a special way, no convergence stopping test would be made until the eighth iteration. Moreover, to compute the initial estimate M'_E, we utilized the fact that $\lim_{n \to \infty} \|\Delta_B^{(n+1)}\| / \|\Delta_B^{(n)}\| = \mu_1^2$ for the Gauss–Seidel process.

To discuss the S and T counters, let us first suppose that $M_{E_s} > \mu_1$ for some s. Then using assumptions and arguments similar to those given in Section 6.7, we obtain the approximation

$$R^{(s, p)} \doteq \frac{|P_{2p, E}(\mu_1)|}{|P_{2p-2, E}(\mu_1)|} = \frac{T_{2p-2}(1/M_E)}{T_{2p}(1/M_E)} \left| \frac{\cos 2p\theta}{\cos 2(p-1)\theta} \right|, \quad (8\text{-}3.53)$$

where θ is given by (6-7.3) with $w(\mu_1) = \mu_1/M_E$. Thus for this case, $R^{(s, p)}$ will oscillate about $[T_{2p-2}(1/M_E)/T_{2p}(1/M_E)] = [r(1 + r^{2p-2})]/[1 + r^{2p}]$. On the other hand, if $M_E < \mu_1$, we have for sufficiently large p that

$$\frac{T_{2p-2}(1/M_E)}{T_{2p}(1/M_E)} < R^{(s, p)} \stackrel{.}{\leq} \frac{P_{2p, E}(\mu_1)}{P_{2p-2, E}(\mu_1)} < 1. \quad (8\text{-}3.54)$$

Thus if $R^{(s, p)}$ does not satisfy (8-3.54), the implication is that $M_E > \mu_1$. If $R^{(s, p)} \geq 1$, the S counter is incremented but no stopping test is performed. If the left inequality in (8-3.54) is not satisfied, the T counter is incremented and the stopping test (8-3.35) is carried out but with $R^{(s, p)}$ replaced by its "theoretical" minimum value, $[r(1 + r^{2p-2})]/[1 + r^{2p}]$.

In the **Parameter Change Test** step, a nonzero product $S * T$ is taken to indicate that $M_E > \mu_1$. When this happens, no new estimation of μ_1 is permitted. If $S * T = 0$ and if the parameter change test (8-3.36) indicates that a new estimate for μ_1 is needed, the required Gauss–Seidel iteration is set up merely by setting $\rho_R = \rho_B = 1.0$ and returning to the **Calculate New Iterate** step.

After the Gauss–Seidel iteration (implied by $\rho_B = 1.0$) is completed, the new estimate M'_E for μ_1 is obtained from (8-3.44). However, a new estimate M'_E is calculated only if $\hat{B}$ satisfies

$$1 < \hat{B}/Q < 1/Q. \tag{8-3.55}$$

This test is required to ensure that $M_E < M'_E < 1$. We have never seen a case for which $\hat{B}$ and Q did not satisfy (8-3.55). However, since there is no guarantee concerning the value of $\hat{B}/Q$ when the 2-norm is used, we include this test merely as a precautionary measure.

We now discuss the steps we take should inequality (8-3.55) not be satisfied. If (a) $M_E < \mu_1$ and if (b) $\|\tilde{\delta}_B^{(s,p)}\|_2/\|\tilde{\delta}_B^{(s,0)}\|_2 \doteq |P_{2p,E}(\mu_1)|$, then we have that

$$\frac{\hat{B}}{Q} \equiv T_{2p}\left(\frac{1}{M_E}\right)\frac{\|\tilde{\delta}_B^{(s,p)}\|_2}{\|\tilde{\delta}_B^{(s,0)}\|_2} \doteq T_{2p}\left(\frac{1}{M_E}\right)|P_{2p,E}(\mu_1)| = T_{2p}\left(\frac{\mu_1}{M_E}\right) > 1. \tag{8-3.56}$$

Now if $(\hat{B}/Q) \leq 1$, then (8-3.56) is not valid which, in turn, implies that assumption (a) and/or assumption (b) is not valid. We assume that $M_E \geq \mu_1$ is the cause of $(\hat{B}/Q) \leq 1$ and instead of calculating a new estimate, we simply set $M'_E = M_E$ and continue. On the other hand, $\hat{B}/Q \geq 1/Q = T_{2p}(1/M_E)$ implies that $\mu_1 \geq 1$ and that the iterative process is diverging. Should this occur, we assume something is wrong and terminate the iterative process.

In Appendix B, we give a Fortran listing of a subroutine, called CCSI, which implements Algorithm 8-3.1 with the exception of the **Calculate New Iterate** portion. The CCSI subroutine is designed for use as a software package to provide the required acceleration parameters and to provide an estimate for the iteration error for the cyclic Chebyshev polynomial method.

Algorithm 8-3.1. An adaptive procedure for the CCSI method using the 2-norm.

Input: $(\zeta, M_E, F, d, \underline{U}_B)$

Initialize:

$n := 0; p := -1; M'_E := M_E; s := 0; \underline{U}_R := \underline{0}.$

Next Iteration:

$n := n + 1; p := p + 1$
If $p = 0$, *then* ⟨Initialize for start of new polynomial⟩

Begin
$s := s + 1; T := 0; S := 0;$
If $M'_E > \tau_s$ *then* $M'_E = \tau_s$; *else* continue;
$M_E := M'_E; \rho_R := 1.0; \rho_B := 2/(2 - M_E^2);$
$r := (1 - \sqrt{1 - M_F^2})/(1 + \sqrt{1 - M_F^2}); p^* := 8;$
If $M_E = 0$ *then* $s := s - 1$; *else* $p^* := [\log d/\log r]$;
If $p^* < 8$ *then* $p^* = 8$; *else* continue;
End
else ⟨Continue polynomial generation⟩
Begin
$\text{DELNO} := \text{DELNP}; \text{ORB} := \rho_B;$
If $\rho_B > 1.0$, *then* $\rho_R := 1/(1 - \frac{1}{4}M_E^2\rho_B); \rho_B := 1/(1 - \frac{1}{4}M_E^2\rho_R);$
               else continue;
End

Calculate New Iterate:

$\underline{U}_R := \rho_R(F_R\underline{U}_B + \underline{c}_R - \underline{U}_R) + \underline{U}_R$
$\underline{\Delta}_B := \rho_B(F_B\underline{U}_R + \underline{c}_B - \underline{U}_B); \text{DELNP} := |\underline{\Delta}_B|_2; \text{DELNE} := \|\underline{\Delta}_B\|_\beta;$
$\underline{U}_B := \underline{\Delta}_B + \underline{U}_B; \text{YUN} := \|\underline{U}_B\|_\eta$

Calculate New Estimate M'_E:

If $p < 3$, *then*
       Begin
       If $p = 0$ *then* $\text{DELNPI} := \text{DELNP}$; *else* continue;
       Go to **Next Iteration**;
       End
       else If $M_E > 0$ *then* continue; *else Go to* **Initial G–S Iteration**;
If $\rho_B = 1.0$, *then* ⟨Compute new estimate for μ_1⟩
       Begin

$$Q := 2r^p/[1 + r^{2p}]; \hat{B} := \left[\frac{2}{2 - M_E^2}\right]\frac{\text{DELNP}}{\text{DELNPI}};$$

       If $\hat{B} < Q$, *then* ⟨Implies current estimate $M_E \geq \mu_1$⟩
              Begin
              $M'_E := M_E; p := -1; Go\ to$ **Next Iteration**;
              End
              else continue;
       If $\hat{B} < 1$ *then* continue; *else* STOP ⟨Possible divergence⟩
       $X := [\frac{1}{2}(1 + r^{2p})(\hat{B} + \sqrt{\hat{B}^2 - Q^2})]^{1/2p};$
       $M'_E := (1/(1 + r))(X + rX^{-1});$
       $p := -1, Go\ to$ **Next Iteration**
       End

else continue;

Stopping Test:

$$R := \left[\frac{2 - \text{ORB}}{2 - \rho_{\text{B}}}\right]\left[\frac{\text{DELNP}}{\text{DELNO}}\right]; \; C := r\left[\frac{1 + r^{2p-2}}{1 + r^{2p}}\right]; \; \beta := \frac{\rho_{\text{B}} - 1}{\text{ORB} - 1};$$

$H := (\beta)(R);$

If $R \geq 1$, *then*

    Begin

    $S := S + 1;$ *Go to* **Next Iteration**;

    End

    else

    Begin

    If $R \geq C$ *then* continue; *else* $T := T + 1; \; H := (\beta)(C);$

    End

If $\dfrac{\text{DELNE}}{\text{YUN}} \leq \zeta(1 - H),$ *then* print final output and STOP $\langle$converged$\rangle$;

                    else continue;

Parameter Change Test:

If $p < \frac{1}{2}p^*,$ *then* *Go to* **Next Iteration**

    else

    Begin

    If $S * T > 0,$ *then* *Go to* **Next Iteration**

        else

        Begin

        If $R \leq [C]^F,$ *then* *Go to* **Next Iteration**

            else

            Begin

            $n := n + 1; \; p := p + 1; \; \rho_{\text{R}} := 1.0;$

            $\rho_{\text{B}} := 1.0;$

            Go to **Calculate New Iterate**

            End

        End

    End

Initial G–S Iteration:

$$R := \frac{\text{DELNP}}{\text{DELNO}}; \; \textit{If } R \geq 1.0, \textit{ then Go to } \textbf{Next Iteration}; \; \textit{else } M'_{\text{E}} := \sqrt{R};$$

If $\dfrac{\text{DELNE}}{\text{YUN}} \leq \zeta(1 - R),$ *then* print final output and STOP $\langle$converged$\rangle$;

            else $p := -1$ and *Go to* **Next Iteration**.

Remark 1. A printout of the ratio $R^{(s, p)}$ $(=R)$ can be useful in appraising the effectiveness of the adaptive procedure. If $R^{(s, p)}$ is a converging sequence in p, the implication is that (8-3.31) is valid and, hence, that $M_E < \mu_1$. If $R^{(s, p)}$ oscillates, the implication is that (8-3.53) is valid and, hence, that $M_E \geq \mu_1$.

Remark 2. Another useful quantity to print is

$$\hat{C}^{(s, p)} = \frac{-\log R^{(s, p)}}{-\log(r(1 + r^{2p-2})/(1 + r^{2p}))}. \tag{8-3.57}$$

Using arguments similar to those given for $C^{(s, p)}$ in Remark 1 of Section 6.4, $\hat{C}^{(s, p)}$ may be given the following interpretation: If $M_E < \mu_1$, the rate of convergence obtained for iteration n using M_E is, at worst, only $\hat{C}^{(s, p)}$ times the optimum convergence rate for iteration n. If $M_E > \mu_1$, $\hat{C}^{(s, p)}$ will be greater than unity and an increasing function of p for those values of p such that the largest zero of $P_{2p, E}(x)$ is less than μ_1. Once p is large enough so that the largest zero of $P_{2p, E}(x)$ becomes larger than μ_1, $\hat{C}^{(s, p)}$ will normally oscillate in sign and will have no meaningful interpretation.

Also, analogous to Section 6.4, the parameter change test (8-3.36) will be satisfied only if $\hat{C}^{(s, p)} \leq F$. Thus as before, the damping factor F may be considered as a convergence criterion on the estimates M_E in the sense that M_E is considered sufficiently close to μ_1 if the convergence rate for iteration n using M_E is at least F times the optimum convergence rate.

We note that $\hat{C}^{(s, p)}$ reflects the convergence rate for iteration n, while $C^{(s, p)}$ of Section 6.4 reflects the average convergence rate.

Remark 3. Since μ_1 may be known for some problems, an option should be added to the procedure of Algorithm 8-3.1 such that the input estimate $M_E > 0$ is never changed. This can be easily done by avoiding the test on τ_1 and setting p^* to a large number.

Remark 4. To ensure that inequality (8-3.46) is satisfied, i.e., that $M_E < M_E' \leq \mu_1$, the quantity $\hat{B}$ used in (8-3.43) must be computed using the S_B-norm. This can be accomplished by computing DELNP $:= \|\Delta_B\|_{S_B}$ for all iterations or only for those iterations used to determine $\hat{B}$. The iteration when $p = 0$ and the Gauss–Seidel iteration determine $\hat{B}$, and both of these iterations are characterized by $\rho_R = 1.0$. Thus one way to modify Algorithm 3.1 such that inequality (8-3.46) is satisfied and such that the computational cost is not increased significantly is to compute DELNP $:= \|\Delta_B\|_{S_B}$ when $\rho_R = 1.0$ and DELNP $:= \|\Delta_R\|_2$ otherwise. For additional remarks concerning the calculation of $\|\Delta_B\|_{S_B}$, see the discussion given in Section 5.6 for the Jacobi method.

B.   The CCG Procedure

The cyclic conjugate gradient (CCG) procedure can be used only when the Jacobi conjugate gradient (J–CG) method (8-1.8) is such that $\gamma_n = 1$ for all n. With the Jacobi matrix B given by (8-1.5), we showed earlier in this section that this condition on γ_n is satisfied provided only that the initial guess vector $u^{(0)}$ be given by (8-3.1) and that the symmetrization matrix W be given by $W = \text{diag}(S_R, S_B)$. Here S_R and S_B are the Cholesky decomposition matrices for D_R and D_B given by (8-2.2).

In this section, we assume that these reasonable choices for $u^{(0)}$ and W are used. Thus the J–CG method can be given by (8-1.8) with $\gamma_{n+1} = 1$ and where $\rho_1 = 1$ and for $n \geq 2$

$$\rho_n = \left[1 - \frac{(W\delta^{(n-1)}, W\delta^{(n-1)})}{(W\delta^{(n-2)}, W\delta^{(n-2)})} \frac{1}{\rho_{n-1}} \right]^{-1}. \tag{8-3.58}$$

Here $\delta^{(n)}$ is defined by (7-4.8) and can be given in its red/black partitioned form as

$$\begin{bmatrix} \delta_R^{(n)} \\ \delta_B^{(n)} \end{bmatrix} = \begin{bmatrix} 0 & F_R \\ F_B & 0 \end{bmatrix} \begin{bmatrix} u_R^{(n)} \\ u_B^{(n)} \end{bmatrix} + \begin{bmatrix} c_R \\ c_B \end{bmatrix} - \begin{bmatrix} u_R^{(n)} \\ u_B^{(n)} \end{bmatrix}. \tag{8-3.59}$$

The CCG procedure is defined by (8-1.11) where, from (8-1.12) and (8-3.58), $\rho_R^{(1)} = \rho_1 = 1$ and for $n \geq 1$

$$\begin{aligned} \rho_B^{(n)} &= \rho_{2n} = \left[1 - \frac{(S_R \delta_R^{(2n-1)}, S_R \delta_R^{(2n-1)})}{(S_B \delta_B^{(2n-2)}, S_B \delta_B^{(2n-2)})} \frac{1}{\rho_R^{(n)}} \right]^{-1}, \\ \rho_R^{(n+1)} &= \rho_{2n+1} = \left[1 - \frac{(S_B \delta_B^{(2n)}, S_B \delta_B^{(2n)})}{(S_R \delta_R^{(2n-1)}, S_R \delta_R^{(2n-1)})} \frac{1}{\rho_B^{(n)}} \right]^{-1}. \end{aligned} \tag{8-3.60}$$

In writing (8-3.60), we used (8-3.3); i.e., we used the fact that $\delta_R^{(n)} = 0$ for n even and that $\delta_B^{(n)} = 0$ for n odd. The difficulty here is in determining the $\delta_R^{(2n-1)}$ and $\delta_B^{(2n)}$ vectors to use in (8-3.60). To compute $\delta_B^{(2n)}$ using (8-3.59) requires $u_R^{(2n)}$, which is not calculated in the CCG process (8-1.11). Similarly, $u_B^{(2n-1)}$ is not available for the calculation of $\delta_R^{(2n-1)}$. However, this difficulty can be circumvented by deriving recurrence formulas for $\delta_R^{(2n-1)}$ and $\delta_B^{(2n)}$.

Let

$$\sigma_B^{(0)} = \delta_B^{(0)}, \qquad \sigma_R^{(0)} = \delta_R^{(0)}(= 0)\dagger,$$

$$\sigma_R^{(n)} = \delta_R^{(2n-1)}, \sigma_B^{(n)} = \delta_B^{(2n)}, \qquad \text{for} \quad n \geq 1.$$

---

† Recall that $\delta_R^{(0)} = 0$ because of the choice (8-3.1) for $u^{(0)}$.

Using the three-term recurrence relation (8-3.4) for $\delta^{(n)}$ together with (8-3.3), we obtain

$$\sigma_R^{(n)} = \rho_R^{(n)} F_R \sigma_B^{(n-1)} + (1 - \rho_R^{(n)})\sigma_R^{(n-1)},$$
$$\sigma_B^{(n)} = \rho_B^{(n)} F_B \sigma_R^{(n)} + (1 - \rho_B^{(n)})\sigma_B^{(n-1)}. \tag{8-3.61}$$

Since $\sigma_R^{(0)} = 0$, it is clear that (8-3.61) can be used to obtain $\sigma_R^{(n)}$ and $\sigma_B^{(n)}$ for all $n \geq 1$ once $\sigma_B^{(0)}$ is determined. Thus with

$$\alpha_R^{(n)} \equiv (S_R \sigma_R^{(n)}, S_R \sigma_R^{(n)}), \qquad \alpha_B^{(n)} \equiv (S_B \sigma_B^{(n)}, S_B \sigma_B^{(n)}), \tag{8-3.62}$$

the acceleration parameters (8-3.60) for $n \geq 1$ may be given by

$$\rho_B^{(n)} = \left[1 - \frac{\alpha_R^{(n)}}{\alpha_B^{(n-1)}} \frac{1}{\rho_R^{(n)}}\right]^{-1}, \qquad \rho_R^{(n+1)} = \left[1 - \frac{\alpha_B^{(n)}}{\alpha_R^{(n)}} \frac{1}{\rho_B^{(n)}}\right]^{-1}. \tag{8-3.63}$$

The CCG process now can be carried out by using (8-1.11) with the ρ parameters given by (8-3.63). However, such a procedure would require matrix–vector multiplications involving F_R and F_B both in the calculation of $\{U_R^{(n)}, U_B^{(n)}\}$ and in the calculation of $\{\sigma_R^{(n)}, \sigma_B^{(n)}\}$. Fortunately, the matrix–vector multiplications in (8-1.11) can be avoided by using the following expression for $U_B^{(n)}$;

$$U_B^{(n)} = \rho_R^{(n)} \rho_B^{(n)} \sigma_B^{(n-1)} + \tilde{\rho}^{(n)}[U_B^{(n-1)} - U_B^{(n-2)}] + U_B^{(n-2)}, \tag{8-3.64}$$

where

$$\tilde{\rho}^{(n)} = 1 + (\rho_B^{(n)}/\rho_B^{(n-1)})(1 - \rho_R^{(n)})(1 - \rho_B^{(n-1)}).$$

To obtain (8-3.64), consider the equations for $U_B^{(n-1)}$, $U_R^{(n)}$, and $U_B^{(n)}$ defined by (8-1.11). These three equations involve $U_B^{(n-2)}$, $U_R^{(n-1)}$, $U_B^{(n-1)}$, $U_R^{(n)}$, and $U_B^{(n)}$. The equation for $U_B^{(n)}$ can be expressed in terms of only $U_B^{(n-1)}$ and $U_B^{(n-2)}$ by using the other two equations to eliminate $U_R^{(n-1)}$ and $U_R^{(n)}$. The expression (8-3.64) then follows by utilizing the additional fact that

$$\sigma_B^{(n-1)} = F_B F_R U_B^{(n-1)} + F_B c_R + c_B - U_B^{(n-1)}. \tag{8-3.65}$$

Thus the CCG procedure can be carried out by using (8-3.61) to obtain $\sigma_R^{(n)}$ and $\sigma_B^{(n)}$, using (8-3.63) to obtain the ρ parameters, and using (8-3.64) to obtain the iterates $U_B^{(n)}$. Once the iterates $U_B^{(n)}$ have converged, the subvector u_R can be obtained easily by using (8-1.3), i.e., using $u_R = F_R u_B + c_R$.

An Overall Computational Algorithm

An overall algorithm for the CCG method is given below as an informal program. The procedure used is based on Eqs. (8-3.61)–(8-3.64). However, in order to avoid the matrix–vector multiplications required in the calculation of $\alpha_R^{(n)}$ and $\alpha_B^{(n)}$, we compute $S_R \sigma_R^{(n)}$ and $S_B \sigma_B^{(n)}$ instead of $\sigma_R^{(n)}$ and $\sigma_B^{(n)}$; i.e., instead of (8-3.61) we use

$$S_R \sigma_R^{(n)} = \rho_R^{(n)} S_R F_R \sigma_B^{(n-1)} + (1 - \rho_R^{(n)}) S_R \sigma_R^{(n-1)},$$
$$S_B \sigma_B^{(n)} = \rho_B^{(n)} S_B F_B \sigma_R^{(n)} + (1 - \rho_B^{(n)}) S_B \sigma_B^{(n-1)}. \tag{8-3.66}$$

By doing this, the number of matrix–vector multiplications required for the CCG procedure is the same as that required for the basic Jacobi method (8-1.4).

For Algorithm 8-3.2, it is assumed that the red/black matrix problem to be solved is given by (8-1.2) and that, as in (8-1.3), $F_R = -D_R^{-1}H$, $F_B = -D_B^{-1}H^T$, $c_R = D_R^{-1}b_R$, and $c_B = D_B^{-1}b_B$. Further, it is assumed that D_R and D_B are factored as in (8-2.2). The underline, $\underline{U}$, is used to indicate more clearly which variables are vectors. The subscripts R and B are used to indicate the red/black subvector partitioning. The notation $S_B \underline{\sigma}_B$, for example, denotes a vector which is equal to $S_B \underline{\sigma}_B$.

Algorithm 8-3.2.   A procedure for the cyclic conjugate gradient (CCG) method.

Input: $(\zeta, \underline{U}_B^{[a]})$

Initialize:

$\underline{U}_R := F_R \underline{U}_B^{[a]} + c_R; \ S_R \underline{\sigma}_R := \underline{0}; \ \underline{U}_B^{[b]} := \underline{0}$
$S_B \underline{\sigma}_B := -(S_B^T)^{-1}(H^T \underline{U}_R + \underline{b}_B) - S_B \underline{U}_B^{[a]}; \ \alpha_B := (S_B \underline{\sigma}_B, S_B \underline{\sigma}_B);$
$\rho_B := 1.0; \ \rho_R := 1.0; \ n := 0.$

Next Iteration:

$n := n + 1; \ \text{ORB} := \rho_B.$

Calculate New Iterate:

$\underline{\sigma}_B := S_B^{-1}(S_B \underline{\sigma}_B); \ \text{DELNE} := \|\underline{\sigma}_B\|_\beta;$
$S_R \underline{\sigma}_R := -\rho_R (S_R^T)^{-1} H \underline{\sigma}_B + (1 - \rho_R) S_R \underline{\sigma}_R; \ \alpha_R := (S_R \underline{\sigma}_R, S_R \underline{\sigma}_R);$
$\rho_B := \left[1 - \dfrac{\alpha_R}{\alpha_B} \dfrac{1}{\rho_R} \right]^{-1}; \ \tilde{\rho} := 1 - \dfrac{\rho_B}{\text{ORB}} (1 - \rho_R)(1 - \text{ORB});$
$\underline{U}_B^{[b]} := \rho_R \rho_B \underline{\sigma}_B + \tilde{\rho}[\underline{U}_B^{[a]} - \underline{U}_B^{[b]}] + \underline{U}_B^{[b]}; \ \text{YUN} := \|\underline{U}_B^{[b]}\|_\eta;$
$\underline{\sigma}_R := S_R^{-1}(S_R \underline{\sigma}_R);$
$S_B \underline{\sigma}_B := -\rho_B (S_B^T)^{-1} H^T \underline{\sigma}_R + (1 - \rho_B) S_B \underline{\sigma}_B; \ \alpha_B := (S_B \underline{\sigma}_B, S_B \underline{\sigma}_B);$
$\rho_R := \left[1 - \dfrac{\alpha_B}{\alpha_R} \dfrac{1}{\rho_B} \right]^{-1};$

$c := a; \ a := b; \ b := c.$ (Relabeling to interchange $\underline{U}^{(a)}$ and $\underline{U}^{(b)}$)

Stopping Test:

Compute M_E (if needed) from the T_n matrix using bisection (see Section 7.5).
If $n \geq 2$, then
 Begin
  If $\dfrac{\text{DELNE}}{\text{YUN}} \leq \zeta(1 - M_E)$, *then*

   Begin
   $\underline{u}_R := F_R \underline{U}_B^{[a]} + \underline{c}_R$;
   Print output and STOP (converged);
   End
   else continue;

 End
 else continue;
Go to **Next Iteration**.

8.4 NUMERICAL RESULTS

In this section we describe results of numerical experiments which were designed to illustrate the effectiveness of the iterative solution procedures given in this chapter. We discuss the behavior of only the CCSI and the RS–CG acceleration procedures. Results for the CCG and the RS–SI methods are not included here since the behavior of these methods can be expected to be similar to that of the RS–CG and CCSI methods, respectively. In fact, as discussed previously in Section 8.1, the u_B subvector iterates of the CCG and the RS–CG methods are identical provided that $u_R^{(0)}$ for the CCG method is given by (8-3.1). The u_B subvector iterates of the CCSI procedure of Algorithm 8-3.1 will differ slightly from the u_B iterates of the RS–SI procedure only because we choose to do a Gauss–Seidel iteration instead of storing the δ_B vector whenever a new estimate M_E' is computed.

Description of Test Problems

The test problems we consider arise from the finite-difference discretization of the elliptic differential equation

$$\frac{\partial}{\partial x}\left(C\frac{\partial \mathcal{U}}{\partial x}\right) + \frac{\partial}{\partial y}\left(C\frac{\partial \mathcal{U}}{\partial y}\right) + F\mathcal{U} = G \qquad (8\text{-}4.1)$$

defined in a rectangular region R and subject to the condition $\mathcal{U}\,(\partial \mathcal{U}/\partial n) = 0$ on the boundary of R. Here $\partial \mathcal{U}/\partial n$ is the outward normal derivative of $\mathcal{U}$ on the boundary. The material coefficients $C > 0$ and $F \leq 0$ are known piece-wise smooth functions of x and y. This problem with $\mathcal{U} = 0$ on the boundaries was discussed previously in Section 1.6. The geometric domain and the mesh subdivisions are given in Fig. 8-4.1.

The finite-difference discretization used is the normal 5-point formula. (See, for example, Varga [1962].) The resulting system of equations is written in matrix form as

$$Au = b. \tag{8-4.2}$$

If F is not zero everywhere or if $\mathcal{U} = 0$ on some part of the boundary, then A is SPD.

For the numerical results given in this chapter and later in Chapter 9, we use a (horizontal) line partitioning (see Sections 1.6 and 1.7) for the unknowns and for the matrix A. Moreover, in this chapter, we always use a red/black ordering of the lines. From the discussion given in Section 1.7, it follows that such an ordering of the unknowns produces a red/black partitioning (8-1.1) for A, where D_R and D_B are block diagonal matrices. Each diagonal sub-matrix $A_{i,i}$ of D_R and D_B is a tridiagonal matrix whose order is either $N - 2$, $N - 1$, or N depending on the boundary conditions imposed on the left and right boundaries.

The specific problems we consider are defined by the data given in Table 8-4.1. The initial guess vectors used are defined in Table 8-4.2. The quantities $I, J, N,$ and M are defined in Fig. 8-4.1.

Results of Numerical Experiments

For the CCSI method, we used Algorithm 8-3.1 with $d = 0.1, F = 0.7$, and an initial guess of 0.0 for M_E. For the RS–CG method, we used Algorithm 7-6.2 as discussed in Section 8.2. For both methods, the ∞-relative norm was

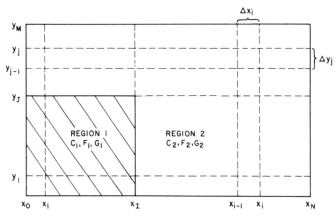

Fig. 8-4.1. Geometric domain.

TABLE 8-4.1

Data for Test Problems

Problem	Boundary conditions				Material data		Mesh data			
	Top	Bottom	Left	Right	Region 1	Region 2	$M = N$	$\Delta x = \Delta y$	I	J
1	$\mathscr{U} = 0$	$\mathscr{U} = 0$	$\mathscr{U} = 0$	$\mathscr{U} = 0$	$C_1 = 1.0$ $F_1 = 0.0$ $G_1 = -9.0$	$C_2 = 1.0$ $F_2 = 0.0$ $G_2 = -1.0$	42	1.0	21	21
2	$\dfrac{\partial \mathscr{U}}{\partial y} = 0$	$\mathscr{U} = 0$	$\dfrac{\partial \mathscr{U}}{\partial x} = 0$	$\mathscr{U} = 0$	$C_1 = 1.0$ $F_1 = 0.0$ $G_1 = -1.0$	$C_2 = 1.0$ $F_2 = 0.0$ $G_2 = -1.0$	42	1.0	21	21
3	$\dfrac{\partial \mathscr{U}}{\partial y} = 0$	$\dfrac{\partial \mathscr{U}}{\partial y} = 0$	$\dfrac{\partial \mathscr{U}}{\partial x} = 0$	$\dfrac{\partial \mathscr{U}}{\partial x} = 0$	$C_1 = 1.0$ $F_1 = -0.1$ $G_1 = -0.1$	$C_2 = 1.0$ $F_2 = -0.1$ $G_2 = -0.1$	42	0.08	14	18
4	$\mathscr{U} = 0$	$\mathscr{U} = 0$	$\mathscr{U} = 0$	$\mathscr{U} = 0$	$C_1 = 0.10$ $F_1 = -1.0$ $G_1 = 0.0$	$C_2 = 0.01$ $F_2 = -1.0$ $G_2 = -1.0$	101	0.01	51	51

TABLE 8-4.2

Initial Guess Vectors

Guess A	Guess B		Guess C
	$u^{(0)} = 0$ except		
	$u^{(0)} = 5$ for $\begin{cases} x_{I+1} \le x \le x_N \\ y_1 \le y \le y_J \end{cases}$		$u^{(0)}$ is a vector whose
$u^{(0)} = 0$			elements are random
			numbers between 0 and 1
	$u^{(0)} = 10$ for $\begin{cases} x_1 \le x \le x_I \\ y_{J+1} \le y \le y_M \end{cases}$		

used in the stopping test. With $\Delta_i^{(n)} = (\Delta_B^{(n)})_i$, the CCSI method was considered converged (see Eq. (8-3.35)) when

$$E_A \equiv \frac{1}{1 - H} \max_i \left| \frac{\Delta_i^{(n)}}{u_i^{(n+1)}} \right| \le \zeta, \tag{8-4.3}$$

where H is defined as in Algorithm 8-3.1. With $\delta_i^{(n)} = (\delta_B^{(n)})_i$, the RS–CG method was considered converged (see Eq. (7-5.2)) when

$$E_A \equiv \frac{1}{1 - \mu_1^2} \max_i \left| \frac{\delta_i^{(n)}}{u_i^{(n+1)}} \right| \le \zeta. \tag{8-4.4}$$

TABLE 8-4.3

Summaries of the Iterative Behavior for the RS–CG and the CCSI Methods for Stopping Criterion of $\zeta = 10^{-5}$

Problem	RS–CG Iter. to converge	Nonadapt. CCSI Iter. to converge	Adaptive CCSI		
			Iter. to converge	Last est. used for μ_1	$\hat{C}^{(s, p)}$ at convergence
1 (Guess A)	47	67 (0.994149)	80	0.994134	0.975
2 (Guess A)	59	119 (0.998533)	140	0.998525	0.941
3 (Guess B)	89	323	344	0.999680	0.996
(Guess C)	78	256 (0.999680)	293	0.999673	0.853
4 (Guess A)	92	125	163	0.998170	1.018
(Guess C)	104	144 (0.998167)	154	0.998241	1.550

The iterative behavior of the CCSI and RS–CG methods for the problems of Table 8-4.1 is summarized in Table 8-4.3. Both adaptive and optimal nonadaptive procedures were used for the CCSI method. For the nonadaptive CCSI procedure, the eigenvalue estimate used for μ_1 is given in parentheses. A stopping criterion of $\zeta = 10^{-5}$ was used for all problems. The quantity $\hat{C}^{(s,\,p)}$ is defined by (8-3.57). Note that this quantity accurately reflected the ratio of actual to optimum convergence rates and was indicative of how closely the last estimate M_E approximated μ_1. For problem 4, $\hat{C}^{(s,\,p)}$ was greater than unity and monotonically increasing with p when convergence occurred. This implied that the last estimate M_E was greater than μ_1 and that $\mu_1 > x^*$, where x^* is the largest zero of $P_{2p,\,E}(x)$ (see discussion following (8-3.57)). For problem 1, it is known (e.g., see Varga [1962]) that $\mu_1 = [\cos \pi/41]/[2 - \cos \pi/41] = 0.994149$.

The typical behavior of $R^{(n)}$, defined by (8-3.32), for the adaptive CCSI procedure is illustrated in Fig. 8-4.2. There, the graph of $R^{(n)}$ versus n is given for problem 3 using vector guess B. The eigenvalue estimates which were used are also given. Gauss–Seidel iterations were done on iterations 4, 9, 14, 20, 28, 40, 60, 75, and 113.

In Table 8-4.4, we give for problem 3 the number of iterations required for convergence when determined by the approximate error measure E_A defined by (8-4.3)–(8-4.4) and when determined by the actual error measure E_T, where

$$E_T \equiv \max_i |(u_i^{(n)} - \bar{u}_i)/\bar{u}_i| \leq \zeta. \tag{8-4.5}$$

Here $\bar{u}_B = (\bar{u}_i)$ is the unique solution to (8-1.14). We remark that for problem 3, $\bar{u}_B$ is known and is unity everywhere. As was true for the simulated iteration results given in Chapters 5 and 7, the approximate error measure E_A closely approximates the true error for the adaptive Chebyshev procedure, but considerably overestimates the true error for the nonadaptive Chebyshev and the conjugate gradient procedures.

The behavior of the actual and estimated errors for the CCSI and RS–CG methods is more clearly indicated by the graphs given in Figs. 8-4.3–8-4.5. In these figures, graphs of $\log_{10} E_A$ and $\log_{10} E_T$ versus n are given for problem 3 using Guess C. Due mainly to oscillations in $R^{(n)}$, the estimated error for the nonadaptive CCSI procedure fluctuated considerably. To simplify the graph in Fig. 8-4.4, the estimated error is not given for all iterations. Actual error estimates are indicated by an x. Any iteration for which an x is not given, the estimated error for that iteration is above the dotted line.

As indicated in Fig. 8-4.5, the use of the difference vector, $\Delta_B^{(n)} \equiv u_B^{(n)} - u_B^{(n-1)}$ can give a reasonable lower bound for the RS–CG iteration error. The difficulty with using $\Delta_B^{(n)}$, however, is in determining how much $\|\Delta_B^{(n)}\|$ underestimates $\|\varepsilon_B^{(n)}\|$. A stopping test which would more accurately measure the

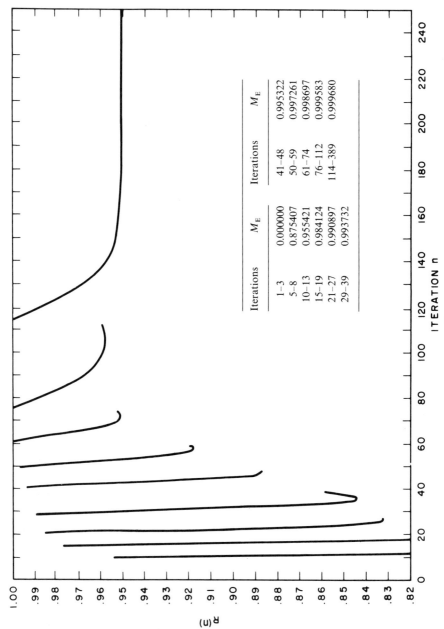

Iterations	M_E	Iterations	M_E
1–3	0.000000	41–48	0.995322
5–8	0.875407	50–59	0.997261
10–13	0.955421	61–74	0.998697
15–19	0.984124	76–112	0.999583
21–27	0.990897	114–389	0.999680
29–39	0.993732		

Fig. 8-4.2. Graph of $R^{(n)}$ versus n for problem 3.

TABLE 8-4.4

*Number of Iterations[a] Required by the CCSI and RS–CG Methods
to Satisfy $E_A \leq \zeta$ and $E_T \leq \zeta$*

Method	$\zeta = 0.1$		$\zeta = 0.01$		$\zeta = 10^{-5}$	
	$n(E_T)$	$n(E_A)$	$n(E_T)$	$n(E_A)$	$n(E_T)$	$n(E_A)$
RS–CG	28	51	40	56	68	78
CCSI (adaptive)	90	92	138	140	292	293
CCSI (nonadaptive)	54	70	98	120	235	256

[a] Problem 3 using Guess C.

iteration error vector for the optimal nonadaptive CCSI procedure possibly could be developed using the relationship (5-7.24). However, difficulties could arise in the use of such a test if the nonadaptive procedure were not optimal; i.e., if the fixed estimate M_E used were not sufficiently close to μ_1.

The problems of Table 8-4.1 were also solved using $\|\Delta_B^{(s,\,p)}\|_{S_B}$ instead† of $\|\Delta_B^{(s,\,p)}\|_2$ in the CCSI procedure of Algorithm 8-3.1. It was found that the iterative behavior of the CCSI iterations is basically independent of whether the 2- or S_B-norm is used to compute $\|\Delta_B^{(s,\,p)}\|$. For the problems considered, the ratio of iterations required for convergence for the two cases varied from 0.95 to 1.05.

Except for problem 4, the number of unknowns N for the problems considered here is small. The iterative behavior of the CCSI method depends strongly on the value‡ of $S(B)$ and only weakly on the size of the matrix problem. However, from Chapter 7, the iterative behavior of the RS–CG method depends on the distribution of all eigenvalues of $F_B F_R$ and on the size N of the matrix problem when N is small. Thus if, N is increased while keeping $S(B)$ fixed, the iterations required for the RS–CG method probably would increase, while the iterations required for the CCSI method probably would remain the same. For the test problems considered here, however, there is no reason to believe that the ratio of RS–CG iterations to CCSI iterations would change significantly with increased N. This follows since, for

† The matrix S_B is defined by (8-2.2).
‡ Roughly, the number of iterations required by the CCSI method is proportional to $1/\sqrt{1 - S(B)}$.

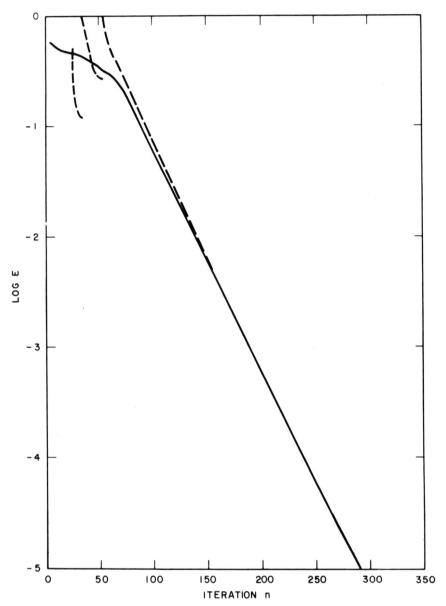

Fig 8-4.3.   Graphs of log E_T and log E_A versus n for problem 3, using the CCSI adaptive procedure. ——, true error E_T; -----, estimated error E_A.

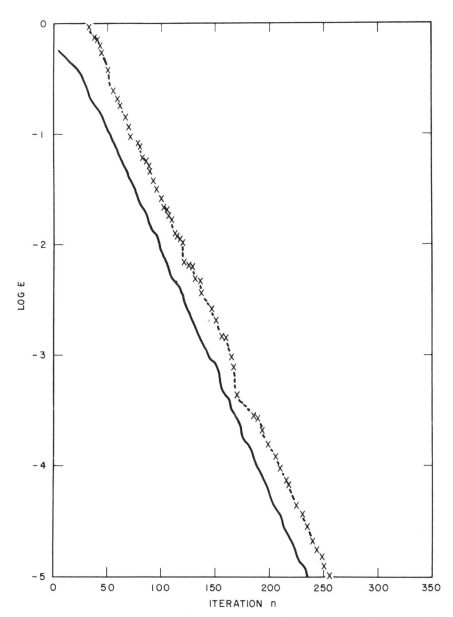

Fig 8-4.4. Graphs of log E_T and log E_A versus n for problem 3, using the CCSI nonadaptive procedure. ——, true error E_T; - - - -, estimated error E_A.

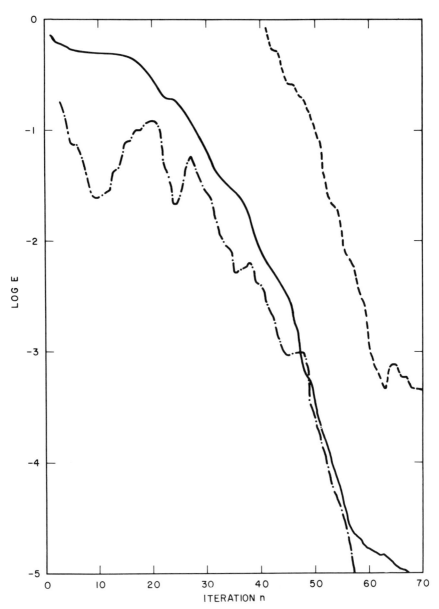

Fig. 8-4.5.   Graphs of log E_T and log E_A versus n for problem 3 for the RS–CG method. ——, true error E_T; -----, estimated error E_A; —·—, $\|u_B^{(n+1)} - u_B^{(n)}\|_{RE}$.

TABLE 8-4.5

Number of Iterations Required by the CCSI and RS–CG Methods for Problem 1^a

| | $\zeta = 10^{-2}$ | | | $\zeta = 10^{-4}$ | | |
| | | CCSI | | | CCSI | |
Problem	RS–CG	Nonadapt.	Adapt.	RS–CG	Nonadapt.	Adapt.
1 (Guess A) $M = 41$ $N = 41$	31	34	47	44	54	68
1 (Guess A) $M = 82$ $N = 82$	64	65	96	92	108	141
1 (Guess A) $M = 164$ $N = 164$	126	132	168	167	226	276

^{a} With Mesh Grids of 41×41, 82×82 and 164×164.

the problems of Table 8-4.1, any increase in N generally causes an accompanying increase in $\mathbf{S}(B)$. To illustrate this, problem 1 was rerun with the mesh increments halved and then halved again; i.e., problem 1 was rerun first with a 82×82 mesh grid and then with a 164×164 mesh grid. The results are given in Table 8-4.5. Note that halving the mesh increments roughly doubled the number of iterations required by both the CCSI and RS–CG methods.

8.5 ARITHMETIC AND STORAGE REQUIREMENTS

Thus far we have compared the RS–CG and CCSI methods relative to the number of iterations required for convergence. However, the cost per iteration also must be considered in the evaluation of any solution method. This we now do for the test problems considered in this section. We assume the matrix problem to be solved is of order N and is expressed in the form (8-1.2). We also assume that each of the D_R and D_B submatrices is of order $N/2$ and that each has been factored as in (8-2.2).

Both methods have the following common requirements:†

storage: H, S_R, S_B, b_R, b_B {$5N$ words}

arithmetic: Matrix–vector multiplications by $D_R^{-1}H$ and $D_B^{-1}H^T$ (8-5.1)
             {$(\div; 2N), (*; 4N), (+; 3N)$}.

† See Section 10.2.

TABLE 8-5.1

Overhead Arithmetic and Storage Requirements for the CCSI
and RS–CG Methods

	Storage	Arithmetic	
Method	Vectors (order $N/2$)	$*$	$+$
CCSI	$\underline{u}_R, \underline{u}_B$	$1\frac{1}{2}N$	$3\frac{1}{2}N$
RS–CG	$\underline{Wu}, \underline{Wu}_\phi$ $\underline{W\delta}, \underline{W\delta}_\phi, \underline{W\theta}$	$3N$	$6N$

The quantities enclosed in parenthesis are the requirements for the 5-point discretization of the elliptic problem (8-4.1). Beyond these common requirements, additional arithmetic and storage are needed to carry out the respective iteration procedures. Based on Algorithms 7-6.2 and 8-3.1, these additional (or overhead) requirements are summarized in Table 8-5.1. The notation used in Table 8-5.1, and in what follows is defined in the appropriate algorithm. For the CCSI method, we assume that only temporary storage is required for the $(\underline{\Delta})_i$ subvectors.† For the RS–CG method, we assume that only one common vector storage array is needed for the storage of the $\underline{\delta}$, $\underline{\delta}_R$, and $\underline{W\theta}$ vectors. The overhead arithmetic for the CCSI method includes the multiplications by the acceleration parameters, the calculation of $\|\underline{\Delta}_B\|_2$, and the additions required for the source vectors $\underline{b}_R$ and $\underline{b}_B$. For the RS–CG method, the overhead arithmetic includes the calculations required to obtain the $\underline{Wu}_\Gamma$ and $\underline{W\delta}_\Gamma$ vectors and $\|\underline{Wu}_\Gamma\|_2^2$ and $\|\underline{W\delta}_\Gamma\|_2^2$. The calculations required for the stopping quantities DELNE and YUN are not included. The arithmetic requirements for the RS–CG method given in Table 8-5.1 minimize the number of multiplications required. Other combinations of the $*$ and $+$ operations are possible. We remark that the RS–CG procedure also may be carried out, with roughly the same overhead requirements, using the classical conjugate gradient formulation (7-4.5) (see, e.g., Chandra [1978].)

Relative to the CCSI method, the RS–CG method requires $4N$ more arithmetic operations per iteration and requires the storage of some extra vectors. The importance of these differences can be measured only in relationship to the common requirements (8-5.1). For example, if the problem size is such that disk bulk storage is needed, then the additional storage requirements and computational complexity of the RS–CG method must be considered.‡

---

† See the discussion on the Jacobi method given in Section 5.6.

‡ Difficulties connected with data flow are discussed in Section 10.2.

For the problem here for which the 5-point discretization formula is used, the CCSI method requires $7\frac{1}{2}N$ multiplications† and $6\frac{1}{2}N$ additions per iteration, while the RS–CG method requires $9N$ multiplications and $9N$ additions. If the calculation time for a multiplication is 1.25 times that for an addition, then the arithmetic work per iteration for the RS–CG method would be approximately 1.25 times that required for the CCSI method. The extra cost for the RS–CG method becomes less important as the amount of common arithmetic increases. For example, the ratio of arithmetic work corresponding to a 9-point discretization formula would be considerably less than the value 1.25 given above for the 5-point formula.

For both methods, the normalization procedure given by Cuthill and Varga [1959] may be used to eliminate the $2N$ divisions (or the related multiplications) in (8-5.1). To describe this normalization procedure, which is carried out prior to the start of the iterative process, we first define the diagonal matrices Λ_R and Λ_B by

$$\Lambda_R \equiv \text{diag}(S_R), \qquad \Lambda_B \equiv \text{diag}(S_B), \tag{8-5.2}$$

where S_R and S_B are defined by (8-2.2). Letting

$$\tilde{u}_R = \Lambda_R u_R, \qquad \tilde{u}_B = \Lambda_B u_B,$$
$$\tilde{b}_R = \Lambda_R b_R, \qquad \tilde{b}_B = \Lambda_B u_B, \tag{8-5.3}$$

we can express the matrix equation (8-1.2) in the form

$$\begin{bmatrix} \tilde{S}_R^T \tilde{S}_R & \tilde{H} \\ \tilde{H}^T & \tilde{S}_B^T \tilde{S}_B \end{bmatrix} \begin{bmatrix} \tilde{u}_R \\ \tilde{u}_B \end{bmatrix} = \begin{bmatrix} \tilde{b}_R \\ \tilde{b}_B \end{bmatrix}, \tag{8-5.4}$$

where $\tilde{S}_R \equiv S_R \Lambda_R^{-1}$, $\tilde{S}_B \equiv S_B \Lambda_B^{-1}$, and $\tilde{H} = \Lambda_R^{-1} H \Lambda_B^{-1}$. The normalized system (8-5.4) now can be solved using either the CCSI or the RS–CG method. Note that the $\tilde{S}_R$ and $\tilde{S}_B$ matrices have unit diagonal elements, which eliminates $2N$ divisions (or multiplications) in the iterative solution procedure. Once convergence has been achieved, u_R and u_B may be obtained easily from $\tilde{u}_R$ and $\tilde{u}_B$ by using (8-5.3).

8.6 COMBINED (HYBRID) CHEBYSHEV AND CONJUGATE GRADIENT ITERATIONS

A combined Chebyshev–conjugate gradient procedure has been suggested by some (see, e.g., Engeli *et al.* [1959] or Axelsson [1977]) as a possible way to reduce computational cost.

---

† We assume the $2N$ divisions in (8-5.1) are replaced by $2N$ multiplications. This can be done by replacing each division (α/s_{ii}) by the multiplication $(\alpha\beta_{ii})$, where $\beta_{ii} = 1/s_{ii}$ and each such β_{ii} is computed prior to the start of the iterative process.

To illustrate the idea involved, consider the solution of the matrix problem (8-1.3) by the RS–CG method. Let $u_B^{(0)}$ be some initial guess vector for the black unknowns and let $\varepsilon_B^{(0)}$ be the associated initial error vector. As in (8-3.27), $\varepsilon_B^{(0)}$ may be expanded in terms of the eigenvectors of $F_B F_R$ as

$$\varepsilon_B^{(0)} = c_1 v_B(1) + c_2 v_B(2) + \cdots + c_{\tilde{N}} v_B(\tilde{N}), \tag{8-6.1}$$

where $\tilde{N}$ is the order of $F_B F_R$. As before, we assume the eigenvector $v_B(i)$ is associated with the eigenvalue λ_i of $F_B F_R$ and that the λ_i are ordered as in (8-3.26), i.e., $0 \le \lambda_{\tilde{N}} \le \cdots \le \lambda_1 = (\mathbf{S}(B))^2$. If $c_i = 0$ for $i \ge l^*$, the RS–CG method (in the absence of rounding errors) will converge in at most l^* iterations.† Thus if a guess vector $u_B^{(0)}$ can be found such that l^* is small, the RS–CG method will converge quickly. The strategy behind the combined procedure is to generate such a guess vector for the conjugate gradient method using the less costly Chebyshev iterations.

For the combined procedure applied to the matrix problem (8-1.3), one first iterates using the CCSI method with $M_E < \mathbf{S}(B)$. After a sufficient number of Chebyshev iterations, the RS–CG iterations are started. Since $M_E < \mathbf{S}(B)$, it follows from (8-3.24), (8-3.27), and (8-3.29) that the error vector of the last CCSI iteration will be dominated by those eigenvectors of $F_B F_R$ associated with eigenvalues $\lambda_i > M_E^2$. Thus the initial error vector for the RS–CG method approximately satisfies

$$\varepsilon_B^{(0)} \doteq c_1 v_B(1) + \cdots + c_{l^*} v_B(l^*), \tag{8-6.2}$$

where $\lambda_{l^*} > M_E^2 \ge \lambda_{l^*+1}$. If l^* is small, only a few RS–CG iterations should be required for convergence.

The data given in Table 8-6.1 illustrates that, indeed, a "smooth"‡ initial guess vector does enhance convergence of the conjugate gradient method. There, we give the number of RS–CG iterations required to solve problem 4 of Section 8.4 using guess vectors A, C, and D. Guess vectors A and C are defined by Table 8-4.2, while D was generated by starting with guess A and then doing 49 Gauss–Seidel iterations before starting the RS–CG procedure. Note that for guess D, the eigenvector coefficient c_i in (8-6.1) associated with the eigenvalue λ_i is $(\lambda_i)^{49}$ times the corresponding c_i associated with guess A. Thus guess D is "smoother" than A. Since the elements of guess C were generated randomly, this guess vector is not likely to be smooth.

The combined Gauss–Seidel, RS–CG procedure is a polynomial method. Hence because of the polynomial minimization property of the conjugate

---

† This follows from the polynomial minimization property of the conjugate gradient method. See Chapter 7.

‡ Here, the term "smooth" implies any guess vector whose associated error vector expansion (8-6.1) is dominated by those eigenvectors associated with the larger eigenvalues of the iteration matrix.

TABLE 8-6.1

*Iterations Required for Different Guess
Vectors*

	RS–CG method iterations	
Problem 4	$\zeta = 10^{-2}$	$\zeta = 10^{-5}$
Guess D	34	69
Guess A	58	92
Guess C	76	104

gradient method, the total number of iterations required by the combined procedure is not likely to be less than that for the RS–CG procedure alone. However, since the cost of each Gauss–Seidel iteration is only 0.6 times that for a RS–CG iteration, the computational cost of the combined procedure could be less. This is not the case for the example given above. Here, the cost for the combined procedure with $\zeta = 10^{-5}$ roughly was equivalent to 99 RS–CG iterations, which is slightly greater than the 92 iterations required by the RS–CG method alone. The cost effectiveness of the combined procedure was improved only slightly when the guess vector for the RS–CG method was generated using the CCSI, instead of the Gauss–Seidel, method. These disappointing results probably are due to the fact that some of the eigenvector terms associated with the smaller eigenvalues in (8-6.1) may be amplified significantly by the conjugate gradient process. The data given below illustrates this behavior.

With $\varepsilon_B^{(0)}$ given by (8-6.1), it follows from (3-2.5) and Theorem 3-2.1 that the error vector for the RS–CG method may be written as

$$\varepsilon_B^{(n)} = Q_n(F_B F_R)\varepsilon_B^{(0)} = c_1 Q_n(\lambda_1)v_B(1) + c_2 Q_n(\lambda_2)v_B(2)$$
$$+ \cdots + c_{\tilde{N}} Q_n(\lambda_{\tilde{N}})v_B(\tilde{N}), \qquad (8\text{-}6.3)$$

where the polynomial $Q_n(x)$ is defined by the recurrence relation (3-2.6), with the parameters ρ_{n+1} and γ_{n+1} determined by the RS–CG procedure. Unlike the corresponding polynomial for the CCSI procedure, $\max|Q_n(x)|$ for $\lambda_{\tilde{N}} \leq x \leq \lambda_1$ need not be bounded by unity. The data of Table 8-6.2 gives some indication of the behavior of $Q_n(x)$ for the three polynomial sequences $\{Q_n(x)\}$ corresponding to the guess vectors A, C, and D for problem 4. In Table 8-6.2, we give $Q_n(x)$ evaluated at selected values of x. The value $x = 0.99634$ corresponds to $\lambda_1 = \mathbf{S}(F_B F_R)$. For guess D, note the exceptionally large values of $Q_n(x)$ at $x = 0.05$. Thus if $v_B(i^*)$ is an eigenvector of $F_B F_R$ with eigenvalue $\lambda_{i^*} \doteq 0.05$, the contribution of $v_B(i^*)$ to $\varepsilon_B^{(n)}$ can become significant

TABLE 8-6.2

Values of the Polynomial $Q_n(x)$ for the RS–CG Method

Problem 4	x	$Q_1(x)$	$Q_{11}(x)$	$Q_{21}(x)$	$Q_{31}(x)$	$Q_{61}(x)$
Guess D	0.99634	0.74×10^0	0.62×10^{-2}	0.43×10^{-4}	-0.38×10^{-4}	-0.13×10^{-4}
	0.9	-0.61×10^1	-0.73×10^0	-0.17×10^0	0.21×10^{-1}	0.23×10^{-4}
	0.05	-0.67×10^2	0.11×10^{12}	-0.88×10^{10}	-0.33×10^9	-0.41×10^6
Guess A	0.9	-0.18×10^1	-0.11×10^0	-0.48×10^{-1}	-0.16×10^{-1}	0.25×10^{-3}
	0.1	-0.25×10^2	-0.26×10^1	-0.22×10^1	-0.57×10^0	-0.30×10^{-3}
Guess C	0.9	0.89×10^0	0.89×10^{-1}	0.35×10^{-1}	0.29×10^{-2}	-0.11×10^{-4}
	0.1	0.71×10^{-2}	0.14×10^{-3}	-0.35×10^{-3}	0.20×10^{-3}	0.63×10^{-6}

even though c_{i*} is very small. The calculations were performed on a computer with a 13-digit mantissa.

Other types of combined or hybrid procedures also have been used. For example, O'Leary [1975] used a combined conjugate gradient–SOR procedure in which the initial conjugate gradient iterations were used as a means to obtain an estimate for the optimal SOR acceleration parameter ω_b. The estimates for ω_b were based on the estimates for $\mathbf{S}(B)$ which can be obtained at little cost from the conjugate gradient iteration data (see Section 7.5).

8.7 PROOFS

In this section we prove Theorem 3.1 and derive the relationships (8-3.33)–(8-3.34) given previously in Section 8.3.

Proof of Theorem 8-3.1

With the eigenvalues of $F_B F_R$ given by (8-3.26), it follows from Definition 6-2.1 that the sequence $\{R_{p,\,E}(y)\}$ is $F_B F_R$-uniformly convergent if $R_{p,\,E}(\lambda_1) \neq 0$ and if

$$\lim_{p \to \infty} \left[|R_{p,\,E}(y)| / |R_{p,\,E}(\lambda_1)| \right] = 0 \qquad \text{for all} \quad y \in [\lambda_{\tilde{N}}, \lambda_t].$$

From (8-2.4) and (8-3.11),

$$R_{p,\,E}(\lambda_1) = R_{p,\,E}(\mu_1^2) = P_{2p,\,E}(\mu_1),$$

where $P_{2p,\,E}(x)$ is given by (5-2.7) with $-m_E = M_E$. Thus $R_{p,\,E}(\lambda_1) > 0$ since $M_E < \mu_1$ implies that $P_{2p,\,E}(\mu_1) > 0$. As the nonzero eigenvalues of $F_B F_R$ are just the squares of the nonzero eigenvalues of B, it follows from (8-1.6) that the α convergence factors (6-2.8) associated with the polynomials $R_{p,\,E}(y)$ can be given by

$$\alpha(2p) \equiv \max_{\lambda_{\tilde{N}} \leq y \leq \lambda_t} \left| \frac{R_{p,\,E}(y)}{R_{p,\,E}(\lambda_1)} \right|$$

$$= \max_{0 \leq x \leq \mu_t} \left| \frac{P_{2p,\,E}(x)}{P_{2p,\,E}(\mu_1)} \right| = \frac{\max_{0 \leq x \leq \mu_t} |T_{2p}(w(x))|}{T_{2p}(w(\mu_1))}, \qquad (8\text{-}7.1)$$

where $w(x) = x/M_E$. Using arguments similar to those given in the proof of Theorem 6-2.4, we can show that

$$\alpha(2p) \leq \frac{T_{2p}(w^*)}{T_{2p}(w(\mu_1))} \qquad \text{and} \qquad \lim_{p \to \infty} \alpha(2p) = 0, \qquad (8\text{-}7.2)$$

where $w^* = 1$ if $\mu_t < M_E$ and $w^* = w(\mu_t)$ if $\mu_t \geq M_E$. Thus the sequence $\{R_{p,E}(y)\}$ is $F_B F_R$-uniformly convergent.

To prove the desired result for the sequence $\{Q_{p,E}(y)\}$, we first show that the polynomial $Q_{p+1,E}(y)$ may be expressed as

$$Q_{p+1,E}(x^2) = P_{2p,E}(x) - P_{2p+2,E}(x). \qquad (8\text{-}7.3)$$

From the three-term recurrence relation (4-2.1) and the definition of $P_{2p,E}(x)$, it follows that

$$T_{2p+2}\left(\frac{1}{M_E}\right)P_{2p+2,E}(x) = 2\left(\frac{x}{M_E}\right)T_{2p+1}\left(\frac{1}{M_E}\right)P_{2p+1,E}(x)$$
$$- T_{2p}\left(\frac{1}{M_E}\right)P_{2p,E}(x). \qquad (8\text{-}7.4)$$

Moreover, from (8-1.12) and (4-3.4),

$$\rho_B^{(p+1)} = \frac{2}{M_E}\frac{T_{2p+1}(1/M_E)}{T_{2p+2}(1/M_E)}.$$

Substituting for $\rho_B^{(p+1)}[xP_{2p+1,E}(x)]$ in (8-3.23) by using the above relations, we may express $Q_{p+1,E}(x^2)$ in the form

$$Q_{p+1,E}(x^2) = \left(\rho_B^{(p+1)} - \frac{T_{2p}(1/M_E)}{T_{2p+2}(1/M_E)}\right)P_{2p,E}(x) - P_{2p+2,E}(x).$$

Equation (8-7.3) now follows since, from (4-3.4) and (4-2.1), $\rho_B^{(p+1)}$ also satisfies

$$\rho_B^{(p+1)} = 1 + \frac{T_{2p}(1/M_E)}{T_{2p+2}(1/M_E)}.$$

Thus from (8-7.3), we have that

$$Q_{p+1,E}(\lambda_1) = Q_{p+1,E}(\mu_1^2) = P_{2p,E}(\mu_1)\left[1 - \frac{P_{2p+2,E}(\mu_1)}{P_{2p,E}(\mu_1)}\right] > 0 \quad (8\text{-}7.5)$$

and that

$$\max_{\lambda_{\tilde{N}} \leq y \leq \lambda_t}\left|\frac{Q_{p+1,E}(y)}{Q_{p+1,E}(\mu_1)}\right| \leq \max_{0 \leq x \leq \mu_t}\left[\frac{|P_{2p,E}(x)| + |P_{2p+2,E}(x)|}{[1 - \mu_1]P_{2p,E}(\mu_1)}\right] \leq \frac{2}{1 - \mu_1}\alpha(2p), \qquad (8\text{-}7.6)$$

where $\alpha(2p)$ is given by (8-7.1). (In (8-7.6), we used the fact that $|P_{2p+2,E}(\mu_1)|/|P_{2p,E}(\mu_1)| \leq \mu_1$ provided $0 < M_E < \mu_1$.) Hence it follows from (8-7.2) that the sequence $\{Q_{p,E}(y)\}$ is also $F_B F_R$-uniformly convergent. ∎

Equation (8-3.33)

If $M_E < \mu_1$ and if p is sufficiently large, from (8-3.31) and (8-7.3), we have

$$\frac{\|\Delta_B^{(s,p)}\|}{\|\Delta_B^{(s,p-1)}\|} \doteq \frac{P_{2p,E}(\mu_1)}{P_{2n-2,F}(\mu_1)} \frac{\Lambda^{(p)}(\mu_1)}{\Lambda^{(p-1)}(\mu_1)}, \tag{8-7.7}$$

where

$$\Lambda^{(p)}(\mu_1) = 1 - [P_{2p+2,E}(\mu_1)/P_{2p,E}(\mu_1)]. \tag{8-7.8}$$

Thus with

$$\tilde{R}^{(s,p)} \equiv \frac{\Lambda^{(p-1)}(\mu_1)}{\Lambda^{(p)}(\mu_1)} \frac{\|\Delta_B^{(s,p)}\|_2}{\|\Delta_B^{(s,p-1)}\|_2}, \tag{8-7.9}$$

we have from (8-7.7) that

$$\tilde{R}^{(s,p)} \doteq P_{2p,E}(\mu_1)/P_{2p-2,E}(\mu_1). \tag{8-7.10}$$

The difficulty with the use of $\tilde{R}^{(s,p)}$ is that the ratio $\Lambda^{(p-1)}(\mu_1)/\Lambda^{(p)}(\mu_1)$ is not known since it is a function of μ_1. However, we show that $R^{(s,p)}$ defined by (8-3.32) is a good approximation to $\tilde{R}^{(s,p)}$ in most cases.

If $M_E < \mu_1$, then $\Lambda^{(p)}(\mu_1)$ is a positive, monotone increasing, convergent sequence of p. Moreover, it is easy to show that

$$0 < \frac{\Lambda^{(p-1)}(\mu_1)}{\Lambda^{(p)}(\mu_1)} < 1, \quad \text{and} \quad \lim_{p \to \infty} \frac{\Lambda^{(p-1)}(\mu_1)}{\Lambda^{(p)}(\mu_1)} = 1. \tag{8-7.11}$$

It is obvious that $[\Lambda^{(p-1)}(M_E)/\Lambda^{(p)}(M_E)]$ also satisfies (8-7.11) and is a good approximation to $[\Lambda^{(p-1)}(\mu_1)/\Lambda^{(p)}(\mu_1)]$ provided M_E is reasonably close to μ_1 and/or p is not small. Thus, since

$$\Lambda^{(p)}(M_E) = 1 - \frac{T_{2p}(1/M_E)}{T_{2p+2}(1/M_E)} = 2 - \rho_B^{(p+1)},$$

we have $R^{(s,p)} \doteq \tilde{R}^{(s,p)}$ and (8-3.33) then follows from (8-7.10).

Equation (8-3.34)

Eliminating $c_{1,s} v_B(1)$ in (8-3.29)–(8-3.30) and using (8-3.11) and (8-7.3), we obtain for sufficiently large p that

$$\tilde{\varepsilon}_B^{(s,p)} \doteq \frac{R_{p,E}(\mu_1^2)}{Q_{p+1,E}(\mu_1^2)} \Delta_B^{(s,p)} = \frac{\Delta_B^{(s,p)}}{1 - \Gamma^{(p+1)}(\mu_1)}, \tag{8-7.12}$$

where $\Gamma^{(p+1)}(\mu_1) \equiv P_{2p+2,E}(\mu_1)/P_{2p,E}(\mu_1)$. Using (8-3.33), we have $\tilde{\varepsilon}_B^{(s,p)} \doteq \Delta_B^{(s,p)}/[1 - R^{(s,p+1)}]$. However, it is not convenient to use $R^{(s,p+1)}$ in the estimation of $\tilde{\varepsilon}_B^{(s,\,p)}$ since $\Delta_B^{(s,\,p+1)}$ would be required. To overcome this difficulty, we use an approximation to $\Gamma^{(p+1)}(\mu_1)$ expressed in terms of $R^{(s,\,p)}$.

To obtain this approximation, we first express $\Gamma^{(p+1)}(\mu_1)$ as

$$\Gamma^{(p+1)}(\mu_1) = [\Gamma^{(p+1)}(\mu_1)/\Gamma^{(p)}(\mu_1)]\Gamma^{(p)}(\mu_1). \tag{8-7.13}$$

Analogous to the behavior of the $\Lambda^{(p)}(\mu_1)$ functions, we have that $\lim_{p \to \infty}(\Gamma^{(p+1)}(\mu_1)/\Gamma^{(p)}(\mu_1)) = 1$ and that the ratio $\Gamma^{(p+1)}(M_E)/\Gamma^{(p)}(M_E)$ closely approximates $\Gamma^{(p+1)}(\mu_1)/\Gamma^{(p)}(\mu_1)$ if M_E is close to μ_1 and/or p is not small. Using this approximation for the ratio in (8-7.13) together with the fact that $\Gamma^{(p+1)}(M_E) = \rho_B^{(p+1)} - 1$, we obtain

$$\Gamma^{(p+1)}(\mu_1) \doteq \frac{\rho_B^{(p+1)} - 1}{\rho_B^{(p)} - 1} \, \Gamma^{(p)}(\mu_1). \tag{8-7.14}$$

But from (8-3.33), $\Gamma^{(p)}(\mu_1) \doteq R^{(s,\,p)}$. The substitution of (8-7.14) into (8-7.12) then gives (8-3.34).

9

Adaptive Procedures
for the Successive
Overrelaxation Method

9.1 INTRODUCTION

The successive overrelaxation (SOR) method, first introduced in Section 2.3, is one of the more efficient and one of the more widely used iterative methods for solving the problem

$$Au = b \tag{9-1.1}$$

when the matrix A is symmetric and positive definite (SPD) and is consistently ordered. The consistently ordered property (defined in Section 9.2) depends on the partitioning which is imposed on A. We show later that any matrix which is partitioned into the red/black form (8-1.1) is consistently ordered. (The converse is not true). For the case when A is partitioned into a red/black form, the SOR, RS–SI, and CCSI methods have the same optimal asymptotic convergence rates, but the RS–SI and CCSI methods have larger average convergence rates (Golub and Varga [1961]). However, sometimes it is not convenient to use a red/black partitioning. In this case, if A is consistently ordered and if optimum iteration parameters are used, then the asymptotic convergence rate for the SOR method is twice as large as that for the Chebyshev acceleration of the Jacobi method given in Chapters 5 and 6.† Analogous to the Chebyshev methods, the SOR method requires the use of an iteration parameter ω which must be properly chosen in order to obtain the greatest rate of convergence. Part of the popularity enjoyed by the SOR method stems from its simplicity of application and from its familiarity in the engineering community.

In Section 9.2, we introduce the important concept of matrices which have Property $\mathscr{A}$ and matrices which are consistently ordered. These properties are defined relative to the block partitioning imposed on the matrix problem (9-1.1) and are necessary for the adaptive determination of the optimum SOR iteration parameter. Section 9.3 contains a brief review of the basic properties of the SOR method.

Like the Chebyshev procedures given in Chapter 6, the SOR adaptive procedures which we give are based on the convergence of the SOR difference vector to an eigenvector of the associated SOR iteration matrix. In Section 9.4, we discuss the convergence properties of the secondary iterative process defined by the sequence of SOR difference vectors. The basic adaptive parameter estimation and stopping procedures are then developed in Section 9.5.

In Section 9.6, we give an overall computational algorithm which includes adaptive procedures for estimating the parameter ω and for terminating the iterative process. For the algorithm given, it is assumed that the coefficient matrix A is SPD, and is consistently ordered relative to the block partitioning imposed. In Section 9.7, we give a computational algorithm which is applicable when the coefficient matrix A is partitioned in the special red/black form. This algorithm utilizes the Rayleigh quotient to obtain estimates for the SOR parameter ω. A useful property of the SOR algorithm given is that no ω estimate can exceed its optimum value. Section 9.8 contains results of numerical experiments. Comments on the relative merits of certain partitionings and certain iterative solution procedures are given in Section 9.9.

The algorithms given in this chapter can be utilized by knowledgeable users of the SOR method without reading all of the preceding material. For example, the use of Algorithm 9-6.1 requires only a familiarity with the concept of a consistent ordering for the coefficient matrix. If desired, the reader may go directly to the section in which the algorithm is given and then refer back to the previous material only if more information is needed.

---

† Similar, but not as precise, statements can be made concerning the relative convergence rates of the CG and SOR methods.

9.2 CONSISTENTLY ORDERED MATRICES AND RELATED MATRICES

In this section we define and give examples of consistently ordered matrices and matrices which have Property $\mathscr{A}$.

Let the $N \times N$ matrix A be partitioned into the form

$$A = \begin{bmatrix} A_{1,1} & \cdots & A_{1,q} \\ \vdots & & \vdots \\ A_{q,1} & \cdots & A_{q,q} \end{bmatrix}, \qquad (9\text{-}2.1)$$

where $A_{i,j}$ is an $n_i \times n_j$ submatrix and $n_1 + \cdots + n_q = N$.

Definition 9-2.1 The $q \times q$ block matrix A of (9-2.1) is said to have *Property* $\mathscr{A}$ if there exists two disjoint nonempty subsets S_R and S_B of $\{1, 2, \ldots, q\}$ such that $S_R \cup S_B = \{1, 2, \ldots, q\}$ and such that if† $A_{i,j} \neq 0$ and $i \neq j$, then $i \in S_R$ and $j \in S_B$ or $i \in S_B$ and $j \in S_R$.

Definition 9-2.2. The $q \times q$ block matrix A of (9-2.1) is said to be *consistently ordered* if for some t there exist disjoint nonempty subsets $S_1, \ldots, S_t$ of $\{1, 2, \ldots, q\}$ such that $\bigcup_{i=1}^{t} S_i = \{1, \ldots, q\}$ and such that if $A_{i,j} \neq 0$ with $i \neq j$ and S_k is the subset containing i, then $j \in S_{k+1}$ if $j > i$ and $j \in S_{k-1}$ if $j < i$.

Note that the Property $\mathscr{A}$ and consistently ordered properties are defined *relative to an imposed partitioning for A*. By letting $S_R = \{S_i : i \text{ odd}\}$ and $S_B = \{S_i : i \text{ even}\}$, it follows that any matrix which is consistently ordered also has Property $\mathscr{A}$. The converse is not always true. However, if the matrix A has Property $\mathscr{A}$ relative to a particular partitioning then by a suitable permutation (reordering) of the given block rows and corresponding block columns of A we can obtain a consistently ordered matrix. (See, e.g., Young [1971].)

It follows easily from the above definitions that any matrix which is partitioned into the red/black form (8-1.1) has Property $\mathscr{A}$ and is consistently ordered. Conversely, as we show later, any matrix A which has Property $\mathscr{A}$ relative to a given partitioning can be cast into a red/black form (8-1.1), where D_R and D_B are block diagonal matrices whose diagonal blocks are consistent with those of the initial partitioning.

If the diagonal submatrices $A_{i,i}$ of A are nonsingular, as is the case when A is SPD, then Property $\mathscr{A}$ is equivalent to the 2-cyclic property given by Varga [1962]. The class of matrices with Property $\mathscr{A}$ was introduced by Young [1950, 1954] for point partitionings and later generalized to block

---

† By $A_{i,j} \neq 0$, it is meant that at least one element of the submatrix $A_{i,j}$ is nonzero.

partitionings by Arms, *et. al.* [1956]. As we shall see in Section 9.3, the consistent ordering property is important for the efficient use of the SOR method. For additional discussion on these properties, see Varga [1962] and Young [1971].

Examples of Matrices Which Have Property $\mathscr{A}$ and Are Consistently Ordered

We first show that any block tridiagonal matrix has Property $\mathscr{A}$ and is consistantly ordered. Let

$$
A = \begin{bmatrix}
A_{1,1} & A_{1,2} & & & \\
A_{2,1} & A_{2,2} & A_{2,3} & & \mathbf{0} \\
& \ddots & \ddots & \ddots & \\
\mathbf{0} & & & & A_{q-1,q} \\
& & & A_{q,q-1} & A_{q,q}
\end{bmatrix},
\tag{9-2.2}
$$

where for convenience we assume q is even. Frequently, the easiest way to determine if a particular partitioning results in a matrix which has Property $\mathscr{A}$ and is consistently ordered is by construction of the subsets given in Definitions 9-2.1 and 9-2.2.† With $S_R = \{1, 3, \ldots, q-1\}$ and $S_B = \{2, 4, \ldots, q\}$, it is obvious that the tridiagonal matrix A of (9-2.2) has Property $\mathscr{A}$. Moreover, the subsets $S_1 = \{1\}$, $S_2 = \{2\}, \ldots, S_q = \{q\}$ satisfy the conditions necessary for A also to be consistently ordered.

We now consider a partitioning for which the matrix A has Property $\mathscr{A}$ but is not consistently ordered. Let

$$
A = \begin{bmatrix}
A_{1,1} & A_{1,2} & 0 & A_{1,4} \\
A_{2,1} & A_{2,2} & A_{2,3} & 0 \\
0 & A_{3,2} & A_{3,3} & A_{3,4} \\
A_{4,1} & 0 & A_{4,3} & A_{4,4}
\end{bmatrix}.
\tag{9-2.3}
$$

We show by construction that the matrix A of (9-2.3) has Property $\mathscr{A}$. Let $1 \in S_R$.

First block row: Since $1 \in S_R$,‡ 2 and 4 must be in S_B. Thus the first block row requires that $S_R = \{1, ?\}$ and $S_B = \{2, 4, ?\}$.

Second block row: Since $2 \in S_B$, 1 and 3 must be in S_R. Thus we have $S_R = \{1, 3\}$ and $S_B = \{2, 4\}$. Since $S_R \cup S_B = \{1, 2, 3, 4\}$, the S_R and S_B

---

† We remark that other procedures exist for determining whether or not a matrix has Property $\mathscr{A}$ and is consistently ordered. Young [1971], for example, gives one such procedure in the form of an algorithm.

‡ By the notation $1 \in S_R$, we mean that the subset S_R contains the number one.

subsets are complete and all that remains is to verify that the necessary conditions are satisfied for the remaining two rows.

Third block row: Since $3 \in S_R$, 2 and 4 must be in S_B, which they are.

Fourth block row: Since $4 \in S_B$, 1 and 3 must be in S_R, which they are.

Thus the matrix A of (9-2.3) has Property $\mathscr{A}$.

We now show by contradiction that the partitioning (9-2.3) is not consistently ordered. Assume the subsets $S_1, \ldots, S_t$ satisfy the necessary condition for A to be consistently ordered. Thus 1 must be in some subset, say S_r.

First block row: Since $1 \in S_r$, 2 and 4 must be in S_{r+1}. Thus we have $S_r = \{1, ?\}, S_{r+1} = \{2, 4, ?\}$.

Second block row: Since $2 \in S_{r+1}$, 1 must be in S_r, which it is, and 3 must be in S_{r+2}. Thus we must have $S_r = \{1\}$, $S_{r+1} = \{2, 4\}$, and $S_{r+2} = \{3\}$.

Third block row: Since $3 \in S_{r+2}$, 2 must be in S_{r+1}, which it is. However, 4 must be in S_{r+3}, which is a contradiction that the subsets be disjoint since 4 is already in S_{r+1}.

Thus (9-2.3) is not a consistently ordered partitioning. The reader should convince himself that a consistent ordering can be obtained either by permuting block rows 1 and 2 and the corresponding block columns 1 and 2 or by permuting block rows 2 and 3 and the corresponding block columns 2 and 3.

We now show that if a matrix A has Property $\mathscr{A}$ relative to a particular partitioning, then by a suitable permutation of the given block rows and corresponding block columns of A we can obtain a red/black partitioning of the form (8-1.1), where D_R and D_B are block diagonal matrices whose diagonal blocks are a permutation of the diagonal submatrices of the original partitioning. Let the partitioned matrix A of (9-2.1) have Property $\mathscr{A}$ and let the subsets S_R and S_B of $\{1, 2, \ldots, q\}$ be given as in Definition 9-2.1. The desired red/black form for A can now easily be obtained through the use of the subsets S_R and S_B. To do this, simply interchange block rows (and corresponding block columns) such that if $i \in S_R$ and $\tilde{i} \in S_B$, then block row i comes before block row $\tilde{i}$.† For example, consider the block tridiagonal matrix A of (9-2.2) where $S_R = \{1, 3, \ldots, q - 1\}$ and $S_B = \{2, 4, \ldots, q\}$. Permuting block rows (and columns) such that block rows of S_R come before those of S_B, we obtain the red/black form

$$A = \begin{bmatrix} D_R & H \\ K & D_B \end{bmatrix}, \qquad (9\text{-}2.4)$$

---

† In solving the matrix problem $Au = b$, similar block interchanges must be made on the rows of b and u. Thus in effect, we are changing the ordering for the partitioned unknown vector u to obtain a red/black partition of the coefficient matrix. See the examples given in Section 1.7.

where

$$
D_R = \begin{bmatrix} A_{1,1} & & & \\ & A_{3,3} & & \text{\large 0} \\ & & \ddots & \\ \text{\large 0} & & & A_{q-1,q-1} \end{bmatrix}, \quad
D_B = \begin{bmatrix} A_{2,2} & & & \\ & A_{4,4} & & \text{\large 0} \\ & & \ddots & \\ \text{\large 0} & & & A_{q,q} \end{bmatrix},
$$

and

$$
H = \begin{bmatrix} A_{1,2} & & \text{\large 0} \\ A_{3,2} & A_{3,4} & \\ \text{\large 0} & \ddots & \ddots \end{bmatrix}, \quad
K = \begin{bmatrix} A_{2,1} & A_{2,3} & & \text{\large 0} \\ & A_{4,3} & A_{4,5} & \\ \text{\large 0} & & \ddots & \ddots \end{bmatrix}.
$$

Note that D_R and D_B are block diagonal matrices whose diagonal blocks are determined by the diagonal submatrices of the original tridiagonal matrix (9-2.2). Thus the D_R and D_B matrices will be "easily invertible" (see Section 1.5) whenever the diagonal blocks $A_{i,i}$ of the original partitioning (9-2.2) are "easily invertible."

A similar permutation of rows (and corresponding columns) based on the S_R and S_B subsets for the matrix A of (9-2.3) gives the red/black form (9-2.4), where, for example,

$$
D_R = \begin{bmatrix} A_{11} & 0 \\ 0 & A_{3,3} \end{bmatrix}, \quad
D_B = \begin{bmatrix} A_{2,2} & 0 \\ 0 & A_{4,4} \end{bmatrix}, \quad \text{and } H = \begin{bmatrix} A_{1,2} & A_{1,4} \\ A_{3,2} & A_{3,4} \end{bmatrix}.
$$

9.3   THE SOR METHOD

In this section, notation is defined and known properties of the SOR method are given.

Let the $N \times N$ matrix problem (9-1.1) be partitioned into the form

$$
\begin{bmatrix} A_{1,1} & A_{1,2} & \cdots & A_{1,q} \\ A_{2,1} & A_{2,2} & \cdots & A_{2,q} \\ \vdots & \vdots & & \vdots \\ A_{q,1} & A_{q,2} & \cdots & A_{q,q} \end{bmatrix}
\begin{bmatrix} U_1 \\ U_2 \\ \vdots \\ U_q \end{bmatrix} =
\begin{bmatrix} F_1 \\ F_2 \\ \vdots \\ F_q \end{bmatrix}, \tag{9-3.1}
$$

where the U_i and F_i represent column matrices of appropriate sizes. The diagonal submatrix $A_{i,i}$ is $n_i \times n_i$ and nonsingular, where $n_1 + n_2 + \cdots + n_q = N$. Relative to the partitioning (9-3.1), the SOR method is defined by

$$
A_{i,i} U_i^{(n+1)} = \omega \left\{ - \sum_{j=1}^{i-1} A_{i,j} U_j^{(n+1)} - \sum_{j=i+1}^{q} A_{i,j} U_j^{(n)} + F_i \right\} + (1 - \omega) A_{i,i} U_i^{(n)}, \tag{9-3.2}
$$

for $i = 1, \ldots, q$. In matrix form, the SOR procedure (9-3.2) may be expressed as

$$Du^{(n+1)} = \omega[C_{L}u^{(n+1)} + C_{U}u^{(n)} + b] + (1 - \omega)Du^{(n)}, \qquad (9\text{-}3.3)$$

where D is the diagonal block matrix of (2-3.11) and C_{L}, C_{U} are the strictly lower and upper triangular matrices, respectively, defined by (2-3.11). The vector $u^{(n)}$ is defined by (2-3.16). In (9-3.2) and (9-3.3), the real number ω is called the *SOR relaxation parameter*. If $\omega = 1$, the SOR method reduces to the Gauss–Seidel method discussed in Section 2.3.

It is easy to show that the SOR error vector $\varepsilon^{(n+1)} \equiv u^{(n+1)} - \bar{u}$ satisfies

$$\varepsilon^{(n+1)} = \mathscr{L}_{\omega}\varepsilon^{(n)}, \qquad (9\text{-}3.4)$$

where $\bar{u}$ is the true solution of (9-1.1) and $\mathscr{L}_{\omega}$ is the SOR iteration matrix given by (2-3.36). From (2-2.4), a necessary and sufficient condition for the convergence of the SOR method is that $S(\mathscr{L}_{\omega}) < 1$. Kahan [1958] has shown that $S(\mathscr{L}_{\omega}) \geq |\omega - 1|$, with equality possible only if all eigenvalues of $\mathscr{L}_{\omega}$ have modulus $|\omega - 1|$. Thus a necessary condition for the convergence of the SOR method is that $0 < \omega < 2$. Further, if A is symmetric and if the block diagonal matrix D is positive definite, then the SOR method converges if and only if $0 < \omega < 2$ and A is positive definite (see, e.g., Varga [1962]).

In what follows, an optimum value of ω, which we denote by ω_{b}, is defined to be any ω which minimizes $S(\mathscr{L}_{\omega})$. The main difficulty in the use of the SOR method is the determination of such an ω_{b}. When the matrix A is consistently ordered, ω_{b} is unique and may be expressed as a function of the spectral radius of the Jacobi iteration matrix associated with (9-3.1). If A is not consistently ordered, however, a prescription for selecting ω_{b} usually is not available.

Let $B = D^{-1}(C_{L} + C_{U})$ be the Jacobi iteration matrix associated with (9-3.1). For the adaptive procedures given in this chapter, it is assumed that

(1) the partitioning for A is such that A is consistently ordered with respect to the partitioning imposed, and that

(2) $(I - B)$ is similar to a SPD matrix; i.e., the associated Jacobi method is symmetrizable (see Definition 2-2.1).

Assumption (2) is sufficient to guarantee the convergence of the SOR method, while assumption (1) is sufficient to ensure the availability of a precise formula for ω_{b}. Note that these assumptions are satisfied if A is a consistently ordered positive definite matrix.

In what follows, let $\{\mu_{i}\}_{i=1}^{N}$ be the set of N eigenvalues for the $N \times N$ matrix B and let $v(i)$ be an eigenvector associated with μ_{i}. Since the Jacobi method is symmetrizable, it follows from Theorem 2-2.1 that the set of eigenvectors $\{v(i)\}_{i=1}^{N}$ may be chosen to be a basis for the associated vector space E^N and that the eigenvalues of B are real and are less than unity. Moreover, since A is consistently ordered, it follows (Young [1971]) that the eigenvalues of B occur in $\pm$ pairs. Thus, the eigenvalues $\{\mu_i\}_{i=1}^{N}$ may be ordered as

$$-\mu_1 < -\mu_2 \leq \cdots \leq -\mu_s < 0 = \cdots = 0 < \mu_s \leq \cdots \leq \mu_2 < \mu_1 < 1.$$
$$(9\text{-}3.5)$$

In order to simplify the notation used in subsequent sections, in (9-3.5) we have assumed that $\mu_1 > \mu_2$. As shown in Section 6.2, there is no loss of generality in this assumption.

Let the set of eigenvalues for $\mathscr{L}_\omega$ be given by $\{\lambda_i\}_{i=1}^{N}$ and let $y(i)$ be an eigenvector associated with λ_i; i.e., $\mathscr{L}_\omega y(i) = \lambda_i y(i)$. When A is consistently ordered, there exists a functional relationship between the eigenvalues of B and those of $\mathscr{L}_\omega$. As in (9-3.5), let the positive eigenvalues of B be denoted by $\mu_1, \mu_2, \ldots, \mu_s$. If $\omega \neq 0$, then for each μ_i, $i = 1, \ldots, s$, the two numbers

$$\lambda_i^+, \lambda_i^- = \left[\frac{\omega \mu_i \pm \sqrt{\omega^2 \mu_i^2 - 4(\omega - 1)}}{2} \right]^2 \qquad (9\text{-}3.6)$$

are eigenvalues of $\mathscr{L}_\omega$. (See, e.g., Young [1971].) The remaining $N - 2s$ eigenvalues of $\mathscr{L}_\omega$ are equal to $1 - \omega$. Using (9-3.6), it can be shown (see, for example, Young [1971]) that

$$\omega_b = \frac{2}{1 + \sqrt{1 - \mu_1^2}} \qquad (9\text{-}3.7)$$

and that

$$S(\mathscr{L}_\omega) = \begin{cases} \left[\dfrac{\omega \mu_1 + \sqrt{\omega^2 \mu_1^2 - 4(\omega - 1)}}{2} \right]^2, & \text{if } 0 < \omega \leq \omega_b, \text{ and} \\ \omega - 1, & \text{if } \omega_b \leq \omega < 2. \end{cases}$$
$$(9\text{-}3.8)$$

From (9-3.8), it is easy to show that $S(\mathscr{L}_\omega) > S(\mathscr{L}_{\omega_b})$ for $\omega \neq \omega_b$. Thus there is a unique value of ω, namely ω_b given by (9-3.7), which minimizes $S(\mathscr{L}_\omega)$. The relationships (9-3.7) and (9-3.8) will be used to determine the adequacy of an approximation ω for ω_b, and for obtaining a new estimate for ω_b, if needed. From (9-3.7), it is obvious that $1 \leq \omega_b < 2$. Thus henceforth, ω is assumed to be in the range $1 \leq \omega < 2$.

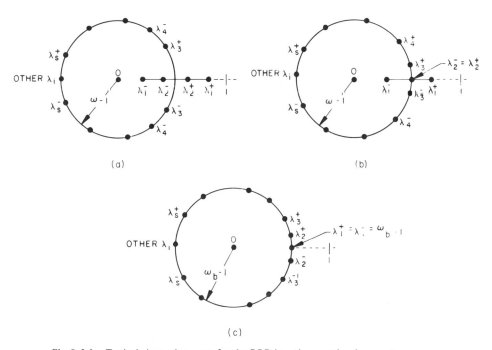

Fig. 9-3.1. Typical eigenvalue maps for the SOR iteration matrix when $\omega \leq \omega_b$.
(a) $2/(1 + \sqrt{1 - \mu_3^2}) < \omega < 2/(1 + \sqrt{1 - \mu_2^2})$, (b) $\omega = 2/(1 + \sqrt{1 - \mu_2^2})$, (c) $\omega = \omega_b = 2/(1 + \sqrt{1 - \mu_1^2})$.

The relation (9-3.6) also gives information concerning the distribution of eigenvalues of $\mathcal{L}_\omega$. After perhaps a small amount of algebraic manipulation, we conclude from (9-3.6) and (9-3.7) the following information:

(a) For $1 < \omega < 2$, $\mathcal{L}_\omega$ has a repeated eigenvalue equal to $(\omega - 1)$ if and only if $\omega = 2/(1 + \sqrt{1 - \mu_i^2})$ for some i. We note that $\omega = 2/(1 + \sqrt{1 - \mu_i^2})$ implies that $\omega^2 \mu_i^2 - 4(\omega - 1) = 0$.

(b) If $\omega \geq \omega_b$, all eigenvalues of $\mathcal{L}_\omega$ lie on a circle of radius $(\omega - 1)$. If $\omega > \omega_b$, no eigenvalues of $\mathcal{L}_\omega$ are real except when some $\mu_i = 0$ and then $\lambda_i = -(\omega - 1)$ is a real eigenvalue of $\mathcal{L}_\omega$. If $\omega = \omega_b$, then $\lambda_1^+ = \lambda_1^- = \omega_b - 1$ is a repeated real eigenvalue of $\mathcal{L}_{\omega_b}$ (see Fig. 9-3.1c).

(c) If $1 < \omega < \omega_b$, the eigenvalues of $\mathcal{L}_\omega$ lie on a circle of radius $\omega - 1$ and on the real line segment $[\lambda_1^-, \lambda_1^+]$. The number of positive eigenvalues of $\mathcal{L}_\omega$ greater than $(\omega - 1)$ equals the number of positive eigenvalues of $\mathcal{L}_\omega$ less than $\omega - 1$. Moreover, from (a) given above, $\mathcal{L}_\omega$ has a repeated eigenvalue equal to $(\omega - 1)$ if and only if $\omega = 2/(1 + \sqrt{1 - \mu_i^2})$ for some i (see Figs. 9-3.1a and 9-3.1b).

For the SOR adaptive procedure, we also will need information concerning eigenvectors of $\mathscr{L}_\omega$. Primarily, we will want to form a basis for the N-dimensional vector space E^N in terms of the eigenvectors of $\mathscr{L}_\omega$ and supplemented by principal vectors† if needed. Fortunately, for the consistently ordered case, a relationship between the eigenvectors of $\mathscr{L}_\omega$ and those of B exists, and this can be used to determine those eigenvectors of $\mathscr{L}_\omega$ which are linearly independent.

From (9-3.5), the Jacobi matrix B has s positive eigenvalues, s negative eigenvalues, and $N - 2s$ zero eigenvalues. Let $v(i)$, $i = 1, 2, \ldots, s$ be the s linearly independent eigenvectors of B corresponding to the s positive eigenvalues of B and let $v(i)$, $i = 2s + 1, \ldots, N$ be the $N - 2s$ linearly independent eigenvectors corresponding to the $N - 2s$ zero eigenvalues of B. The following theorem relates the eigenvectors of $\mathscr{L}_\omega$ to those of B.

Theorem 9-3.1   Let the matrix A of (9-3.1) be consistently ordered and let $\omega > 1$. For each μ_i, $i = 1, \ldots, s$, $(\lambda_i^+, \lambda_i^-)$ of (9-3.6) are two eigenvalues of $\mathscr{L}_\omega$ and

(a)   if $\mu_i^2\omega^2 - 4(\omega - 1) \neq 0$, the eigenvectors of $\mathscr{L}_\omega$ corresponding to $(\lambda_i^+, \lambda_i^-)$ may be expressed by $y^+(i) = E([\lambda_i^+]^{1/2})v(i)$ and $y^-(i) = E([\lambda_i^-]^{1/2})v(i)$, where $E(x)$ is a diagonal matrix whose diagonal elements are certain powers of x.

(b)   if $\mu_i^2\omega^2 - 4(\omega - 1) = 0$, $y^+(i) = E([\lambda_i^+]^{1/2})v(i)$ is an eigenvector of $\mathscr{L}_\omega$ corresponding to $\lambda_i^+ = \lambda_i^-$ and $p(i) = (2/[\lambda_i^+]^{1/2})F([\lambda_i^+]^{1/2})v(i)$ is the associated principal vector of grade 2 associated† with the repeated eigenvalue $\lambda_i^+ = \lambda_i^-$. If $E_j(x)$ is the jth diagonal element of $E(x)$, then F is a diagonal matrix whose jth element is $dE_j(x)/dx$.

The remaining $N - 2s$ eigenvalues of $\mathscr{L}_\omega$ are equal to $(1 - \omega)$ with the associated eigenvectors $y(i) = E([1 - \omega]^{1/2})v(i), i = 2s + 1, \ldots, N$. Further, the N vectors $y^+(i)$, $y^-(i)$, $p(i)$, $y(i)$ defined above are linearly independent and, hence, form a basis for the associated vector space E^N.

Proof.   For a proof of this theorem, see Young [1950] or Ref. 13 in Hageman and Kellogg [1968].

Thus if $\mu_i^2\omega^2 - 4(\omega - 1) \neq 0$ for all i (as is the case for the example given in Fig. 9-3.1a), the eigenvectors of $\mathscr{L}_\omega$ include a basis for the N-dimensional vector space. However, if $[\mu_i^2\omega^2 - 4(\omega - 1)] = 0$ for some i (as is the case for the examples given in Figs. 9-3.1b and 9-3.1c), then the eigenvectors of $\mathscr{L}_\omega$ must be supplemented by the associated principal vectors in order to form a basis. If $\mu_i^2\omega^2 - 4(\omega - 1) = 0$, then the eigenvector

† For a discussion on principal vectors, see Section 1.3.

$y^+(i)$ of $\mathcal{L}_\omega$ with eigenvalue λ_i^+ and the associated principal vector $p(i)$ satisfy (see Section 1.3)

$$\mathcal{L}_\omega^n p(i) = (\lambda_i^+)^n p(i) + n(\lambda_i^+)^{n-1} y^+(i). \qquad (9\text{-}3.9)$$

The diagonal matrix $E(x)$ used to obtain eigenvectors of $\mathcal{L}_\omega$ from those of B is a function of the so-called "ordering vector" (see Young [1971]) which exists for any consistent ordering. Since the explicit form of $E(x)$ will be used only for problems with a red/black partitioning, no general discussion of $E(x)$ will be given. The matrix $E(x)$ may easily be defined for the red/black or general tridiagonal partitioning cases. If the matrix A of (9-3.1) is the block tridiagonal matrix of (9-2.2), then $E(x)$ may be given by

$$E(x) = \begin{bmatrix} I_1 & & & 0 \\ & xI_2 & & \\ & & \ddots & \\ 0 & & & x^{q-1}I_q \end{bmatrix}, \qquad (9\text{-}3.10)$$

where I_j is the identity matrix of the same order as $A_{j,j}$. If A is partitioned in the special red/black form

$$A = \begin{bmatrix} A_{1,1} & A_{1,2} \\ A_{2,1} & A_{2,2} \end{bmatrix},$$

then (9-3.10) reduces to

$$E(x) = \begin{bmatrix} I_1 & 0 \\ 0 & xI_2 \end{bmatrix}. \qquad (9\text{-}3.11)$$

9.4 EIGENVECTOR CONVERGENCE OF THE SOR DIFFERENCE VECTOR

Let the difference vector for the SOR method of (9-3.4) be defined by†

$$\Delta^{(n)} \equiv u^{(n)} - u^{(n+1)} = -[\mathcal{L}_\omega u^{(n)} + k - u^{(n)}] \qquad (9\text{-}4.1)$$

and let the error vector be defined, as before, by

$$\varepsilon^{(n)} \equiv u^{(n)} - \bar{u}. \qquad (9\text{-}4.2)$$

† The vector $\Delta^{(n)}$ defined by (9-4.1) is the negative of the more conventional definition of "difference" vector. Since we will be concerned only with the norm of $\Delta^{(n)}$, the sign is immaterial. We also note that the difference vector (9-4.1) is identical, except for sign, to the pseudoresidual vector $\delta^{(n)}$ associated with the SOR method.

It is easy to show that the error and difference vectors for the SOR method satisfy

$$\varepsilon^{(n)} = \mathscr{L}_\omega \varepsilon^{(n-1)} = \mathscr{L}_\omega^n \varepsilon^{(0)},$$

$$\Delta^{(n)} = \mathscr{L}_\omega \Delta^{(n-1)} = \mathscr{L}_\omega^n \Delta^{(0)}, \quad \text{and} \tag{9-4.3}$$

$$\Delta^{(n)} = (I - \mathscr{L}_\omega)\varepsilon^{(n)}.$$

We assume throughout this section that the eigenvalues of the associated Jacobi matrix B are given by (9-3.5) and that $1 < \omega < \omega_b$. Thus without loss of generality, the eigenvalues of $\mathscr{L}_\omega$ may be ordered (see Figs. 9-3.1a and 9-3.1b) as

$$0 < \lambda_N \leq \cdots < |\lambda_p| = \cdots = |\lambda_{g+1}| \leq \lambda_g \leq \cdots \leq \lambda_2 < \lambda_1 < 1, \tag{9-4.4}$$

where, for $i = g + 1, \ldots, p$, λ_i satisfies $|\lambda_i| = \omega - 1$ but $\lambda_i \neq \omega - 1$. As before, we let $y(i)$ denote an eigenvector of $\mathscr{L}_\omega$ associated with λ_i. We now consider the two cases: (1) $\lambda_g > |\lambda_{g+1}| = \omega - 1$ and (2) $\lambda_g = |\lambda_{g+1}| = \omega - 1$. For the first case, the set of eigenvectors for $\mathscr{L}_\omega$ includes a basis for the associated vector space. The first case is illustrated by the eigenvalue distribution given in Fig. 9-3.1a. For the second case, the set of eigenvectors of $\mathscr{L}_\omega$ must be supplemented by a principal vector to form a basis. The second case is illustrated by the eigenvalue distribution given in Fig. 9-3.1b.

Case 1: $\lambda_g > |\lambda_{g+1}| = \omega - 1$

Since no eigenvalue of $\mathscr{L}_\omega$ equals $(\omega - 1)$ for this case, we have that $\mu_i^2 \omega^2 - 4(\omega - 1) \neq 0$ for all i. Hence from Theorem 9-3.1, the set of eigenvectors $\{y(i)\}$ for $\mathscr{L}_\omega$ includes a basis for E^N. Thus we can write $\varepsilon^{(0)}$ in the form†

$$\varepsilon^{(0)} = c_1 y(1) + \sum_{i=2}^N c_i y(i). \tag{9-4.5}$$

Multiplying (9-4.5) by $(I - \mathscr{L}_\omega)$ and using (9-4.3), we can express $\Delta^{(0)}$ in the related form

$$\Delta^{(0)} = d_1 y(1) + \sum_{i=2}^N d_i y(i), \tag{9-4.6}$$

---

† We assume throughout this section that $c_1 \neq 0$. There is no real loss of generality in this assumption. However, if $c_1 = c_2 = \cdots = c_k = 0$ and $c_{k+1} \neq 0$, then the assumption that $\omega < \omega_b$ in the theorems which follow should be interpreted as meaning $\omega < \tilde{\omega}_b$, where $\tilde{\omega}_b = 2/(1 + \sqrt{1 - \mu_{k+1}^2})$.

where $d_i = (1 - \lambda_i)c_i$. As in Chapter 6, the contamination factor K associated with the initial difference vector is defined by

$$K = \left[\sum_{i=2}^{N} \|d_i y(i)\|_\psi \right] \Big/ \|d_1 y(1)\|_\psi, \qquad (9\text{-}4.7)$$

where the norm ψ is determined by the context in which K is used. Recall that the contamination factor K gives a measure of how closely $\Delta^{(0)}$ approximates the eigenvector $y(1)$ with eigenvalue λ_1. If $K = 0$, then $\Delta^{(0)}$ (and also $\varepsilon^{(0)}$) is an eigenvector of $\mathscr{L}_\omega$.

We now have

Theorem 9-4.1.   If $1 < \omega < \omega_b$ and if $[\mu_i^2 \omega^2 - 4(\omega - 1)] \neq 0$ for all i, then for sufficiently large n and for any vector norm, we have that

$$\frac{\|\Delta^{(n)}\|}{1 - \lambda_1} [1 - \alpha_n] \leq \|\varepsilon^{(n)}\| \leq \frac{\|\Delta^{(n)}\|}{1 - \lambda_1} [1 + \alpha_n], \qquad (9\text{-}4.8)$$

where with K defined by (9-4.7)

$$\alpha_n = \left| \frac{\lambda_2}{\lambda_1} \right|^n \left[\frac{K}{1 - |\lambda_2/\lambda_1|^n K} \right]. \qquad (9\text{-}4.9)$$

In addition,

$$\lambda_1 \left[\frac{1 - \beta_n}{1 + \beta_{n-1}} \right] \leq \frac{\|\Delta^{(n)}\|}{\|\Delta^{(n-1)}\|} \leq \lambda_1 \left[\frac{1 + \beta_n}{1 - \beta_{n-1}} \right], \qquad (9\text{-}4.10)$$

where

$$\beta_n = |\lambda_2/\lambda_1|^n K. \qquad (9\text{-}4.11)$$

Proof.   See Section 9.10.

Case 2: $\lambda_g = |\lambda_{g+1}| = \omega - 1$

We now consider the case where the set of eigenvectors for $\mathscr{L}_\omega$ does not include a basis. From Theorem 9-3.1, this occurs when ω is such that $\omega - 1 = \frac{1}{4}\mu_i^2 \omega^2$ for some i. When this condition happens, $\omega - 1$ is a repeated eigenvalue of $\mathscr{L}_\omega$; i.e., $\lambda_g = \lambda_{g-1} = \omega - 1$ in (9-4.4). Moreover, a principal vector is associated with the repeated eigenvalue $\omega - 1$ of $\mathscr{L}_\omega$. Since we are assuming $\omega < \omega_b$ in this section, $\omega - 1 = \frac{1}{4}\mu_2^2 \omega^2$ is the largest† eigenvalue of $\mathscr{L}_\omega$ for which a principal vector can be associated. For reasons which will become obvious later, this is the worst case for our purposes. For this

---

† If ω is such that $\omega - 1 = \frac{1}{4}\mu_1^2\omega^2$, then $\omega = \omega_b$, which violates the assumption that $\omega < \omega_b$.

"worst" case (i.e., $\omega - 1 = \frac{1}{4}\mu_2^2\omega^2$), it follows from Fig. 9-3.1b that the ordering (9-4.4) for the eigenvalues of $\mathscr{L}_\omega$ can be written as†

$$0 < \lambda_N < |\lambda_{N-1}| = \cdots = |\lambda_4| = \lambda_3 = \lambda_2 < \lambda_1 < 1,$$

where $\lambda_2 = \omega - 1$ and where $\lambda_i \neq \omega - 1$ for $i \geq 4$. Since $[\mu_i^2\omega^2 - 4(\omega - 1)] \neq 0$ for $i = 1$ and for $i \geq 4$, it follows from Theorem 9-3.1 that a basis for E^N can be given by the set of vectors $\{y(1), p(2), y(3), \ldots, y(N)\}$, where $y(i)$ denotes an eigenvector of $\mathscr{L}_\omega$ associated with the eigenvalue λ_i and where $p(2)$ is a principal vector of $\mathscr{L}_\omega$ associated with the repeated eigenvalue $\lambda_2 = \lambda_3$. In terms of this basis, we can express $\varepsilon^{(0)}$ in the form

$$\varepsilon^{(0)} = c_1 y(1) + c_2 p(2) + c_3 y(3) + \sum_{i=4}^{N} c_i y(i). \tag{9-4.12}$$

Multiplying (9-4.12) by $I - \mathscr{L}_\omega$ and using (9-4.3) together with the fact that $\mathscr{L}_\omega p(2) = \lambda_2 p(2) + y(3)$, we can express $\Delta^{(0)}$ in the related form

$$\Delta^{(0)} = d_1 y(1) + d_2 p(2) - c_2 y(3) + \sum_{i=3}^{N} d_i y(i), \tag{9-4.13}$$

where $d_i = (1 - \lambda_i)c_i$.

For this case, we define the contamination factor to be

$$K' = \left[\|c_2 y(3)\|_\psi + \|d_2 p(2)\|_\psi + \sum_{i=3}^{N} \|d_i y(i)\|_\psi \right] \Big/ \|d_1 y(1)\|_\psi. \tag{9-4.14}$$

We now have

Theorem 9-4.2. If $1 < \omega < \omega_b$ and if $\mu_2^2\omega^2 - 4(\omega - 1) = 0$, then for n sufficiently large and for any vector norm, we have that

$$\frac{\|\Delta^{(n)}\|}{1 - \lambda_1}[1 - \alpha_n] \leq \|\varepsilon^{(n)}\| \leq \frac{\|\Delta^{(n)}\|}{1 - \lambda_1}[1 + \alpha_n], \tag{9-4.15}$$

where with K' defined by (9-4.14)

$$\alpha_n = \frac{\left(\dfrac{\lambda_2}{\lambda_1}\right)^n K' + n\left(\dfrac{\lambda_2}{\lambda_1}\right)^{n-1}\left(\dfrac{\lambda_1 - \lambda_2}{1 - \lambda_2}\right)\left(\dfrac{1}{\lambda_1}\right)\dfrac{\|d_2 y(3)\|}{\|d_1 y(1)\|}}{1 - \left(\dfrac{\lambda_2}{\lambda_1}\right)^n K' + n\left(\dfrac{\lambda_2}{\lambda_1}\right)^{n-1}\left(\dfrac{1}{\lambda_1}\right)\dfrac{\|d_2 y(3)\|}{\|d_1 y(1)\|}}. \tag{9-4.16}$$

Moreover,

$$\lambda_1\left[\frac{1 - \beta_n}{1 + \beta_{n-1}}\right] \leq \frac{\|\Delta^{(n)}\|}{\|\Delta^{(n-1)}\|} \leq \lambda_1\left[\frac{1 + \beta_n}{1 - \beta_{n-1}}\right], \tag{9-4.17}$$

---

† Here, we implicitly assume that $\mu_3 \neq \mu_2$ in (9-3.5). For reasons similar to those given in Section 6.2, there is no loss of generality in this assumption.

where

$$\beta_n = \left(\frac{\lambda_2}{\lambda_1}\right)^n K' + n\left(\frac{\lambda_2}{\lambda_1}\right)^{n-1}\left(\frac{1}{\lambda_1}\right)\frac{\|d_2 y(3)\|}{\|d_1 y(1)\|}. \tag{9-4.18}$$

Proof.   See Section 9.10.

If $1 < \omega < \omega_b$, then $|(\lambda_2/\lambda_1)| < 1$ and $\lim_{n\to\infty} n|(\lambda_2/\lambda_1)|^{n-1} = 0$. Thus for both Case 1 and Case 2 conditions, we deduce from Theorems 9-4.1 and 9-4.2 that for any vector norm

$$\lim_{n\to\infty} [\|\Delta^{(n)}\|/\|\Delta^{(n-1)}\|] = \lambda_1, \tag{9-4.19}$$

$$\lim_{n\to\infty} [\|\varepsilon^{(n)}\|/\|\Delta^{(n)}\|] = 1/(1 - \lambda_1). \tag{9-4.20}$$

The relationships (9-4.19) and (9-4.20) serve as a basis for the parameter estimation and stopping procedures developed in the next section.

The behavior of the sequences $\{\alpha_n\}$ and $\{\beta_n\}$ gives a good indication of the rates at which the ratios $\|\varepsilon^{(n)}\|/\|\Delta^{(n)}\|$ and $\|\Delta^{(n)}\|/\|\Delta^{(n-1)}\|$ converge to $1/(1 - \lambda_1)$ and λ_1, respectively. If $1 < \omega < \omega_b$, both sequences $\{\alpha_n\}$ and $\{\beta_n\}$ converge to zero at a rate governed by $|\lambda_2/\lambda_1|^n$ for Case 1 conditions, but at a slower rate governed by $n[|\lambda_2/\lambda_1|]^{n-1}$ for Case 2 conditions. As the estimate ω approaches $\omega_b -$, $|\lambda_2/\lambda_1|$ approaches $1 -$. Thus the ratios of (9-4.19) and (9-4.20) will converge slowly to λ_1 and $1/(1 - \lambda_1)$ whenever ω closely approximates ω_b. Even for this case, however, the use of $\|\Delta^{(n)}\|/\|\Delta^{(n-1)}\|$ and $\|\Delta^{(n)}\|/(1 - \lambda_1)$ as approximations to λ_1 and $\|\varepsilon^{(n)}\|$, respectively, are reasonable whenever the contamination factor K or K' is small.

9.5   SOR ADAPTIVE PARAMETER AND STOPPING PROCEDURES

To clearly indicate the dependence on the ω used, notation similar to that given previously for successive Chebyshev polynomial generation will be used. We let

$\mathcal{L}_{\omega_s}$ denote the SOR iteration matrix currently being used, where the subscript s on ω implies this is the sth estimate for ω.

n denote the current iteration step.

q denote the last iteration on which the estimate ω_{s-1} was used. For the initial estimate where $s = 1$, we let $q = 0$.

$p \equiv n - q$ denote the number of iterations since ω was last changed.†

---

† For example, if ω_s was first used to compute $u^{(m)}$, then $q = n - 1$. The value of p corresponding to iteration n is then $p = 1$.

From the above definitions and from (9-4.3), we have $\varepsilon^{(n)} = \varepsilon^{(p+q)} = \mathscr{L}_{\omega_s}^p \varepsilon^{(q)}$. To clearly indicate the relationship between n, p, and ω_s, we express $\varepsilon^{(n)} = \varepsilon^{(p+q)}$ as $\varepsilon^{(s,\,p)}$. Thus we have

$$\varepsilon^{(n)} \equiv \varepsilon^{(s,\,p)} = [(\mathscr{L}_{\omega_s})^p]\varepsilon^{(s,\,0)} = [(\mathscr{L}_{\omega_s})^p]\varepsilon^{(q)}. \tag{9-5.1}$$

Similarly, we write $\Delta^{(n)} = \Delta^{(p+q)}$ as $\Delta^{(s,\,p)}$. Thus $\Delta^{(s,\,0)} = \Delta^{(q)}$ and

$$\Delta^{(n)} \equiv \Delta^{(s,\,p)} = [(\mathscr{L}_{\omega_s})^p]\Delta^{(s,\,0)} = [(\mathscr{L}_{\omega_s})^p]\Delta^{(q)}. \tag{9-5.2}$$

Using the subscript s or superscript (s) to indicate a dependence on ω_s, we let $\{\lambda_i^{(s)}\}_{i=1}^N$ denote the set of eigenvalues for $\mathscr{L}_{\omega_s}$ and let $y_s(i)$ denote an eigenvector of $\mathscr{L}_{\omega_s}$ corresponding to the eigenvalue $\lambda_i^{(s)}$. We assume that the eigenvalues $\lambda_i^{(s)}$ are ordered as in (9-4.4).

With the basic assumption that $\omega_s < \omega_b$, the adaptive ω estimation procedure given in this section is designed to handle the worst circumstance for the Case 2 condition discussed in Section 9.4. Recall that, for our purposes, the worst Case 2 condition is when $(\omega_s - 1) = \frac{1}{4}(\mu_2^2\omega_s^2)$ and that for this occurrence, the matrix $\mathscr{L}_{\omega_s}$ has a principal vector of grade two associated with the repeated eigenvalue $\lambda_2^{(s)} = \lambda_3^{(s)}$. We denote this principal vector by $p_s(2)$. From the discussion given in Section 9.4, a basis for E^N can be given by the set of vectors $\{y_s(1), p_s(2), y_s(3), \ldots, y_s(N)\}$. In terms of this set, we may express the vector $\varepsilon^{(s,\,0)}$ in the expanded form

$$\varepsilon^{(s,\,0)} = c_1^{(s)}y_s(1) + c_2^{(s)}p_s(2) + \sum_{i=3}^N c_i^{(s)}y_s(i). \tag{9-5.3}$$

As in Section 9.4, we can obtain from (9-5.3) the related expansion for $\Delta^{(s,\,0)}$:

$$\Delta^{(s,\,0)} = d_1^{(s)}y_s(1) + d_2^{(s)}p_s(2) - c_2^{(s)}y_s(3) + \sum_{i=3}^N d_i^{(s)}y_s(i), \tag{9-5.4}$$

where $d_i^{(s)} = (1 - \lambda_i^{(s)})c_i^{(s)}$.

From (9-4.14), the contamination factor associated with $\Delta^{(s,\,0)}$ is

$$K_s' = \frac{\|c_2^{(s)}y_s(3)\|_\psi + \|d_2^{(s)}p_s(2)\|_\psi + \sum_{i=3}^N \|d_i^{(s)}y_s(i)\|_\psi}{\|d_1^{(s)}y_s(1)\|_\psi}. \tag{9-5.5}$$

Numerical results indicate that the contamination factors K_s' generally decrease with s when the corresponding $p^{(s)}$ is sufficiently large. However, since the set of eigenvectors for $\mathscr{L}_{\omega_s}$ vary with s, it is difficult to prove that this always must be true. This fact coupled with the additional complications that the set of eigenvectors for $\mathscr{L}_{\omega_s}$ need not include a basis and that some eigenvalues of $\mathscr{L}_{\omega_s}$ are complex make any adaptive procedure for the SOR method more difficult than that given previously for the Chebyshev procedures.

Equation (9-3.8) and the results of Theorem 9-4.2 serve as a basis for the SOR adaptive procedure to be given. Three important ingredients of the adaptive procedure are (a) the test to determine if a new approximation ω for ω_b is needed, (b) the method used to obtain a new estimate for ω, and (c) the criteria used to determine when the iteration error is sufficiently small. However, because of the complications given above, the procedures we will use for (a)–(c) are not valid for all iterations. Thus additional tests are needed to indicate those iterations for which the procedures used for (a)–(c) are likely to be valid. The tests we use for this purpose, which we call *strategy sufficient conditions*, are discussed below. We remark that the adaptive procedure given here differs from those given by Carré [1961] and Kulsrud [1961] primarily because of the strategy sufficient conditions imposed. Numerical results (e.g., see Ref. 7 in Hageman and Porsching [1975]) clearly indicate that some tests of this nature are needed.

Parameter Change Test

The largest possible (or optimum) convergence rate for the SOR method is that obtained when $\omega = \omega_b$ and is $-\log S(\mathscr{L}_{\omega_b}) = -\log(\omega_b - 1)$. The asymptotic convergence rate using ω_s is $-\log S(\mathscr{L}_{\omega_s})$. The estimate ω_s is assumed to be satisfactory if the convergence rate using ω_s is greater than F times the optimum rate, where $F < 1$. Thus ω_s is satisfactory if

$$-\log S(\mathscr{L}_{\omega_s}) \geq F[-\log(\omega_b - 1)] \tag{9-5.6}$$

or equivalently if

$$S(\mathscr{L}_{\omega_s}) \leq (\omega_b - 1)^F. \tag{9-5.7}$$

If $\omega_s < \omega_b$, then the inequality $S(\mathscr{L}_{\omega_s}) \leq (\omega_s - 1)^F$ implies that (9-5.7) and (9-5.6) are also satisfied. Also, if $\omega_s < \omega_b$ and if p is sufficiently large, then from (9-4.19) we have approximately that

$$\frac{\|\Delta^{(s,\,p)}\|}{\|\Delta^{(s,\,p-1)}\|} \doteq \lambda_1 = S(\mathscr{L}_{\omega_s}). \tag{9-5.8}$$

Thus letting

$$R^{(s,\,p)} \equiv \frac{\|\Delta^{(s,\,p)}\|_2}{\|\Delta^{(s,\,p-1)}\|_2}, \tag{9-5.9}$$

we take ω_s to be a satisfactory estimate for ω_b if

$$R^{(s,\,p)} \leq (\omega_s - 1)^F \tag{9-5.10}$$

As before, we refer to F as the damping factor. Typically, F is chosen to lie in the interval $[0.65, 0.8]$.

Obtaining a New Estimate for ω_b

If $\omega_s \le \omega_b$, then from (9-3.8), μ_1 satisfies

$$S(\mathscr{L}_{\omega_s}) = \{\tfrac{1}{2}[\omega_s \mu_1 + \sqrt{\omega^2 \mu_1^2 - 4(\omega_s - 1)}]\}^2. \qquad (9\text{-}5.11)$$

Using (9-5.8) to approximate $S(\mathscr{L}_{\omega_s})$, we take as the new estimate μ_1' for μ_1, the largest real x which satisfies the SOR *equation*

$$\{\tfrac{1}{2}[\omega_s x + \sqrt{\omega_s^2 x^2 - 4(\omega_s - 1)}]\}^2 = R^{(s, p)}. \qquad (9\text{-}5.12)$$

Given μ_1', the new estimate ω' for ω_b is then obtained by

$$\omega' = 2/[1 + \sqrt{1 - (\mu_1')^2}]. \qquad (9\text{-}5.13)$$

Note that the solution μ_1' to (9-5.12) is simply

$$\mu_1' = [R^{(s, p)} + \omega_s - 1]/\omega_s [R^{(s, p)}]^{1/2}. \qquad (9\text{-}5.14)$$

Iteration Stopping Test

If $\omega_s < \omega_b$ and if p is sufficiently large, then from (9-4.20) we have approximately that

$$\|\varepsilon^{(s, p)}\|_\beta \doteqdot \frac{1}{1 - \lambda_1}\|\Delta^{(s, p)}\|_\beta, \qquad (9\text{-}5.15)$$

where $\|\cdot\|_\beta$ is any vector norm. If $\bar{u}$ is the solution to (9-1.1), then using (9-5.8) to approximate λ_1 and using $\|u^{(n+1)}\|_\eta$ to approximate $\|\bar{u}\|_\eta$, we obtain the approximation

$$\frac{\|\varepsilon^{(s, p)}\|_\beta}{\|\bar{u}\|_\eta} \doteqdot \frac{1}{1 - R^{(s, p)}} \frac{\|\Delta^{(s, p)}\|_\beta}{\|u^{(n+1)}\|_\eta}. \qquad (9\text{-}5.16)$$

Thus for the iteration termination test, we use†

$$\frac{1}{1 - H} \frac{\|\Delta^{(s, p)}\|_\beta}{\|u^{(n+1)}\|_\eta} \le \zeta, \qquad (9\text{-}5.17)$$

where ζ is stopping criterion number and where

$$H \equiv \max[\omega_s - 1, R^{(s, p)}]. \qquad (9\text{-}5.18)$$

If $R^{(s, p)} \ge 1$, no test for convergence is made.

The quantity H instead of $R^{(s, p)}$ is used in (9-5.17) as a precautionary step in case $\omega_s \ge \omega_b$. If $\omega_s \ge \omega_b$, we show in the next section that the quantity

---

† For a discussion concerning the β- and η-norms, see the comments given following Eq. (5-4.32).

$R^{(s, p)}$ normally will oscillate about $(\omega_s - 1)$. Thus on some iterations, $(1 - R^{(s, p)})$ may be considerably larger than $[1 - \mathbf{S}(\mathscr{L}_{\omega_s})]$. The use of H instead of $R^{(s, p)}$ prevents any exceptionally large value of $(1 - R^{(s, p)})$ from being used in the stopping test.

Strategy Sufficient Conditions

The approximations used in the above tests and the approximations used in the equations for computing the new estimate ω' are reasonable provided $\omega_s < \omega_b$ and provided p is sufficiently large. One of the more difficult problems associated with the SOR adaptive procedure is to determine conditions which ensure that p is sufficiently large. For the discussion which follows, we continue to assume that ω_s satisfies the Case 2 condition of $\omega_s - 1 = \frac{1}{4}\mu_2^2\omega_s^2 = \lambda_2$.

We first examine the behavior of $R^{(s, p)} = \|\Delta^{(s, p)}\|_2/\|\Delta^{(s, p-1)}\|_2$ in more detail. With $\Delta^{(s, 0)}$ given by (9-5.4), we show later in Section 9.10 that $\Delta^{(s, p)}$ can be expressed (see Eq. (9-10.10)) in the form

$$\Delta^{(s, p)} = \lambda_1^p d_1^{(s)} y_s(1) + p(\lambda_2)^{p-1} d_2^{(s)} y_s(2)$$

$$+ \lambda_2^p[d_2^{(s)}p_s(2) - c_2^{(s)}y_s(3)] + \sum_{i=3}^{N} \lambda_i^p y_s(i). \qquad (9-5.19)$$

Because of the $p\lambda_2^{p-1}$ term, $\|\Delta^{(s, p)}\|_2$ may be an increasing function of p initially. Thus the ratio $R^{(s, p)}$ may be greater than unity for small p. For $\lambda_2 \ (= \omega_s - 1)$ close to unity, a typical graph of $R^{(s, p)}$ versus p is given in Fig. 9-5.1. The first iteration, $\hat{p}$, for which $R^{(s, p)}$ is less than unity can be as large as $(\omega_s - 1)/(2 - \omega_s)$. This is the value of p which maximizes $p(\omega_s - 1)^{p-1}$.

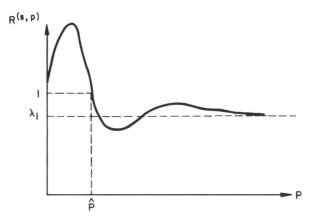

Fig. 9-5.1. Typical behavior for $R^{(s, p)}$.

To compensate for this possible behavior, we will require that p and $R^{(s,p)}$ satisfy certain conditions before permitting a new estimate for ω_b to be calculated. The conditions we impose are that

(1) p must be greater than p^*, where p^* is the smallest integer greater than 5 which satisfies

$$p^*(\omega_s - 1)^{p^* - 1} \leq PSP, \tag{9-5.20}$$

and that

(2) $R^{(s,p)}$ must satisfy

$$(-10)RSP \leq (R^{(s,p-1)} - R^{(s,p)}) \leq RSP. \tag{9-5.21}$$

Here PSP and RSP are strategy parameters. Numerical results indicate that the values $PSP = 0.5$ and $RSP = 0.0001$ are appropriate.† Condition (1) is imposed so that $R^{(s,p)}$ is well into its decreasing tail (see Fig. 9-5.1). Condition (2) may be considered as a type of convergence test on $R^{(s,p)}$. It prevents a new ω estimate from being computed while $R^{(s,p)}$ is still changing significantly. Most often, it is the test (9-5.21) which dictates when a new ω is to be computed. The significance of the test (9-5.21) is discussed in more detail in Section 9.10.

In an attempt to ensure that $\omega_s < \omega_b$, upper bounds are imposed on the ω_s. If the strategy parameters τ_s satisfy

$$\tau_1 = 1.6, \qquad \tau_2 = 1.8, \qquad \tau_3 = 1.90, \qquad \tau_4 = 1.95,$$
$$\tau_5 = 1.975, \qquad \tau_6 = 1.985, \qquad \tau_7 = 1.990, \qquad \tau_8 = 1.995,$$

and

$$\tau_s = 1.995 \quad \text{for} \quad s \geq 8, \tag{9-5.22}$$

then ω_s is required to satisfy $\omega_s \leq \tau_s$.

9.6   AN OVERALL COMPUTATIONAL ALGORITHM

In this section we describe a computational procedure for the SOR method based on the above discussion. The overall algorithm is given below as an informal program. We assume that the matrix problem to be solved is partitioned in the form (9-3.1) and that the *coefficient matrix A is consistently ordered* relative to the partitioning imposed (see Section 9.2). To ensure convergence of the SOR process, it is also assumed that A is SPD *or*

---

† Alternative suggested values to use for RSP are given in (9-10.24).

that the Jacobi method associated with the matrix problem (9-3.1) is symmetrizable (see Definition 2-2.1).

The following input is required.

ζ the stopping criterion number ζ used in (9-5.17).

ω the initial estimate for ω_b. For Algorithm 9-6.1, ω must satisfy $1.0 \leq \omega < 2.0$, If ω_b is not known, it is usually best to let $\omega = 1.0$ or some value close to 1.0.

F the damping factor used in the parameter change test (9-5.10). Typically, F should satisfy $0.65 \leq F \leq 0.8$.

PSP, RSP strategy parameters used for the tests (9-5.20) and (9-5.21). Suggested values are $PSP = 0.5$ and $RSP = 0.0001$.

$\underline{u}_\Gamma$ the initial guess vector.

The control variables and counters used are

n counter for the current iteration step.

p counter for the number of iterations using the current value of ω.

s counter for the number of times ω is changed.

τ_s upper bound for the sth estimate ω_s for ω_b. Suggested values for τ_s are given in (9-5.22).

p^* required number of iterations using the current value of ω before the parameter change test is made. p^* is given by (9-5.20).

$\hat{p}$ required number of iterations using the current value of ω before any convergence testing is done. $\hat{p}$ is given by (9-6.2) and is discussed later.

S counter used to determine if possibly $\omega_s \geq \omega_b$.

The underline, $\underline{u}$, is used to indicate more clearly which variables are vectors. At the beginning of iteration n, $\underline{u}_\Gamma$ is the vector iterate $u^{(n-1)}$. At the end of iteration n, $\underline{u}_\Gamma$ is the vector iterate $u^{(n)}$. The ith block of unknowns in the partitioned vector $\underline{u}_\Gamma$ is denoted by $(\underline{u}_\Gamma)_i$. Note that the ith block of unknowns, $(\underline{u}_\Gamma)_i$, in $\underline{u}_\Gamma$ is updated as soon as the improved values are available. Since the subvectors $\underline{u}_T$ and $(\underline{\Delta})_i$ are used only to obtain $(\underline{u}_\Gamma)_i$, only $2\bar{n}$ words of temporary storage is required for the $\underline{u}_T$ and $(\underline{\Delta})_i$ subvectors. Here, $\bar{n} = \max_{1 \leq i \leq q} n_i$, where n_i is the order of the $(\underline{u}_\Gamma)_i$ block of unknowns.

For a discussion on the solution of the subsystem $A_{i,i}(\underline{u}_T)_i = \underline{y}_i$, where $\underline{y}_i$ is known, see Sections 2.3 and 5.6.

If the input value for ω equals 1.0, an initial estimate for ω is obtained by doing four Gauss–Seidel† iterations initially. It can be shown that $\lim_{n \to \infty} [\|\Delta^{(n)}\|_2 / \|\Delta^{(n-1)}\|_2] = \mu_1^2$ for the Gauss–Seidel process. Thus at the

---

† Recall that the SOR method reduces to the Gauss–Seidel method when $\omega = 1.0$.

completion of the Gauss–Seidel iterations we obtain an estimate for ω_b using the formula

$$\omega' = \frac{2}{1 + \sqrt{1 - \|\Delta^{(3)}\|_2 / \|\Delta^{(2)}\|_2}}. \tag{9-6.1}$$

Algorithm 9-6.1.   An adaptive procedure for the SOR method.

Input: $(\zeta, \omega, F, PSP, RSP, \underline{u}_\Gamma)$

Initialize:

$n := 0; p := -1; \omega' := \omega; s := 0; R := 1.0; \text{DELNP} := 1.0.$

Next Iteration:

$n := n + 1; p := p + 1; \text{DELNO} := \text{DELNP};$
If $p = 0$, *then* ⟨Initialize for use of new ω estimate⟩
        Begin
        $S := 0; s := s + 1; \hat{p} := 3; p^* := 6; \omega := \omega';$
        If $\omega' > \tau_s$ *then* $\omega := \tau_s;$ *else* continue;
        If $(\omega - 1)/(2 - \omega) > \hat{p}$ *then* $\hat{p} := (\omega - 1)/(2 - \omega);$ *else* continue;
        If $p^*(\omega - 1)^{p^*-1} > PSP$ *then* $p^* := p^* + 1$ and repeat *If* (this line);
                                else continue;
        If $\omega = 1.0$ *then*
                        Begin
                        $\hat{p} := 2; p^* := 3; s := s - 1;$
                        End
                        else continue;
        End
        else continue;

Calculate New Iterate:

Do (1), (2), and (3) for $i = 1, 2, \ldots, q$

(1)   Solve for $\underline{u}_\Gamma$, where $A_{i,i}\underline{u}_\Gamma = -\sum_{j=1}^{i-1} A_{i,j}(\underline{u}_\Gamma)_j - \sum_{j=i+1}^{q} A_{i,j}(\underline{u}_\Gamma)_j + \underline{F}_i$

(2) $(\Delta)_i := \omega[(\underline{u}_\Gamma)_i - \underline{u}_\Gamma]$
(3) $(\underline{u}_\Gamma)_i := (\underline{u}_\Gamma)_i - (\Delta)_i$
End of *Do*
Remark: While doing the above *Do* Loop, also compute

$$\text{DELNP} := \|\Delta\|_2; \quad \text{DELNE} := \|\Delta\|_\beta; \quad \text{and} \quad \text{YUN} := \|\underline{u}_\Gamma\|_\eta.$$

Stopping Test:

$RO := R$; $R := \text{DELNP}/\text{DELNO}$; $H := R$;
If $p < \hat{p}$ *then Go to* **Next Iteration**; *else* continue;
If $R \geq 1.0$ *then Go to* **Next Iteration**; *else* continue;
If $R \leq \omega - 1$, *then*

$\qquad$ *Begin*
$\qquad H := \omega - 1$; $S := S + 1$;
$\qquad$ *End*
$\qquad$ *else* continue;
If $\text{DELNE}/\text{YUN} \leq \zeta(1 - H)$, *then* print final output and STOP (converged);
$\qquad\qquad\qquad\qquad$ *else* continue;

Parameter Change Test:

If $p < p^*$ *then Go to* **Next Iteration**; *else* continue;
If $S > 0$ *then Go to* **Next Iteration**; *else* continue;
If $\omega = 1$, *then*

$\qquad$ *Begin*
$\qquad \omega' := 2/[1 + \sqrt{1 - R}]$;
$\qquad p := -1$
$\qquad$ *Go to* **Next Iteration**
$\qquad$ *End*
$\qquad$ *else* continue;
If $R < (\omega - 1)^F$ *then Go to* **Next Iteration**; *else* continue;
$\text{DELR} := (RO - R)$
If $(-10.0) * (RSP) \leq \text{DELR} \leq RSP$ *then* continue;

$\qquad\qquad\qquad$ *else*
$\qquad\qquad\qquad$ *Begin*
$\qquad\qquad\qquad$ *If* $s > 2$ *then Go to* **Next Iteration**;
$\qquad\qquad\qquad\qquad$ *else* continue;
$\qquad\qquad\qquad$ *If* $R < (\omega - 1)^{0.1}$ *then Go to*
$\qquad\qquad\qquad\qquad\qquad\qquad$ **Next Iteration**;
$\qquad\qquad\qquad\qquad\qquad$ *else* continue;
$\qquad\qquad\qquad$ *End*
$\mu' := [R + \omega - 1]/[\omega\sqrt{R}]$; $\omega' := 2/[1 + \sqrt{1 - (\mu')^2}]$; $p := -1$;
Go to **Next Iteration**.

When a new estimate for ω is used, no convergence testing is done until $\hat{p}$ iterations have been done, and no parameter change testing is done until p^* iterations are done. As discussed previously, p^* is the smallest integer greater than 5 which satisfies (9-5.20). The value used for $\hat{p}$ is

$$\hat{p} = \max[3, (\omega_s - 1)/(2 - \omega_s)]. \qquad (9\text{-}6.2)$$

The parameter $\hat{p}$ is used primarily to avoid any use of $R^{(s, p)}$ for small p when its behavior is sometimes erratic. Note that the value $(\omega_s - 1)/(2 - \omega_s)$ for $\hat{p}$ approximates the point p in Fig. 9-5.1, where $R^{(s, p)}$ first becomes less than unity.

The adaptive parameter estimation procedures developed in Section 9.5 are based on the assumption that each estimate ω_s is less than ω_b. However, it may happen that some estimate ω_s turns out to be greater than ω_b. We show below that whenever $\omega_s > \omega_b$, then the ratio $R^{(s, p)}$ defined by (9-5.9) is likely to exhibit an oscillatory behavior about the value $\omega_s - 1$. The counter S is used in the algorithm to detect this occurrence. When a new estimate ω_s is first used, the counter S is set to zero. Then, in the **Stopping Test** portion of the algorithm, the counter S is incremented by one whenever, for $p \geq \hat{p}$, $R^{(s, p)}$ is less than $\omega_s - 1$.

If $\omega_s > \omega_b$, all eigenvalues λ_j of $\mathscr{L}_{\omega_s}$ lie on a circle of radius $\omega_s - 1$ and can be expressed as $\lambda_j = (\omega_s - 1)e^{i\theta_j}$. Note that $\theta_j \neq 0$ for all j since no eigenvalue of $\mathscr{L}_\omega$ can equal $(\omega_s - 1)$ when $\omega_s > \omega_b$. For this case, from Eqs. (9-10.2) and (9-10.3) given in Section 9.10, $\varepsilon^{(s, p)}$ and $\Delta^{(s, p)}$ may be expressed as

$$\varepsilon^{(s, p)} = (\omega_s - 1)^p \{2\mathrm{Re}[e^{i(p\theta_1)}c_1 y(1)] + \cdots\} \tag{9-6.3}$$

and

$$\Delta^{(s, p)} = (\omega_s - 1)^p \{2\mathrm{Re}[e^{i(p\theta_1)}(1 - \lambda_1)c_1 y(1)] + \cdots\}. \tag{9-6.4}$$

From (9-6.4), it is obvious that the ratio $R^{(s, p)}$ will show oscillations about $\omega_s - 1$. For any such oscillations, the S counter in the adaptive procedure will be nonzero for sufficiently large p. Thus $S \neq 0$ is used as a signal that ω_s is probably greater than ω_b. When this occurs, no new estimation of ω is permitted; i.e., ω_s is used until convergence is achieved. Moreover, we continue to use (9-5.17) for the iteration stopping test. Even though the assumptions used in the development of (9-5.17) are not valid when $\omega_s \geq \omega_b$, the numerical results given in Section 9.8 indicate that this stopping test is reasonably accurate even when $\omega_s \geq \omega_b$.

In Appendix C, we give a Fortran listing of a subroutine, called SOR, which implements Algorithm 9-6.1 with the exception of the **Calculate New Iterate** portion. The SOR subroutine is designed for use as a software package to provide the required acceleration parameters and to provide an estimate of the iteration error for the SOR method.

Remark 1. Printing the ratio $R^{(s, p)} = \|\Delta^{(s, p)}\|_2/\|\Delta^{(s, p-1)}\|_2$ can be useful in appraising the effectiveness of the adaptive procedure. For example, if $R^{(s, p)}$ is a converging sequence, the implication is that $\omega_s < \omega_b$. If $R^{(s, p)}$ oscillates about $\omega_s - 1$, the implication is that $\omega_s > \omega_b$.

Another useful quantity to print is

$$C^{(s,\,p)} = [-\log R^{(s,\,p)}]/[-\log(\omega_s - 1)], \qquad (9\text{-}6.5)$$

which, for p sufficiently large, is the approximation for the ratio of actual to optimum convergence rates for iteration n. From (9-5.6)–(9-5.10), ω will be changed only if $C^{(s,\,p)} < F$.

Remark 2.   Since ω_b may be known for some problems, an option should be added to the procedure of Algorithm 9-6.1 to permit the input ω to be used for all iterations. This can be done easily by avoiding the test on τ_1 and setting p^* to a large number.

Remark 3.   Reid [1966] suggested that the Rayleigh quotient with respect to the Jacobi iteration matrix B be used to approximate μ_1. Using the notation of Eq. (9-3.3), the Rayleigh quotient (see Section 1.3) with respect to B satisfies

$$(v,(C_L + C_U)v)/(v, Dv) \le \mu_1 \qquad (9\text{-}6.6)$$

with equality if $v = v(1)$, the eigenvector of B associated with μ_1. Reid [1966] suggested that the relationship (see Theorem 9-3.1) between the eigenvectors of $\mathscr{L}_\omega$ and those of B be used to get an estimate for $v(1)$ and then use this estimate in (9-6.6) to obtain a lower bound for μ_1. If $\omega_s < \omega_b$ and if p is sufficiently large, $\Delta^{(s,\,p)}$ may be taken as an estimate for $y_s(1)$. Thus from Theorem 9-3.1,

$$v^{(s,\,p)} = E^{-1}(\sqrt{\lambda_1^+})\Delta^{(s,\,p)} \qquad (9\text{-}6.7)$$

is then an estimate for $v(1)$. In (9-6.7), λ_1^+ is approximated by $R^{(s,\,p)}$. However, numerical results using the Rayleigh quotient (9-6.6) have been somewhat disappointing for the general consistent ordering case. (For example, see Hageman and Porsching [1975, Ref. 7].) The reason for this is believed due to the fact that the matrix $E^{-1}(\sqrt{\lambda_1^+})$ and, hence, the approximation $v^{(s,\,p)}$ depend strongly on the estimate for λ_1^+ for the general consistent ordering case. However, for problems with red/black partitionings, the use of a Rayleigh quotient approximation is often very effective. We discuss this application in the next section.

The use of the Rayleigh quotient (9-6.6) requires some additional calculations and, possibly, some additional storage. Its main advantage is that the resulting ω estimate is guaranteed to be less than ω_b.

9.7   THE SOR METHOD FOR PROBLEMS WITH RED/BLACK PARTITIONINGS

In this section, we describe a special SOR algorithm to solve matrix problems which are partitioned in the red/black form

$$\begin{bmatrix} D_R & H \\ H^T & D_B \end{bmatrix} \begin{bmatrix} u_R \\ u_B \end{bmatrix} = \begin{bmatrix} b_R \\ b_B \end{bmatrix}. \tag{9-7.1}$$

We assume that the coefficient matrix in (9-7.1) is SPD. The adaptive parameter procedure used in this special algorithm utilizes a Rayleigh quotient. Because of this, any ω estimate obtained is guaranteed to be less than ω_b. We show that the Rayleigh quotient is effective for the problem (9-7.1) primarily because of the simple relationships which exist between the eigenvectors of the SOR, the Gauss–Seidel, and the Jacobi iteration matrices.†

The SOR method associated with (9-7.1) can be written as

$$\begin{aligned} D_R u_R^{(n+1)} &= \omega[-Hu_B^{(n)} + b_R] + (1-\omega)D_R u_R^{(n)} \\ D_B u_B^{(n+1)} &= \omega[-H^T u_R^{(n+1)} + b_B] + (1-\omega)D_B u_B^{(n)}. \end{aligned} \tag{9-7.2}$$

We assume that the difference and error vectors are partitioned in a red/black form consistent with (9-7.2). The notation of Section 9.5 is used to indicate successive ω estimates; i.e., $\Delta^{(n)}$ is expressed by $\Delta^{(n)} \equiv \Delta^{(s,\,p)}$.

Let B, $\mathscr{L}_1$, and $\mathscr{L}_{\omega_s}$ denote, respectively, the Jacobi, Gauss–Seidel, and SOR iteration matrices associated with the partitioning (9-7.1). If $\omega_s < \omega_b$, each of these iteration matrices has a positive eigenvalue which is equal to its spectral radius. Letting $v(1)$, $z(1)$, and $y(1)$ denote eigenvectors of B, $\mathscr{L}_1$, and $\mathscr{L}_{\omega_s}$, respectively, corresponding to the eigenvalues $S(B)$, $S(\mathscr{L}_1)$ and $S(\mathscr{L}_{\omega_s})$, we may write the respective eigenequations for B, $\mathscr{L}_1$, and $\mathscr{L}_{\omega_s}$ in the forms

$$B\begin{bmatrix} v_R(1) \\ v_B(1) \end{bmatrix} = S(B)\begin{bmatrix} v_R(1) \\ v_B(1) \end{bmatrix}, \qquad \mathscr{L}_1\begin{bmatrix} z_R(1) \\ z_B(1) \end{bmatrix} = S(\mathscr{L}_1)\begin{bmatrix} z_R(1) \\ z_B(1) \end{bmatrix}, \qquad \text{and}$$

$$\mathscr{L}_{\omega_s}\begin{bmatrix} y_R(1) \\ y_B(1) \end{bmatrix} = S(\mathscr{L}_{\omega_s})\begin{bmatrix} y_R(1) \\ y_B(1) \end{bmatrix}. \tag{9-7.3}$$

---

† Because of larger average convergence rates, the RS–SI and CCSI methods of Chapter 8 are slightly more efficient than the SOR method for red/black partitioned problems. Thus if possible, the RS–SI or CCSI method, rather than the SOR method, should be used for the special matrix problem (9-7.1).

Using the partitioned forms (8-1.5) and (9-7.6) for B and $\mathscr{L}_1$, respectively, we have by direct calculation that $z_R(1)$ and $z_B(1)$ can be given by $z_R(1) = v_R(1)$ and $z_B(1) = [S(\mathscr{L}_1)]^{1/2}v_B(1)$. Moreover, from Theorem 9-3.1, the eigenvector $y(1)$ of $\mathscr{L}_{\omega_s}$ and the eigenvector $v(1)$ of B satisfy $y(1) = E([S(\mathscr{L}_{\omega_s})]^{1/2})v(1)$, where $E(x)$ is given by (9-3.11) for the red/black partitioned problem. Thus the eigenvectors $v(1)$, $z(1)$, and $y(1)$ of B, $\mathscr{L}_1$, and $\mathscr{L}_{\omega_s}$, respectively, can be chosen to satisfy

$$\begin{bmatrix} v_R(1) \\ v_B(1) \end{bmatrix} = \begin{bmatrix} z_R(1) \\ [\sqrt{S(\mathscr{L}_1)}]^{-1}z_B(1) \end{bmatrix} = \begin{bmatrix} y_R(1) \\ [\sqrt{S(\mathscr{L}_{\omega_s})}]^{-1}y_B(1) \end{bmatrix}$$

or equivalently, since eigenvectors may be multiplied by an arbitrary non-zero constant, the eigenvectors of B, $\mathscr{L}_1$, and $\mathscr{L}_{\omega_s}$ can be chosen to satisfy

$$\begin{bmatrix} v_R(1) \\ v_B(1) \end{bmatrix} = \begin{bmatrix} \sqrt{S(\mathscr{L}_1)}z_R(1) \\ z_B(1) \end{bmatrix} = \begin{bmatrix} \sqrt{S(\mathscr{L}_{\omega_s})}y_R(1) \\ y_B(1) \end{bmatrix}. \tag{9-7.4}$$

Thus the black subvector part for both the $v(1)$ and $z(1)$ eigenvectors has the same shape as $y_B(1)$.

If $\omega_s < \omega_b$, we show later in Section 9.10 (see Lemmas 9-10.1 and 9-10.2) that $\lim_{p \to \infty} \{\Delta^{(s,p)}/[S(\mathscr{L}_{\omega_s})]^p\} = y(1)$, where $y(1)$ is an eigenvector of $\mathscr{L}_{\omega_s}$ corresponding to the eigenvalue $S(\mathscr{L}_{\omega_s})$. Thus for sufficiently large p, we have approximately that

$$\Delta_B^{(s,p)} = y_B(1) = z_B(1) = v_B(1). \tag{9-7.5}$$

We now use the fact that $\Delta_B^{(s,p)}$ approximates $z_B(1)$ to define a Rayleigh quotient which bounds $S(\mathscr{L}_1)$.

The Gauss–Seidel iteration matrix associated with the partitioning (9-7.1) is

$$\mathscr{L}_1 = \begin{bmatrix} 0 & -D_R^{-1}H \\ 0 & D_B^{-1}H^TD_R^{-1}H \end{bmatrix}. \tag{9-7.6}$$

Thus the nonzero eigenvalues of $\mathscr{L}_1$ are the same as those for $D_B^{-1}H^TD_R^{-1}H$. Moreover, from (9-7.3), $(D_B^{-1}H^TD_R^{-1}H)z_B(1) = S(\mathscr{L}_1)z_B(1)$. The matrix $D_B^{-1}H^TD_R^{-1}H$ is similar to the symmetric matrix $D_B^{-1/2}H^TD_R^{-1}HD_B^{-1/2}$. With respect to this symmetric matrix, the Rayleigh quotient (see (1-3.5)) for any nonzero vector w satisfies

$$\frac{(w, (D_B^{-1/2}H^TD_R^{-1}HD_B^{-1/2})w)}{(w, w)} \le S(\mathscr{L}_1) \tag{9-7.7}$$

or equivalently†

$$\frac{(\Delta_B^{(s,\,p)},\,(H^T D_R^{-1} H)\Delta_B^{(s,\,p)})}{(\Delta_B^{(s,\,p)},\,D_B \Delta_B^{(s,\,p)})} \le \mathsf{S}(\mathscr{L}_1), \tag{9-7.8}$$

with equality if $\Delta_B^{(s,\,p)} = z_B(1)$. Thus if $\Delta_B^{(s,\,p)}$ closely approximates $z_B(1)$, the Rayleigh quotient (9-7.8) will give an accurate estimate for $\mathsf{S}(\mathscr{L}_1) = \mathsf{S}^2(B)$. Note that the Rayleigh quotient (9-7.8) is not dependent on estimates for $\mathsf{S}(\mathscr{L}_{\omega_s})$ as is the case for a general consistent ordering (see Eqs. (9-6.6)–(9-6.7)).

If the matrices D_R and D_B have been factored (see Section 2.3) as

$$D_R = S_R^T S_R \quad \text{and} \quad D_B = S_B^T S_B, \tag{9-7.9}$$

then (9-7.8) can be expressed in the computationally simpler form

$$\frac{([(S_R^T)^{-1} H\Delta_B^{(s,\,p)}],\,[(S_R^T)^{-1} H\Delta_B^{(s,\,p)}])}{([S_B \Delta_B^{(s,\,p)}],\,[S_B \Delta_B^{(s,\,p)}])} \le \mathsf{S}(\mathscr{L}_1). \tag{9-7.10}$$

If $\Delta_B^{(s,\,p)}$ is stored, the vectors $(S_R^T)^{-1} H\Delta_B^{(s,\,p)}$ and $S_B \Delta_B^{(s,\,p)}$ can be computed by slightly modifying the computational process of (9-7.2). The major disadvantage in the use of (9-7.10) is that the $\Delta_B^{(s,\,p)}$ vector must be stored. We now present a Rayleigh quotient procedure which does not require storage of the $\Delta_B^{(s,\,p)}$ vector but does require that two Gauss–Seidel iterations be done every time a new estimate for ω is to be calculated.

If $\Delta^{(0)}$ and $\Delta^{(1)}$ are the difference vectors (9-4.1) for two Gauss–Seidel iterations, then since $\Delta^{(1)} = \mathscr{L}_1 \Delta^{(0)}$ we have

$$\Delta_B^{(1)} = D_B^{-1} H^T D_R^{-1} H\Delta_B^{(0)}. \tag{9-7.11}$$

Thus, using (9-7.9) we have that the ratio

$$\frac{\|\Delta_B^{(1)}\|_{S_B}^2}{\|\Delta_B^{(0)}\|_{S_B}^2} = \frac{(\Delta_B^{(0)},\,(H^T D_R^{-1} H D_B^{-1} H^T D_R^{-1} H)\Delta_B^{(0)})}{(\Delta_B^{(0)},\,D_B \Delta_B^{(0)})}, \tag{9-7.12}$$

is equivalent (see Eqs. (9-7.7) and (9-7.8)) to the Rayleigh quotient with respect to the symmetric matrix $[(S_B^T)^{-1} H^T D_R^{-1} H D_B^{-1} H^T D_R^{-1} H(S_B)^{-1}]$, which is similar to $(D_B^{-1} H^T D_R^{-1} H)^2$. Thus

$$\|\Delta_B^{(1)}\|_{S_B}^2 / \|\Delta_B^{(0)}\|_{S_B}^2 \le \mathsf{S}^2(\mathscr{L}_1) \tag{9-7.13}$$

with equality if $\Delta_B^{(0)} = z_B(1)$. The only cost in using (9-7.13), in addition to the Gauss–Seidel iterations, is that the S_B vector norm must be used. Note that the Rayleigh quotient (9-7.7) is applied to $D_B^{1/2} H^T D_R^{-1} H D_B^{1/2}$, while that of (9-7.13) is applied to $(D_B^{1/2} H^T D_R^{-1} H D_B^{1/2})^2$.

† Replace w in (9-7.7) by $D_B^{1/2}\Delta_B^{(s,\,p)}$.

An adaptive procedure utilizing the Rayleigh quotient (9-7.13) is given in Algorithm 9-7.1. This procedure differs from that given in Algorithm 9-6.1 in the following ways:

(1)   The control parameter g has been added. $g \neq 0$ implies the two Gauss–Seidel iterations to compute (9-7.13) are being done.

(2)   The norm $\|\Delta_B^{(n)}\|_{S_B}$ is computed if $\omega = 1.0$.

(3)   Since any estimate ω' must satisfy $\omega' \leq \omega_b$, the upper bounds τ_s imposed previously and the control word S have been removed.

The adaptive procedure of Algorithm 9-6.1 also may be used when the coefficient matrix A is partitioned in a red/black form. However, to utilize the special eigenvector relationship (9-7.4) most fully, $\|\Delta_B^{(s,\,p)}\|_2$ rather than $\|\Delta^{(s,\,p)}\|_2$ should be used in the adaptive parameter procedure.

For Algorithm 9-7.1, we assume that the matrix problem to be solved is partitioned into the red/black form

$$
\begin{bmatrix}
A_{1,1} & & & A_{1,l+1} & \cdots & A_{1,q} \\
 & \ddots & 0 & \vdots & & \vdots \\
0 & & A_{l,l} & A_{l,l+1} & \cdots & A_{l,q} \\
\hline
A_{l+1,1} & \cdots & A_{l+1,l} & A_{l+1,l+1} & & 0 \\
\vdots & & \vdots & & \ddots & \\
A_{q,1} & \cdots & A_{q,l} & 0 & & A_{q,q}
\end{bmatrix}
\begin{bmatrix}
U_1 \\ U_2 \\ \vdots \\ U_l \\ \hline U_{l+1} \\ \vdots \\ U_q
\end{bmatrix}
=
\begin{bmatrix}
F_1 \\ F_2 \\ \vdots \\ F_l \\ \hline F_{l+1} \\ \vdots \\ F_q
\end{bmatrix}, \quad (9\text{-}7.14)
$$

where $l \geq 1$ and $q \geq 2$. Moreover, we assume each diagonal submatrix $A_{i,i}$ has been factored (see Sections 2.3 and 5.6) as

$$
A_{i,i} = S_i^T S_i,
$$

where S_i is an upper triangular matrix. If the D_R and D_B submatrices in (9-7.1) are not of the block diagonal form given in (9-7.14), suitable adjustments must be made in the **Calculate New Iterate** portion of the algorithm.

Algorithm 9-7.1.   An adaptive procedure for problems with red/black partitionings.

Input: $(\zeta, \omega, F, PSP, RSP, \underline{u}_\Gamma)$

Initialize: Same as Algorithm 9-6.1.

Next Iteration: Same as Algorithm 9-6.1 except replace $S := 0$ by $g := 0$ when $p = 0$.

Calculate New Iterate:

Do steps (1)–(3) for $i = 1, \ldots, l$

(1)   Solve for $\underline{u}_T$, where $A_{i,i}\underline{u}_T = -\sum_{j=l+1}^{q} A_{i,j}(\underline{u}_\Gamma)_j + \underline{F}_i$

(2) $(\underline{\Delta})_i := \omega[(\underline{u}_\Gamma)_i - \underline{u}_T]$
(3) $(\underline{u}_\Gamma)_i := (\underline{u}_\Gamma)_i - (\underline{\Delta})_i$
End of *Do*.
$\|\underline{\Delta}_B\| := 0.0$
Do steps (4)–(6) for $i = l + 1, \ldots, q$

(4)   Solve for $\underline{u}_T$, where $A_{i,i}\underline{u}_T = -\sum_{j=1}^{l} A_{i,j}(\underline{u}_\Gamma)_j + \underline{F}_i$

(5) $(\underline{\Delta})_i := \omega[(\underline{u}_\Gamma)_i - \underline{u}_T]$
      If $\omega = 1.0$ *then* $\|\underline{\Delta}_B\| := \|\underline{\Delta}_B\| + (S_i(\underline{\Delta})_i, S_i(\underline{\Delta})_i)$;
                  else $\|\underline{\Delta}_B\| := \|\underline{\Delta}_B\| + ((\underline{\Delta})_i, (\underline{\Delta})_i)$
(6) $(\underline{u}_\Gamma)_i := (\underline{u}_\Gamma)_i - (\underline{\Delta})_i$;
End of *Do*.
DELNP $:= [\|\underline{\Delta}_B\|]^{1/2}$
   Remark.   While doing the above *Do* loop (or loops), also compute the appropriate quantities DELNE $:= \|\underline{\Delta}\|_\beta$ and YUN $:= \|\underline{u}_\Gamma\|_\eta$ for use in the convergence test.

Stopping Test:

$RO := R$; $R := \text{DELNP}/\text{DELNO}$; $H := R$;
If $p < \hat{p}$ *then Go to* **Next Iteration**; *else* continue;
If $R \geq 1.0$ *then Go to* **Next Iteration**; *else* continue;
If $g = 1$ *then* $p^* := p$ and *Go to* **Parameter Change Test**; *else* continue;
If $R \leq \omega - 1$ *then* $H := \omega - 1$; *else* continue;
If DELNE/YUN $\leq \zeta(1 - H)$ *then* print final output and STOP (converged);
else continue;

Parameter Change Test:

If $p < p^*$ *then Go to* **Next Iteration**; *else* continue;
If $\omega = 1$ *then*
        Begin
        $\omega' := 2/[1 + \sqrt{1 - R}]$; $p := -1$;
        Go to **Next Iteration**;
        End
        else continue;
If $R < (\omega - 1)^F$ *then Go to* **Next Iteration**; *else* continue;
DELR $:= RO - R$

If $(-10.0)(RSP) \leq$ DELR $\leq$ *RSP then* continue;
                else
                Begin
                If $s > 2$ *then Go to* **Next Iteration**;
                         else continue;
                If $R < (\omega - 1)^{0.1}$ *then Go to*
                             Next Iteration;
                             else continue;
                End
$\omega := 1.0$; $g := 1$; $p := 0$; $\hat{p} := 1$; $p^* := 1$; $n := n + 1$;
Go to **Calculate New Iterate**

9.8 NUMERICAL RESULTS

In this section we describe results of numerical experiments which were designed to illustrate the effectiveness of the adaptive procedures discussed in this chapter. To do this, three computer programs were written to solve the two-dimensional elliptic test problems described previously in Section 8.4. The three programs and the solution procedures utilized are

SOR.NO Uses the one-line SOR iteration method with a natural ordering (see Sections 1.6 and 1.7) of the mesh lines. The adaptive procedure used is that given in Algorithm 9-6.1.

SOR.RB Uses the one-line SOR iteration method with a red/black ordering (see Sections 1.6 and 1.7) of the mesh lines. The adaptive procedure used is that given in Algorithm 9-6.1 except that DELNP $:= \|\Delta_{\mathrm{B}}^{(s,p)}\|_2$ is used instead of DELNP $:= \|\Delta^{(s,p)}\|_2$.

SOR.RB/RQ Uses the one-line SOR iteration method with a red/black ordering of the mesh lines. The adaptive procedure used is that given in Algorithm 9-7.1.

The basic difference between the SOR.NO and SOR.RB programs is the line ordering used. In addition, to utilize the special eigenvector relationship (9-7.4) valid for problems with a red/black partitioning, the SOR.RB program uses $\|\Delta_{\mathrm{B}}^{(s,p)}\|_2$ rather than $\|\Delta^{(s,p)}\|_2$ in the adaptive parameter procedure. The SOR.RB and SOR.RB/RQ programs differ only in the fact that the SOR.RB/RQ program utilizes the Rayleigh quotient (9-7.13) in the estimation of ω_b. Only for the SOR.RB/RQ program is it known mathematically that any estimate ω must satisfy $\omega \leq \omega_b$.

For all three programs, the stopping test used was (9-5.17) using the relative error norm. That is, with $u_i^{(n)} \equiv (u^{(n)})_i$ and $\zeta = 10^{-6}$, the problem

TABLE 9-8.1

Summaries of the Line SOR Iterative Behavior for the Test Problems of Section 8.4

| | Nonadaptive | | | | Adaptive | | | | | | | |
| | SOR.NO | | SOR.RB | | SOR.NO | | | SOR.RB | | | SOR.RB/RQ | | |
Problem	Iter.	Value for fixed ω	Iter.	Value for fixed ω	Iter.	Last estimate for ω	$C^{(s,p)}$ at convergence	Iter.	Last estimate for ω	$C^{(s,p)}$ at convergence	Iter.	Last estimate for ω	$C^{(s,p)}$ at convergence
1 (Guess A)	92	1.8050	89	1.8050	118	1.7988	0.752	103	1.8052	0.927	101	1.8033	0.856
2 (Guess A)	168	1.8976	156	1.8976	169	1.9000	—	174	1.8976	0.978	178	1.8971	0.920
3 (Guess B)	370	1.9507	364	1.9507	419	1.9527	—	452	1.9500	0.835	445	1.9506	0.939
(Guess C)	336	1.9507	322	1.9507	392	1.9500	0.842	385	1.9500	0.833			
4 (Guess A)	187	1.8859	150	1.8859	189	1.8880	—	192	1.8862	0.955	191	1.8857	0.927
(Guess C)	201	1.8859	161	1.8859	254	1.8759	0.597	211	1.8857	0.926	214	1.8859	0.935

was considered converged when

$$E_A \equiv \frac{1}{1 - H} \max_i \left| \frac{u_i^{(n+1)} - u_i^{(n)}}{u_i^{(n+1)}} \right| < \zeta, \qquad (9\text{-}8.1)$$

where H is defined by (9-5.18). For all problems, the strategy parameter values used were $F = 0.75$, $PSP = 0.5$, and $RSP = 0.0001$. The initial estimate for ω was always taken to be 1.0.

Table 9-8.1 contains summaries of the iterative behavior for each of the three programs in solving the test problems of Table 8-4.1. The guess vectors used are defined in Table 8-4.2. The "Iter" column gives the iterations required to satisfy (9-8.1) with $\zeta = 10^{-6}$. The values of ω and $C^{(s, p)}$ (defined by (9-6.5)) when convergence was achieved are given in columns headed by "Last estimate for ω" and "$C^{(s, p)}$ value," respectively. A blank entry for $C^{(s, p)}$ indicates that the values for $C^{(s, p)}$ have an oscillatory behavior. The best estimates for the spectral radius of the Jacobi iteration matrix B and the optimum acceleration factor ω_b for each problem are given in Table 9-8.2. Table 9-8.2 also gives the number of iterations required for convergence and the value of $C^{(1, p)}$ at convergence when a fixed $\omega = 1.8$ was used. In Table 9-8.2, guess A was used for problems 2 and 4, while guess C was used for problem 3.

Concerning the behavior of these problems, we make the following observations:

1. For most problems, the ratio $C^{(s, p)}$ of actual to optimum convergence rate at convergence was considerably greater than the F factor of 0.75. For the SOR.NO program, the $C^{(s, p)}$ factor at convergence for problem 4 was considerably less than 0.75. This was caused by the fact that $R^{(s, p)}$ was less than $(\omega - 1)$ on one iteration. Recall from Section 9.6 that $R^{(s, p)} < \omega_b$ is used as a signal to indicate that $\omega \geq \omega_b$ and that no new estimation of ω is permitted when this occurs.

TABLE 9-8.2

Additional Iteration Data

| Problem | Best Estimate | | $\omega = 1.8$ Fixed | |
	For $S(B)$	For ω_b	Iter	$C^{(1, p)}$
1	0.994149	1.805022		
2	0.998533	1.897284	495	0.126
3	0.999680	1.950664	1164	0.026
4	0.998167	1.885878	420	0.161

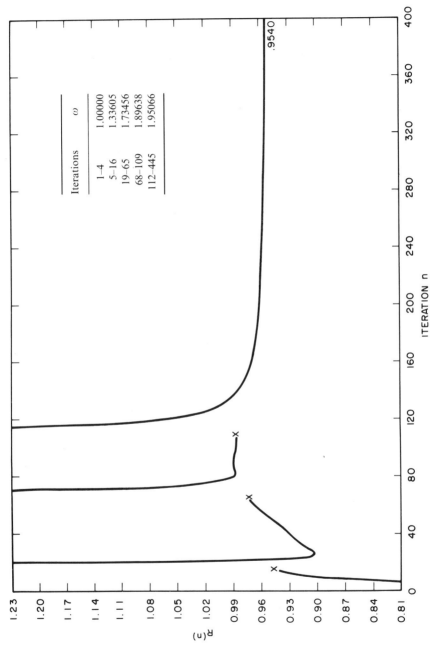

Iterations	ω
1–4	1.00000
5–16	1.33605
19–65	1.73456
68–109	1.89638
112–445	1.95066

.9540

Fig. 9-8.1. Graph of $R^{(n)}$ versus n for problem 3 using the SOR.RB/RQ program.

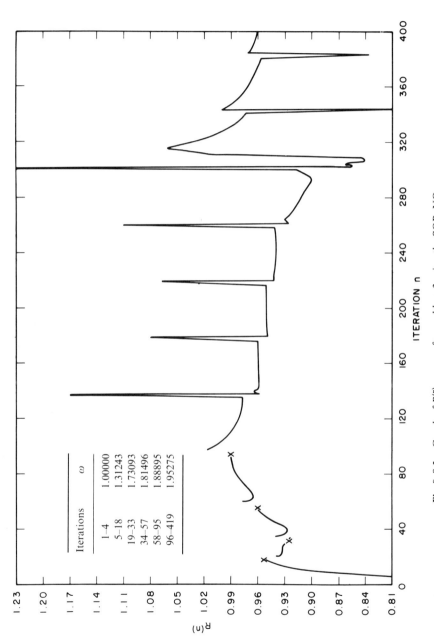

Iterations	ω
1–4	1.00000
5–18	1.31243
19–33	1.73093
34–57	1.81496
58–95	1.88895
96–419	1.95275

Fig. 9-8.2. Graph of $R^{(n)}$ versus n for problem 3 using the SOR. NO program.

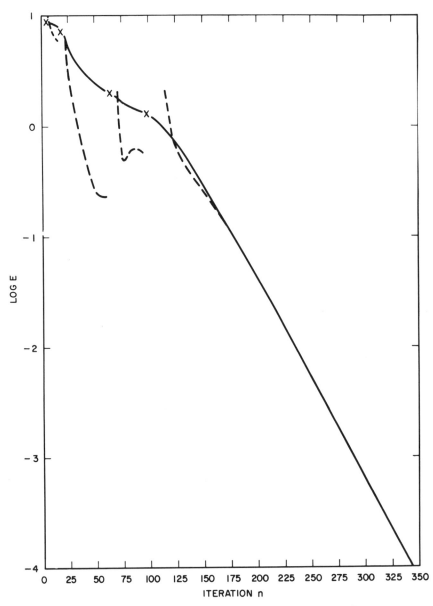

Fig. 9-8.3. Graphs of log E_T and log E_A versus n for problem 3 using the adaptive SOR procedure. ———, true error E_T; ---, estimated error E_A; ×, iteration at which new estimate for ω is used.

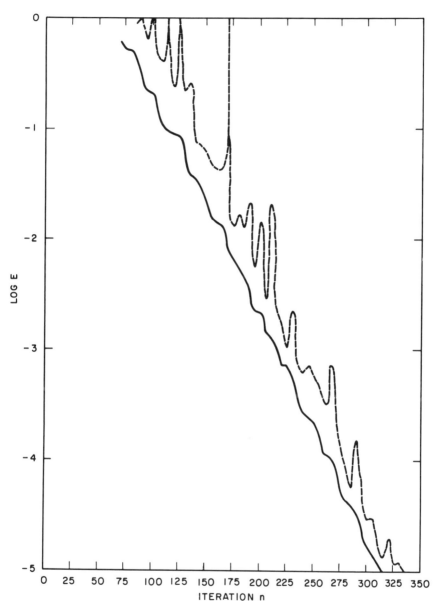

Fig. 9-8.4.   Graphs of log E_T and log E_A versus n for problem 3 with fixed $\omega = 1.9506552$.
————, true error E_T; ———, estimated error E_A.

2.   The SOR.NO program slightly overestimated ω_b for three of the four problems. However, note that this improved convergence and verified the well-known fact that it is beneficial to slightly overestimate ω_b.

3.   Relative to the two different line orderings, the behavior of the adaptive process generally is more consistent for problems with a red/black partitioning. Moreover, with $\omega = \omega_b$ fixed, the iterations required for convergence using a red/black line ordering is consistently less than that when the natural line ordering is used.

4.   For the problems considered here, there appears to be little difference between the procedures used in the SOR.RB and SOR.RB/RQ programs. The advantage of the SOR.RB/RQ procedure is that the sequence of estimates $\{\omega'\}$ generated is guaranteed to converge to ω_b. The disadvantage of the Rayleigh quotient procedure is that the S_B-norm of Δ_B is required for some iterations.

Graphs of $R^{(n)}$ versus iterations are given in Figs. 9-8.1 and 9-8.2 for problem 3. The graph of Fig. 9-8.1 was obtained using the SOR.RB/RQ program and illustrates the behavior of $R^{(n)}$ when the ω used is always less than ω_b. The graph of Fig. 9-8.2 was obtained using the SOR.NO program and illustrates the behavior of $R^{(n)}$ when the ω used becomes larger than ω_b.

For problem 3, the solution $\bar{u}$ is known and is unity everywhere. Thus for this problem we may compare the actual error

$$E_T \equiv \max_i |[u_i^{(n+1)} - \bar{u}_i]/\bar{u}_i|, \qquad (9\text{-}8.2)$$

with the estimate given by (9-8.1). The graphs of Figs. 9-8.3 and 9-8.4 give an indication of the behavior of the actual error and the estimated error as a function of the iteration n. The data for Fig. 9-8.3 was obtained from the SOR.RB program, while that for Fig. 9-8.4 was obtained from the SOR.NO program. Guess B was used for both programs. The graphs of Fig. 9-8.3 clearly indicate that the estimated error is not very accurate for small n. In order for the estimated error to accurately measure the actual error at convergence, the stopping criteria number ζ must be sufficiently small. Generally, ζ should be less than 0.05 or 0.01.

9.9   ON THE RELATIVE MERITS OF CERTAIN PARTITIONINGS AND CERTAIN ITERATIVE PROCEDURES

In this and previous chapters, we have discussed convergence rates and have described computational algorithms for various iterative procedures. The reader faced with solving an actual problem must decide which solution scheme is "best" suited for his particular application. Because of the many nonmathematical factors involved in the use of iterative methods and because of the diversity of problems to be solved, no one iterative procedure is best for all situations. From a practical point of view, the best solution method is one that accomplishes the job with a minimum total cost. The cost here includes computer cost and the man-hour cost to develop and program the solution scheme. If only a small amount of computer time is involved, the solution method selected should be one that works well, not necessarily the best, and one that can be implemented easily. However, for large scientific problems which saturate or nearly saturate the capabilities of the available computer, details of implementation become much more important. For example, we show in Chapter 10 that the computer times required to solve a particular problem can vary by as much as a factor of eight for different implementations of the iterative method applied. For large scale scientific computations, the most effective iterative procedures often are those which converge at a reasonable rate (not necessarily the fastest rate) and which can be especially tailored to the architectural features of the computer at hand.

Among the important computer characteristics which must be considered in the selection and implementation of an iterative procedure are those involving memory capacity, computation rate of the arithmetic unit(s), structure (serial, parallel, pipelined, etc.) of the arithmetic unit(s), and data transfer rate from secondary storage devices (tape, disk, etc.). Advanced scientific computers today are capable of executing in excess of 20 million floating point operations per second. However, only those algorithms which "fit" the architecture of the computer and are carefully implemented will execute with high arithmetic efficiency. (See, for example, Buzbee et al. [1977] and Pfeifer [1963]). Since a detailed discussion of computer architectures transcends the material given in this book, we will discuss only general implementation aspects. The remarks we make in this section and later in Chapter 10 concerning execution efficiencies of certain iterative procedures should be taken as observations, some of which may not be valid for all computer architectures.

From the above discussion, it is apparent that the selection and application of iterative procedures to large scale scientific problems is an art as well as a science. Because of this, we can offer the reader only a feeling for what can be done, what he can expect, and what he should look for in the utilization of certain iteration procedures. In Chapters 10 and 11, we do this by discussing specific examples. In this section, we discuss the significance of the imposed partitioning on the utilizations of certain iterative procedures.

Suppose that we seek to solve the matrix equation $Au = b$, where A is a sparse SPD matrix, which is partitioned in the form

$$A = \begin{bmatrix} A_{1,1} & A_{1,2} & \cdots & A_{1,q} \\ A_{2,1} & A_{2,2} & \cdots & A_{2,q} \\ \vdots & \vdots & & \vdots \\ A_{q,1} & A_{q,2} & \cdots & A_{q,q} \end{bmatrix}. \tag{9-9.1}$$

We first discuss relative merits of the SOR, Chebyshev, and CG procedures applied to the Jacobi method under varying assumptions concerning the partitioning (9-9.1). Later, we will discuss other, perhaps less obvious, ways in which the imposed partitioning can affect the behavior of certain iterative procedures.

Since the matrix A is SPD, the Jacobi matrix is symmetrizable for any partitioning imposed on A. Thus without additional assumptions, the Chebyshev and CG methods of Chapters 5–7 may be used to accelerate the Jacobi method corresponding to (9-9.1). (As before, we denote these two acceleration methods by J–SI and J–CG). However, for the more effective polynomial acceleration methods of Chapter 8, we require that the partitioning for A be red/black. For the SOR procedures described in this chapter, we require that A be consistently ordered relative to the partitioning imposed. We discuss the red/black case first.

Solution Methods for the Red/Black Partitioning Case

For problems of this type, the RS–CG (or equivalently the CCG) method and the CCSI (or equivalently the RS–SI) method should be considered as the principal candidates for a solution scheme.†

The choice between the RS–CG method and the CCSI method is not always apparent. The decision as to which procedure is more appropriate can be made by weighing the positive (and negative) features of each method relative to problem(s) to be solved. Important features of both methods are summarized below.

(a)   The RS–CG method converges faster, sometimes significantly faster, than the CCSI method.

(b)   The iteration parameters for the RS–CG method are generated automatically during the iteration process while the optimum parameters for the CCSI method are functions of the spectral radius, $S(B)$, of the Jacobi matrix B.

---

† As noted previously, for problems with a red/black partitioning, the SOR method is slightly less effective than the CCSI method; their asymptotic convergence rates are the same but the CCSI method has the larger average convergence rate.

(c) The CCSI method requires less storage and less work per iteration than the RS–CG method. Moreover, for the CCSI method, unlike the RS–CG method, it is possible to skip the computation of the pseudoresidual vector and its norm for many iterations.

(d) The iteration error can be measured accurately in any norm when the adaptive CCSI procedure is used. Procedures for accurately measuring the iteration error remain unknown for the RS–CG and optimal nonadaptive CCSI methods.

(e) The use of a symmetrization matrix W can be avoided in the CCSI procedure whereas W (or at least $W^T W$) must be used in the RS–CG procedure.

(f) Unlike the RS–CG method, the iteration parameters for iteration $n + 1$ of the CCSI procedure can be computed before the completion of iteration n. (This feature is required in the implementation of the "concurrent" iteration procedure described later in Section 10.2).

Items (a) and (b) are positive features of the RS–CG method, while items (c)–(f) are positive features of the CCSI method. The positive features of the RS–CG method usually overshadow those of the CCSI method if the use of a symmetrization matrix W is computationally convenient and if data flow from secondary storage devices is not an important consideration. If either of these conditions is not valid, the positive features of the CCSI method increase in importance. Reasons for this are given later in Section 10.2 and Chapter 11.

Solution Methods for the Consistently Ordered (but Not Red/Black) Case

Sometimes programming and data alignment considerations make it more convenient to use a partitioning which is consistently ordered but not red/black. For this case, the principal choice is between the SOR method and the J–CG method. The J–SI method is not included here since the SOR method not only converges twice as fast as the J–SI method but also requires less storage and less work per iteration.

The SOR method is likely to be superior to the J–CG method for most problems of this type. This follows by reinterpreting items (a)–(f) given above. With the SOR and J–CG methods replacing the CCSI and RS–CG methods, respectively, items (c)–(f) remain valid and are positive features of the SOR method. However, since the ratio of J–CG to SOR convergence rates for this case is known only to be greater than one-half, item (a) need not be valid.

Solution Methods for the General Partitioning Case

Here the principal choice is between the J–SI method and the J–CG method. The SOR method may also be used; however, the usefulness of the SOR method is limited for problems with a general partitioning because of uncertainties concerning its convergence rate and because of uncertainties concerning a precise prescription for the optimum relaxation factor.

The choice between the J–CG and J–SI methods should be made after considering the positive (or negative) features of each method. Features (a)–(f) given above and the comments made there concerning the CCSI and RS–CG methods also are valid, respectively, for the J–SI and J–CG methods. The J–SI method here, however, has the additional negative feature that the smallest eigenvalue $m(B)$ of the iteration matrix B is required as well as the largest eigenvalue $M(B)$.

Other Partitioning Aspects

Thus far we have discussed relative merits of certain iterative procedures, given a partitioning. Now, we wish to comment briefly on the relative merits of certain partitionings, given a problem.

Convergence Rates as a Function of Partitioning

To illustrate the sensitivity of convergence rates to the partitioning imposed, we again consider the solution of the matrix problem $Au = b$ which results from a discretization of the elliptic differential Eq. (8-4.1) over the rectangular mesh subdivision of Fig. 8-4.1. Here, however, we assume the discretization is based on a 9-point stencil instead of the 5-point stencil used previously (see Fig. 10-2.1). We assume only that the 9-point discretization formula used is such that the resulting coefficient matrix A is SPD. The structure of the nonzero elements of the matrix A corresponding to a natural ordering of the unknowns (see Section 1.6) is illustrated in Fig. 9-9.1. It is easy to show that the coefficient matrix A for the 9-point discretization case has Property $\mathscr{A}$ relative to a line partitioning but not relative to a point partitioning. We now wish to examine the convergence rates of certain Chebyshev acceleration procedures as a function of these two partitionings.

We denote the Jacobi iteration matrices corresponding to the point and line partitionings, respectively, by B^P and B^L. Also, as before, we let $m(B)$ and $M(B)$ denote, respectively, the algebraically smallest and largest eigenvalues of a Jacobi iteration matrix B. In what follows, we assume that $M(B^P)$

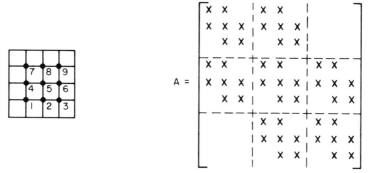

Fig. 9-9.1.   Coefficient matrix A produced by a 9-point discretization stencil for the mesh grid shown.

and $M(B^L)$ are close to unity. Thus for $\delta(> 0)$ small and for $x_L \delta(> 0)$ small, we can write

$$M(B^P) \equiv 1 - \delta \qquad \text{and} \qquad M(B^L) \equiv 1 - x_L \delta. \qquad (9\text{-}9.2)$$

Here, the parameter x_L depends on the specific problem being solved and is normally greater than unity. As A has Property $\mathscr{A}$ relative to a line partitioning, we have from (9-3.5) that $m(B^L) = -M(B^L) = -1 + x_L \delta$. For the point partitioning case, utilizing the structure of the nonzero elements of B^P and using the fact that A is SPD, we can show† that $m(B^P) > -3$. Thus we can write

$$m(B^L) = -1 + x_L \delta \qquad \text{and} \qquad m(B^P) = -3 + \Lambda, \qquad (9\text{-}9.3)$$

where $\Lambda > 0$.

From (4-2.22), the asymptotic convergence rate for the Chebyshev acceleration method applied to B^P is $\bar{R}_\infty(P_n(B^P)) = -\tfrac{1}{2} \log \bar{r}$, where $\bar{r}$ is defined by (4-2.19). Since $M(B^P) = 1 - \delta$, where δ is small, it follows (see (4-4.11)) that approximately $\bar{R}_\infty(P_n(B^P)) \doteq 2[\delta/(1 + \delta - m(B^P))]^{1/2}$. Similarly, for the Chebyshev acceleration method applied to B^L, we have that $\bar{R}_\infty(P_n(B^L)) \doteq 2[x_L \delta/(1 + x_L \delta - m(B^L))]^{1/2}$. Thus substituting $m(B^P)$ and $m(B^L)$ from (9-9.3) into these expressions, we have approximately that

$$\bar{R}_\infty(P_n(B^L))/\bar{R}_\infty(P_n(B^P)) \doteq \{x_L[\tfrac{1}{2}(4 + \delta - \Lambda)]\}^{1/2}. \qquad (9\text{-}9.4)$$

In addition, if a red/black line partitioning is used, the CCSI method may be applied‡ to B^L. Letting $\bar{R}_\infty(\text{CCSI}(B^L))$ denote the asymptotic convergence

---

† We omit a formal proof.

‡ Since A does not have Property $\mathscr{A}$ relative to a point partitioning, it is not possible to obtain a red/black partitioning (8-1.1), where D_R and D_B are diagonal matrices. Thus the CCSI method cannot be utilized for the point partitioning case.

rate of the CCSI method, we have from Chapter 8 that $\bar{R}_\infty(\text{CCSI}(B^L)) = 2\bar{R}_\infty(P_n(B^L))$, which when combined with (9-9.4) gives

$$\bar{R}_\infty(\text{CCSI}(B^L))/\bar{R}_\infty(P_n(B^P)) \doteq 2(x_L)^{1/2}[\tfrac{1}{2}(4 + \delta - \Lambda)]^{1/2}. \quad (9\text{-}9.5)$$

Assuming that Λ as well as δ are small, we see from (9-9.5) that the best Chebyshev procedure for a line partitioning converges roughly $2(2x_L)^{1/2}$ times faster than the best Chebyshev procedure for a point partitioning. Note that a significant part of this improvement is due to the fact that eigenvalue domain for B^L is smaller than that for B^P. We remark that the problem dependent parameter x_L frequently lies in the interval $[1, 4]$.

The evaluation of any solution procedure must also include considerations concerning the computational cost per iteration. For the example given above, the cost per iteration for the CCSI procedure utilizing the (Property $\mathscr{A}$) line partitioning is no greater than that for the Chebyshev procedure utilizing the (non-Property $\mathscr{A}$) point partitioning. We note, however, that this need not be the case for other problems. See, e.g., Section 10.3.

Similar, but not as precise, remarks can be made for the corresponding CG methods applied to B^L and B^P and for the corresponding SOR procedures based on the line and point partitionings.

Considerations in Choosing a Partitioning with Property $\mathscr{A}$

In the above example, we illustrated the sensitivity of convergence rates relative to partitionings with and without Property $\mathscr{A}$ and showed that iterative procedures which utilize partitionings with Property $\mathscr{A}$ generally have the more favorable convergence rates. Often, many feasible partitionings exist for which the coefficient matrix A has Property $\mathscr{A}$; moreover, each of these partitionings possibly can lead to a different convergence rate for the iterative method under consideration. The one-line methods used in Sections 8.4 and 9.8 were obtained by partitioning the unknowns by single horizontal lines. However, other partitionings could have been used; for example, we obtain a coefficient matrix with Property $\mathscr{A}$ if the unknowns are partitioned by l successive horizontal lines or by l successive vertical lines. Usually, the number of iterations required for convergence decreases as the block size increases. Unfortunately, in passing to larger blocks, the cost per iteration increases also.

To illustrate this, consider the numerical solution of Poisson's equation (1-7.1) in a square region with a uniform mesh of size h. We assume, as in Section 8.4, that the discretization is based on the normal 5-point formula. If, as before, we let $\bar{R}_\infty(\text{CCSI}(B^{lL}))$ denote the asymptotic convergence rate of the CCSI method for the l-line partitioning problem, Parter [1961] has shown for $l \geq 1$ that

$$\bar{R}_\infty(\text{CCSI}(B^{(l+1)L}))/\bar{R}_\infty(\text{CCSI}(B^{lL})) \sim [(l+1)/l]^{1/2}, \qquad h \to 0. \quad (9\text{-}9.6)$$

For this model problem, we have that the asymptotic convergence rate of the one-line CCSI method is roughly $\sqrt{\frac{1}{2}}$ times that for the two-line method which in turn is $\sqrt{\frac{2}{3}}$ times that for the three-line method. However, the computational cost per iteration for the three-line CCSI method is 1.28 times that for the two-line method which in turn is 1.2 times that for the one-line CCSI method.[†] Thus considering only computational requirements and convergence rates, the computer time required to solve this model problem is minimized for $l = 2$.

In some problems, geometrical and material properties result in a situation where the coefficients in the discretization formula are much larger in one direction than those in other directions. In this situation, the orientation of a partitioning in the direction of the large coefficients is particularly effective.[‡] The reason for this is that, for such a partitioning, the large coefficients enter into the diagonal submatrices $A_{i,i}$ of A, while the weak coefficients enter into the off-diagonal submatrices of A. More details concerning this subject can be found in Wachspress [1966] and in Nakamura [1977 Ref. 1, p. 137].

9.10 PROOFS OF THEOREMS AND DISCUSSION OF THE STRATEGY CONDITION (9-5.21)

In this section we first give proofs of Theorems 9-4.1 and 9-4.2. We then discuss the significance of the strategy test (9-5.21) and give alternative suggested values to use for the associated parameter RSP.

Proof of Theorem 9-4.1

We first show

Lemma 9-10.1. If $1 < \omega < \omega_b$ and if $\mu_i^2 \omega^2 - 4(\omega - 1) \neq 0$ for all i, then

$$\varepsilon^{(n)} = \frac{\Delta^{(n)}}{1 - \lambda_1} - \lambda_2^n \left\{ \sum_{i=2}^{N} \left(\frac{\lambda_i}{\lambda_2} \right)^n \left(\frac{\lambda_1 - \lambda_i}{1 - \lambda_1} \right) c_i y(i) \right\}, \qquad (9\text{-}10.1)$$

[†] We assume the Cuthill–Varga normalization procedure described in Section 8.5 is used to eliminate the required divisions.

[‡] For example, if the coefficients $N_{i,j}$ and $S_{i,j}$ in the discretization formula (1-6.4) are larger than the $E_{i,j}$ and $W_{i,j}$ coefficients, then a partitioning by vertical mesh lines should be used.

and

$$\Delta^{(n)} = \lambda_1^n \left\{ d_1 y(1) + \left(\frac{\lambda_2}{\lambda_1}\right)^n \sum_{i=2}^{N} \left(\frac{\lambda_i}{\lambda_2}\right)^n d_i y(i) \right\}, \qquad (9\text{-}10.2)$$

where $d_i = (1 - \lambda_i)c_i$.

Proof of Lemma.   From (9-4.3) and (9-4.5), we have

$$\varepsilon^{(n)} = \lambda_1^n c_1 y(1) + \lambda_2^n \sum_{i=2}^{N} \left(\frac{\lambda_i}{\lambda_1}\right)^n c_i y(i) \qquad (9\text{-}10.3)$$

and

$$\frac{\Delta^{(n)}}{1 - \lambda_1} = \lambda_1^n c_1 y(1) + \lambda_2^n \sum_{i=2}^{N} \left(\frac{\lambda_i}{\lambda_2}\right)^n \left(\frac{1 - \lambda_i}{1 - \lambda_1}\right) c_i y(i). \qquad (9\text{-}10.4)$$

Equation (9-10.1) follows by substituting $\lambda_1^n c_1 y(1)$ from (9-10.4) into (9-10.3). Equation (9-10.2) follows directly from (9-4.3) and (9-4.6). This completes the proof of Lemma 9-10.1.

Turning now to the proof of Theorem 9-4.1, we have from (9-10.1) that

$$\|\varepsilon^{(n)}\| \le \frac{\|\Delta^{(n)}\|}{1 - \lambda_1} \left[1 + \frac{|\lambda_2|^n \sum_{i=2}^{N} |\lambda_i/\lambda_2|^n |\lambda_1 - \lambda_i| \cdot \|c_i y(i)\|}{\|\Delta^{(n)}\|} \right]. \qquad (9\text{-}10.5)$$

and from (9-10.2) that

$$\|\Delta^{(n)}\| \ge \lambda_1^n \left\{ \|d_1 y(1)\| - \left|\frac{\lambda_2}{\lambda_1}\right|^n \sum_{i=2}^{N} \left|\frac{\lambda_i}{\lambda_2}\right|^n \|d_i y(i)\| \right\}. \qquad (9\text{-}10.6)$$

For n sufficiently large, it follows from the definition (9-4.7) of K that $\sum_{i=2}^{N} |\lambda_i/\lambda_2|^n |\lambda_1 - \lambda_i| \cdot \|c_i y(i)\| < K \|d_1 y(1)\|$ and from (9-10.6) that $\|\Delta^{(n)}\| \ge \lambda_1^n \|d_1 y(1)\| [1 - |\lambda_2/\lambda_1|^n K] > 0$. Using these inequalities in (9-10.5), we obtain the upper bound given in (9-4.8). The lower bound given in (9-4.8) follows similarly by writing $\|\varepsilon^{(n)}\|$ as

$$\|\varepsilon^{(n)}\| \ge \frac{\|\Delta^{(n)}\|}{1 - \lambda_1} \left[1 - \frac{|\lambda_2|^n \sum_{i=2}^{N} |\lambda_i/\lambda_2|^n |\lambda_1 - \lambda_i| \cdot \|c_i y(i)\|}{\|\Delta^{(n)}\|} \right]. \qquad (9\text{-}10.7)$$

The inequalities of (9-4.10) follow easily from (9-10.2) since for n sufficiently large we have

$$\left[1 - \left|\frac{\lambda_2}{\lambda_1}\right|^n K \right] \|d_1 y(1)\| \lambda_1^n \le \|\Delta^{(n)}\| \le \lambda_1^n \|d_1 y(1)\| \left[1 + \left|\frac{\lambda_2}{\lambda_1}\right|^n K \right]. \quad \blacksquare$$

$$(9\text{-}10.8)$$

Proof of Theorem 9-4.2

We first show

Lemma 9-10.2. If $1 < \omega < \omega_b$ and if $[\mu_2^2 \omega^2 - 4(\omega - 1)] = 0$, then

$$\varepsilon^{(n)} = \frac{\Delta^n}{1 - \lambda_1} - \frac{1}{1 - \lambda_1}\left\{ n\lambda_2^{n-1}(\lambda_1 - \lambda_2)c_2\, y(3) \right.$$

$$\left. + \lambda_2^n\left[-c_2\, y(3) + (\lambda_1 - \lambda_2)c_2\, p(2) + \sum_{i=3}^{N} \left(\frac{\lambda_i}{\lambda_2}\right)^n (\lambda_1 - \lambda_i)c_i\, y(i) \right] \right\} \tag{9-10.9}$$

and

$$\Delta^{(n)} = \lambda_1^n d_1 y(1) + n(\lambda_2)^{n-1} d_2\, y(3)$$

$$+ \lambda_2^n\left\{ d_2\, p(2) - c_2\, y(3) + \sum_{i=3}^{N} \left(\frac{\lambda_i}{\lambda_2}\right)^n d_i\, y(i) \right\}, \tag{9-10.10}$$

where $d_i = (1 - \lambda_i)c_i$.

Proof of Lemma. From (9-3.9), (9-4.3), and (9-4.12), we have

$$\varepsilon^{(n)} = \lambda_1^n c_1 y(1) + n\lambda_2^{n-1} c_2\, y(3) + \left\{ \lambda_2^n c_2\, p(2) + \sum_{i=3}^{N} \lambda_i^n c_i\, y(i) \right\} \tag{9-10.11}$$

and

$$\Delta^{(n)} = \lambda_1^n(1 - \lambda_1)c_1 y(1) + n\lambda_2^{n-1}(1 - \lambda_2)c_2\, y(3) - \lambda_2^n c_2\, y(3)$$

$$+ \left\{ \lambda_2^n(1 - \lambda_2)c_2\, p(2) + \sum_{i=3}^{N} \lambda_i^n(1 - \lambda_i)c_i\, y(i) \right\}. \tag{9-10.12}$$

Substituting for $\lambda_1^n c_1 y(1)$ in (9-10.11) from (9-10.12), we obtain (9-10.9). The expression (9-10.10) for $\Delta^{(n)}$ follows directly from (9-4.5), (9-4.6), and (9-10.12), which completes the proof of Lemma 9-10.2.

We now turn to the proof of Theorem 9-4.2. Since $c_i = d_i/(1 - \lambda_i)$, from (9-10.9) we have

$$(1 - \lambda_1)\|\varepsilon^{(n)}\| \underset{(\geq)}{\overset{\leq}{}} \|\Delta^{(n)}\| \underset{(-)}{\overset{+}{}} \left\{ n\lambda_2^{n-1}\left(\frac{\lambda_1 - \lambda_2}{1 - \lambda_2}\right)\|d_2\, y(3)\| + \lambda_2^n\|d_1 y(1)\|K' \right\}. \tag{9-10.13}$$

The inequalities of (9-4.15) then follow by dividing (9-10.13) by $\|\Delta^{(n)}\|$ and noting that

$$\|\Delta^{(n)}\| \geq \lambda_1^n\|d_1 y(1)\|\left\{ 1 - \left[n\left(\frac{\lambda_2}{\lambda_1}\right)^{n-1}\left(\frac{1}{\lambda_1}\right)\frac{\|d_2\, y(3)\|}{\|d_1 y(1)\|} + \left(\frac{\lambda_2}{\lambda_1}\right)^n K' \right] \right\}. \tag{9-10.14}$$

The inequalities of (9-4.17) follow from (9-10.14) and

$$\|\Delta^{(n)}\| \leq \lambda_1^n \|d_1 y(1)\| \left\{ 1 + \left[n\left(\frac{\lambda_2}{\lambda_1}\right)^{n-1} \left(\frac{1}{\lambda_1}\right) \frac{\|d_2 y(3)\|}{\|d_1 y(1)\|} + \left(\frac{\lambda_2}{\lambda_1}\right)^n K' \right] \right\}. \quad \blacksquare$$

(9-10.15)

Significance of the Strategy Test (9-5.21)

For our discussion here, we use the notation of Section 9.5 and assume that the eigenvalues of B and of $\mathscr{L}_{\omega_s}$ are given by (9-3.5) and (9-4.4), respectively. We let μ_1' be the estimate (9-5.14) for μ_1 obtained using $R^{(s, p)}$ and let ω' be the associated estimate (9-5.13) for ω_b. For notational convenience, we let $S \equiv S(\mathscr{L}_{\omega_s})$, $R \equiv R^{(s, p)}$, and

$$\Delta R \equiv R^{(s, p-1)} - R^{(s, p)}. \tag{9-10.16}$$

If I_ω is the number of iterations required for convergence using ω, then for μ_1 and μ_1' close to unity, the ratio $(I_{\omega'}/I_{\omega_b})$ approximately satisfies (see Hageman and Kellogg [1968]).

$$I_{\omega'}/I_{\omega_b} \lesssim 1 + \delta, \tag{9-10.17}$$

where

$$\delta^2 = |\mu_1 - \mu_1'|/(1 - \mu_1'). \tag{9-10.18}$$

Intuitive arguments will now be given to show that a relationship exists between δ and the strategy parameter RSP used in (9-5.21). We continue to assume that μ_1 and μ_1' are close to unity.

Since μ_1' is close to unity, from (9-5.13) we have approximately that $(1 - \mu_1') \doteq \frac{1}{8}(2 - \omega')^2$. Using this approximation together with S from (9-5.11) and μ_1' from (9-5.14), we have

$$\frac{|\mu_1 - \mu_1'|}{1 - \mu_1'} \doteq 8 \frac{|S - R|}{(2 - \omega')^2} \left\{ \frac{1 + \sqrt{RS} - \omega_s}{\omega_s \sqrt{RS}[\sqrt{R} + \sqrt{S}]} \right\}. \tag{9-10.19}$$

If $\omega_s < \omega_b$, the eigenvalue error $|S - R|$ may be approximated by (see, e.g., Hageman and Kellogg [1966]).

$$|S - R| \doteq |\delta R|/(1 - \bar{\sigma}), \tag{9-10.20}$$

where $\bar{\sigma} = |\lambda_2|/S$, Here λ_2 is the second largest eigenvalue of $\mathscr{L}_{\omega_s}$ given by (9-4.4). From Fig. 9-3.1, it is clear that $(\omega_s - 1)/S \leq \bar{\sigma} < 1$ whenever $1 \leq \omega_s < \omega_b$. Thus $0 < 1 - \bar{\sigma} \leq (1 + S - \omega_s)/S$ and we may write

$$1 - \bar{\sigma} = \alpha[(1 + S - \omega_s)/S], \tag{9-10.21}$$

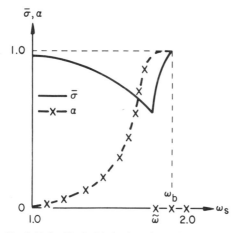

Fig. 9-10.1. Typical behavior of $\bar{\sigma}$ and α versus ω_s.

where $0 < \alpha \leq 1$. With $\tilde{\omega} \equiv 2/(1 + \sqrt{1 - \mu_2^2})$, the behavior† of $\bar{\sigma}$ and α versus ω_s is indicated by the graphs of Fig. 9-10.1. Note that $\alpha = 1$ for $\tilde{\omega} \leq \omega_s < \omega_b$. Substituting (9-10.21) and (9-10.20) into (9-10.19), we have approximately that

$$\frac{|\mu_1 - \mu_1'|}{1 - \mu_1'} \doteqdot \left[\frac{8S}{\omega_s \sqrt{RS}(\sqrt{R} + \sqrt{S})} \left(\frac{1 + \sqrt{RS} - \omega_s}{1 + S - \omega_s} \right) \right] \frac{|\Delta R|}{\alpha(2 - \omega')^2}. \tag{9-10.22}$$

The expression in brackets is about 2.0. Thus if test (9-5.21) is satisfied, the δ in (9-10.17) is approximately equal to $[1/(2 - \omega')][2RSP/\alpha]^{1/2}$.

On the other hand, if (9-10.17) is to be satisfied for a fixed $\hat{\delta}$, then RSP should approximately satisfy

$$RSP \doteqdot \tfrac{1}{2}\alpha(2 - \omega')^2 \delta^2. \tag{9-10.23}$$

Thus in order to satisfy (9-10.17) for a fixed δ, RSP should be a function of $(2 - \omega')^2$ when μ_1 and μ_1' are close to unity. For the test problems given in Section 9.8, RSP was taken to be

$$RSP = \begin{cases} 0.0001 & \text{if } \omega_s \leq 1.9 \\ (0.01)(2 - \omega_s)^2 & \text{if } \omega_s > 1.9. \end{cases} \tag{9-10.24}$$

The expression $(0.01)(2 - \omega_s)^2$ in (9-10.24) corresponds to choosing $\delta = 0.2$, $\alpha = 0.5$, and using ω_s in place of ω'. To account for the fact that $\alpha = 0.5$ is

† From (9-3.6), it is clear that $\bar{\sigma} = [\mu_2/\mu_1]^2$ when $\omega_s = 1$. As ω_s increases from unity to $\tilde{\omega}$, $\bar{\sigma}$ decreases from $[\mu_2/\mu_1]^2$ to $[(\tilde{\omega} - 1)/S(\mathcal{L}_\tilde{\omega})]$. The value of $\bar{\sigma}$ then increases to unity as ω_s increases from $\tilde{\omega}$ to ω_b.

inadequate for most problems when ω_s is considerably less than ω_b, RSP in (9-10.24) is fixed at 0.0001 when $\omega_s \leq 1.9$.

If $\omega_s \ll \omega_b$, the convergence rate of $u^{(n)}$ to $\bar{u}$ is considerably smaller than the optimum rate. Moreover, since $\bar{\sigma}$ is usually close to unity for this case, the convergence rate of $R^{(s, p)}$ to $S(\mathcal{L}_{\omega_s})$ also is small. This implies that a significant number of iterations, which are converging at a slow rate, may be required before (9-5.21) is satisfied and ω_s is changed. We try to avoid this pitfall in Algorithms 9-6.1 and 9-7.1 by bypassing the test (9-5.21) whenever the following two conditions are satisfied: (1) $s \leq 2$ and (2) the ratio of actual to optimum convergence rates is less than 0.1; i.e., when $R^{(s, p)} \geq (\omega_s - 1)^{0.1}$.

10

The Use of Iterative Methods in the Solution of Partial Differential Equations

10.1 INTRODUCTION

Most of the large sparse systems which are solved by iterative methods arise from discretizations of partial differential equations. In the previous chapters we have discussed convergence properties of various iterative methods and have given algorithms for estimating acceleration parameters and for terminating each iterative process. These aspects, based on mathematical properties of the method considered, are essential for the efficient use of iterative methods. However, since the total computer time depends on the work required per iteration as well as the number of iterations, it is clear that factors other than those associated with the iterative convergence rate need to be considered. In this chapter we discuss other problem aspects, such as mesh structure and discretization stencils, which affect the efficiency of iterative methods in the solution of multidimensional boundary-value problems. Our goal is to present an overview and to give a general flavor for some of the factors which affect solution costs. Since the best numerical procedures depend on many factors, no hard and fast recommendations will be given. We merely seek to illustrate some of the factors which affect the efficiencies of certain iterative solution methods.

Given a physical problem, one first needs a *mathematical model* whose solution in some sense approximates the solution of the physical problem. Next, a *numerical model* of the mathematical problem is needed. The solution to the numerical model only approximates the mathematical model solution, but should be such that the error in this approximation theoretically can be made arbitrarily small. The final step is to *solve the numerical model*. Again, because of the solution method and the finite word length of digital computers, the computer solution only approximates the solution to the numerical model. Given the numerical solution τ, the engineer or scientist must consider the question: Does τ adequately represent the solution of the physical problem? Since complete theoretical results usually are not available for general practical problems, the answer to this difficult and important question can only be answered by carefully designed numerical and physical experiments. Thus a computer program must be constructed for generating the numerical solution before any answer to this question can be obtained. However, many factors need to be considered in the *design* of such a program.

To illustrate some of these factors, we consider the problem of determining the temperature distribution in a conducting solid. A general mathematical model for this problem can be expressed as

$$\rho(r, t) \frac{\partial T(r, t)}{\partial t} = \nabla \cdot k(r, t, T)\nabla T(r, t) + g(r, t), \qquad (10\text{-}1.1)$$

where r is the spatial vector (x, y, z), ∇ the spatial gradient vector, t time, T the unknown temperature distribution, ρ the specific heat, k the thermal conductivity, and g the heat source. We assume appropriate initial and boundary conditions are given. At a point r on the boundary S of the geometric domain D, the temperature satisfies one of the following conditions:

(1) $T(r, t) = f(r, t)$, where $f(r, t)$ is a known function , or
(2) $-k(r, t, T) \, \partial T(r, t)/\partial n = h(r, t, T)[T(r, T) - T_0(r, t)]^{p(r)}$, where h is the film coefficient, T_0 is the sink temperature, $p(r) \geq 0$, and $\partial T/\partial n$ is the derivative of T in the direction of the outward normal to S.

Before progressing to the next step of determining the numerical model, one needs to be more specific concerning the generality of the mathematical model. For example, how many spatial dimensions are needed? What type of boundary conditions are required? Can the thermal conductivity k be assumed to vary just with time and space? Is only the steady-state solution desired?

Obviously, one would like to solve the most general problem which is practically possible. The practicality criterion must be based on cost, which can be divided into the following categories: (a) initial mathematical and programming development cost and (b) computer solution cost. We will discuss only computer solution cost. Usually, the more flexible and general the program is, the greater are its solution costs. Thus in order to keep solution costs reasonable, one is often forced to make compromises concerning the flexibility and generality of the program. In order to make these compromises, one must decide whether a particular feature or generality is worth the cost, i.e., one must weigh *generality value* versus *cost*. The value of a particular generality is a function of the problem or class of problems to be solved, and, thus, the final judgement concerning value versus cost usually should rest with the engineer wishing a solution to a physical problem or class of problems. In order to make this judgement, however, one needs to obtain some estimate for the cost. For example, what is the cost for solving a three-dimensional problem? What is the cost for solving a two-dimensional problem?

Three basic parts of a computer program to solve a boundary-value problem are the mesh generation, the discretization, and the solution of the matrix problem. The mesh generation and the discretization parts basically determine the accuracy or value of the numerical solution, while the matrix solution part determines most of the computer solution cost. Unfortunately, the three parts are not independent of each other. Increased generality in the mesh generation and discretization parts can significantly increase the matrix solution cost. Thus solution cost can only be given as a function of the mesh decomposition and discretization methods under consideration.

The factors which most affect matrix solution costs are

(1) Total arithmetic operations required,
(2) Storage requirements, and
(3) Overhead due to data transmission and to logical opera- (10-1.2) tions associated with the implementation of the solution method.

The relative importance of these factors is, of course, a strong function of the computer at hand and the type and size of problems one is trying to solve. The most efficient solution procedures usually are those which minimize storage and arithmetic requirements. However, recent advances in computer technology have been such that logical operations and data transmission cannot be neglected.

We will not attempt to analyze the general nonlinear model given by (10-1.1). Instead, we will discuss iterative solution costs relative to the more modest linear† mathematical model in two and three space dimensions.

---

† That is, we assume the thermal conductivity k, is not a function of temperature and the exponent $p(r)$ in boundary condition (2) is either 0 or 1.

10.2   THE TIME-INDEPENDENT TWO-DIMENSIONAL PROBLEM

In this section we consider the numerical solution of the time-independent, two-dimensional problem

$$-\frac{\partial}{\partial x}\left(k\frac{\partial T}{\partial x}\right) - \frac{\partial}{\partial y}\left(k\frac{\partial T}{\partial y}\right) + \gamma T = g \qquad (10\text{-}2.1)$$

in a bounded connected region R with boundary S. Here $k > 0$, $\gamma \geq 0$, and g are material properties of the specific problem to be solved and are assumed to be only *piecewise smooth functions of* x and y.

The general steps needed to solve this problem numerically are

(1)   Choose a suitable mesh subdivision of the geometric domain.

(2)   Choose a procedure—called discretization—whereby an approximate solution to the problem (10-2.1) can be represented as the solution to the system of equations

$$\sum_{j=1}^{N} a_{ij}\tau_j = g_i, \qquad i = 1, \ldots, N,$$

or equivalently in matrix form

$$A\tau = g. \qquad (10\text{-}2.2)$$

(3)   Choose a suitable direct or iterative method to obtain the solution to the matrix problem.

In step (2), τ represents the numerical approximation for T and *we always assume A is a large sparse, positive definite* matrix.

For the mesh generation part, the types of mesh elements which we consider are rectangles, quadrilaterals (or triangles), and curved four-sided elements (see Fig. 10-2.1). Another important aspect of the mesh procedure is the way the mesh elements are put together or what we call the *global mesh structure*. The types of global mesh structures considered will be given later.

For the discretization procedure, we assume that each τ_j can be associated with a point of a mesh element and that the coupling $a_{ij} \neq 0$ only if τ_i and τ_j are associated with points of the same element.† If $a_{ij} \neq 0$, we say node i is coupled to node j. We consider 3-, 4-, and 9-node couplings for rectangular elements, 4- and 9-node couplings for quadrilaterals, and 9-node couplings for curved elements.‡ If τ_P is the unknown associated with element point P,

---

† This assumption insures that A is a sparse matrix.

‡ For reasons of simplicity, we consider only four-sided elements here. However, most of what is said in this chapter is also valid for triangular and general three-sided elements.

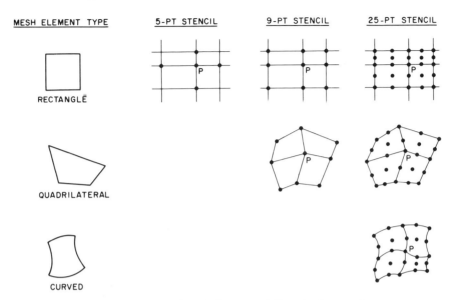

Fig. 10-2.1.   Types of mesh elements and discretization stencils.

we indicate the j's such that $a_{Pj} \neq 0$ by a stencil. Examples of 5-, 9-, and 25-point stencils for the various mesh elements are given in Fig. 10-2.1. We assume only that the coefficient matrix A resulting from either the 5-, 9-, or 25-point discretization formula is SPD.

For the mesh generation procedure, we say the global mesh structure is *regular* if the geometric domain is subdivided into a $p \times q$ mesh element array; i.e., the domain is divided by q rows with p elements per row. See Fig. 10-2.2. A regular mesh structure implies that the discretization equations and unknowns may be ordered such that the nonzero entries of the matrix A lie in a predictable pattern. For example, if a 5-point stencil is used for the mesh element array of Fig. 10-2.2a, the unknowns and equations may be ordered such that $a_{ij} \neq 0$ only if $|i - j| = 0, 1,$ or p. Thus in the solution process, no directory is needed to determine the nonzero elements of A. We say the global mesh structure is *semiregular* if the geometric domain is subdivided into a row and column mesh element array but with a variable number of mesh elements per row. See Figs. 10-2.3 and 10-2.4b. To determine the nonzero elements in A for semiregular mesh structures, some information concerning the row element structure is needed. However, no general directory is required. A global mesh structure which is neither regular nor semiregular is said to be *helter-skelter*. A helter-skelter mesh structure implies that the nonzero entries of A may be scattered rather haphazardly throughout the matrix

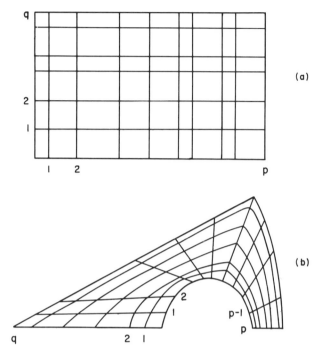

Fig. 10-2.2.   Examples of regular global mesh structures.

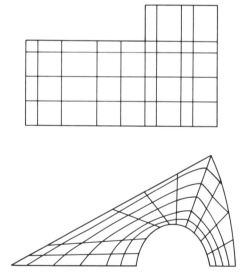

Fig. 10-2.3.   Examples of semiregular global mesh structures.

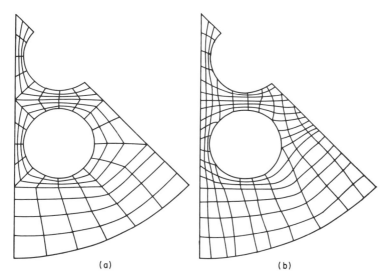

Fig. 10-2.4. Examples of helter-skelter and semiregular global mesh structures. (a) helter-skelter global structure. (b) semiregular global structure.

Thus some type of directory is needed almost at every point i to determine the j's such that $a_{ij} \neq 0$.

A helter-skelter mesh subdivision is useful in accommodating the mesh to an irregular geometric domain. Semiregular mesh subdivisions may also be used for irregular domains but usually require more elements. See Fig. 10-2.4b. For regular and semiregular mesh structures, any element point is common to at most four mesh elements. For example, the configuration given in Fig. 10-2.5, where element point P is common to six elements is possible only for a helter-skelter mesh structure.

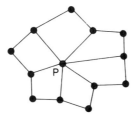

Fig. 10-2.5. 13-point discretization stencil which may result from a helter-skelter global mesh structure.

A Basic Problem

To establish a basic problem which can be used for comparison purposes, we first consider the solution of (10-2.1) in a square region R with $T = 0$ on the boundary. We assume that the regular mesh structure given in Fig. 10-2.6 is imposed on R and that the standard 5-point discretization formula is used.† We now wish to consider the iterative solution‡ of this problem using a line partitioning of the unknowns.

Let T_i denote the column vector whose elements are the unknowns on line i of Fig. 10-2.6. Using a red/black ordering for the line unknowns,§ we can express the matrix equation (10-2.2) in the partitioned form

$$\begin{bmatrix} A_{2,2} & & 0 & \vdots & A_{1,2}^T & A_{2,3} & & 0 \\ & A_{4,4} & & \vdots & & A_{3,4}^T & A_{4,5} & \\ 0 & & \ddots & \vdots & & 0 & & \ddots \\ \cdots & \cdots & \cdots & \cdots & \cdots & \cdots & \cdots & \cdots \\ A_{1,2} & & 0 & \vdots & A_{1,1} & & 0 & \\ A_{2,3}^T & A_{3,4} & & \vdots & & A_{3,3} & & \\ 0 & \ddots & & \ddots & \vdots & 0 & & \ddots \end{bmatrix} \begin{bmatrix} T_2 \\ T_4 \\ \vdots \\ \cdots \\ T_1 \\ T_3 \\ \vdots \end{bmatrix} = \begin{bmatrix} G_2 \\ G_4 \\ \vdots \\ \cdots \\ G_1 \\ G_3 \\ \vdots \end{bmatrix}, \quad (10\text{-}2.3)$$

where the elements of the submatrix $A_{i,j}$ are the couplings of the unknowns from line i to those on line j. Because of matrix symmetry, $A_{i,j} = A_{j,i}^T$. The CCSI method (see Chapter 8) relative to the partitioning (10-2.3) can then be given by

$$T_i^{(n+1)} = \rho_R^{(n+1)}[\hat{T}_i^{(n+1)} - T_i^{(n)}] + T_i^{(n)}, \qquad (10\text{-}2.4a)$$

where

$$A_{i,i}\hat{T}_i^{(n+1)} = -A_{i-1,i}^T T_{i-1}^{(n)} - A_{i,i+1}T_{i+1}^{(n)} + G_i, \qquad i = 2, 4, \ldots,$$

and

$$T_i^{(n+1)} = \rho_B^{(n+1)}[\hat{T}_i^{(n+1)} - T_i^{(n)}] + T_i^{(n)}, \qquad (10\text{-}2.4b)$$

where

$$A_{i,i}\hat{T}_i^{(n+1)} = -A_{i-1,i}^T T_{i-1}^{(n+1)} - A_{i,i+1}T_{i+1}^{(n+1)} + G_i, \qquad i = 1, 3, \ldots.$$

Here $\rho_R^{(n+1)}$ and $\rho_B^{(n+1)}$ are the Chebyshev parameters defined by (8-3.7).

---

† The test problems used in Sections 8.4 and 9.8 are similar to the basic problem considered here.

‡ Direct methods can also be used to solve the system of equations resulting from discretizations of multidimensional, boundary-value problems. Many efficient direct methods exist (Reid [1977], George [1973], and Irons [1970]) and should be considered as a possible solution method. Among the factors to be considered in deciding between direct and iterative methods are "storage" and "work," i.e. number of arithmetic operations. For many problems, there is a "crossover point" in the number of unknowns above which a "good" iterative method becomes more cost effective than a "good" direct method.

§ For convenience later, here we have taken T_i to be a red line of unknowns if i is even.

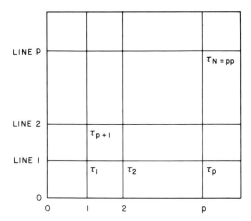

Fig. 10-2.6.   Regular rectangular mesh subdivision.

Each iteration of the CCSI method requires solutions to the subsystems

$$A_{i,i}\,\hat{T}_i^{(n+1)} = y_i, \qquad i = 1, \dots, p, \qquad (10\text{-}2.5)$$

where y_i is known. Since most of the arithmetic cost per iteration is involved in the formulation of the y_i vectors and in the solution of the subsystems (10-2.5), it is important that these calculations be done efficiently. For the problem considered here, each $A_{i,i}$ is a positive definite tridiagonal matrix so that the subsystems (10-2.5) can be solved efficiently, for example, using the Cholesky decomposition procedure described in Section 2.3.

We now wish to discuss in some detail the implementation, relative to cost, of the CCSI method. Since the cost of evaluating the required Chebyshev parameters has been discussed previously in Section 8.5, we ignore this aspect in this chapter. Much of what we say concerning the computational aspects of the CCSI method applies equally well to the SOR, RS–SI, RS–CG, and CCG methods since these methods and the CCSI method differ primarily in the way the acceleration parameters are calculated.

We first determine the storage requirements. From Section 2.3, the tridiagonal system (10-2.5) can be solved by factoring each $A_{i,i}$ as $A_{i,i} = S_i^T S_i$ and then solving for $\hat{T}_i^{(n+1)}$ by using (2-3.25) and (2-3.26). Here S_i is an upper bidiagonal matrix, which has at most two nonzero elements per row. We assume that the $S_i^T S_i$ decomposition of $A_{i,i}$ is done prior to the start of the iterative process and that the nonzero elements of S_i, rather than the elements of $A_{i,i}$, are stored. Each $A_{i,i+1}$ matrix has at most one nonzero element per row. If N is the order of the matrix A, then the storage requirements η for the CCSI method are $3N$ words for the S_i and $A_{i,i+1}$ matrices† and $2N$ words for

---

† The nonzero elements of $A_{i-1,i}^T$ are easily obtained from $A_{i-1,i}$.

the $T^{(n)}$ and g vectors. Thus

$$\eta = 5N. \tag{10-2.6}$$

We measure the arithmetic operations θ by the total number of multi-plications and divisions† required to solve the problem. We express θ as

$$\theta = FI_\alpha \alpha, \tag{10-2.7}$$

where F is the number of multiplications and divisions required per iteration, I_α is the number of iterations required to reduce the iteration error by 0.1, and α is the number of significant digits (to the base 10) required of the solution. Forming the y_i vectors and solving for $\hat{T}_i^{(n)}$ in (10-2.5) requires a total of $4N$ multiplications and $2N$ divisions. Thus with the $\rho^{(n+1)}$ multiplication in (10-2.4), we have $F = 7N$. (As noted previously in Section 8.5, η can be reduced to $4N$ and F to $5N$ by using the change of variable method suggested by Cuthill and Varga [1959].) For most physical problems of interest, α will lie in the range from 2 to 6. The I_α factor in (10-2.7) is determined by the rate of convergence of the CCSI method, which in general depends on the mesh element size‡ as well as the physical data of the problem. Thus θ is often difficult to predict since it depends on I_α, which is usually unknown.

For the problem considered here, it is reasonable to assume that the spectral radius, $M(B^L)$, of the Jacobi iteration matrix B^L associated with the line partitioning (10-2.3) can be expressed approximately in the form§

$$M(B^L) \doteq 1 - [w/(p + 1)]^2. \tag{10-2.8}$$

Here p is defined in Fig. 10-2.6 and w is a variable which depends on the physical data of the problem but is independent of p. We assume that $(w/(p + 1))^2$ is small. (The variable w usually lies in the interval $[0.1, 20]$). Using the approximation (10-2.8) for $M(B^L)$, we can express (see Section 9.9) the asymptotic convergence rate, $\bar{R}_\infty(CCSI(B^L))$, of the CCSI method in the

---

† For reasons of simplicity, we combine the multiplication and division operations here. As noted in Section 8.5, most divisions may be replaced by a multiplication, if desired. Since an addition is usually associated with each multiplication, the number of additions required is about the same as the number of multiplications.

‡ For some physical problems, the CCSI convergence rate does not depend significantly on mesh element size. We shall discuss two such problems in the next chapter.

§ For the model problem (i.e., Eq. (10-2.1) with $k = 1$ and constant γ) in the unit square, it can be shown (see, e.g., Varga [1962]) that

$$M(B^L) = \frac{\cos(\pi/(p + 1))}{2 - \cos(\pi/(p + 1)) + \gamma/(2(p + 1)^2)} \doteq 1 - \frac{(\pi^2 + \gamma/2)}{(p + 1)^2}$$

approximate form: $\bar{R}_\infty(\text{CCSI}(B^L)) \doteq 2\sqrt{2}(w/(p+1))$. Now, from (2-2.12), $I_\alpha = (\log 10)/\bar{R}_\infty(\text{CCSI}(B^L))$. Thus I_α may be approximated roughly by

$$I_\alpha \doteq \frac{\log 10}{2\sqrt{2}}\left(\frac{p+1}{w}\right) \doteq 8\left(\frac{p}{w}\right). \tag{10-2.9}$$

Since $F = 7N = 7p^2$, we have approximately that

$$\theta \doteq 5.6p^3(\alpha/w). \tag{10-2.10}$$

The estimates (10-2.6) and (10-2.10) give some indication of the storage and arithmetic factors which affect solution costs. The remaining factor listed in (10-1.2), overhead due to data transmission and logical operations, is more difficult to measure but is as important as the other factors. Modern computers have reached a point where logical operations and data transmission may be more time consuming than that required for the arithmetic operations.

We indicate overhead costs by what we call arithmetic efficiency, which is defined by

$$\text{Arith. Effic.} \equiv \frac{\text{Time to perform multiplications, divisions, and additions}}{\text{Total matrix solution time}}.$$

We include additions here since additions and multiplications require about the same time on some computers. However, the numerator does not include "fetch" and "store" operations.

To estimate overhead costs, we calculate the arithmetic efficiency for three programs written for the CDC-7600 computer to solve the matrix problem (10-2.3) by using the line CCSI method. The computer configuration used was roughly that given in Fig. 10-2.7.

Program A was compiled with the CDC FORTRAN IV RUN compiler. The storage and arithmetic requirements are as given above; i.e., $5N$ words of storage and $5N$ multiplication, $2N$ divisions, and $6N$ additions per iteration are required.

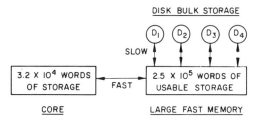

Fig. 10-2.7. CDC-7600 computer configuration. Each disk has 2.0×10^7 words of storage and a streaming transfer rate of 5×10^{-6} sec/word. Multiply time $= 1.37 \times 10^{-7}$ sec: add time $= 1.10 \times 10^{-7}$ sec.

Program B was also compiled using the RUN compiler but with some parts of the program written in machine assembly language.† The Cuthill–Varga change of variable method (See Section 8.5) is used to eliminate the $2N$ divisions and to reduce storage. Thus $4N$ words of storage, $5N$ multiplications, and $6N$ additions per iteration‡ are required.

In addition to the machine language programming and the Cuthill–Varga method used in Program B, *Program C* utilizes a concurrent iteration procedure§ (e.g., see Pfeifer [1963]) which enables much of disk data transfer time for large problems to be covered by central processor calculations. The concurrent iteration precedure will be described later.

The overhead efficiencies for these programs are indicated by the data given in Table 10-2.1. For problem 1, $p = 100$, $N = 10^4$, and all data is stored in fast memory. For problem 2, $p = 240$, $N = 5.76 \times 10^4$, and all data is stored in disk bulk storage.‖ The time given in the table is for 100 iterations and does not include any adaptive parameter or stopping calculations since these calculations were purposely bypassed. For Program A, the division time was assumed to be 5.48×10^{-7} sec.

For problem 1, the matrix solution time for Program A was about eight times that for Program B. The elimination of the $2N$ divisions in Program B produced a factor of two improvement and the use of machine language programming another factor of four. The difference in solution times between Program B and Program C is due to the additional overhead connected with the implementation of the concurrent iteration procedure.

For problem 2, one disk was used to read the previous iterate, T^n, one disk to write the result of the current iterate, T^{n+1}, and the remaining two disks were used to store the matrix coefficients and the vector g. Thus for all three programs, at least one disk required the storage of $2N$ words. The transfer time for these words, which usually must be moved to the central processor every iteration, is $(5 \times 10^{-6})(11.52 \times 10^4) = 0.576$ sec. Thus for this case, each iteration will require at least 0.576 sec.

---

† Machine language programming is done to utilize the parallel or independent function units on the CDC–7600 which enables one to fetch, multiply, and store at the maximum central processor speed.

‡ The initialization and storage costs of the change of variable method are minor and are ignored here.

§ The use of concurrent iterations apparently was first suggested in the late 1950s by S. W. Dunwell of the IBM Corporation.

‖ For Programs B and C, the storage required for problem 2 is $4N$ or 2.3×10^5 words which would seem to fit in large fast memory. However, because of additional storage requirements of these programs, disk bulk storage was used.

TABLE 10-2.1

Arithmetic Efficiencies for Three Programs to Solve the Basic Problem.[a]

	Problem 1: $p = 100$, $N = 10,000$			Problem 2: $p = 240$, $N = 57,600$		
	Time (Sec)			Time (Sec)		
	$*, \div, +$	Matrix solution	Arithmetic efficiency	$*, \div, +$	Matrix solution	Arithmetic efficiency
Program A	2.44	14.5	0.17	14.05	87.0	0.16
Program B	1.345	1.68	0.80	7.75	59.6	0.13
Program C	1.345	2.634	0.51	7.75	12.47	0.62

[a] The matrix solution time indicated is for 100 iterations of the line CCSI method.

For Program A, the central processor time required per iteration† for problem 2 is about 0.8 sec. Thus the disk transfer time is covered by central processor calculations and the arithmetic efficiency is not reduced even though disk storage is used. For the more efficient Program B, however, the central processor time per iteration is about 0.097 sec. Thus the slower disk transfer rate will determine the calculation time. This is reflected in the large reduction of the arithmetic efficiency for Program B when disk storage is required. The concurrent iteration procedure used in Program C permits calculations to be basically carried out at the faster central processor speed independent of the type of storage used.

Now, we wish to illustrate the concurrent iteration procedure for the CCSI method (10-2.4). We assume the indexing of lines is as given in Fig. 10-2.6. Let the data for the first 12 lines, for example, be in large fast memory. Using this data, we can calculate iterate 1 for the red lines 2, 4, ..., 12 and for the black lines 1, 3, ..., 11. Now iterate 2 can be calculated for red lines 2, 4, ..., 10 and for the black lines 1, 3, ..., 9. Using only the data for the first 12 lines, we can continue this process to obtain finally iterate 6 for lines 1 and 2, iterate 5 for lines 3 and 4, etc. After this "triangle" of iterates (see Fig. 10-2.8) is completed, the data for lines 1 and 2 are replaced in fast memory by that for lines 13 and 14. We then can calculate iterate 1 for lines 13 and 14, iterate 2 for lines 11 and 12, ..., and iterate 6 for lines 3 and 4. The data for lines 3 and 4 are then replaced by that for lines 15 and 16, and we can calculate one additional iteration for lines 5–16. Continuing this process over all lines, we do six *concurrent* iterations while moving the coefficient data from

† Since the central processor time for problem 1 is basically the matrix solution time, an estimate of the central processor time per point per iteration can be obtained from the data of problem 1.

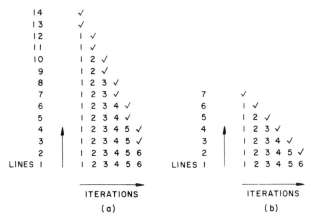

Fig. 10-2.8.   Concurrent iterations. (a) Concurrent CCSI iterations. (b) Concurrent SOR iterations.

disk to large fast memory only once. Moreover, if buffering capabilities are available, much of the time required to transfer this data from disk can be covered by central processor calculations (Pfeifer [1963]). For example, while calculating iterate 1 for lines 13 and 14, ..., and iterate 6 for lines 3 and 4 in the above, we can transfer the data for lines 15 and 16 from disk to a buffered area in fast memory. Thus all (or almost all) of the data for lines 15 and 16 will be available in fast memory when needed. Treating the data transfer for other lines similarly, we can cover most of the disk data transfer time by central processor calculations.

In general, data for $2(C + 1)$ lines must be stored in fast memory in order to do C concurrent CCSI iterations with data buffering. We remark that concurrent SOR iterations also may be carried out. In fact, data for only $C + 1$ lines is required in fast memory in order to do C concurrent iterations of the SOR method when a natural line ordering is used. The case $C = 6$ is illustrated in Fig. 10-2.8. Since the values for the conjugate gradient parameters γ_{n+1} and ρ_{n+1} (see (7-4.9)) depend on results obtained from iteration n, the direct utilization of the concurrent iteration procedure for the conjugate gradient method is not possible.

From the data of Table 10-2.1, it is clear that overhead costs can be significant,† and that effort made to minimize data transfer and logical operations

---

† For the problems of Table 10-2.1, Program C is five to seven times fater than Program A but the differences in computer time are relatively small. Thus Program C would have to be used more than a few times in order for the savings in computer cost to offset the greater initial programming cost of Program C. However, for more general problems (such as three-dimensional or time-dependent problems) where solution times are in minutes or hours, a factor of 5 reduction in solution cost becomes significant almost immediately.

can have a large payoff. We remark that the arithmetic efficiency improvements obtained in Programs B and C relied heavily on the regular mesh structure of the problems. If the precise location of the nonzero elements in the coefficient matrix could not be assumed in the programming of the solution method, an arithmetic efficiency greater than 0.1 would be difficult to achieve.†

Problems Using 9- and 25-Point Stencils

We now consider iterative solution costs when the more general 9- or 25-point stencil of Fig. 10-2.1 is used in the discretization of (10-2.1). We continue to assume that the regular mesh structure of Fig. 10-2.6 is imposed on the solution region and that $T = 0$ on the boundary.

For the 9-point discretization problem, using the same line partitioning and the same line ordering as for the 5-point problem, we again obtain the red/black partitioned form (10-2.3). For this case, both the diagonal submatrix $A_{i,i}$ and the off-diagonal submatrix $A_{i,i+1}$ are tridiagonal.

For discretization procedures which yield 25-point stencils, some preliminary remarks are needed before solution aspects for this problem can be discussed. The 25-point stencil given in Fig. 10-2.1 is applicable for any unknown τ_P associated with an element point P, which is common to four mesh elements. However, when P is common to fewer than four elements, a stencil with fewer points will result. This follows from our previous assumption (see discussion following (10-2.2)) that the unknowns τ_P and τ_j are coupled only if P and j are points of the same element. Stencils for τ_P when P is common to one mesh element and when P is common to two mesh elements are illustrated in Fig. 10-2.9.

Strang and Fix [1973] note that the unknowns τ_P associated with element points interior to mesh elements can be eliminated at little cost prior to the start of the solution process. This early elimination of unknowns, which is known as *static condensation*, also simplifies the stencils for the remaining unknowns. For example, the 25-point stencil of Fig. 10-2.1, after static condensation, reduces to a stencil with only 21 points (see Fig. 10-2.9). In what follows, we ignore the small cost involved in this early elimination of unknowns.

In order to obtain a convenient red/black partition for problems which utilize the 21-point stencil, we partition the unknowns by "mesh element lines": Let T_i denote the column vector whose elements are the unknowns on the "element-line" i given in Fig. 10-2.10.‡ Such a partitioning leads, as before.

---

†For example, the effectiveness of machine language programming would be significantly reduced for this case.

‡ Note that "element-line" $p + 1$ is a special case.

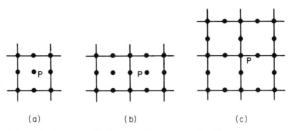

Fig. 10-2.9.   Discretization stencils for τ_P. (a) Element point P common to one mesh element. (b) Element point P common to two mesh elements. (c) Element point P common to four elements after static condensation.

to a matrix equation of the form (10-2.3), but now the diagonal submatrices $A_{i,i}$ have a half-bandwidth† of four instead of one as before. Thus solutions for the subsystems $A_{i,i} \hat{T}_i = y_i$ will be more costly to obtain for the 21-point stencil case. Each row of the off-diagonal submatrices $A_{i,i+1}$ has, on the average, 6.5 nonzero elements.

In Table 10-2.2, estimates for the multiplication and storage requirements of the block CCSI solution method are given for the various stencils which may be used in the discretization of (10-2.1). The last row of the table gives the requirements for the 21-point case assuming that the solution region has been subdivided into only $\frac{1}{2}p \times \frac{1}{2}p$ mesh elements. Note that the arithmetic requirements for this case are roughly the same as those for the five- and nine-point cases when a $p \times p$ mesh subdivision is used. The ratios of the multiplication and storage requirements for the 9- and 25-point problems to the requirements of the 5-point problem are given in Table 10-2.3. We have assumed that the rate of convergence for the block CCSI method is independent of the discretization formula used. Numerical experience (Abu-Shumays and Hageman [1975]) indicates that this is true for problems for which the 5- and 9-point stencils are used, but we have no data concerning the truth of this for the 25-point case.

The value of the 25-point stencil lies in two areas:

(1)   Use with curved mesh elements to describe problem geometry if curved boundaries are involved.

(2)   Use as a higher order discretization formula if the solution is sufficiently smooth.

If an accurate description of the geometric domain is important, then one is forced to pay the price in solution costs. However, if rather gross approximations are used to model other parts of the physical problem, then some

† The matrix $H = (h_{i,j})$ is said to have a half-bandwidth of b if $h_{i,j} = 0$ whenever $|j - i| > b$.

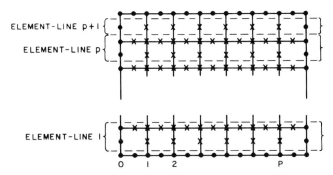

Fig. 10-2.10. An "element-line" partitioning for a 21-point discretization approximation.

thought should be given as to the worth of an accurate description of the geometric domain.

The value of the 25-point stencil as a higher order discretization formula is often an open question. Singularities in low-order derivatives of the solution occur in most problems where corners exist on the boundaries and/or where the physical parameters (k, γ, g) of Eq. (10-2.1) are not smooth. Near singular points, there is no reason to expect a "high-order" 25-point discretization formula to be more accurate than a "low-order" 9-point formula (Babuska and Kellogg [1973]). For problems with few singular points, the accuracy of high-order methods can be maintained through the use of special techniques. For example, around the singular points, a suitable mesh refinement (Birkhoff *et al.* [1974]) or special discretization formulas (Strang and Fix [1973]) could be used. However, for more general problems, it is difficult and costly to account for each singular point individually.

TABLE 10-2.2

*Multiplications and Storage Requirements of the CCSI Solution Method as a
Function of Discretization Stencil*

$p \times p$ Mesh elements	F	I_α	θ	Storage, η
5-Point stencil				
($\doteq p^2$ unknowns)	$7\,p^2$	$0.8\ (p/w)$	$5.6\ \alpha\ (p^3/w)$	$5\,p^2$
9-Point stencil				
($\doteq p^2$ unknowns)	$11\,p^2$	$0.8\ (p/w)$	$8.8\ \alpha\ (p^3/w)$	$7\,p^2$
21-Point stencil				
($\doteq 3p^2$ unknowns)	$60\,p^2$	$0.8\ (p/w)$	$48.0\ \alpha\ (p^3/w)$	$34\,p^2$
21-Point stencil[a]				
($\doteq \frac{3}{4}p^2$ unknowns)	$15\,p^2$	$0.4\ (p/w)$	$6.0\ \alpha\ (p^3/w)$	$8.5\,p^2$

[a] $\frac{1}{2}p \times \frac{1}{2}p$ mesh elements.

TABLE 10-2.3

*Ratios of Multiplication and Storage Requirements for
the 9- and 21-Point Stencils to the Requirements of the
5-Point Stencil*

		Block CCSI	
		Arith., θ	Storage, η
$(p \times p)$ Elements $\begin{cases} \dfrac{\text{9-point}}{\text{5-point}} \\ \dfrac{\text{21-point}}{\text{5-point}} \end{cases}$		1.6	1.4
		8.6	6.8
$\dfrac{\{(\tfrac{1}{2}p \times \tfrac{1}{2}p) \text{ Elements}\}\ \text{21-point}}{\{(p \times p) \text{ Elements}\}\ \text{5-point}}$		1.1	1.7

Often, a concern to the user is the error caused by the discretization of the
mathematical model. For most practical problems there is no easy way to
estimate pointwise this discretization error. Usually, error estimates can only
be obtained (Zienkiewicz [1973]) by analyzing the solutions from successively
finer mesh subdivisions. Previous information may beneficially be used when
analyzing discretization errors by successive mesh subdivisions. For example,
a good initial approximation for a refined mesh problem can be obtained by
simple interpolation of the coarse mesh solution. Moreover, the optimum
acceleration parameters for the refined mesh problem usually can be easily
estimated from the coarse mesh problem. We note that at least two mesh
subdivisions are necessary to get useful information concerning the discretization error. Thus the cost for doing numerical discretization error analysis can
be high and is often impossible due to size restrictions of the program being
used.

Global Mesh Structure Effects

Thus far we have discussed the solution cost factors given in (10-1.2) relative
to various discretization formulas only for regular mesh structures. The
generalization of the above estimates to the semiregular mesh structure case is
straightforward. However, some care should be taken for the semiregular
mesh case in estimating the data transmission overhead cost factor for large
problems where disk storage is required. Since the amount of data required
for each line may be different, the use of concurrent iterations may be more
complicated.

For a helter-skelter global mesh structure, block CCSI solution methods†
may be used to solve the matrix equation but usually at a considerably greater
cost than that for a problem with a regular mesh structure. The greater cost
is due to increased arithmetic and storage requirements and increased overhead due to logical operations and data transfer. The efficiency of a particular
program is often determined by how well the regularities of the numerical pro
cedure can be adapted to the characteristics of the computer at hand. The
fewer regularities which a numerical procedure possesses, the more difficult
it is to make use of the characteristics of the computer. For problems with a
helter-skelter global mesh element structure, the effectiveness of iterative
solution methods are reduced considerably.

10.3 THE TIME-INDEPENDENT THREE-DIMENSIONAL PROBLEM

In this section we consider the numerical solution of the time-independent,
three-dimensional problem

$$-\frac{\partial}{\partial x}\left(k\frac{\partial T}{\partial x}\right) - \frac{\partial}{\partial y}\left(k\frac{\partial T}{\partial y}\right) - \frac{\partial}{\partial z}\left(k\frac{\partial T}{\partial z}\right) + \gamma T = g \qquad (10\text{-}3.1)$$

with $T = 0$ on the surface boundaries. The mesh element types and discretization stencils considered are given in Fig. 10-3.1. The global mesh element
structure is restricted to be of the regular type given in Fig. 10-3.2.

A one-line partitioning of the unknowns leads to a red/black partitioning
of the coefficient matrix only for the 7- and 11-point stencils corresponding to
elements 3.1a and 3.1b. For the 27-point stencil corresponding to element
3.1c, a one-plane partitioning of the unknowns is needed to produce a convenient red/black partitioning for the coefficient matrix. In Table 10-3.1, we
give the arithmetic and storage requirements for the block CCSI solution
method as a function of the mesh element used in the discretization of (10-3.1).
We have assumed that $q \le p \le r$. The increased arithmetic and storage
requirements for the one-plane CCSI method is caused by the increase in the
half-bandwidth of the subsystem of equations which must be solved at each
step of the iterative process. The asymptotic rates of convergence for the CCSI
method were obtained by assuming that the spectral radii μ_1 of the associated
Jacobi iteration matrices to be of the form $\mu_1 \doteq 1 - (w/q)^2$. The subscript on

---

† For example, the Cuthill–McKee [1969] reordering algorithm developed to minimize
bandwidth, may be used to obtain a block red/black partitioning of the coefficient matrix.

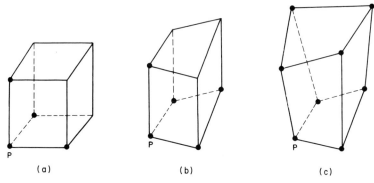

(a)   (b)   (c)

Fig 10-3.1. Types of mesh elements and discretization formulas for three-dimensional problems. (a) 4-node rectangular prism element, 7-point discretization stencil. (b) 5-node quadrilateral prism element, 11-point discretization stencil. (c) 8-node hexagonal element, 27-point discretization stencil.

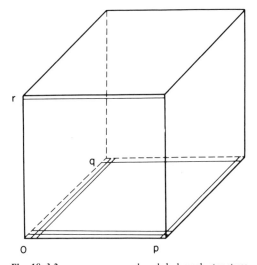

Fig. 10-3.2. $p \times q \times r$ regular global mesh structure.

the w variable in Table 10-3.1 indicates the dependence† of this variable on the partitioning used.

The semilog graphs of θ and η versus p given in Figs. 10-3.3 and 10-3.4 clearly indicate the magnitude of the storage and computational requirements for the three-dimensional problem. The graphs are based on the assumption that $p = q = r$, that $\alpha = 5$, and that $w_L = w_P = 0.7$. The right margin of

---

† For the model problem ($k = 1.0$, $\gamma = 0.0$, and $p = q = r$) in the unit cube, $w_L^2 \doteq \frac{3}{4}\pi^2$ and $w_P^2 \doteq \frac{3}{2}\pi^2$.

TABLE 10-3.1

Multiplication and Storage Requirements of the Block CCSI Solution Method as a Function of Discretization Stencil.[a]

	I_z	F	θ	Storage, η
7-Point stencil (line partitioning)	$\dfrac{0.8q}{w_L}$	$9pqr$	$\dfrac{7.2\alpha}{w_L}pq^2r$	$6pqr$
11-Point stencil (line partitioning)	$\dfrac{0.8q}{w_L}$	$13pqr$	$\dfrac{10.4\alpha}{w_L}pq^2r$	$8pqr$
27-Point stencil (plane partitioning)	$\dfrac{0.8q}{w_P}$	$2(p+9)pqr$	$\dfrac{1.6\alpha(p+9)}{w_P}pq^2r$	$(p+11)pqr$

[a] The quantities I_z, F, θ, and η are defined by (10-2.6)–(10-2.7).

Fig. 10-3.3 gives the time required just to carry out the required number of multiplications on a CDC-7600 computer.

For the computer configuration of Fig. 10-2.7, the storage limitation is 8×10^7 words. Thus if $p = q = r$, the storage limitation restricts p to be less than 90 when the 27-point stencil and the one-plane CCSI solution method are used. Moreover, the effectiveness of the iterative solution method is reduced considerably when a 27-point discretization formula is used. The reduced effectiveness is caused not only by the increased arithmetic and storage requirements but also by the possibly even greater increase in the logical and data transmission overhead associated with the implementation of the method.

For most three-dimensional problems of modest size, some data must be stored in disk bulk storage. We illustrate this for the CDC-7600 configuration of Figure 10-2.7. With $p = q = r$, for example, an upper bound on p before disk bulk storage is required is $p = 20$ and $p = 29$ for the plane and line partitionings, respectively. Since the disk streaming transfer time (assuming the use of four disks in Fig. 10-2.7) is ten times larger than the multiplication time, many of calculations would be done at the slower bulk storage transfer rate. The use of the concurrent iteration procedure discussed in Section 10.2 permits some of the bulk data transfer time to be covered by central processor calculations. The difficulty here is that large fast memory might not be large enough to hold the data required. In Table 10-3.2, we give the number of concurrent iterations which can be done as a function of p for the one-plane SOR and the one-line SOR methods using a natural (non–red/black) ordering for the partitioned unknowns. For the one-plane SOR and one-line SOR methods, respectively, we assume that $(p^3 + 10p^2)$ and $(8p^2)$ words per plane are required. To do C concurrent SOR iterations, $(C + 1)$ planes of data are required in the large fast memory. For the red/black ordering required by the

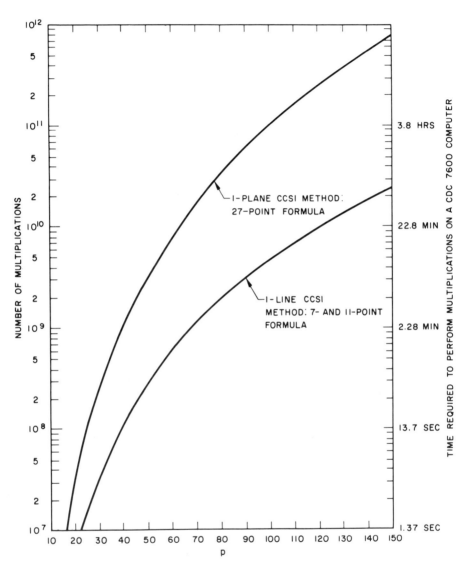

Fig. 10-3.3. Number of multiplications versus p for three-dimensional problems with $N = p^3$ unknowns.

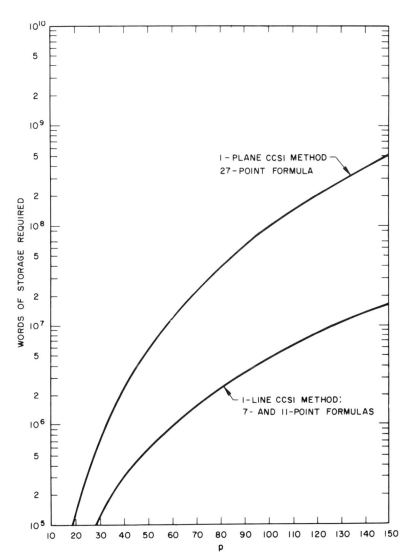

Fig. 10-3.4. Storage requirements versus p for three-dimensional problems with $N = p^3$ unknowns.

TABLE 10-3.2

Number of Concurrent Iterations as a Function of p
for the One-Plane SOR and One-Line SOR
Methods[a]

	No. of concurrent iterations	
$p = q = r$	One-plane SOR	One-line SOR
10	124	155
20	20	76
30	6	32
34	4	28
35	3	22
37	2	20
41	1	18
56		8
60		6
70		6
72		4

[a] The plane method assumes a 27-point discretization formula while the line SOR method is for an 11-point formula.

CCSI methods, $2(C + 1)$ planes of data are required in the large fast memory in order to do C concurrent iterations. Thus the number of concurrent iterations which may be done using the CCSI method is roughly one-half that given in Table 10-3.2 for the SOR method. It is for this reason that the SOR method using a natural ordering for the partitioned unknowns often is used instead of the CCSI method in the solution of three-dimensional problems.

Taking into account the three cost factors given in (10-1.2), Hageman [1975] gives estimates for the solution times of the various methods for different values of p. His estimates are given in Table 10-3.3 and are based on the assumptions that[†] $p = q = r$, that the computer configuration of Fig. 10-2.7 is used, and that concurrent iterations are used. The estimates given are rough and are based on many assumptions. However, they do indicate the magnitude of the computing time required to solve three-dimensional problems.

Based on Figs. 10-3.3 and 10-3.4 and the estimates of Table 10-3.3, it is clear that the solution time required for a 7- or 11-point discretization system is significantly less than that required for a 27-point system. The difficulty is that the 7- or 11-point discretization formula is usually applicable only for

---

† If p and q in Fig. 10-3.2 are small relative to r, the computational cost for the plane method (relative to the line method) is considerably reduced.

TABLE 10-3.3

Estimates for the Solution Time Required by Different Solution Methods on a CDC–7600 Computer

	Solution time (hr)	
$p = q = r$	27-Point formula one-plane SOR	7- or 11-Point formula one-line SOR
19	0.06	0.02
34	0.96	0.17
40	4.26	0.32
41	11.20	0.33
50		0.84

those problems whose geometry can be described using rectangular or quadrilateral prism mesh elements.

We remark that partitionings other than plane should also be considered in the solution of the 27-point discretization problem. Consider, for example, a partitioning of the unknowns by lines. Since such a partitioning produces a coefficient matrix which does not have Property $\mathscr{A}$, neither the CCSI method or the SOR method† is applicable for this case. However, other iterative procedures can be used. One such method is the J–SI method (see Section 9.9). The plane CCSI method discussed above converges faster, perhaps three to six times faster,‡ than the line J–SI method; however, the total computer time required by the line solution method may be less than that for the plane method since less storage and less arithmetic per iteration are required.

10.4   THE TIME-DEPENDENT PROBLEM

In this section we consider the numerical solution of the time-dependent linear problem

$$\rho \frac{\partial T}{\partial t} = \nabla \cdot k \, \nabla T + g \qquad (10\text{-}4.1)$$

in two or three space dimensions. We assume that $T = 0$ on the boundary and that some initial condition T_0 is given. We also assume that $\rho > 0$ and $k > 0$.

---

† The SOR method can be used but a precise prescription for ω_b would not be available.

‡ This estimate is based on arguments similar to those given in Section 9.9 to obtain the ratio (9-9.5).

Using one of the spatial discretization methods discussed previously, we can express (see, e.g., Varga [1962]) the space–time continuous problem (10-4.1) in the time continuous matrix differential form

$$C \frac{d\tau}{dt} = -A(t)\tau + g(t), \tag{10-4.2}$$

where, for simplicity, we assume that the matrix C is independent of t. Except for the time variable, the matrix $A(t)$ and the vectors $\tau(t)$ and $g(t)$ are defined as in (10-2.2). To discretize relative to time, we integrate (10-4.2) from t_n to t_{n+1}. Using the approximation

$$\int_{t_n}^{t_{n+1}} W\tau \, dt \doteq (t_{n+1} - t_n)[\theta W(t_{n+1})\tau(t_{n+1}) + (1 - \theta)W(t_n)\tau(t_n)]$$

for terms on the right side of (10-4.2), we obtain

$$\left[\frac{C}{\Delta t_{n+1}} + \theta A(t_{n+1})\right]\tau_{n+1} = \left[\frac{C}{\Delta t_{n+1}} - (1 - \theta)A(t_n)\right]\tau_n$$
$$+ [\theta g(t_{n+1}) + (1 - \theta)g(t_n)]. \tag{10-4.3}$$

Here τ_n represents the numerical approximation for $\tau(t_n)$, $\Delta t_{n+1} = t_{n+1} - t_n$, and $0 \le \theta \le 1$ is the integration weight factor. The backward-difference method is obtained with $\theta = 1$ and the central-difference (Crank–Nicolson) method corresponds to $\theta = \frac{1}{2}$. For sufficiently small Δt, the central-difference method is considered to be the most accurate. From (10-4.3), it is seen that an equation of the type

$$[(C/\Delta t_{n+1}) + \theta A(t_{n+1})]\tau_{n+1} = s_{n+1}, \tag{10-4.4}$$

where s_{n+1} is a known vector, must be solved for τ_{n+1} at each time step.

The solution cost for the time-dependent problem (10-4.1) depends on the number of time steps and the work required per time step. The number of time steps required is a function of the time increments, Δt, used[†] and is strongly problem dependent. Thus we will discuss only computational cost per time step here. Moreover, in what follows, we assume that $\theta > 0$.

If we let

$$\tilde{A}_{n+1} = (C/\Delta t_{n+1}) + \theta A(t_{n+1}), \tag{10-4.5}$$

then for each time step the equation

$$\tilde{A}_{n+1}\tau_{n+1} = s_{n+1} \tag{10-4.6}$$

---

† The selection of Δt required to achieve a certain degree of accuracy is a difficult problem. Often numerical experimentation or an adaptive procedure is required to determine adequate values for Δt. See, for example, Gear [1971].

must be solved for τ_{n+1}. The known vector s_{n+1} in (10-4.6) is given by the right side of Eq. (10-4.3). Since the matrix A corresponds to the discretization of $-\nabla \cdot k\nabla$ and the matrix C corresponds to the discretization of γ in the time-independent equations (10-2.1) or (10-3.1), the matrix equation (10-4.6) is of the form considered previously in Sections 10.2 and 10.3. Thus the computational cost factors for solving (10-4.6) by the block CCSI method are the same as those given previously.

The number of CCSI iterations required to solve (10-4.6) for τ_{n+1} is usually less than that required for the time-independent problems, however. The reasons for this are

(1)   For many problems, a stopping criterion number of 10^{-4} or 10^{-5} for the iterative process is adequate for reasonable accuracy. Moreover, if τ does not change rapidly in the time interval (t_n, t_{n+1}), τ_n is a good approximation for τ_{n+1} so that the relative error of the initial guess is already small.
(2)   The number of spatial unknowns for time-dependent problems is usually kept modest in order that the total computer time required be reasonable.
(3)   The convergence rate of the CCSI method is now also a function of Δt and θ, where the number of iterations required for a fixed degree of accuracy decreases† as Δt and θ decrease from ∞ and 1, respectively.
(4)   The estimate for the optimum acceleration parameters usually needs little modification in going from time t_n to time t_{n+1}.

One of the more difficult problems in the use of an iterative solution process to solve (10-4.6) is the value ζ to use in the iteration stopping test (8-3.35). If ζ is picked too large, the global error may be adversely affected; on the other hand, if ζ is picked too small, the number of iterations per time step would be unnecessarily large. Obviously, the optimum value of ζ is a function of Δt: perhaps, a good choice for ζ would be $\frac{1}{10}$ the value of the local time discretization error for that Δt. Unfortunately, this discretization error usually is not known a priori. However, in order to control the time step size dynamically, methods (see, e.g., Hindmarsh [1974] and Byrne and Hindmarsh [1975]) have been developed to estimate the local (time) truncation error from the numerical solution. If this is done, then ζ can be made a function of the criterion imposed on the local truncation error.

For time-dependent problems, the matrix‡ $\tilde{A}_{n+1}$ and the vector s_{n+1} must be recalculated for each time step. The cost of doing these calculations may be as great or greater than that required for the solution of (10-4.6). Other than

---

† This is especially beneficial in the solution of problems where the use of a small Δt is required.

‡ We assume that the thermal conductivity k in (10-4.1) varies with time.

careful programming, little can be done to reduce the cost involved in the calculation of $\tilde{A}_{n+1}$.

From (10-4.3), the calculation of the source vector s_{n+1} involves the matrix–vector product

$$[(C/\Delta t_{n+1}) - (1 - \theta)A(t_n)]\tau_n. \tag{10-4.7}$$

Since it is not possible to obtain $A(t_n)$ or C from $\tilde{A}_n$, it is clear that either the matrix C and/or the matrix $A(t_n)$ must be available to compute s_{n+1}. One of the more efficient ways to compute s_{n+1} is to make use of the fact that

$$\tilde{A}_n \tau_n \equiv [(C/\Delta t_n) + \theta A(t_n)]\tau_n = s_n. \tag{10-4.8}$$

Thus if the matrix C and the vectors s_n and $g(t_n)$ are available from time step n, then s_{n+1} may be expressed as†

$$
s_{n+1} = \left[1 + \frac{\Delta t_{n+1}}{\Delta t_n}\left(\frac{1-\theta}{\theta}\right)\right]\frac{C}{\Delta t_{n+1}}\,\tau_n - \left(\frac{1-\theta}{\theta}\right)s_n
$$
$$
+ \left[\theta g(t_{n+1}) + (1 - \theta)g(t_n)\right]. \tag{10-4.9}
$$

The arithmetic required to compute s_{n+1} by using (10-4.9) is less than that of one CCSI iteration. However, additional storage is required since the matrix C and the vector $g(t_n)$ must be stored during time step n.

If the two-dimensional 5-point stencil or the three-dimensional 7-point stencil is used in the spatial discretization of (10-4.1), C is a diagonal matrix so that the additional storage required is small. However, for other discretization formulas discussed in Sections 10.2 and 10.3, the nonzero elements of C are likely to be structured the same as for $A(t_n)$ or $\tilde{A}_n$. For this case, the additional words of storage required to store C can be significant.

---

† Since τ_n is obtained by iteration, Eq. (10-4.8) will not be satisfied exactly. Because of this, the vector s_{n+1} given by (10-4.9) will not exactly equal the right side of (10-4.3). The effect of this on global error is not known to the authors.

CHAPTER

11

Case Studies

11.1 INTRODUCTION

In this chapter, we describe the use of iterative methods in the solution of the few-group diffusion, the monoenergetic transport, and the nonlinear network flow multidimensional boundary-value problems. The goal here is not to give particular solution methods for these problems but rather to illustrate the versatility of iterative methods and to illustrate some problem aspects which must be considered in their use.

In the previous chapters, numerical results were given to illustrate the behavior of a particular numerical method. The method was given and then the problem to be solved was chosen. Of course, the problem was selected so that the use of the numerical method was justified theoretically. In this chapter, we consider the reverse situation: we are given the problem and must select a solution method. Often, for this case, theoretical results may be applicable only for certain model problems, or they may involve assumptions which are not valid for some problems of interest, or they may be valid only when some variable is sufficiently small. Thus in the solution of practical problems, mathematical rigor often must be complemented by sound mathematical reasoning and carefully designed numerical experimentation.

When an a priori rigorous mathematical justification for a particular solution procedure does not exist, an a posteriori numerical verification is needed. This verification can be made by printing, during the solution process, selected numerical data† which indicates a posteriori whether or not the assumptions made are valid for the particular problem being solved. Should the method fail, the behavior of the numerical data often will indicate which assumption is not valid. Frequently, relatively easy modifications then can be made in the numerical procedure to correct the difficulty. For the three problems studied in this chapter, we state what assumptions are being made in the numerical procedure and why these assumptions are reasonable.

In Section 11.2, we discuss the numerical solution of the two-group diffusion eigenvalue problem. The numerical procedure described illustrates (1) the use of inner–outer (or multistage) iterations and (2) the use of Chebyshev acceleration in the solution of eigenvalue problems.

In Section 11.3, we discuss the numerical solution of the inhomogeneous monoenergetic transport equation in x–y geometry. The numerical procedure illustrates (1) the use of Chebyshev acceleration when the associated Jacobi iteration matrix may have complex eigenvalues, (2) the use of a Chebyshev acceleration procedure whose convergence rate is almost independent of the spatial discretization, and (3) another use of inner–outer iterations.

In Section 11.4, we discuss the numerical solution of nonlinear network flow problems. The numerical procedure illustrates (1) the use of the SOR method to solve systems of nonlinear equations and (2) the use of inner–outer iterations in which the block SOR method is used for the outer iterations and Newton's method is used for the inner iterations.

11.2   THE TWO-GROUP NEUTRON DIFFUSION PROBLEM

In the following discussion on the numerical solution of the two-group neutron diffusion problem of reactor physics, we illustrate the use of inner or multistage iterative processes and the use of Chebyshev polynomials to accelerate the convergence of the power eigenvalue iterative process.

Statement of the Continuous Problem

The two-group diffusion problem (e.g., see Bell and Glasstone [1970]) is to determine the largest (in modulus) eigenvalue Λ_1 and the corresponding eigenvector, $(\varphi_1(r),\ \varphi_2(r))$, satisfying the coupled differential system

$$-\nabla \cdot D_1(r)\nabla\varphi_1(r) + \sigma_1(r)\varphi_1(r) = \Lambda^{-1}[\sigma_{11}^f(r)\varphi_1(r) + \sigma_{12}^f \varphi_2(r)],$$

$$-\nabla \cdot D_2(r)\nabla\varphi_2(r) + \sigma_2(r)\varphi_2(r) - \sigma^s(r)\varphi_1(r) = \qquad\qquad (11\text{-}2.1)$$
$$\Lambda^{-1}[\sigma_{21}^f(r)\varphi_1(r) + \sigma_{22}^f(r)\varphi_2(r)].$$

---

† For example, see the remarks given at the end of Sections 6.4, 8.3, and 9.6.

Here $\varphi_g(r)$, $g = 1$, 2, denotes the neutron flux in an energy interval E_g, r denotes spatial dependence, and ∇ is the gradient operator. The coefficients $D_g(r)$, $\sigma_g(r)$, $\sigma^s(r)$, and $\sigma^f_{gg'}(r)$ are physically meaningful nonnegative functions of r. The functions $D_g(r)$ and $\sigma_g(r)$ are assumed to be strictly positive.

We assume the solution region to be a rectangle R in the x–y plane and that

(a)   the solution region R is composed of a finite number of simply connected rectangular subregions, in each of which the coefficients $D_g(r)$, etc., possess sufficiently high derivatives,

(b) $\varphi_g(r)$ and $D_g(r)\,\partial\varphi_g(r)/\partial n$ are continuous across region interfaces ($\partial\varphi/\partial n$ is the derivative of φ in the direction of the normal to the region interface),

(c)   on the boundary Γ of R, the homogeneous boundary condition

$$\alpha(r)\varphi_g(r) + \beta(r)\frac{\partial\varphi_g(r)}{\partial n} = 0$$

holds, where $\alpha(r) \geq 0$, $\beta(r) \geq 0$, $\alpha(r) + \beta(r) > 0$ and where $\partial\varphi_g/\partial n$ is the derivative in the direction of the outward normal to the boundary.

For the two-group approximation, the continuous neutron energy range is divided into the two intervals E_1 and E_2. The equations given in (11-2.1) are conservation equations for neutrons in the energy intervals (groups) E_1 and E_2 at a spatial point r. The $(-\nabla \cdot D_g \nabla\varphi_g)$ term gives the net loss due to diffusion, the $(\sigma_g \varphi_g)$ term gives the loss of neutrons in group g due to neutron collisions with material nuclei, and the $(\sigma^f_{g1}\varphi_1 + \sigma^f_{g2}\varphi_2)$ term gives the neutrons born into group g from fissioning. The $(\sigma^s\varphi_1)$ term in the second equation of (11-2.1) gives the neutrons which are scattered from group 1 into group 2 through a nonabsorbing collision. The eigenvalue Λ_1 indicates whether there is a net loss or gain of neutrons and that the eigenvector gives the neutron scalar flux shape. Habetler and Martino [1961] have shown that the largest (in modulus) eigenvalue Λ_1 of (11-2.1) is positive and that if $\Lambda_k(\neq \Lambda_1)$ is any other eigenvalue of (11-2.1), then $\Lambda_1 > |\Lambda_k|$. Moreover, they show that the eigenvector $(\varphi_1(r), \varphi_2(r))$ corresponding to Λ_1 can be chosen to be positive almost everywhere.

Statement of the Discrete Problem

To obtain the discrete analog of (11-2.1), we subdivide the rectangular region R into a $p \times q$ mesh element array (see Fig. 10-2.2a) such that all internal interfaces and external boundaries lie exactly on mesh lines. The coupled differential equations (11-2.1) are then approximated by a coupled system of linear equations obtained by standard finite difference or finite element discretization techniques. The resulting coupled linear system may be expressed in matrix form as

$$M\Phi = \lambda^{-1}F\Phi, \qquad (11\text{-}2.2)$$

where

$$M = \begin{bmatrix} A_1 & 0 \\ -R_1 & A_2 \end{bmatrix}, \quad F = \begin{bmatrix} F_{1,1} & F_{1,2} \\ F_{2,1} & F_{2,2} \end{bmatrix}, \quad \text{and} \quad \Phi = \begin{bmatrix} \varphi_1 \\ \varphi_2 \end{bmatrix}.$$

Here φ_g, $g = 1, 2$ is a vector of order N whose components are the approximations for $\varphi_g(r)$ at prescribed spatial points and λ is the discrete approximation for Λ. The matrix–vector products $A_g\varphi_g$, $F_{gg'}\varphi_{g'}$, and $R_1\varphi_1$ are the discrete analogs of $(-\nabla \cdot D_g(r)\nabla\varphi_g(r) + \sigma_g\varphi_g(r))$, $\sigma^f_{gg'}\varphi_{g'}(r)$, and $\sigma^s\varphi_1(r)$, respectively. The matrices A_1 and A_2 are symmetric and positive definite. Thus M is nonsingular.

The discrete problem is to determine the largest (in modulus) eigenvalue λ_1 and the corresponding eigenvector Φ_1 of (11-2.2). For the standard finite difference discretization† of (11-2.1), Birkhoff and Varga [1958] have shown that the discrete problem (11-2.2) possesses a unique positive, largest (in modulus) eigenvalue λ_1 and that the eigenvector Φ_1 corresponding to λ_1 can be chosen to have positive elements. As the spatial mesh increments approach zero, the assumption is that the fundamental solution (λ_1 and Φ_1) of (11-2.2) approaches the fundamental solution of the continuous problem (11-2.1).

Numerical Solution of the Discrete Problem

We assume that $\lambda_1 > |\lambda_k|$ for $k \neq 1$. Thus the eigenvalues of the $2N \times 2N$ matrix $M^{-1}F$ can be ordered as

$$\lambda_1 > |\lambda_2| \geq |\lambda_3| \geq \cdots \geq |\lambda_{2N}|. \qquad (11\text{-}2.3)$$

Moreover, we let Φ_i be the eigenvector associated with λ_i, i.e., $\lambda_i\Phi_i = M^{-1}F\Phi_i$.

The well-known power method (Wilkinson [1965]) may be used to obtain iteratively λ_1 and Φ_1. Given an arbitrary (positive) guess vector $\Phi(0)$ and guess eigenvalue $\lambda(0)$, the power method generates successive estimates for λ_1 and Φ_1 by the process

$$\Phi(l + 1) = \lambda(l)^{-1}M^{-1}F\Phi(l),$$
$$\lambda(l + 1) = \lambda(l)(e^T F\Phi(l + 1)/e^T F\Phi(l)), \qquad (11\text{-}2.4)$$

---

† The nonnegative properties of certain matrices assumed by Birkhoff and Varga need not be valid for some discretization techniques.

where $e^T \equiv (1, 1, \ldots, 1)$. If $(11\text{-}2.3)$ is valid,† the power method is a convergent process, i.e., $\lim_{l \to \infty} \lambda(l) = \lambda_1$ and $\lim_{l \to \infty} \Phi(l) = \Phi_1$. Moreover, the rate at which $\Phi(l)$ converges to Φ_1 depends (Wilkinson [1965]) on how well separated the fundamental eigenvalue λ_1 is from the other eigenvalues of $M^{-1}F$, i.e., the convergence rate depends directly on the *dominance ratio* ρ, where

$$\rho \equiv |\lambda_2|/\lambda_1 < 1. \qquad (11\text{-}2.5)$$

The smaller the ratio, the faster the convergence. When ρ is close to unity, the power method converges very slowly. We now describe how Chebyshev polynomials can be used to accelerate the basic power method.

To use Chebyshev acceleration, we make the following additional assumptions:

(a)   the eigenvalues $\{\lambda_k\}_1^{2N}$ of $M^{-1}F$ are real, and
(b)   the set of eigenvectors $\{\Phi_k\}$ for $M^{-1}F$ includes a basis $(11\text{-}2.6)$
      for the associated vector space E^{2N}.

From assumption (b), the initial guess $\Phi(0)$ can be expanded in terms of the eigenvectors of $M^{-1}F$, i.e., we can express $\Phi(0)$ in the form

$$\Phi(0) = \sum_{k=1}^{2N} c_k \Phi_k \qquad (11\text{-}2.7)$$

for some constants c_k. We assume $c_1 \neq 0$. For expository purposes, we now assume the eigenvalue estimates $\lambda(0), \lambda(1), \ldots$ in $(11\text{-}2.4)$ are approximately equal to λ_1. The iterate $\Phi(l)$ in $(11\text{-}2.4)$ then may be expressed approximately as

$$\Phi(l) \doteq \frac{M^{-1}F}{\lambda_1} \Phi(l-1) \doteq \left(\frac{M^{-1}F}{\lambda_1}\right)^l \Phi(0) = c_1 \Phi_1 + \sum_{k=2}^{2N} \left(\frac{\lambda_k}{\lambda_1}\right)^l c_k \Phi_k. \qquad (11\text{-}2.8)$$

Thus in l power iterations, the most slowly decaying contribution (i.e., Φ_2) to the error is multiplied by the lth power of the dominance ratio ρ. We note that the l power iterations here are equivalent to applying the matrix operator $(M^{-1}F/\lambda_1)^l$ to $\Phi(0)$. The power method is a special case of the general polynomial procedure

$$\Phi(l) = P_l\left(\frac{M^{-1}F}{\lambda_1}\right)\Phi(0) = P_l(1)c_1 \Phi_1 + \sum_{k=2}^{2N} P_l\left(\frac{\lambda_k}{\lambda_1}\right)c_k \Phi_k. \qquad (11\text{-}2.9)$$

Here $P_l(z)$ is a polynomial of degree l in z.

---

† As mentioned above, Birkhoff and Varga have shown $(11\text{-}2.3)$ to be valid under certain conditions.

If $P_l(z)$ could be chosen such that $P_l(1) = 1$ and $P_l(\lambda_k/\lambda_1) = 0$ for $k = 2, \ldots,$ $2N$, then $\Phi(l) = c_1\Phi_1$. Choosing such a $P_l(z)$ is not practical. Instead, we choose $P_l(z)$ such that $P_l(1) = 1$ and such that the maximum of $|P_l(z)|$ is minimized for $\lambda_{2N}/\lambda_1 \leq z \leq \lambda_2/\lambda_1 (= \rho)$. From Theorem 4-2.2, such a polynomial can be given (Flanders and Shortley [1950]) by

$$P_l(z) = T_l\left(\frac{2z - \rho - b}{\rho - b}\right) \bigg/ T_l\left(\frac{2 - \rho - b}{\rho - b}\right), \qquad (11\text{-}2.10)$$

where $b = \lambda_{2N}/\lambda_1$ and $T_l(x)$ is the Chebyshev polynomial given by (4-2.1). For computational purposes, using the recurrence relation (4-2.1) and using estimates $\lambda(l)$ in place of λ_1, we can express (Varga [1961]) the iterates $\Phi(l)$ of (11-2.9) in the recurrence form:

$$\hat{\Phi}(l + 1) = \lambda(l)^{-1}M^{-1}F\Phi(l),$$

$$\Phi(l + 1) = \Phi(l) + \alpha_{l+1}[\hat{\Phi}(l + 1) - \Phi(l)]$$
$$+ \beta_{l+1}[\Phi(l) - \Phi(l - 1)], \qquad (11\text{-}2.11)$$

$$\lambda(l + 1) = \lambda(l)(e^{\mathrm{T}}F\Phi(l + 1)/e^{\mathrm{T}}F\Phi(l)),$$

where $\alpha_1 = 2/(2 - \rho - b)$, $\beta_1 = 0$, and for $l \geq 1$

$$\alpha_{l+1} = \frac{4}{\rho - b}\frac{T_l[(2 - \rho - b)/(\rho - b)]}{T_{l+1}[(2 - \rho - b)/(\rho - b)]},$$
$$\beta_{l+1} = \frac{T_{l-1}[(2 - \rho - b)/(\rho - b)]}{T_{l+1}[(2 - \rho - b)/(\rho - b)]}. \qquad (11\text{-}2.12)$$

As for the inhomogeneous fixed source problems discussed in Chapter 4, the effectiveness of the Chebyshev method (11-2.11) depends upon accurate estimates of certain eigenvalues bounds, which for this case are $\rho = \lambda_2/\lambda_1$ and $b = \lambda_{2N}/\lambda_1$. In most diffusion programs, the assumption[†] is made that the eigenvalues of $M^{-1}F$ are nonnegative and, thus, $b = 0$ is used. With this assumption, i.e., $b = 0 \leq \lambda_{2N}/\lambda_1$, effective adaptive procedures for estimating ρ may be developed by comparing the observed[‡] convergence rate with the theoretical rate. Such procedures (e.g., see Ref. 9 in Hageman and Kellogg [1968]) are similar to those given in Chapters 5 and 6 and will not be described here.

[†] For the model problem, it can be shown (Wachspress [1966]) that the eigenvalues are nonnegative. However, this has not been proven for general heterogeneous problems. In fact, Kristiansen [1963] gives an example for which $M^{-1}F$ has a negative eigenvalue.

[‡] Apparently, the idea of obtaining improved estimates for ρ by comparing the observed Chebyshev convergence rate with the theoretical rate was first suggested in the mid-1950s by E. L. Wachspress. See Ref. 81 in Wachspress [1966].

Note that we have no mathematical proof that the Chebyshev-acceleration method outlined above will work for all problems. However, the method has been used successfully in many programs. This success indicates that the assumptions of $b = 0 \le \lambda_{2N}/\lambda_1$, of (11-2.3), and of (11-2.6) are satisfied (or at least nearly satisfied). There is nothing wrong with using numerical methods based on (reasonable) assumptions, provided sufficient numerical testing is done and provided these assumptions are kept clearly in mind. Moreover, it is often possible (and advisable) during the solution process to print selected numerical data which indicate a posteriori whether or not the assumptions made are valid for the particular problem being solved. For example, for the Chebyshev procedure (11-2.11), the behavior of the residual vector–norm ratio

$$H(l) \equiv \frac{\|\hat{\Phi}(l) - \Phi(l-1)\|}{\|\hat{\Phi}(l-1) - \Phi(l-2)\|} \tag{11-2.13}$$

can be used for this purpose. See Ref. 9 in Hageman and Kellogg [1968]. If the method fails and it can be determined which assumption is not valid, corrective modifications often can be made. For the Chebyshev procedure (11-2.11), for example, changes to account for negative eigenvalues (i.e., $b < 0$) can be easily made. The other assumptions are much more crucial. However, even these assumptions may often be relaxed somewhat by appropriate modifications of the Chebyshev procedure. We will discuss one such modification in Section 11.3.

Inner Iterations

We first discuss the implementation and the effects of inner iterations relative to the power method (11-2.4). In order to carry out the power iterations numerically (and also the Chebyshev iterations), a matrix equation of the form

$$M\Phi(l) = (\lambda(l-1))^{-1}F\Phi(l-1) \tag{11-2.14}$$

must be solved for $\Phi(l)$ every iteration. From the definition (11-2.2) of the matrix M, $\Phi(l)$ may be obtained by solving successively the system of group equations

$$A_1\varphi_1(l) = \frac{1}{\lambda(l-1)}[F_{1,1}\phi_1(l-1) + F_{1,2}\phi_2(l-1)],$$

$$A_2\varphi_2(l) = R_1\varphi_1(l) + \frac{1}{\lambda(l-1)}[F_{2,1}\varphi_1(l-1) + F_{2,2}\varphi_2(l-1)]. \tag{11-2.15}$$

Thus $\Phi(l)$ can be determined if we can solve the matrix equations

$$A_g \varphi_g(l) = b_g, \qquad g = 1, 2, \tag{11-2.16}$$

where b_g is a known vector and A_g is the discrete analog of $(-\nabla \cdot D_g \nabla + \sigma_g)$. Recall that A_g is assumed to be a symmetric positive definite matrix. The iterative procedures (11-2.4) and (11-2.11) assume that the $\varphi_g(l)$ of (11-2.15) can be obtained exactly using some direct inversion procedure. However, for most two- and three-dimensional problems, the use of a direct method to solve (11-2.16) is impractical. Thus the solutions $\varphi_g(l)$ to the group equations (11-2.15) are usually only approximated by some convergent iterative process. The iterations used to obtain these approximations are called *inner iterations*. The main iterations (11-2.4) or (11-2.11) are called *outer iterations*.

Since the matrices A_1 and A_2 are SPD, most of the iterative methods discussed in the previous chapters could be used as the inner iteration method. Let $\varphi_g^{(m)}(l)$ be the approximation to $\varphi_g(l)$ in (11-2.16) after m inner iterations. From Eqs. (2-2.6) or (3-2.5), we may express the error vector $\varepsilon_g^{(m)}(l) \equiv \varphi_g^{(m)}(l) - \varphi_g(l)$ as

$$\varepsilon_g^{(m)}(l) = Q_g^{(m)} \varepsilon_g^{(0)}(l), \tag{11-2.17}$$

where $Q_g^{(m)}$ is the convergent iteration error matrix associated with the inner iteration method.† If the inner iteration process is started with $\varphi_g^{(0)}(l) = \varphi_g(l-1)$, it easily follows from (11-2.16) and (11-2.17) that

$$\varphi_g^{(m)}(l) = Q_g^{(m)} \varphi_g(l-1) + (I - Q_g^{(m)}) A_g^{-1} b_g. \tag{11-2.18}$$

We now let $\tilde{\varphi}_g(l)$, $g = 1, 2$, denote the outer iteration iterates obtained from the power method (11-2.4) when m_1 and m_2 inner iterations are done in groups 1 and 2, respectively. From (11-2.18), instead of satisfying (11-2.15), the fluxes $\tilde{\varphi}_g(l)$ satisfy the pseudosystem of group equations‡

$$\tilde{A}_1 \tilde{\varphi}_1(l) = \tilde{A}_1 Q_1^{(m_1)} \tilde{\varphi}_1(l-1) + \frac{1}{\lambda(l-1)}[F_{1,1} \tilde{\varphi}_1(l-1) + F_{1,2} \tilde{\varphi}_2(l-1)],$$

$$\tilde{A}_2 \tilde{\varphi}_2(l) = \tilde{A}_2 Q_2^{(m_2)} \tilde{\varphi}_2(l-1) + R_1 \tilde{\varphi}_1(l) \tag{11-2.19}$$

$$+ \frac{1}{\lambda(l-1)}[F_{2,1} \tilde{\varphi}_1(l-1) + F_{2,2} \tilde{\varphi}_2(l-1)],$$

† For the Chebyshev or CG methods, $Q_g^{(m)}$ would be some polynomial of degree m. For the SOR method, $Q_g^{(m)} = \mathscr{L}_{\omega_g}^m$.

‡ For a derivation of (11-2.19) and for additional discussions on the interrelations between inner and outer iterations, see Wachspress [1966] and Ref. 160 in Cuthill [1964].

where $\tilde{A}_g = A_g(I - Q_g^{(m_g)})^{-1}$. Equivalently, with $\tilde{\Phi}(l) \equiv (\tilde{\varphi}_1(l)^{\mathrm{T}}, \tilde{\varphi}_2(l)^{\mathrm{T}})^{\mathrm{T}}$, we may express (11-2.19) in the form

$$\tilde{\Phi}(l) = [I - \tilde{M}^{-1}M(I - \lambda(l-1)^{-1}M^{-1}F)]\tilde{\Phi}(l-1), \quad (11\text{-}2.20)$$

where

$$\tilde{M} = \begin{bmatrix} A_1(I - Q_1^{(m_1)})^{-1} & 0 \\ -R_1 & A_2(I - Q_2^{(m_2)})^{-1} \end{bmatrix}.$$

Note that $\tilde{M}$ approaches M as the number of inner iterations in each group approaches infinity. We always assume that the same number of inner iterations m_g is performed in group g every outer iteration. If m_g were allowed to vary with l, then the matrix $\tilde{M}$ would also vary with l.

Thus when inner iterations are done, the power method is being applied to the problem

$$\gamma\tilde{\Phi} = [I - \tilde{M}^{-1}M(I - M^{-1}F/\lambda)]\tilde{\Phi} \quad (11\text{-}2.21)$$

instead of the problem (11-2.2). If $\tilde{\Phi}$ satisfies (11-2.21) with $\lambda = \lambda_1$ and $\gamma = 1.0$, then $\tilde{\Phi}$ also satisfies $\tilde{\Phi} = (M^{-1}F/\lambda_1)\tilde{\Phi}$, which implies that $\tilde{\Phi} = \Phi_1$. For this case then, the desired solution of (11-2.2) is also a solution of (11-2.21).

For expository purposes we assume, as was done in (11-2.8), that $\lambda = \lambda_1$ in (11-2.21). We let $\{\gamma_k\}_{k=1}^{2N}$ be the set of eigenvalues for the $2N \times 2N$ matrix $[I - \tilde{M}^{-1}M(I - M^{-1}F/\lambda_1)]$. If an infinite number of inner iterations is done in each group, then $\tilde{M} = M$ and $\gamma_k = \lambda_k/\lambda_1$. Thus from (11-2.3), the γ_k for this case may be ordered as

$$1 = \gamma_1 > |\gamma_2| \geq \cdots \geq |\gamma_{2N}|. \quad (11\text{-}2.22)$$

If the number of inner iterations done in each group is sufficiently large but finite, then (11-2.22) still holds but $\gamma_k \neq \lambda_k/\lambda_1$ for $k \neq 1$. Henceforth, we assume (11-2.22) is satisfied. If $\lambda(l-1) = \lambda_1$, condition (11-2.22) guarantees the convergence of the power iterates $\tilde{\Phi}(l)$ in (11-2.20) to Φ_1. Moreover, the rate at which $\tilde{\Phi}(l)$ converges to Φ_1 depends on the dominance ratio $\tilde{\rho}$, where

$$\tilde{\rho} \equiv |\gamma_2|/\gamma_1 = |\gamma_2|. \quad (11\text{-}2.23)$$

How close $\tilde{\rho}$ is to ρ depends on the number of inner iterations done. We discuss this aspect later.

To apply Chebyshev acceleration to (11-2.21), we assume as before that the set of eigenvectors for $[I - \tilde{M}^{-1}M(I - M^{-1}F/\lambda_1)]$ includes a basis for E^{2N}. But we must be more careful concerning our assumption for the domain containing the eigenvalues $\{\gamma_k\}$. If the Chebyshev method is used for the inner iterations, it can be shown (Wachspress [1966]) for the model problem that the matrix $[I - \tilde{M}^{-1}M(I - M^{-1}F/\lambda_1)]$ may have small complex eigenvalues as indicated in Fig. 11-2.1. In Fig. 11-2.1, δ is the maximum of $S(Q_1^{(m_1)})$ and $S(Q_2^{(m_2)})$, where $S(Q_g^{(m_g)})$ is the spectral radius of the inner iteration error matrix $Q_g^{(m)}$ given in (11-2.17). Generally, the presence of complex eigenvalues significantly reduces the effectiveness of the Chebyshev-acceleration procedure. (See, e.g., Ref. 9 in Hageman and Kellogg [1968]). However, for the eigenvalue domain of Fig. 11-2.1 with δ not large, we can virtually eliminate the negative effects due to complex eigenvalues by doing a few power iterations before starting the Chebyshev procedure.

To see why this is so, let the initial guess $\tilde{\Phi}(0)$ be expanded in terms of the eigenvectors $\tilde{\Phi}_k$ of $[I - \tilde{M}^{-1}M(I - M^{-1}F/\lambda_1)]$ as $\tilde{\Phi}(0) = \sum_{k=1}^{2N} c_k \tilde{\Phi}_k$. We assume $\tilde{\Phi}_k$ is associated with the eigenvalue γ_k. After p initial power iterations (11-2.20) are calculated, we obtain $\tilde{\Phi}(p) = \sum_{k=1}^{2N} c_k(\gamma_k)^p \tilde{\Phi}_k$. If Chebyshev acceleration is started on iteration $p + 1$ then, analogous to (11-2.9), we have for $l \geq 1$ that

$$\tilde{\Phi}(l + p) = \tilde{P}_l\left[I - \tilde{M}^{-1}M\left(I - \frac{M^{-1}F}{\lambda_1}\right)\right]\tilde{\Phi}(p),$$

where

$$\tilde{\Phi}(p) = \sum_{k=1}^{2N} c_k(\gamma_k)^p \tilde{\Phi}_k. \qquad (11\text{-}2.24)$$

Here $\tilde{P}_l(z)$ is a normalized Chebyshev polynomial which we define later. Note that any eigenvector in the expansion of $\tilde{\Phi}(0)$ associated with a complex eigenvalue has already been multiplied by a factor less than $(\delta)^p$ before Chebyshev acceleration has begun. Normally then, for δ less than 0.1 and for $p = 4$ or 5, eigenvector modes associated with small complex eigenvalues need not be reduced further by Chebyshev acceleration. On the other hand, Chebyshev acceleration should not significantly amplify these eigenvector modes either. Numerical evidence indicates that harmful amplification does not occur if we determine $\tilde{P}_l(z)$ based on the assumptions that the eigenvalues of $[I - \tilde{M}^{-1}M(I - M^{-1}F/\lambda_1)]$ are real and that they lie in the

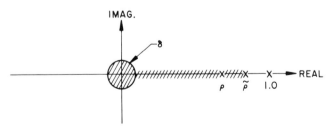

Fig. 11-2.1.   Eigenvalue domain of $[I - \tilde{M}^{-1}M(I - M^{-1}F/\lambda_1)]$ for the model problem.

interval $[-\delta, \tilde{\rho}]$. With these assumptions, analogous to (11-2.10), we define $\tilde{P}_l(z)$ by

$$\tilde{P}_l(z) = T_l\!\left(\frac{2z - \tilde{\rho} + \delta}{\tilde{\rho} + \delta}\right)\!\Big/ T_l\!\left(\frac{2 - \tilde{\rho} + \delta}{\tilde{\rho} + \delta}\right).$$

We remark that undesirable amplification may occur with $\tilde{P}_l(z)$ defined this way if δ is too large. See Table 11-2.1.

Computational Aspects

In summary, when inner iterations are required the following procedure is used:

(1)   A fixed number of inner iterations is done in each group every outer iteration. The number of inner iterations m_g done in group g is chosen such that the error reduction achieved by the inner iterations is less than δ.

(2)   The initial four outer iterations are done using the power method.

(3)   Chebyshev acceleration is started on outer iteration 5 using (11-2.11) with $\tilde{\Phi}(l)$ replacing $\Phi(l)$. The parameters α_l and β_l are computed using (11-2.12) with $b = -\delta$ and with $\tilde{\rho}$ estimated adaptively.

The data given in Table 11-2.1 illustrate the effect of δ on the dominance ratio $\tilde{\rho}$ and the number of outer iterations required for convergence for a typical problem. The line cyclic Chebyshev method was used for the inner iterations and the value of $\tilde{\rho}$ was obtained numerically when the problem was solved using only power iterations. By decreasing the number of inner iterations (increasing δ), we affect the outer iterations in two ways. First, as illustrated in Table 11-2.1, the dominance ratio $\tilde{\rho}$ increases. The increase in the number of outer iterations *due solely* to a larger value for $\tilde{\rho}$ is usually compensated for by the fewer number of inner iterations performed. This is

TABLE 11-2.1

Effect of Inner Iterations on Outer Iterations

Inner iter. error red., δ	No. of inner iterations per outer		Dominance ratio, $\tilde{\rho}$	No. of outer iterations	
	Group 1	Group 2		Power method	Chebyshev accel.
0.01	21	16	0.895	39	14
0.15	11	8	0.901	41	15
0.25	9	7	0.911	45	18
0.5	7	6	0.922	51	Did not converge

illustrated by the data given in Table 11-2.1 for the nonaccelerated power method. The second effect of decreasing the number of inner iterations is that the magnitudes of the negative and complex eigenvalues are usually increased. As indicated by the data of Table 11-2.1, this can seriously reduce the effectiveness of Chebyshev acceleration if the inner iteration convergence criterion δ is too large. More detailed numerical studies indicate that δ in the range 0.075–0.15 is usually sufficient for the Chebyshev acceleration not to be adversely affected by the inner iterations.

We note that the eigenvalue bounds $\tilde{\rho}$ and $\tilde{b} = -\delta$ required by the Chebyshev procedure are not fixed if the number of inner iterations is allowed to vary with l.† The adaptive procedure for estimating $\tilde{\rho}$ is more effective when the same number of inner iterations is used every outer iteration. This is especially helpful for slowly converging problems, i.e., when ρ is close to unity.

For the inner iteration method, the line CCSI method has been found to be more effective than the line SOR method. For the problems considered in Chapters 8 and 9, the CCSI method required 4–16% fewer iterations than did the line SOR method. However, when used as the inner iteration method here, the advantage of the CCSI method was even greater. Numerical results indicate that the total number of inner iterations required using the CCSI method was often 25–40% less than that required when the SOR method was used as the inner iteration method. Except possibly for problems with Neumann boundary conditions, Wang [1977] found the CG method‡ to be less effective than the SOR method. He attributes this to the combined facts that the convergence requirement for the inner iterations is rather loose ($\delta = 0.05$) and that CG acceleration is a nonlinear process which converges more slowly during the initial iterations.

We now turn our attention to the problem of determining the number of inner iterations to be done. For each group equation (11-2.16), we want to determine m_g such that an error reduction of δ is achieved by the m_g inner iterations. If $\varphi_g(l)$ is the unique solution to (11-2.16) and if the error vector for inner iteration m is $\varepsilon_g^{(m)}(l) \equiv \varphi_g^{(m)}(l) - \varphi_g(l)$, then using the notation of (11-2.17) we want m_g to be chosen such that

$$\|\varepsilon_g^{(m_g)}(l)\| = \|Q_g^{(m_g)}\varepsilon_g^{(0)}(l)\| \le \delta\|\varepsilon_g^{(0)}(l)\|. \tag{11-2.25}$$

Thus we want to pick m_g to be the smallest integer satisfying

$$\|Q_g^{(m_g)}\| \le \delta. \tag{11-2.26}$$

† For this case, $\tilde{\rho}$ and $\tilde{b}$ are not well defined.
‡ Wang applied CG acceleration to an approximate factorization method described in Section 2.5.

Any convenient norm may be used to determine m_g in (11-2.26). For the line CCSI method, for example, it is convenient to require that (11-2.25) hold only for the "black" unknowns. For this case, it follows from Chapter 8 that there exists a symmetrization matrix W such that

$$\|Q_g^{(m_g)}\|_W = P_{2m_g, \mathrm{E}}(\mathsf{S}(B_g)), \qquad (11\text{-}2.27)$$

where, from (8-3.10), $P_{2m_g, \mathrm{E}}(x) = T_{2m_g}(x/M_\mathrm{E})/T_{2m_g}(1/M_\mathrm{E})$ and where $\mathsf{S}(B_g)$ is the spectral radius of the line Jacobi iteration error matrix B_g. If $M_\mathrm{E} = \mathsf{S}(B_g)$ is used in the CCSI iterations, it follows from (4-2.20) that $P_{2m_g, \mathrm{E}}(\mathsf{S}(B_g)) = 2\bar{r}^{m_g}/(1 + \bar{r}^{2m_g})$. Thus if $\mathsf{S}(B_g)$ is known, m_g may be obtained by picking m_g to be the smallest integer satisfying

$$2(\bar{r})^{m_g}/(1 + (\bar{r})^{m_g}) \le \delta. \qquad (11\text{-}2.28)$$

Here, $\bar{r} = (1 - \sqrt{1 - \mathsf{S}(B_g)^2})/(1 + \sqrt{1 - \mathsf{S}(B_g)^2})$.

Two methods may be used to obtain an accurate estimate for the spectral radius $\mathsf{S}(B_g)$ required in (11-2.28). One method is to estimate $\mathsf{S}(B_g)$, while carrying out the inner iterations for the first outer iteration, using a Chebyshev adaptive procedure, such as that given previously in Chapter 8. (To insure that an accurate estimate for $\mathsf{S}(B_g)$ is obtained, a reasonably tight inner iteration convergence criteria should be used for the first outer iteration). A second method is to obtain an estimate for $\mathsf{S}(B_g)$ by a separate calculational procedure performed prior to the start of the outer iterations. For example, an a priori estimate for $\mathsf{S}(B_g)^2$ may be obtained from the power iteration method applied to the line Gauss–Seidel iteration error matrix $\mathscr{L}_1$. (Recall that $\mathsf{S}(\mathscr{L}_1) = \mathsf{S}(B)^2$ for the red/black matrix problem.) Hageman and Kellogg [1968] consider this second approach and discuss the use of Chebyshev polynomials to accelerate the convergence of the power iterations.

When inner iterations are performed, we again have no a priori gurantee that inner–outer iteration process is a convergent procedure. However, as discussed previously, a posteriori confirmations exist which indicate whether or not the iteration process worked as *designed* (e.g., see Ref. 1 in Nakamura [1977], p. 137).

The number of mesh points used for the spatial discretization has a significant effect on the number of inner iterations required but only a minor effect on the number of outer iterations. The reason for this lack of sensitivity in the outer iteration convergence rate is that the outer iterations could have been formulated first in terms of the continuous problem (11-2.1) and then discretized spatially. Thus at least for sufficiently small mesh spacings, the spatially discretized outer iterations converge at a rate determined by the convergence rate of the outer iterations defined in terms of the continuous problem.

11.3 THE NEUTRON TRANSPORT EQUATION
IN $x - y$ GEOMETRY

In this section, we discuss the numerical solution of the inhomogeneous, monoenergetic neutron transport equation in $x-y$ geometry. The numerical procedure described illustrates the use of Chebyshev acceleration when the associated Jacobi iteration matrix may have complex eigenvalues, the use of a Chebyshev acceleration procedure whose convergence rate is almost independent of the spatial discretization, and the use of inner iterations.

Statement of the Continuous Problem

In $x-y$ geometry, the monoenergetic neutron transport equation with isotropic scattering and isotropic source can be written as (see Davison and Sykes [1957])

$$\Omega \cdot \text{grad } N + \sigma N = \frac{\sigma^s}{2\pi} \int_0^{2\pi} \int_0^{\pi/2} N \sin \theta \, d\theta \, d\varphi + S, \qquad (11\text{-}3.1)$$

where

(a) $\Omega = (\Omega_x, \Omega_y, \Omega_z) = (\sin \theta \cos \varphi, \sin \theta \sin \varphi, \cos \theta)$, $0 \le \varphi \le 2\pi$, $0 \le \theta \le \pi/2$, is the unit direction vector.
(b) $N(x, y, \Omega)$ is the neutron flux at the spatial point (x, y) in the direction Ω, and grad N is the spatial gradient vector of N,
(c) $S(x, y)$ is the isotropic source, and,
(d) $\sigma(x, y)$ and $\sigma^s(x, y)$ are bounded positive functions of x, y which satisfy, for positive constants c and c_1, $\sigma(x, y) \ge c_1 > 0$ and

$$\sigma^s(x, y)/\sigma(x, y) \le c < 1. \qquad (11\text{-}3.2)$$

Equation (11-3.1) represents a neutron conservation law which basically states that the total derivative in the direction Ω of the neutron flux at position (x, y) per unit area per unit solid angle (leakage) equals the neutrons introduced (S plus σ^s terms) in that direction less the number which are removed (σ term) by collisions. The σ^s term in (11-3.1) represents those neutrons scattered by collisions into the direction Ω from all other solid angle elements.

We consider solutions to (11-3.1) in a rectangular region $R = \{(x, y) | 0 \le x \le a, 0 \le y \le b\}$ subject to vacuum (no neutrons entering R) or reflecting (mirror reflection) boundary conditions. We assume that σ and σ^s are piecewise continuous functions of x, y and that $N(r + s\Omega, \Omega)$ is a continuous function of s for all Ω and all spatial points r in R. It can be shown that (11-3.1) has a unique solution when S is square integrable (e.g., see Kellogg [1969]).

For x–y geometry problems, $N(x, y, \Omega_x, \Omega_y, \Omega_z) = N(x, y, \Omega_x, \Omega_y, -\Omega_z)$. Thus we seek $N(x, y, \Omega)$ only for those directions Ω with $\Omega_z \geq 0$. To define an iterative procedure to solve (11-3.1), we split the unknowns $N(x, y, \Omega)$ into two parts. For any direction $\Omega = (\Omega_x, \Omega_y, \Omega_z)$ with $\Omega_y \geq 0$, $\Omega_z \geq 0$, we define the associated direction $\tilde{\Omega} \equiv (\Omega_x, -\Omega_y, \Omega_z)$ and the new variables†

$$\psi(x, y, \Omega) \equiv N(x, y, \Omega),$$
$$\eta(x, y, \Omega) \equiv N(x, y, \tilde{\Omega}). \tag{11-3.3}$$

In terms of these new variables, the transport equation (11-3.1) may be written in the "partitioned" operator matrix form

$$\begin{bmatrix} D_\psi & -G \\ -F & D_\eta \end{bmatrix} \begin{bmatrix} \psi(x, y, \Omega) \\ \eta(x, y, \Omega) \end{bmatrix} = \begin{bmatrix} S \\ S \end{bmatrix} \tag{11-3.4}$$

for $\Omega \in \mathscr{S}$, where $\mathscr{S} \equiv \{\Omega = (\Omega_x, \Omega_y, \Omega_z) | \Omega_y \geq 0 \text{ and } \Omega_z \geq 0\}$. Here,

$$G\eta = \frac{\sigma^s}{2\pi} \int_0^\pi \int_0^{\pi/2} \eta \sin\theta \, d\theta \, d\varphi, \qquad F\psi = \frac{\sigma^s}{2\pi} \int_0^\pi \int_0^{\pi/2} \psi \sin\theta \, d\theta \, d\varphi,$$

$$D_\psi \psi = \Omega \cdot \operatorname{grad} \psi + \sigma\psi - F\psi, \quad \text{and} \tag{11-3.5}$$

$$D_\eta \eta = \tilde{\Omega} \cdot \operatorname{grad} \eta + \sigma\eta - G\eta.$$

Along the $x = 0$ and $x = a$ boundaries, both η and ψ satisfy either vacuum or reflecting conditions. For $y = 0$, ψ satisfies

$$\psi(x, 0, \Omega) = L\eta(x, 0, \Omega), \tag{11-3.6}$$

and for $y = b$, η satisfies

$$\eta(x, b, \Omega) = T\psi(x, b, \Omega), \tag{11-3.7}$$

where L and T are either zero or unity (zero implies a vacuum boundary and unity implies a reflecting boundary condition).

If n is the iteration index, the Gauss–Seidel iteration method corresponding to the partitioning (11-3.4) is defined by

$$\begin{bmatrix} D_\psi & 0 \\ -F & D_\eta \end{bmatrix} \begin{bmatrix} \psi^{(n)} \\ \eta^{(n)} \end{bmatrix} = \begin{bmatrix} 0 & G \\ 0 & 0 \end{bmatrix} \begin{bmatrix} \psi^{(n-1)} \\ \eta^{(n-1)} \end{bmatrix} + \begin{bmatrix} S \\ S \end{bmatrix}, \tag{11-3.8}$$

where

$$\psi^{(n)}(x, 0, \Omega) = L\eta^{(n-1)}(x, 0, \Omega),$$
$$\eta^{(n)}(x, b, \Omega) = T\psi^{(n)}(x, b, \Omega). \tag{11-3.8a}$$

---

† In the x–y plane ψ corresponds to those neutrons traveling in the positive y direction, while η corresponds to neutrons traveling in the negative y direction.

Along $x = 0$ and $x = a$, we always assume $\psi^{(n)}$ and $\eta^{(n)}$ satisfy the specified boundary conditions. The iterative process is initiated by guessing $G\eta^{(0)}$ everywhere and, if $L = 1$, by guessing $\eta^{(0)}$ along the $y = 0$ boundary.

The corresponding Jacobi iteration method may be similarly defined.

Kellogg [1969] has shown that the iteration procedure (11-3.8) is well defined and converges to the unique solution of (11-3.4) provided the eigenvalues λ of the Gauss–Seidel iteration operator† satisfy $|\lambda| < 1$. Davis and Hageman [1969] show that, indeed, $|\lambda| < 1$. Moreover, if a vacuum condition exists along the $y = 0$ boundary $(L = 0)$, they show that the eigenvalues λ are real and satisfy

$$-c/(4 - c) \le \lambda \le c/(4 - 3c), \qquad (11\text{-}3.9)$$

where $c < 1$ is defined in (11-3.2). If all four boundaries are reflecting, some λ may be complex.

It can be shown that every nonzero eigenvalue λ_B of the Jacobi iteration operator B is related to an eigenvalue λ of the Gauss–Seidel operator by

$$\lambda_B^2 = \lambda. \qquad (11\text{-}3.10)$$

The relationship (11-3.10) is not surprising since the splitting (11-3.4) is the continuous analog of the red/black partitioning for linear matrix problems (see Chapter 8). Chebyshev acceleration may be applied to the Gauss–Seidel method‡ or to the Jacobi method. Of course, any Chebyshev procedure used here must allow for the possibility of complex eigenvalues. We will discuss only the Chebyshev acceleration of the Jacobi method.

Chebyshev Acceleration When the Eigenvalues of the Iteration Operator Are Complex

Let $\tilde{\psi}^{(n)}$ and $\tilde{\eta}^{(n)}$ be the iterates of the Jacobi method corresponding to (11-3.4), and let B denote the corresponding Jacobi iteration operator. We

---

† The eigenvalues and eigenfunctions for the Gauss–Seidel iteration operator satisfy the equations

$$\lambda \begin{bmatrix} D_\psi & 0 \\ -F & D_\eta \end{bmatrix} \begin{bmatrix} \psi \\ \eta \end{bmatrix} = \begin{bmatrix} 0 & G \\ 0 & 0 \end{bmatrix} \begin{bmatrix} \psi \\ \eta \end{bmatrix}$$

subject to the boundary conditions

$$\lambda\psi(x, 0, \Omega) = L\eta(x, 0, \Omega), \qquad \eta(x, b, \Omega) = T\psi(x, b, \Omega).$$

‡ Chebyshev acceleration may also be applied to the reduced system (see Section 8.2) obtained from (11-3.4).

assume that

(a) the set of eigenfunctions for B is complete, and

(b) the eigenvalues of B lie in or on an ellipse as shown in Fig. 11-3.1, where $e < S(B)$.†

(11-3.11)

For polynomial acceleration, we let $P_n(z) = \sum_{i=0}^{n} \alpha_{n,i} z^i$ be a real polynomial of degree n such that $P_n(1) = 1$ and then generate the sequences $\{\psi^{(n)}\}$ and $\{\eta^{(n)}\}$ by

$$\begin{bmatrix} \psi^{(n)} \\ \eta^{(n)} \end{bmatrix} = \sum_{i=0}^{n} \alpha_{n,i} \begin{bmatrix} \tilde{\psi}^{(n)} \\ \tilde{\eta}^{(n)} \end{bmatrix}. \qquad (11\text{-}3.12)$$

As in Section 4.2, the error function $\varepsilon^{(n)} \equiv [\psi^{(n)} - \psi, \eta^{(n)} - \eta]^T$ of the $\psi^{(n)}$ and $\eta^{(n)}$ sequences may be written as $\varepsilon^{(n)} = P_n(B)\varepsilon^{(0)}$, where $P_n(B) = \sum_{i=0}^{n} \alpha_{n,i} B^i$ is a matrix polynomial. The associated real polynomial $P_n(z)$ which has the least maximum modulus over the ellipse of Fig. 11-3.1 can be

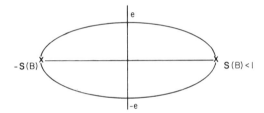

Fig. 11-3.1. Eigenvalue domain for B.

expressed (Clayton [1963]) in terms of Chebyshev polynomials and is given by

$$P_n(z) \equiv \frac{T_n(z/\bar{\sigma})}{T_n(1/\bar{\sigma})}, \qquad (11\text{-}3.13)$$

where $T_n(w)$ is given by (4-2.1) and

$$\bar{\sigma} = \sqrt{S(B)^2 - e^2}.$$

With $P_n(z)$ given by (11-3.13), using the recurrence relation (4-2.1), we can express (see Theorem 3-2.1) the iterates $\psi^{(n)}$ and $\eta^{(n)}$ of (11-3.12) in the recurrence form‡

$$\begin{bmatrix} \psi^{(n)} \\ \eta^{(n)} \end{bmatrix} = \rho^{(n)} \left\{ B \begin{bmatrix} \psi^{(n-1)} \\ \eta^{(n-1)} \end{bmatrix} + \begin{bmatrix} D_\psi & 0 \\ 0 & D_\eta \end{bmatrix}^{-1} \begin{bmatrix} S \\ S \end{bmatrix} - \begin{bmatrix} \psi^{(n-2)} \\ \eta^{(n-2)} \end{bmatrix} \right\} + \begin{bmatrix} \psi^{(n-2)} \\ \eta^{(n-2)} \end{bmatrix},$$

(11-3.14)

---

† It follows from (11-3.10) that the eigenvalues of B are contained in some ellipse about the origin. The assumption that $e < S(B)$ will be discussed later.

‡ Equation (11-3.14) also follows from the discussion given in Section 12.2. See Eq. (12-2.10).

where $\psi^{(n)}$ and $\eta^{(n)}$ satisfy the boundary conditions (11-3.8a) with $\eta^{(n)}(x, b, \Omega) = T\psi^{(n-1)}(x, b, \Omega)$ replacing the second equation. The parameters $\rho^{(n)}$ are defined by $\rho^{(1)} = 1$ and for $n > 1$

$$\rho^{(n)} = \frac{2}{\bar{\sigma}} \frac{T_{n-1}(1/\bar{\sigma})}{T_n(1/\bar{\sigma})}.$$

As for the Chebyshev methods of Chapter 4, $\rho^{(n)}$ may equivalently be defined recursively by Eqs. (4-2.15). Also, as before, $\lim_{n \to \infty} \rho^{(n)} \equiv \rho^{(\infty)} = \bar{r} + 1$, where

$$\bar{r} = \frac{1 - \sqrt{1 - \bar{\sigma}^2}}{1 + \sqrt{1 - \bar{\sigma}^2}}. \tag{11-3.15}$$

If $\mathcal{D}$ denotes the domain bounded by the ellipse of Fig. 11-3.1, then $\max_{z \in \mathcal{D}} |P_n(z)| = P_n(S(B))$. Moreover, it can be shown (e.g., see Wachspress [1966]) that

$$\lim_{n \to \infty} [P_n(S(B))]^{1/n} = \left[\left(\frac{S(B) + e}{S(B) - e}\right)\bar{r}\right]^{1/2}. \tag{11-3.16}$$

The right side of (11-3.16) gives the expected average error reduction factor for the Chebyshev iterations (11-3.14). The corresponding asymptotic convergence rate is

$$R_\infty(P_n(B)) = -\log\left\{\left(\frac{S(B) + e}{S(B) - e}\right)\bar{r}\right\}^{1/2}. \tag{11-3.17}$$

Note that (11-3.17) reduces to (4-2.22) when $e = 0$.

As e approaches $S(B)$, it follows from (11-3.15) and (11-3.17) that $R_\infty(P_n(B))$ approaches $-\log S(B)$,† which is the expected convergence rate, $R(B)$, of the Jacobi method. Thus as the ellipse in Fig. 11-3.1 approaches a circle, the Chebyshev process does not converge much faster than the Jacobi method. In fact, if $S(B)$ is close to unity, then for any $e > \sqrt{1 - S(B)}$ we have (see, e.g., Ref. 9 in Hageman and Kellogg [1968])

$$R_\infty(P_n(B)) < R_\infty(B)/e. \tag{11-3.18}$$

Thus the acceleration obtained by the Chebyshev method may be small if e is not small.

The related cyclic Chebyshev method (see Section 8.3) for the "red/black" partitioned problem (11-3.4) is defined by

$$\psi^{(n)} = \rho_R^{(n)}[D_\psi^{-1}(G\eta^{(n-1)} + S) - \psi^{(n-1)}] + \psi^{(n-1)},$$
$$\eta^{(n)} = \rho_B^{(n)}[D_\eta^{-1}(F\psi^{(n)} + S) - \eta^{(n-1)}] + \eta^{(n-1)}, \tag{11-3.19}$$

† Also see Section 12.2.

where $\psi^{(n)}$ and $\eta^{(n)}$ satisfy the boundary conditions (11-3.8a). The parameters $\rho_R^{(n)}$ and $\rho_B^{(n)}$ are given by (8-3.7) with $\bar{\sigma}$ replacing M_E. The asymptotic convergence rate of the cyclic Chebyshev method is twice that of the Chebyshev method (11-3.14). The difficult task of obtaining estimates for $\bar{\sigma}$ will be discussed later.

For general problems, it is not possible to carry out the iteration methods defined above in terms of continuous operators. Thus we resort to a numerical approximation by discretizing the angular and spatial variables and using the discrete analogue of the continuous iterative method. The fact that the iterative method used is a discrete analog of a convergent nondiscretized iterative method suggests that the rate of convergence of the discretized iterations will not be greatly affected by mesh sizes. The numerical results given later indicate that this is indeed true.

Angular Discretization

Let the integral $\int_0^{2\pi} \int_0^{\pi/2} N \sin\theta \, d\theta \, d\varphi$ be approximated by the quadrature formula $\sum_{p=1}^{P} w_p N(x, y, \Omega^p)$, where $\sum_{p=1}^{P} w_p = 2\pi$ and $N(x, y, \Omega^p)$ is the neutron flux at position (x, y) in the discrete direction Ω^p. We now let $N^p(x, y)$ be the approximation to $N(x, y, \Omega^p)$ which satisfies the following approximate form of the transport equation (11-3.1):

$$\Omega^p \cdot \text{grad } N^p + \sigma N^p = \frac{\sigma^s}{2\pi} \sum_{q=1}^{P} w_q N^q + S, \qquad p = 1, \ldots, P. \qquad (11\text{-}3.20)$$

The P coupled differential equations (11-3.20) are called the *discrete ordinate equations*.

The weights w_p and discrete directions Ω^p are determined (e.g., see Abu-Shumays [1977]) by the quadrature formula used. We require only that the quadrature formula be compatible with reflecting boundary conditions. That is, we require that there be exactly $P/4(\equiv \tau)$ directions $\Omega^q, q = 1, \ldots, \tau$, with positive x- and y-direction cosines (Ω_x^q, Ω_y^q) and for every such direction there must be three associated directions $\Omega^{q+\tau}, \Omega^{q+2\tau}, \Omega^{q+3\tau}$ whose x- and y-direction cosines are, respectively, $(-\Omega_x^q, \Omega_y^q), (\Omega_x^q, -\Omega_y^q)$, and $(-\Omega_x^q, -\Omega_y^q)$. We also require that $w_q = w_{q+\tau} = w_{q+2\tau} = w_{q+3\tau}$. With this understanding, the discrete angle approximation to the ψ and η variables of (11-3.3) are

$$\begin{aligned}
\psi^q(x, y) &\equiv N^q(x, y), & q &= 1, 2, \ldots, 2\tau, \\
\eta^q(x, y) &\equiv N^{q+2\tau}(x, y), & q &= 1, 2, \ldots, 2\tau.
\end{aligned} \qquad (11\text{-}3.21)$$

In terms of the ψ^q and η^q discrete angular variables, (11-3.4) can be written as

$$\begin{bmatrix} D_{\psi,\Omega} & -G_\Omega \\ -F_\Omega & D_{\eta,\Omega} \end{bmatrix} \begin{bmatrix} \psi_\Omega(x, y) \\ \eta_\Omega(x, y) \end{bmatrix} = \begin{bmatrix} S_\Omega \\ S_\Omega \end{bmatrix}, \qquad (11\text{-}3.22)$$

where ψ_Ω and η_Ω are the $2\tau \times 1$ vectors with components $\psi^q(x, y)$ and $\eta^q(x, y)$, respectively. $D_{\psi,\Omega}$, $D_{\eta,\Omega}$, G_Ω, and F_Ω are $2\tau \times 2\tau$ matrices whose elements are semidiscrete operators,† which for $q = 1, \ldots, 2\tau$ are defined by

$$(G_\Omega \eta_\Omega)_q = \frac{\sigma^s}{2\pi} \sum_{k=1}^{2\tau} w_k \eta^k(x, y), \qquad (F_\Omega \psi_\Omega)_q = \frac{\sigma^s}{2\pi} \sum_{k=1}^{2\tau} w_k \psi^k(x, y),$$

$$(D_{\psi,\Omega}\psi_\Omega)_q = \Omega^q \cdot \text{grad } \psi^q(x, y) + \sigma\psi^q(x, y) - (F_\Omega \psi_\Omega)_q,$$

$$(D_{\eta,\Omega}\eta_\Omega)_q = \Omega^{q+2\tau} \cdot \text{grad } \eta^q(x, y) + \sigma\eta^q(x, y) - (G_\Omega \eta_\Omega)_q.$$

(11-3.23)

Along the $x = 0$ and $x = a$ boundaries, ψ^q and η^q satisfy either vacuum or reflecting conditions. For $y = 0$, we have $\psi^q(x, 0) = L\eta^q(x, 0)$ and for $y = b$, $\eta^q(x, b) = T\psi^q(x, b)$. L and T have the same meaning as before. Madsen [1971] shows that the discrete ordinate problem (11-3.22) has a unique solution which is also nonnegative if $S_\Omega \geq 0$.

Iterative solution methods for (11-3.22) analogous to those given for solving (11-3.4) are easily defined. It can be shown (Davis and Hageman [1969]) that eigenvalues of the Gauss–Seidel and Jacobi iteration operators for the discrete ordinate problem (11-3.22) possess the same basic properties as was shown for (11-3.4). That is, the Gauss–Seidel and Jacobi methods converge and the relationships (11-3.9)–(11-3.10) remain valid for the discrete ordinate equations (11-3.22).

Spatial Discretization

To obtain the discrete spatial problem, we first impose a mesh of horizontal lines on the rectangle R and use "upstream" differencing to approximate the y-derivative in $\Omega \cdot \text{grad } N = \Omega_x(\partial N/\partial x) + \Omega_y(\partial N/\partial y)$. Thus along the mesh line y_j of Fig. 11-3.2, we use the approximations‡

$$\Omega_y^q \frac{\partial \psi^q(x, y_j)}{\partial y} \doteq \frac{\Omega_y^q}{\Delta y_j} [\psi_j^q(x) - \psi_{j-1}^q(x)], \qquad q = 1, \ldots, 2\tau,$$

(11-3.24)

$$\Omega_y^{q+2\tau} \frac{\partial \eta^q(x, y_j)}{\partial y} \doteq -\frac{\Omega_y^q}{\Delta y_{j+1}} [\eta_{j+1}^q(x) - \eta_j^q(x)], \qquad q = 1, \ldots, 2\tau,$$

where $\psi_j^q(x)$ and $\eta_j^q(x)$ are approximations to $\psi^q(x, y_j)$ and $\eta^q(x, y_j)$. In what follows, we let $\psi_{\Omega, j}(x)$ and $\eta_{\Omega, j}(x)$ denote the $2\tau \times 1$ vectors with components $\psi_j^q(x)$ and $\eta_j^q(x)$, $q = 1, \ldots, 2\tau$.

---

† Continuous in the x and y variables.

‡ Recall that ψ corresponds to neutrons traveling in the positive y-direction, while η corresponds to those neutrons traveling in the negative y-direction. We also use the fact that $\Omega_y^{q+2\tau} = -\Omega_y^q$ in (11-3.24).

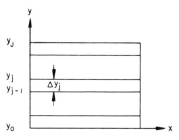

Fig. 11-3.2. y mesh grid.

In terms of the semidiscrete variables $\psi_j^q(x)$ and $\eta_j^q(x)$, (11-3.22) can be written as

$$\begin{bmatrix} D_{\psi,\Omega,y} & -G_{\Omega,y} \\ -F_{\Omega,y} & D_{\eta,\Omega,y} \end{bmatrix} \begin{bmatrix} \psi_{\Omega,y}(x) \\ \eta_{\Omega,y}(x) \end{bmatrix} = \begin{bmatrix} S_{\Omega,y} \\ S_{\Omega,y} \end{bmatrix}, \tag{11-3.25}$$

where the $2\tau J \times 1$ vectors $\psi_{\Omega,y}(x)$ and $\eta_{\Omega,y}(x)$ may be expressed in the $J \times 1$ block form

$$\psi_{\Omega,y}(x) = \begin{bmatrix} \psi_{\Omega,1}(x) \\ \vdots \\ \psi_{\Omega,j}(x) \\ \vdots \\ \psi_{\Omega,J}(x) \end{bmatrix}, \quad \eta_{\Omega,y}(x) = \begin{bmatrix} \eta_{\Omega,J-1}(x) \\ \vdots \\ \eta_{\Omega,j}(x) \\ \vdots \\ \eta_{\Omega,0}(x) \end{bmatrix}. \tag{11-3.26}$$

The $2\tau J \times 2\tau J$ matrices $D_{\psi,\Omega,y}$ and $G_{\Omega,y}$ may be expressed in the $J \times J$ block forms

$$D_{\psi,\Omega,y} = \begin{bmatrix} \tilde{D}_{\psi,\Omega,1} & & & 0 \\ -H_{\psi,\Omega,1} & \tilde{D}_{\psi,\Omega,2} & & \\ & & \ddots & \ddots & \\ 0 & & & -H_{\psi,\Omega,J-1} & \tilde{D}_{\psi,\Omega,J} \end{bmatrix},$$

$$G_{\Omega,y} = \begin{bmatrix} & & \tilde{G}_{\Omega,0} \\ & 0 & \\ & \ddots & \\ \tilde{G}_{\Omega,J-1} & & 0 \end{bmatrix}. \tag{11-3.27}$$

Here $\tilde{D}_{\psi,\Omega,j}$, $H_{\psi,\Omega,j}$, and $\tilde{G}_{\Omega,j}$ are $2\tau \times 2\tau$ matrices whose elements are semi-discrete operators,† which, for $q = 1, 2, \ldots, 2\tau$, are defined by

$$(\tilde{D}_{\psi,\Omega,j}\psi_{\Omega,j}(x))_q = \Omega_x^q \frac{\partial \psi_j^q(x)}{\partial x} + \left(\frac{\Omega_y^q}{\Delta y_j} + \sigma\right)\psi_j^q(x) - \frac{\sigma^s}{2\pi}\sum_{k=1}^{2\tau} w_k \psi_j^k(x),$$

$$j = 1, \ldots, J, \quad (11\text{-}3.28)$$

$$(H_{\psi,\Omega,j}\psi_{\Omega,j}(x))_q = (\Omega_y^q/\Delta y_{j+1})\psi_j^q(x), \quad \text{for} \quad j = 1, \ldots, J-1, \quad (11\text{-}3.29)$$

$$(\tilde{G}_{\Omega,j}\eta_j(x))_q = (\sigma^s/2\pi)\sum_{k=1}^{2\tau} w_k \eta_j^k(x), \quad \text{for} \quad j = 1, \ldots, J-1, \quad (11\text{-}3.30)$$

and

$$(\tilde{G}_{\Omega,0}\eta_0(x))_q = (\sigma^s/2\pi)\sum_{k=1}^{2\tau} w_k \eta_0^k(x) + L(\Omega_y^q/\Delta y_1)\eta_0^q(x). \quad (11\text{-}3.31)$$

The additional term in (11-3.31) is obtained by using the boundary condition (11-3.6) to replace $\psi_0^q(x)$ in (11-3.24) for $j = 1$.

The matrices $D_{\eta,\Omega,y}$ and $F_{\Omega,y}$ in (11-3.25) are similar to $D_{\psi,\Omega,y}$ and $G_{\Omega,y}$ given above and will not be given explicitly.

Again, iterative solution methods for (11-3.25) analogous to those given for solving the continuous problem (11-3.4) can be easily defined, and, again, it can be shown that the convergence properties and the eigenvalue relationships (11-3.9)–(11-3.10) are retained. Moreover, it can be shown that $\psi_{\Omega,y}(x)$ and $\eta_{\Omega,y}(x)$ are nonnegative if $S_{\Omega,y} \geq 0$.

The x variable may be discretized in a manner similar to that used for y; i.e., the approximation for $\partial N/\partial x$ is based on the direction in which the neutron N is moving. However, since the computational process may be adequately described using (11-3.25), the details of the x-variable discretization will not be given.

The spatial discretization given here is called the step model by Carlson [1963]. Other more accurate approximations exist (Lathrop [1969a]). However, the solution positivity (Lathrop [1969b]) and some iterative convergence properties (Davis *et al.* [1967]) need not hold for the more accurate approximations.

The Discrete Cyclic Chebyshev Method

For the Ω and y discrete problem (11-3.25), the cyclic Chebyshev method (11-3.19) may be expressed in the form:

$$D_{\psi,\Omega,y}\tilde{\psi}_{\Omega,y}^{(n)}(x) = G_{\Omega,y}\eta_{\Omega,y}^{(n-1)}(x) + S_{\Omega,y}, \quad (11\text{-}3.32a)$$

$$\psi_{\Omega,y}^{(n)}(x) = \rho_R^{(n)}[\tilde{\psi}_{\Omega,y}^{(n)}(x) - \psi_{\Omega,y}^{(n-1)}(x)] + \psi_{\Omega,y}^{(n-1)}(x), \quad (11\text{-}3.32b)$$

† Continuous in the x variable.

$$D_{\eta,\Omega,y}\tilde{\eta}^{(n)}_{\Omega,y}(x) = F_{\Omega,y}\psi^{(n)}_{\Omega,y}(x) + S_{\Omega,y}, \qquad (11\text{-}3.32c)$$

$$\eta^{(n)}_{\Omega,y}(x) = \rho^{(n)}_B[\tilde{\eta}^{(n)}_{\Omega,y}(x) - \eta^{(n-1)}_{\Omega,y}(x)] + \eta^{(n-1)}_{\Omega,y}(x). \qquad (11\text{-}3.32d)$$

Since $D_{\psi,\Omega,y}$ (given by (11-3.27)) is a block lower triangular matrix, the vector $\tilde{\psi}^{(n)}_{\Omega,y}(x)$ can be determined by solving successively the line equations

$$\tilde{D}_{\psi,\Omega,j}\tilde{\psi}_{\Omega,j}(x) = b_{\eta,\Omega,j}, \qquad j = 1, \ldots, J, \qquad (11\text{-}3.33a)$$

where $b_{\eta,\Omega,j}$ is a known vector. Similarly, $\tilde{\eta}^{(n)}_{\Omega,y}(x)$ may be determined by solving successively the equations

$$\tilde{D}_{\eta,\Omega,j}\tilde{\eta}_{\Omega,j}(x) = b_{\psi,\Omega,j}, \qquad j = J-1, \ldots, 0. \qquad (11\text{-}3.33b)$$

For general problems, the line equations (11-3.33) are best solved by some iterative process. As before, these iterations are called *inner iterations*, while the main iterations (11-3.32) are called *outer iterations*. For the numerical results given later, the one-dimensional discrete analog of the Gauss–Seidel method (11-3.8) is used as the inner iteration method† We omit the details. Henceforth, we assume the x variable has also been discretized in (11-3.32) and (11-3.33).

We assume the eigenvalues and eigenfunctions of the Jacobi iteration matrix $B_{\Omega,y,x}$ corresponding to (11-3.25) satisfy conditions (a) and (b) of (11-3.11). For the spatial discretization presented here, it can be shown that condition (b) holds for any combination of boundary conditions. Nothing is known concerning the validity of condition (a).

We also assume that a fixed number of inner iterations is done for each line every outer iteration. The number of inner iterations m_j done for line j is chosen such that the error reduction achieved by the inner iterations is less than δ. As discussed in the previous section, the utilization of inner iterations results in the outer iterations being carried out with a modified iteration matrix. With inner iterations, the Jacobi matrix B_δ can be written as (Nichols [1973])

$$B_\delta = B_{\Omega,y,x} + T_\delta(I - B_{\Omega,y,x}). \qquad (11\text{-}3.34)$$

Here, T_δ is the iteration error operator of the inner iteration process with an error reduction criterion of δ and $B_{\Omega,y,x}$ is the Jacobi iteration matrix when the discretized line equations (11-3.33) are solved by some direct method. Obviously, the spectral radius, $S(B_\delta)$, of B_δ will depend on the value of δ.

Procedure for Estimating Acceleration Parameters

If the eigenvalues of B_δ are assumed to lie in the ellipse of Fig. 11-3.1, the cyclic Chebyshev parameters $\rho^{(n)}_R$ and $\rho^{(n)}_B$ are functions of $\bar{\sigma} = (S(B_\delta)^2 - e^2)^{1/2}$.

---

† For most problems, the convergence of the Gauss–Seidel method in solving the line equations is sufficiently rapid that no acceleration is deemed necessary.

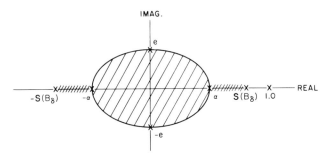

Fig. 11-3.3.   Eigenvalue domain for B_δ with $e < \alpha < S(B_\delta)$.

To numerically estimate both $S(B_\delta)$ and e is a difficult task. We try to avoid this difficulty by using a procedure which, in essence, attempts to estimate only $S(B_\delta)$ numerically. A brief description of this procedure and the assumptions made are given below.

Assume now the eigenvalues of B_δ are contained in the domain given in Fig. 11-3.3. (We discuss this assumption later.) Now suppose that the Chebyshev procedure (11-3.32) is used with the estimate $\bar{\sigma}_E = (S_E(B_\delta)^2 - e_0^2)^{1/2}$, where the approximations $S_E(B_\delta)$ and e_0 are assumed to satisfy

$$e \le e_0 < S_E(B_\delta), \tag{11-3.35a}$$

$$\alpha \le S_E(B_\delta) < S(B_\delta). \tag{11-3.35b}$$

If $\mathscr{D}$ denotes the domain bounded by the ellipse of Fig. 11-3.3 and if (11-3.35) is satisfied, it can be shown that

$$\max_{z \in \mathscr{D}} |P_{n,E}(z)| \le P_{n,E}(S_E(B_\delta)) < P_{n,E}(S(B_\delta)), \tag{11-3.36}$$

where $P_{n,E}(z) = T_n(z/\bar{\sigma}_E)/T_n(1/\bar{\sigma}_E)$. With the difference vector for the black unknowns η defined by

$$\Delta_B^{(n)} \equiv \eta_{\Omega,y,x}^{(n)} - \eta_{\Omega,y,x}^{(n+1)}, \tag{11-3.37}$$

it follows from (11-3.36) that the results given for $\Delta_B^{(n)}$ in Chapter 8 also are valid for the $\Delta_B^{(n)}$ of (11-3.37). (It is only necessary to replace M_E and μ_1 in Chapter 8 by $\bar{\sigma}_E$ and $\bar{\sigma} \equiv (S(B_\delta)^2 - e_0^2)^{1/2}$, respectively.) Thus if the conditions (11-3.35) are satisfied, we can obtain new estimates $\bar{\sigma}_E'$ using the procedures of Chapter 8.† We note that the solution to the Chebyshev

---

† Note that we are not attempting to obtain the optimum $\bar{\sigma} = (S(B_\delta)^2 - e^2)^{1/2}$. Instead we seek to obtain the pseudo-optimum $\bar{\sigma} = (S(B_\delta)^2 - e_0^2)^{1/2}$ when $e_0 > e$ is fixed.

equation (8-3.40) gives the new estimate $\bar{\sigma}'_E$ for $\bar{\sigma}$. Although the new approximation $S'_E(B_\delta)$ is not needed explicitly in the generation of the new Chebyshev polynomial, it can be obtained by $S'_E(B_\delta) = [(\bar{\sigma}'_E)^2 + e_0^2]^{1/2}$.

The question now is how to choose an e_0 to use in (11-3.35). For problems with a vacuum boundary condition, it is relatively easy to obtain such an e_0. For problems with all reflecting conditions, the task can be considerably more difficult.

If a vacuum boundary condition exists, we may without loss of generality assume it exists at the $y = 0$ boundary. For this case, it follows from (11-3.9) and (11-3.10) that the eigenvalues λ_B of the Jacobi iteration operator for (11-3.8) are either real or purely imaginary. Moreover, the real eigenvalues satisfy $\lambda_B^2 \le c/(4 - 3c)$ and the purely imaginary eigenvalues satisfy $|i\lambda_B|^2 \le c/(4 - c)$. It is reasonable to expect that the eigenvalues of $B_{\Omega, y, x}$ (the discrete approximation to B) satisfy these same conditions. Moreover, from (11-3.34), if δ is small then B_δ will closely approximate $B_{\Omega, y, x}$. Consequently, it is reasonable to assume that the eigenvalues of B_δ satisfy these same conditions. Thus for problems with a vacuum boundary, $e_0^2 = c/(4 - c)$ can be used in (11-3.35). For reasons given later, we will not use this estimate in the computational procedure.

For problems with all reflecting boundary conditions, it can be shown (for the difference equations given here) that the eigenvalues of $B_{\Omega, y, x}$ are contained in a domain given by Fig. 11-3.3. However, depending on problem conditions, e may now take on any value between 0 and $S(B_\delta)$. For these problems, we start the Chebyshev process using some e_0 and proceed assuming that the conditions (11-3.35) are satisfied. If the Chebyshev process appears to be divergent, we terminate the generation of the Chebyshev process, increment e_0 by 0.1 and start a new Chebyshev polynomial. If e_0 becomes greater than 0.75, we cease using Chebyshev acceleration and continue the iterations using the Gauss–Seidel method ($\rho_R = \rho_B = 1.0$), which is guaranteed to converge. As noted previously, the acceleration achieved by the Chebyshev process is small if e is large. Thus little is lost by using the Gauss–Seidel method when e is large.

This rather arbitrary procedure for obtaining estimates for e is not very appealing. However, no completely satisfactory method is known to the authors. (For other approaches to this problem, see Wrigley [1963], Wachspress [1966], and Manteuffel [1978].) Fortunately, many transport problems have at least one vacuum boundary. Moreover, even for problems with all reflecting boundaries, the complex eigenvalues are frequently small in magnitude.

Several CG variants were tried by Lewis [1977] in solving the one-dimensional transport problem. He found the CG-convergence behavior to be poor and erratic.

Computational Aspects

For the numerical results given below, the Chebyshev outer iterations (11-3.32) were carried out using the following procedure:

(1)   Five Gauss–Seidel iterations ($\rho_R = \rho_B = 1.0$) are carried out. An estimate, $S_E(B_\delta)$, for $S(B_\delta)$ is obtained from these iterations.†

(2)   Chebyshev acceleration is started using the estimates $S_E(B_\delta)$ and e_0. Initially, $e_0 = 0.0$. The adaptive procedure of Chapter 8 is used to update $S_E(B_\delta)$ and to terminate the outer iterations.

(3)   If the Chebyshev process appears to be diverging, the Chebyshev procedure of item (2) is terminated and e_0 is incremented by 0.1; i.e., $e_0 = e_0^{old} + 0.1\ (\geq 0.4)$.‡ If $e_0 < 0.75$, go to (1). If $e_0 \geq 0.75$, the Chebyshev process is terminated and the solution procedure is continued using only Gauss–Seidel iterations.

In the above procedure, we have taken the initial estimate e_0 for e to be zero which, based on the previous discussion, may not seem appropriate at first glance. However, by (11-2.24) and the comments made there, the initial Gauss–Seidel iterations carried out can greatly reduce the effect of small complex eigenvalues on the effective convergence rate of the Chebyshev method. Thus when the complex eigenvalues of B_δ are small (in magnitude), $e_0 = 0.0$ is often an adequate estimate for e.§ Numerical results indicate that the initial estimate $e_0 = 0.0$ works well for almost all problems with a vacuum boundary condition and for many of the problems with all reflecting boundary conditions.

If the behavior of the Chebyshev process is not as expected, we assume this is caused by the fact that $e_0 < e$. Thus we increase e_0. However, since the Chebyshev adaptive estimate for $S(B_\delta)$ using an $e_0 < e$ is likely to be in error, we again do Gauss–Seidel iterations in order to get a new estimate for $S(B_\delta)$ before restarting the Chebyshev process. This strategy for obtaining estimates for e undoubtedly could be improved. However, since $e_0 = 0.0$ works well for most problems of present interest, no concentrated effort has been made to do this.

Since $\psi^{(n)}$ appears in the Eq. (11-3.32c) for $\tilde{\eta}^{(n)}$ only through boundary conditions and the $(F_{\Omega, y, x}\psi^{(n)})$ term, the extrapolation (11-3.32b) of $\psi^{(n)}$

---

† For the eigenvalue domain of Fig. 11-3.3, the Gauss–Seidel method is a convergent process, and the difference vector of (11-3.37) satisfies $\lim_{n \to \infty} \|\Delta_B^{(n)}\|/\|\Delta_B^{(n-1)}\| = S(B_\delta)^2$.

‡ No nonzero e_0 less than 0.4 is used.

§ From (11-3.16), the acceleration achieved by the Chebyshev process is reduced considerably when $e > 0$. Thus by using the estimate $e_0 = 0.0$, we maximize the convergence rate provided no detrimental effects result from its use.

needs to be done only for these terms. Thus instead of (11-3.32b), we do

$$(F_{\Omega, y, x} \psi^{(n)}_{\Omega, y, x}) = \rho^{(n)}_{R}[(F_{\Omega, y, x} \tilde{\psi}^{(n)}_{\Omega, y, x}) - (F_{\Omega, y, x} \psi^{(n-1)}_{\Omega, y, x})] + (F_{\Omega, y, x} \psi^{(n-1)}_{\Omega, y, x}).$$

(11-3.38)

When (11-3.38) is used instead of (11-3.32b), less storage is required since $(F_{\Omega, y, x} \psi_{\Omega, y, x})$ does not depend on Ω. We remark that $\psi^{(n)}_{\Omega, y, x}$ must be extrapolated at the $y = b$ and $x = 0$ boundaries† if reflecting conditions exist along these boundaries. Similar remarks are valid for the extrapolation (11-3.32d) for $\eta^{(n)}$.

For the numerical results given below, a fixed number of inner iterations is done for each outer iteration. The number of inner iterations for line j is chosen such that the error reduction achieved is less than δ. Moreover, the "weighted diamond" spatial discretization method (Lathrop [1969a]) was used instead of the step model described above.

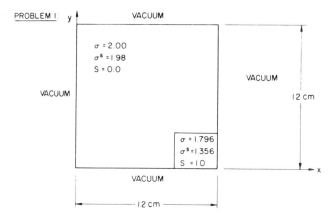

PROBLEM 2 : SAME AS PROBLEM I EXCEPT THAT REFLECTING BOUNDARY CONDITIONS ARE IMPOSED ALONG ALL FOUR BOUNDARIES.

Fig. 11-3.4. Description of two test problems.

To illustrate the effect of the line inner iterations on the convergence rate of the outer iterations, we consider the solution of problem 1 defined in Fig. 11-3.4 using four discrete angles and using 24 spatial mesh points in each of the x and y directions; i.e., $\Delta x = \Delta y = 0.5$. The problem was solved three times with different values for the inner iteration convergence criterion δ. The results are given in Table 11-3.1. We note that the red/black cyclic Chebyshev analysis becomes mathematically rigorous only as the iteration error matrix T_{δ} in (11-3.34) approaches the null matrix. This probably accounts for the

† The extrapolation of ψ at the $x = 0$ boundary is required since inner iterations are done.

TABLE 11-3.1

*Effect of Line Inner Iterations on the Outer Iteration Convergence Rate for
Problem 1 with Four Angles and $\Delta x = \Delta y = 0.5$*

	Inner iteration convergence criterion, δ		
	$\delta = 0.25$	$\delta = 0.05$	$\delta = 0.0006$
Estimate for $S(B_\delta)$ from the Gauss–Seidel iterations	0.9700	0.96205	0.9588
Estimate for $S(B_\delta)$ from the Chebyshev iterations	0.9843	0.9672	0.9588
No. of Chebyshev iterations for convergence	68	49	44
No. of Gauss–Seidel iterations for convergence	222	181	168

fact that the estimates for $S(B_\delta)$ determined adaptively from the Chebyshev
iterations become less accurate for the larger values of δ.† When the outer
iterations are accelerated by the cyclic Chebyshev method, an inner iteration
δ in the range 0.025–0.075 appears to be adequate.

To illustrate the nondependence of the outer iteration convergence rate on
the discretization mesh, problem 1 was solved with a varying number of
discrete angles and spatial mesh points. The results are given in Table 11-3.2.

TABLE 11-3.2

*Nondependence of Convergence Rate on Number of Angles and Spatial Mesh Points for Problem
1 with $\delta = 0.05$*

	4 angles 24 × 24 mesh	12 angles 24 × 24 mesh	24 angles 24 × 24 mesh	12 angles 48 × 48 mesh
Estimate for $S(B_\delta)$ from the Gauss–Seidel iterations	0.9621	0.9625	0.9608	0.9610
No. of Gauss–Seidel iterations for convergence	181	182	175	176

This nondependence of convergence rate on the discretization mesh may be
used to reduce the computing time required to solve large, slowly converging

---

† When the number of inner iterations is finite, the relationship (8-1.10) between the J–SI
and CCSI iterates is not strictly valid. Thus the cyclic Chebyshev equation (8-3.40) also is not
strictly valid. Hence the accuracy of the adaptive estimates for $S(B_\delta)$ is limited. Because of this,
the damping factor F should not be too large unless δ is sufficiently small. For δ in the range
0.025–0.075, F between 0.65 and 0.7 seems appropriate. (Recall F, introduced in Section 5.4,
controls the accuracy required of the estimates for $S(B_\delta)$.)

TABLE 11-3.3

The Effect of $e \neq 0$ on the Convergence Rate of Chebyshev Acceleration for Problem 1 with Four Angles and
$\Delta x = \Delta y = 0.05$

	$e_0 = 0.0$	$e_0 = 0.2$	$e_0 = 0.5$	$e_0 = 0.6$	$e_0 = 0.75$	(Gauss–Seidel) $e_0 = \rho$
Estimate for $S(B_\delta)$ from the Chebyshev iterations	0.9672	0.9642	0.9692	0.944	0.943	0.9672
No. of Chebyshev iterations for convergence	49	65	105	126	150	181

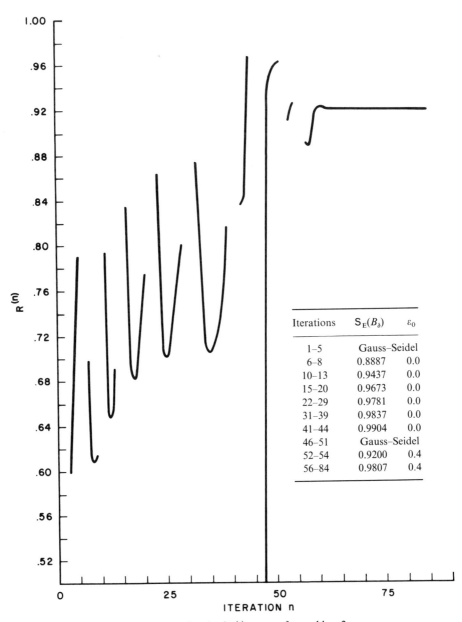

Iterations	$S_E(B_\delta)$	ε_0
1–5	Gauss–Seidel	
6–8	0.8887	0.0
10–13	0.9437	0.0
15–20	0.9673	0.0
22–29	0.9781	0.0
31–39	0.9837	0.0
41–44	0.9904	0.0
46–51	Gauss–Seidel	
52–54	0.9200	0.4
56–84	0.9807	0.4

Fig. 11-3.5.   Graph of $R^{(n)}$ versus n for problem 2.

problems. For example, such a problem could first be solved with a reduced number of angles and spatial mesh points to obtain the optimum acceleration parameters and a reasonable guess for $(F_{\Omega, y, x} \eta_{\Omega, y, x})$. Such procedures have been used to reduce computer solution time with moderate success.

To illustrate that a nonzero e can have a significant effect on the convergence rate, problem 1 was solved using different (fixed) values for e_0 in the Chebyshev process. The results are given in Table 11-3.3. The seemingly bad estimates for $S(B_\delta)$ with $e_0 = 0.6$ and $e_0 = 0.75$ are due to the Chebyshev strategy. With $F = 0.65$, new estimates for $S(B_\delta)$ are calculated only if the actual convergence rate is less than 65% of the estimated theoretical rate (See Eq. (8-3.36)). As e_0 increases, the convergence rate becomes less dependent on the estimate for $S(B_\delta)$ and, thus, less accurate estimates are required to satisfy the parameter test (8-3.36).

The behavior of the Chebyshev process when relatively large complex eigenvalues are present is illustrated in Fig. 11-3.5. There, a plot of $R^{(n)}$ versus n is given for problem 2 of Fig. 11-3.4. (Recall from Chapter 8 that $R^{(n)}$ is the approximation to the error reduction achieved on iteration n (see Eq. (8-3.33)). On outer iteration 45, it was determined that the Chebyshev process was not converging properly. e_0 was then increased to 0.4 and five Gauss–Seidel iterations were done to get a new estimate for $S(B_\delta)$. The Chebyshev process was then restarted on iteration 52 and continued without interruption until convergence was achieved.

Problem 2 was also solved using $e_0 = 0.3$ and $e_0 = 0.4$ as the initial estimates for e. The estimates for $S(B_\delta)$ and e used in the successive Chebyshev polynomial generation are given in Table 11-3.4. On the iterations not given in Table 11-3.4, a Gauss–Seidel iteration, which is needed to obtain a new estimate for $S(B_\delta)$, is being carried out (see Section 8.3). Surprisingly, as shown

TABLE 11-3.4

Estimates for $S(B_\delta)$ and e for Problem 2 Using Initial Estimates of $e_0 = 0.3$ and $e_0 = 0.4$

Iterations	$S_E(B_\delta)$	e_0	Iterations	$S_E(B_\delta)$	e_0
1–5	Gauss–Seidel		1–5	Gauss–Seidel	
6–8	0.8887	0.3	6–8	0.8887	0.4
10–13	0.9404	0.3	10–13	0.9387	0.4
15–20	0.9632	0.3	15–23	0.9612	0.4
22–44	0.9730	0.3	25–111	0.9726	0.4
46–64	0.9813	0.3	113–128	0.9823	0.4
66–71	0.9856	0.3			
73–77	0.9940	0.3			
79–83	Gauss–Seidel				
84–110	0.9807	0.4			

in Fig. 11-3.5, the fewest number of iterations were required when the initial estimate $e_0 = 0.0$ was used. This can be attributed to the fact that the error reduction factors for eigenvector modes associated with the larger real eigenvalues of B_δ are smaller for iterations with $e_0 = 0.0$ then for iterations with $e_0 \neq 0.0$.

11.4   A NONLINEAR NETWORK PROBLEM

In this section, we discuss an iterative solution procedure for solving the systems of nonlinear equations which arise in the network representation of nonlinear flow problems. The numerical procedure illustrates the use of an inner–outer iteration process using the (nonlinear) block SOR method for the outer iterations and the Newton method for the inner iterations. We also discuss such problem areas as the choice for the initial guess and the selection of overrelaxation parameters.

Statement of Problem

The problems we consider are special cases of the network problems studied by Birkhoff and Kellogg [1966], Porsching [1971], and Rheinboldt [1970].

We assume that we are given a network consisting of a set of nodes $I = \{1, \ldots, M\}$ and a set of links, $\Lambda = \{i, j\} \subset I \times I$, joining these nodes. We assume the network is symmetric in the sense that $(i, j) \in \Lambda$ implies $(j, i) \in \Lambda$ and that the network is connected; i.e., for any $i, j \in I$, there is a path of links of the form $(i, i_1)(i_1, i_2) \cdots (i_m, j)$. For every link $(i, j) \in \Lambda$, we are given a conductance function $\varphi_{i,j}(s, t)$ with the following properties:

(P1) $\varphi_{i,j}(s, t)$ is a continuous function of s and t.
(P2) $\varphi_{i,j}(s, t)$ is strictly isotone in s; i.e., if $s' < s$. then $\varphi_{i,j}(s', t) < \varphi_{i,j}(s, t)$.
(P3) $\varphi_{i,j}(s, t)$ is strictly antitone in t; i.e., if $t' < t$, then $\varphi_{i,j}(s, t') > \varphi_{i,j}(s, t)$.
(P4) $\varphi_{i,j}(s, t) + \varphi_{j,i}(t, s) = 0$.
(P5) $\varphi_{i,j}(s, t)$ becomes unbounded as s or $t \to \pm\infty$.

If one interprets s and t as temperatures at nodes i and j, respectively, then $\varphi_{i,j}(s, t)$ may be regarded as the directed heat flow from node i to node j through the connecting link (i, j). With this interpretation, properties (P2) and (P3) imply that this directed heat flow decreases when the temperature at the downstream node is decreased but increases when the upstream temperature is decreased. Property (P4) states that the directed heat flow from node i to node j is the negative of the flow from node j to node i. For hydraulic networks, the state variables s and t represent nodal pressures.

If $u_1, u_2, \ldots, u_M$ are the state variables at the nodes of I, then the net efflux from node i is defined to be

$$f_i(u_1, \ldots, u_M) = \sum_k \varphi_{i,k}(u_i, u_k), \qquad (11\text{-}4.1)$$

where, for notational purposes, we define $\varphi_{i,k}(u_i, u_k) \equiv 0$ if $(i, k) \notin \Lambda$. If the u_i are assigned known values on a nonvoid set of boundary nodes, $N + 1$, $N + 2, \ldots, M$, then the network problem of interest here is to determine the state variables $u_1, \ldots, u_M$ which satisfy the boundary conditions and for which the net efflux at all interior nodes equals a prescribed value. Thus we seek $u_1, \ldots, u_M$ which satisfy

$$\begin{aligned} f_i(u_1, \ldots, u_M) &= q_i, & i &= 1, \ldots, N, \\ u_i &= q_i, & i &= N + 1, \ldots, M, \end{aligned} \qquad (11\text{-}4.2)$$

where the q_i are prescribed constants. The system of equations (11-4.2) may also be written as

$$Fu = 0, \qquad (11\text{-}4.3)$$

where $u = (u_1, u_2, \ldots, u_N)^T$ is the unknown state vector and F is the continuous mapping $F: R^N \to R^N$ with the component functions $(f_i(u_1, \ldots, u_M) - q_i)$, $i = 1, \ldots, N$. Here R^N is the N-dimensional linear space of real column vectors.

Birkhoff and Kellogg [1966] have shown that a unique solution of the network problem (11-4.2) always exists. Their result is valid for more general boundary conditions than the "Dirichlet" condition used here. In what follows, we let u^* denote the unique solution of (11-4.2).

The Nonlinear Block Successive Overrelaxation (NBSOR) Method

We first define what is meant by a partitioning of a nonlinear system of equations into block subfunctions. Let $n_1, \ldots, n_q$ be q positive integers such that $\sum_{k=1}^q n_k = N$, and consider R^N as a q-fold Cartesian product $R^{n_1} \times R^{n_2} \times \cdots \times R^{n_q}$. If P_k is the natural projection operator of R^N onto R^{n_k}, then the kth block of $Fy = 0$ is defined by the n_k equations

$$P_k F(P_1 y, P_2 y, \ldots, P_q y) = 0. \qquad (11\text{-}4.4)$$

We say F is block diagonal solvable if for any $y \in R^N$ and any k, $1 \le k \le q$, the system of n_k equations

$$P_k F(P_1 y, \ldots, P_{k-1} y, z, P_{k+1} y, \ldots, P_q y) = 0 \qquad (11\text{-}4.5)$$

has a unique solution $z \in R^{n_k}$. It follows from a result† of Rheinboldt [1970b] that the network problem (11-4.3) is block diagonal solvable.

If l is the iteration index, the NBSOR iterates $u(l)$ are defined by the following procedure: For $k = 1, 2, \ldots, q$, determine $z \in R^{n_k}$ such that

$$P_k F(P_1 u(l), \ldots, P_{k-1} u(l), z, P_{k+1} u(l - 1), \ldots, P_q u(l - 1)) = 0 \quad (11\text{-}4.6a)$$

and set

$$P_k u(l) = \omega z + (1 - \omega) P_k u(l - 1), \quad (11\text{-}4.6b)$$

where ω is the overrelaxation parameter. Since Fu is block diagonal solvable. the NBSOR process is well defined. If all block sizes are unity, i.e., $q = N$, we refer to (11-4.6) as the nonlinear point successive overrelaxation (NPSOR) method. The nonlinear block Gauss–Seidel (NBGS) method is defined by (11-4.6) when $\omega = 1$. The modifications to (11-4.6) required to define the nonlinear block Jacobi (NBJ) method are straightforward.

For the network problem considered here, it is known (Rheinboldt [1970b]) that the NBGS and the NBJ methods are globally convergent; i.e., these methods converge to u^* for any starting vector $u(0) \in R^N$. Unfortunately, these global convergence results do not in general extend to the truly overrelaxed ($\omega > 1$) process. Nor is there any global theory for choosing ω in order to optimize the rate of convergence. However, as we shall see, often more can be said about the local convergence of the overrelaxed process.

Local Convergence of the NBSOR Method

For the rest of this section, we assume the mapping (11-4.3) is continuously differentiable in some neighborhood $H_0 \subset R^N$ of the solution u^*. Thus the $N \times N$ Jacobian matrix $F'(u)$,

$$F'(u) \equiv \begin{bmatrix} \partial f_1/\partial u_1 & \cdots & \partial f_1/\partial u_N \\ \vdots & & \vdots \\ \partial f_N/\partial u_1 & \cdots & \partial f_N/\partial u_N \end{bmatrix}, \quad (11\text{-}4.7)$$

corresponding to the mapping F is well defined for any $u \in H_0$. We partition $F'(u)$ in a form compatible with the subfunction partitioning (11-4.4): i.e.,

$$F'(u) = \begin{bmatrix} F'_{1,1} & \cdots & F'_{1,q} \\ \vdots & & \vdots \\ F'_{q,1} & \cdots & F'_{q,q} \end{bmatrix}, \quad (11\text{-}4.8)$$

---

† The results by Rheinboldt [1970b] are based on an analysis of M-functions. Previously, Rheinboldt [1970a] showed that the network considered here gives rise to M-functions.

where $F'_{i,j}$ is a $n_i \times n_j$ submatrix and the $n_k, k = 1, \ldots, q$, are defined by the partitioning (11-4.4). If $D(u) = \text{diag}(F'_{i,i})$ is the diagonal block matrix of (11-4.8), then $F'(u)$ may be written as

$$F'(u) = D(u) - L(u) - U(u), \tag{11-4.9}$$

where L and U are strictly lower and upper triangular matrices, respectively. We assume $D(u)$ is nonsingular for any $u \in H_0$. Thus we can define the u-dependent matrices

$$\mathscr{L}_\omega(u) \equiv [D(u) - \omega L(u)]^{-1}[\omega U(u) + (1 - \omega)D(u)],$$
$$B(u) \equiv D(u)^{-1}[L(u) + U(u)], \tag{11-4.10}$$

If $F'(u^*)$ is regarded as a coefficient matrix, then $\mathscr{L}_\omega(u^*)$ and $B(u^*)$ in (11-4.10) are the linear SOR and Jacobi iteration matrices corresponding to the partitioning (11-⁴.8) of $F'(u^*)$.

It follows from a result of Ortega and Rheinboldt [1970] that if† $\mathbf{S}(\mathscr{L}_\omega(u^*))$ is less than unity, then there is a neighborhood of u^*, say $H(u^*)$, such that the NBSOR iterates converge to u^* for any starting guess $u(0) \in H(u^*)$. Moreover, they show that the least possible asymptotic convergence rate for these NBSOR iterations is $-\log \mathbf{S}(\mathscr{L}_\omega(u^*))$. Thus asymptotically the optimum ω for the NBSOR method may be taken to be that ω which minimizes $\mathbf{S}(\mathscr{L}_\omega(u^*))$.

If $F'(u)$ is consistently ordered and if the eigenvalues of $B(u^*)$ are real with $\mathbf{S}(B(u^*)) < 1$, it is known (see Chapter 9) that $\mathbf{S}(\mathscr{L}_\omega(u^*)) < 1$ for $1 \le \omega < 2$ and that $\mathbf{S}(\mathscr{L}_\omega(u^*))$ is minimized for $\omega = \omega_b$, where

$$\omega_b = \frac{2}{1 + \sqrt{1 - \mathbf{S}(B(u^*))^2}}. \tag{11-4.11}$$

Thus if (11-4.8) is a consistently ordered partitioning for $F'(u^*)$ and if $F'(u^*)$ is SPD, it follows that the NBSOR method is convergent provided the starting guess $u(0)$ is sufficiently close to u^*. Moreover, to maximize the asymptotic rate of convergence, ω should approach ω_b as the iterations become large. Henceforth we assume that $F'(u^*)$ is SPD and that the partitioning (11-4.4) is chosen such that the corresponding partitioning (11-4.8) for $F'(u)$ is consistently ordered.

Let $\Delta(l) \equiv u(l) - u(l + 1)$ and $\varepsilon(l) \equiv u(l) - u^*$ denote the difference and error vector, respectively, for the NBSOR method. If $u(l)$ and $u(l + 1)$ are

---

† Recall that $\mathbf{S}(G)$ denotes the spectral radius of the matrix G.

sufficiently close to u^*, using the mean value theorem and the continuity of the elements of $F'(u)$, we can show that $\Delta(l)$ and $\varepsilon(l)$ approximately satisfy

$$\Delta(l) \doteq \mathscr{L}_\omega(u^*)\Delta(l-1),$$

$$\varepsilon(l) \doteq \mathscr{L}_\omega(u^*)\varepsilon(l-1), \quad \text{and} \qquad (11\text{-}4.12)$$

$$\varepsilon(l) \doteq (I - \mathscr{L}_\omega(u^*))^{-1}\Delta(l).$$

Thus the $\Delta(l)$ and $\varepsilon(l)$ vectors approximately satisfy the same relationships as do the Δ and ε vectors for the linear SOR method (see (9-4.3)). Thus the adaptive procedure for estimating ω_b and for estimating the iteration error given in Chapter 9 can also be used for the NBSOR method when the iterates $u(l)$ are sufficiently close to u^*.

The difficulty in using the above results is that the initial guess $u(0)$ for the NBSOR iterates must already be sufficiently close to u^*. Utilizing the global convergence of the NBGS method, Hageman and Porsching [1975] give an a posteriori numerical procedure for obtaining such a $u(0)$. Basically, their procedure is

(1)   Do 15 NBGS iterations ($\omega = 1.0$). Let $S_E(B)$ be the estimate for $S(B(u^*))$ obtained from these iterations.†

(2)   Start the NBSOR iterations using the initial estimate $\omega^{(0)} = 2/[1 + \sqrt{1 - S_E(B)^2}]$. The adaptive procedure of Chapter 9 is used to update the estimates for ω and to estimate the iteration error.

(3)   If the NBSOR process appears to be diverging‡, go to (1).

Inner Iterations

To carry out the NBGS or NBSOR process (11-4.6), nonlinear subsystems of the form

$$G_k(z) \equiv P_k F(P_1 u(l), \ldots, P_{k-1} u(l), z, P_{k+1} u(l-1), \ldots, P_q u(l-1)) = 0$$

$$(11\text{-}4.13)$$

must be solved for z. Unless the system of n_k equations, $G_k(z) = 0$, is of an extremely simple nature, it is not possible to solve for the solution vector z in closed form. Consequently, some iterative process must be used. As before, the iterations used to solve these subsystems are called *inner iterations*, while the primary iterations of (11-4.6) are called *outer iterations*.

---

† Since $F'(u^*)$ is consistently ordered and positive definite, it follows that $S(B(u^*))^2 = S(\mathscr{L}_1(u^*))$ and, from (11-4.12), that $\lim_{l \to \infty} \|\Delta(l)\|/\|\Delta(l-1)\| = S(\mathscr{L}_1(u^*))$. Thus we take $S_E(B)^2 = \|\Delta(14)\|/\|\Delta(13)\|$.

‡ For example, divergence is implied if the ratio $R^{(s,p)}$ defined by (9-5.9) is consistently greater than unity.

Some methods which may be used to solve the subsystem $G_k(z) = 0$ are:

(i)   The NPGS method. The global convergence of the NPGS inner iterations follows from Rheinboldt [1970b].

(ii)   Linearization methods. If the elements of the Jacobian matrix $G'_k(z)$ corresponding to $G_k(z)$ can be easily determined or approximated, one can attempt to solve the subsystems using any one of the variants of Newton's method.

(iii)   Combinations of (i) and (ii). For instance, the globally convergent NPGS method may be used to obtain an approximate solution which is then refined by Newton's method.

We now discuss the inner iteration strategy given by Hageman and Porsching [1975] which uses (iii) above, with safeguards added to improve reliability. We assume that the elements of the Jacobian matrix $G'_k(z)$ can be easily determined and that the Newton method is locally convergent. Moreover, we assume that $P_k u(l)$ is used as the initial guess for inner iterations to solve the subsystem

$$G_k(z) = 0 \qquad (11\text{-}4.14)$$

on outer iteration $(l + 1)$.

Newton's method is appealing because of its quadratic convergence properties. As the NBSOR iterates $u(l)$ converge to the solution u^*, the initial guesses available for Newton's method also converge to the solution of (11-4.14). Thus the most recent NBSOR iteration result $P_k u(l)$ should be an adequate guess for Newton's method in solving (11-4.14) on outer iteration $(l + 1)$ provided l is large enough. However, for l small, the guess $P_k u(l)$ may cause divergence of the Newton iterates. We correct for this by using the globally convergent NPGS method to improve $P_k u(l)$ before starting the Newton process.

Specifically, in solving (11-4.14) on outer iteration $(l + 1)$, two NPGS iterations are performed prior to the start of the Newton iterations if

$$\|u(l) - u(l - 1)\|_2/\|u(l)\|_2 \geq \alpha, \qquad (11\text{-}4.15)$$

where α is an inner iteration strategy parameter. For the numerical examples given later, we picked $\alpha = 0.001$. If after the preliminary NPGS iterations are done the Newton method fails to converge, we neglect the Newton results and solve (11-4.14) by the NPGS method. No preliminary NPGS iterations are performed if inequality (11-4.15) is not satisfied. If the Newton method fails to converge for this case, we retreat, do two NPGS iterations and then attempt to use Newton's method again.

The inner iterations for solving (11-4.14) are considered converged when the estimate for the ratio of final error to initial error is less than β. Specifically, if $P_k u(l)$ is the initial guess for the solution z, we try to do enough inner iterations so that the final approximation $\tilde{z}$ for z satisfies†

$$\|\tilde{z} - z\|_2 / \|P_k u(l) - z\|_2 \leq \beta. \tag{11-4.16}$$

For the inner iteration procedures discussed previously in Sections 14.2 and 14.3, we assumed that the number of inner iterations required to achieve the error reduction (11-4.16) could be numerically predicted a priori. Thus the same (fixed) number of inner iterations could be done every outer iteration. Because the subsystems in this section are nonlinear, the number of inner iterations required for (11-4.16) to be satisfied is likely to vary with the outer iteration. Thus the convergence test (11-4.16) is used here to terminate the inner iterations.

The Newton and NPGS inner iteration procedures for outer iteration $l + 1$ are summarized below.

Newton Method.   If $G_k'(z)$ is the Jacobian matrix for $G_k(z)$, the Newton iterations are defined by

$$d^{(m+1)} \equiv z^{(m+1)} - z^{(m)} = -[G_k'(z^{(m)})]^{-1} G_k(z^{(m)}), \tag{11-4.17}$$

where m is the Newton iteration index and $z^{(0)}$ is taken to be $P_k u(l)$ or the result of the preliminary NPGS iterations. The process (11-4.17) is considered converged when $\|d^{(m+1)}\|_2 / \|d^{(1)}\|_2 \leq \beta$, and is considered to have failed on iteration $m + 1 > 3$ if $\|d^{(m+1)}\|_2 / \|d^{(1)}\|_2 \geq 2.0$ or $m + 1 = 15$.

NPGS Method.   Let $G_k(z) = (g_1(z), \ldots, g_{n_k}(z))^T$, where $z = (z_1, \ldots, z_{n_k})$. The NPGS iterates are defined by the procedure: For $p = 1, \ldots, n_k$, determine $z_p^{(m+1)}$ such that

$$g_p(z_1^{(m+1)}, \ldots, z_{p-1}^{(m+1)}, z_p^{(m+1)}, z_{p+1}^{(m)}, \ldots, z_{n_k}^{(m)}) = 0, \tag{11-4.18}$$

where m is the iteration index and $z^{(0)} = P_k u(l)$. The NPGS process (11-4.18) is terminated if $\|z^{(m+1)} - z^{(m)}\|_2 / \|z^{(1)} - z^{(0)}\|_2 \leq \beta$ or if $m + 1 = \tilde{M}$, where $\tilde{M} = 2$ if the NPGS process is being used to provide a guess for the Newton iterations and $\tilde{M} = 40$ otherwise. Note that single nonlinear equations of the form (11-4.18) must be solved at each step of the NPGS process. These single equations may be approximated by bisection, regula falsi, or Newton iterations. Here again only reasonable accuracy is required. (For the numerical examples considered later, the bisection method is used.)

In the terminology of Ortega and Rheinboldt [1970], the inner–outer iteration process described here could be classified as a variable m-step SOR–Newton method. Analogously, we could also define the variable m-step

---

† Numerical studies indicate that a value of $\beta = 0.01$ is appropriate.

Chebyshev–Newton and the variable m-step CG–Newton methods. Based on the results of the previous chapters for linear problems, it seems likely that SOR–Newton would be the least effective of the three methods.† The use of the CG method seems especially appealing. However, numerical results have been obtained only for the SOR–Newton method.

A Numerical Example

We now discuss the numerical solution of a particular network problem. The connected network is taken to be the 57 node double ladder shown in Fig. 11-4.1. The state variables u_i are assigned fixed values at boundary nodes

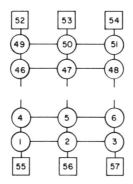

Fig. 11-4.1.   Double ladder network.

52–57. The conductance function for a link joining node i to node j is of the form

$$\varphi_{i,j}(u_i, u_j) = \gamma_{ij}|u_i - u_j|^{\alpha_{ij}} \operatorname{sgn}(u_i - u_j), \qquad (11\text{-}4.19)$$

where the constants γ_{ij} and α_{ij} satisfy $\alpha_{ij} = \alpha_{ji} > 0$ and $\gamma_{ij} = \gamma_{ji} > 0$. It is easy to see that the $\varphi_{i,j}$ of (11-4.19) satisfy properties (P1)–(P5). We consider the following three problems:

Problem 1 $\alpha_{ij} = 0.5$ for all links of the network,

$$\gamma_{ij} = \begin{cases} 1 & \text{for the risers} \\ 10^{-0.5} & \text{for the left rungs} \\ 10^{-1} & \text{for the right rungs} \end{cases}$$

$$u_{52} = 0.0, \qquad u_{53} = 2.0, \qquad u_{54} = 4.0$$

$$u_{55} = 5.08, \qquad u_{56} = 16.8, \qquad u_{57} = 27.8$$

† We assume that one of the red/black ordering variants for the Chebyshev and CG methods is used.

Problem 2
$$\alpha_{ij} = \begin{cases} 3.0 & \text{for the risers} \\ 0.7 & \text{for the left rungs} \\ 2.0 & \text{for the right rungs} \end{cases}$$

$$\gamma_{ij} = \quad 1.0 \qquad \text{for all links}$$

$$u_{52} = 1.0, \qquad u_{53} = 2.0, \qquad u_{54} = 3.0$$

$$u_{55} = 4.0, \qquad u_{56} = 5.0, \qquad u_{57} = 6.0$$

Problem 3 $\alpha_{ij} = \quad 0.5 \qquad$ for all links

$$\gamma_{ij} = \begin{cases} 1.0 & \text{for the risers} \\ 10.0 & \text{for the left rungs} \\ 20.0 & \text{for the right rungs} \end{cases}$$

$$u_{52} = 0.0, \qquad u_{53} = 1.0, \qquad u_{54} = 0.0$$
$$u_{55} = 5.08, \qquad u_{56} = 9.08, \qquad u_{57} = 4.08.$$

All problems were solved using the NBSOR method (11-4.6) with each block in the partitioning being associated with a horizontal line of nodes. Thus for the network of Fig. 11-4.1, there are three equations in each of the 17 subfunction blocks. For each problem the corresponding Jacobian matrix $F'(u)$ is positive definite in some neighborhood about the solution and is consistently ordered relative to the line partitioning chosen for $Fu = 0$.

As discussed previously, NBGS iterations are done initially and possibly later if the NBSOR iterations appear to be diverging. The adaptive procedure used to obtain estimates for ω and to estimate the iteration error is that given by Hageman and Porsching [1975]. Their procedure is basically that of Algorithm 9-6.1 with $F = 0.5$ as the damping factor and, instead of (9-5.22), they use the following smaller upper bounds on the estimates ω.

$$\tau_1 = 1.33, \qquad \tau_2 = 1.5, \qquad \tau_3 = 1.74, \qquad \tau_4 = 1.8, \qquad \tau_5 = 1.85,$$
$$\tau_6 = 1.90, \qquad \tau_7 = 1.96, \qquad \tau_8 = 1.97, \qquad \tau_9 = 1.98, \qquad \tau_{10} = 1.985,$$

and

$$\tau_s = 1.99 \qquad \text{for} \quad s > 10. \tag{11-4.20}$$

$\zeta = 10^{-3}$ was used as the outer iteration convergence criterion.

The relationships (11-4.12), which are important in the estimation of ω, are only valid for $u(l)$ sufficiently close to u^* and even then are only approximate. The smaller (more conservative) bounds τ_s of (11-4.20) are used to prevent the estimates for ω from possibly becoming too large before $u(l)$ is sufficiently close to u^*. In addition to the approximate nature of (11-4.12), the effect† of inner iterations may prevent the theoretical convergence rate predicted by linear theory from being achievable. Thus the accuracy of the adaptive estimates for ω_b may be limited. Because of this, a relatively loose convergence criterion should be used for the ω estimates. In the adaptive procedure here, this is accomplished by using the relatively small value of 0.5 for F (see Sections 5.4 and 9.5).

The inner iteration procedure used is the combination of NPGS and Newton iterations described previously. $\alpha = 0.001$ and $\beta = 0.01$ were used for the strategy parameters in (11-4.15) and (11-4.16). We remark that the partial derivatives of the $\varphi_{i,j}(s, t)$ in (11-4.19) with respect to s and t are easily calculated. Moreover, the Jacobian matrices $G'_k(z)$ for the subfunctions $G_k(z)$ are tridiagonal. Thus the system of linear equations associated with Newton's method (11-4.17) may be solved easily.

A flat guess was used for all problems: 27.0 for problem 1, 7.0 for problem 2, and 10.0 for problem 3. Most problems were solved three times: once using the adaptive procedure to estimate ω and twice using ω fixed. Fixed values of $\omega = 1.0$ (NBGS) and $\omega \lesssim \omega_b$ were used.‡ The iteration data is summarized in Table 11-4.1. For all problems considered, no NBGS iterations were required once the truly overrelaxed process was started. Figure 11-4.2 presents a graph of $R(l)$ versus l for the NBSOR method applied to problem 1 when solved using the adaptive procedure to estimate ω. Here, as in Chapter 9, $R(l) \equiv \|\Delta(l)\|_2 / \|\Delta(l - 1)\|_2$.

The inner iteration procedure worked well for these problems. For problem 1, the average number of Newton iterations/line varied from 3 for the initial NBSOR iterations to 2 for the last 30 NBSOR iterations. The two preliminary NPGS iterations were terminated on outer iteration 81. Problems 2 and 3 behaved similarly. However, for problem 3, the Newton method failed to converge for several lines when the preliminary NPGS iterations were terminated on outer iteration 82. The second Newton attempt, however, was always successful.

In order to verify that the subfunction systems were being solved with sufficient accuracy, problem 1 was rerun using a tighter inner iteration convergence criterion β in (11-4.16). Changing β from 0.01 to 0.0001 resulted in practically no change in the outer iterations. However, the number of Newton inner iterations almost doubled for the initial outer iterations.

---

† When the number of inner iterations is finite, the relationships resulting from the consistent ordering property are no longer strictly valid. For example, Eq. (9-3.8) is only approximately satisfied when inner iterations are done. The approximate nature of (9-3.8), in turn, implies that the theoretical optimum SOR convergence rate can only be approximated by $-\log(\omega_b - 1)$.

‡ Even when a fixed $\omega \neq 1.0$ was used, the first 15 iterations were carried out using the NBGS method.

TABLE 11-4.1

Block Method Iteration Results

Method Problem	NBGS ($\omega = 1$)			NBSOR with ω strategy		NBSOR with ω fixed	
	Iter.	Last est. for $S(\mathscr{L}_1)$	ω	Iter.	Last ω used	Iter.	ω
1	644	0.9849	1.786	95	1.788	65	1.76
2	269	0.9619	1.675	63	1.682	48	1.66
3	368	0.9718	1.713	82	1.723		

As Ortega and Rheinboldt [1970] note, the *asymptotic* convergence rate of the NBSOR method is not enhanced by doing more than one Newton iteration. Thus as convergence is approached, one Newton iteration is likely to be sufficient. However, because of the convergence test used, a minimum of two Newton iterations were required for the numerical examples here. We remark that the Newton convergence test could easily be modified to permit only one Newton iteration when the iterates are sufficiently close to the solution. Since the quotient $\|d^{(2)}\|/\|d^{(1)}\|$ used in the Newton convergence test measures the

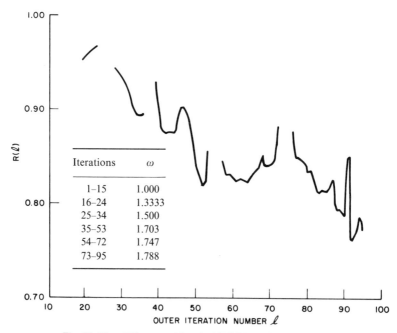

Iterations	ω
1–15	1.000
16–24	1.3333
25–34	1.500
35–53	1.703
54–72	1.747
73–95	1.788

Fig. 11-4.2. $R(l)$ versus l for the NBSOR method: Problem 1.

error reduction of the first Newton iteration, one could use this quotient to indicate when one Newton iteration is sufficient. For example, if $\|d^{(2)}\|/\|d^{(1)}\|$ ≤ 0.01 for all lines for two successive NBSOR iterations, then one Newton iteration could be done for subsequent NBSOR iterations.

These problems were also solved using the NPSOR method; i.e., each block in the SOR process (11-4.6) is now associated with a single node instead of a line of nodes. The iteration data is summarized in Table 11-4.2. Computer

TABLE 11-4.2
Point Method Iteration Results

Method Problem	NPGS ($\omega = 1$)			NPSOR with ω strategy		NPSOR with ω fixed	
		Last est. for					
	Iter.	$S(\mathscr{L}_1)$	ω	Iter.	Last ω used	Iter.	ω
1	930	0.9881	1.803	122	1.809	92	1.78
2	>999 (18,000)	0.9988	1.934	525	1.965	Did not converge	1.946
3	>999 (900,000)	0.9992	1.945	>1482 (2400)	1.990		

time limitations made it unfeasible to run some of these problems to convergence. However, estimates for the number of iterations required for convergence are given in parenthesis. As expected, the point methods required more iterations for convergence than did the block methods. The significant increase for problems 2 and 3 is caused by the strong coupling† between nodes on a horizontal node line. For these problems, the coupling between nodes on a horizontal line is much larger than the coupling between nodes on a vertical line. Thus, the interblock couplings (see Section 9.9) for the line block method are much smaller than those for the point method. In this connection, we note that the convergence rates for the NBSOR method would have been significantly reduced had we chosen the blocks to conform to vertical lines of nodes.

Note that problem 2 did not converge when the NPSOR method was used with $\omega = 1.946$ fixed. However, when the initial guess was changed from a flat value of 7.0 to a flat value of 5.0, convergence was achieved in 262 iterations.

The numerical procedures described in this section have also been used to solve the mildly nonlinear elliptic boundary value problems considered by Ortega and Rockoff [1966]. Numerical results for these problems are given by Hageman and Porsching [1975].

---

† As determined by the appropriate element in the Jacobian matrix $F'(u^*)$.

The Nonsymmetrizable Case

12.1 INTRODUCTION

We now will consider various procedures for accelerating basic iterative methods which are not symmetrizable. Suppose that the basic method

$$u^{(n+1)} = Gu^{(n)} + k \qquad (12\text{-}1.1)$$

is used to solve the $N \times N$ nonsingular matrix problem

$$Au = b. \qquad (12\text{-}1.2)$$

For the acceleration procedures given previously in Chapters 5–8, we assumed that the basic iterative method was symmetrizable. By Theorem 2-2.1, this implies in particular that (a) the eigenvalues of G are real, (b) the algebraically largest eigenvalue of G is less than unity, and (c) the set of eigenvectors for G includes a basis for the associated vector space E^N. However, it is not uncommon to encounter problems for which (12-1.1) is nonsymmetrizable and for which one or more of the properties (a)–(c) are not valid for the matrix G. Because of the possible presence of complex eigenvalues and because of the possible eigenvector deficiency, the nonsymmetrizable case is essentially more difficult than the symmetrizable case.

Previously, in Section 6.8, we discussed some of the difficulties which are encountered when property (b) or property (c) is not valid. In this chapter, we will be concerned with the case where the matrix G has complex eigenvalues.

In place of property (b), we will require that the real part of any eigenvalue μ of G satisfy

$$\text{Re}(\mu) < 1. \tag{12-1.3}$$

We will continue to assume that the matrix A is nonsingular and that the set of eigenvectors for G includes a basis for E^N.[†] In addition, we assume that the matrices A and G are *real* and that (2-2.1) holds, i.e., that G and k can be expressed in the form

$$G = I - Q^{-1}A, \qquad k = Q^{-1}b \tag{12-1.4}$$

for some nonsingular splitting matrix Q.

The complex eigenvalue case may arise for a problem where A is symmetric but where the splitting matrix Q is not symmetric. Most complex eigenvalue cases, however, arise from the application of a standard iterative method such as the RF, Jacobi, or SSOR method to a linear system where the coefficient matrix A is not symmetric.[‡]

One possible approach for solving nonsymmetric linear systems is to consider the equivalent symmetric system

$$A^T A u = A^T b. \tag{12-1.5}$$

Since $A^T A$ is SPD for A nonsingular, the techniques discussed in Chapters 4–7 can now be used to solve the system (12-1.5). However, the condition number of $A^T A$ generally will be much greater than that of A. Because of this,[§] an accelerated procedure for solving (12-1.5) in many cases would be no faster than an unaccelerated procedure for solving the nonsymmetric system (12-1.2). In addition, the cost per iteration would be greater for the system (12-1.5) than for (12-1.2). Thus the use of (12-1.5) is not recommended except in special cases.

In Section 12.2 we will show how Chebyshev acceleration can sometimes be applied effectively in the nonsymmetrizable case. In Section 12.3 we describe three generalizations of conjugate gradient (CG) acceleration for nonsymmetrizable iterative methods. Each of the methods we present may be considered as a "generalized CG-acceleration procedure" in the sense that each procedure reduces to CG acceleration in the symmetrizable case. In Section 12.4 we discuss an acceleration procedure based on Lanczos method. While

---

[†] For discussions of the case where G has an eigenvector deficiency, see Section 6.8 and also Manteuffel [1975].

[‡] We remark that the RF, Jacobi, and SSOR methods are symmetrizable even if A is not SPD provided there exists a corresponding nonsingular block diagonal matrix Σ such that $\Sigma A \Sigma^{-1}$ is SPD. For the RF method, Σ need not be a block diagonal matrix; it is sufficient that Σ be nonsingular.

[§] For example, the effect of condition number on the convergence rate of the RF method can be obtained from Eq. (2-3.6).

the procedures described in Sections 12.3–12.4 have been found to work well in some preliminary experiments, theoretical results are lacking except in special cases. Further research and numerical experimentation are needed.

In Section 12.5 we consider a procedure which can sometimes be used if the matrix A of (12-1.2) is "positive real" in the sense that $A + A^T$ is SPD. For such problems, it is often convenient to use a basic iterative method corresponding to the splitting matrix $Q = \frac{1}{2}(A + A^T)$. (Thus Q is the symmetric part of A.) The corresponding iterative method was considered by Concus and Golub [1976a] and by Widlund [1978] and is referred to as the "GCW method." Although the iterative method is not symmetrizable, the eigenvalues of its iteration matrix G are purely imaginary. In such cases, one can apply either Chebyshev or CG acceleration and analyze the behavior of the resulting procedure in terms of the theory for the symmetrizable case.

In Section 12.6 we describe the application of some of the methods to solve a nonsymmetric linear system corresponding to a partial differential equation of convection-diffusion.

12.2   CHEBYSHEV ACCELERATION

In this section we consider Chebyshev polynomial acceleration of the basic method (12-1.1) which is applicable when the eigenvalues of the iteration matrix G are known to lie within or on an ellipse in the complex plane.

From (3-2.5), the error vector associated with a polynomial method applied to (12-1.1) can be expressed in the form

$$\varepsilon^{(n)} = Q_n(G)\varepsilon^{(0)}, \tag{12-2.1}$$

where $Q_n(G) = \alpha_{n,0} I + \alpha_{n,1} G + \cdots + \alpha_{n,n} G^n$. As in Chapter 4, we require that $Q_n(1) = 1$. Since the matrix G is real, we also require that the coefficients $\alpha_{n,j}$ be real in order to avoid calculations with complex numbers. Let $\mathcal{D}$ be a convex region of the complex plane which contains the eigenvalues of G. We define the *virtual spectral radius of* $Q_n(G)$ with respect to $\mathcal{D}$ to be

$$\bar{S}_{\mathcal{D}}(Q_n(G)) \equiv \max_{z \in \mathcal{D}} |Q_n(z)|. \tag{12-2.2}$$

The subscript $\mathcal{D}$ is used to indicate that the virtual spectral radius is now a function of the domain $\mathcal{D}$ which is chosen to enclose the spectrum of G. We also define $\bar{S}_{\mathcal{D}}(Q_n(z))$ to be the right side of (12-2.2). Analogous to (3-2.17), we define the *asymptotic virtual rate of convergence* of the polynomial method (12-2.1) for this case to be

$$\bar{R}_{\mathcal{D},\infty}(Q_n(G)) = \lim_{n \to \infty} [-(1/n)\log \bar{S}_{\mathcal{D}}(Q_n(G))]. \tag{12-2.3}$$

When $\mathscr{D}$ is chosen to be an ellipse, the polynomial $Q_n(G)$ which maximizes $\bar{R}_{\mathscr{D},\infty}(Q_n(G))$ can often be given in terms of Chebyshev polynomials.

Suppose the eigenvalues of G are known to lie in the region $\mathscr{D}$, whose boundary is the ellipse

$$[(x - d)/a]^2 + (y/b)^2 = 1, \qquad (12\text{-}2.4)$$

where the constants a, b, and d are real and where

$$a + d < 1. \qquad (12\text{-}2.5)$$

Because any complex eigenvalues of the real matrix G must occur in complex conjugate pairs and because of the condition (12-1.3), neither of the above conditions on the constants a, b, and d is restrictive.

With $a \neq b$, consider the normalized Chebyshev polynomial

$$P_n(z) \equiv \frac{T_n[(z - d)/c]}{T_n[(1 - d)/c]}, \qquad (12\text{-}2.6)$$

where $T_n(w)$ is the Chebyshev polynomial (4-2.1) and where

$$c^2 = a^2 - b^2. \qquad (12\text{-}2.7)$$

Since a and b are real, the constant c is either real or purely imaginary. For either case, $P_n(z)$ is a real polynomial. This follows since $T_n(w)$ is an odd (even) polynomial for n odd (even). We remark that the points $d + c$ and $d - c$ are the foci of the ellipse (12-2.4). See Fig. 12-2.1.

If the foci of the ellipse (12-2.4) are real, we have the following result due to Clayton [1963].

Theorem 12-2.1. Let $\mathscr{D}$ be the region enclosed by the ellipse (12-2.4), where $d + a < 1$ and where $b < a$. If $\mathscr{S}_n$ is the set of all real polynomials $Q_n(z)$ of degree n or less satisfying $Q_n(1) = 1$, then the polynomial $P_n(z)$ of

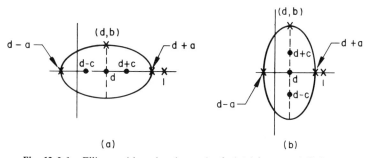

Fig. 12-2.1. Ellipses with real and complex foci. (a) $b < a$ and (b) $b > a$.

(12-2.6) is the unique polynomial in the set $\mathscr{S}_n$ which for any $Q_n(z) \in \mathscr{S}_n$ satisfies

$$\bar{S}_\mathscr{D}(P_n(z)) \le \bar{S}_\mathscr{D}(Q_n(z)).$$

Manteuffel [1977] has shown that the above result cannot be extended to the case where c is purely imaginary. However, he shows that the result of Theorem 12-2.1 is asymptotically true, i.e., if $d + a < 1$ and if $b > a$, then $\lim_{n \to \infty} [\bar{S}_\mathscr{D}(P_n(z))]^{1/n} \le \lim_{n \to \infty} [\bar{S}_\mathscr{D}(Q_n(z))]^{1/n}$ for any $Q_n(z) \in \mathscr{S}_n$. Moreover, Manteuffel notes that this asymptotic behavior is achieved very quickly.

As in Chapter 4, we can use (4-2.1) to express $P_n(z)$ in the recursion form

$$P_0(z) = 1, \qquad P_1(z) = \gamma z - \gamma + 1,$$
$$P_{n+1}(z) = \rho_{n+1}(\gamma z + 1 - \gamma)P_n(z) + (1 - \rho_{n+1})P_{n-1}(z), \qquad n \ge 1, \tag{12-2.8}$$

where

$$\gamma = \frac{1}{1 - d}, \qquad \rho_{n+1} = 2\left(\frac{1 - d}{c}\right) \frac{T_n[(1 - d)/c]}{T_{n+1}[(1 - d)/c]}. \tag{12-2.9}$$

It now follows from Theorem 3-2.1 that the iterates for the polynomial method (12-2.1) based on $P_n(z)$ can be expressed in the three-term form

$$u^{(n+1)} = \rho_{n+1}(\gamma \delta^{(n)} + u^{(n)}) + (1 - \rho_{n+1})u^{(n-1)}, \tag{12-2.10}$$

where $\delta^{(n)} = Gu^{(n)} + k - u^{(n)}$. Again, using (4-2.1), we can express the parameters ρ_n in the more computationally convenient form

$$\rho_{n+1} = \begin{cases} 1, & \text{if } n = 0, \\ (1 - \tfrac{1}{2}\sigma^2)^{-1}, & \text{if } n = 1, \\ (1 - \tfrac{1}{4}\sigma^2 \rho_n)^{-1}, & \text{if } n \ge 2, \end{cases} \tag{12-2.11}$$

where

$$\sigma^2 = c^2/(1 - d)^2. \tag{12-2.12}$$

Since $P_n(z)$ is analytic, we have by the maximum modulus principle that

$$\max_{z \in \mathscr{D}} |P_n(z)| = \max_{z \in E(d, a, b)} |P_n(z)|,$$

where $E(d, a, b)$ is the ellipse (12-2.4). Now, using the fact that a maximum of $|P_n(z)|$ over $E(d, a, b)$ occurs at $z = d + a$, we obtain (see, e.g., Wachspress [1966]), after some algebraic manipulation, that

$$\bar{R}_{\mathscr{D}, \infty}(P_n(G)) = -\log\left[\frac{a + b}{1 - d + \sqrt{(1 - d)^2 - c^2}}\right]. \tag{12-2.13}$$

When the basic method (12-1.1) is symmetrizable, we have that $b = 0$. For this case, the ellipse (12-2.4) reduces to the interval $[d - a, d + a]$ and $\bar{R}_{\mathscr{D}, \infty}(P_n(G))$ reduces to

$$\bar{R}_{\mathscr{D}, \infty}(P_n(G)) = -\log[(1 - \sqrt{1 - \sigma^2})/(1 + \sqrt{1 - \sigma^2})]^{1/2}, \quad (12\text{-}2.14)$$

where $\sigma = a/(1 - d)$. This is consistent with the results of Chapter 4. To show this, we first note that with $b = 0$ we have that $M(G) \leq d + a$ and $m(G) \geq d - a$. Now, from (4-3.23), we have that the convergence rate for the Chebyshev procedure of Chapter 4 with $M_E = d + a$ and with $m_E = d - a$ satisfies

$$\bar{R}_\infty(P_{n, E}(G)) \geq -\log[(1 - \sqrt{1 - \sigma_E^2})/(1 + \sqrt{1 - \sigma_E^2})]^{1/2} = \bar{R}_{\mathscr{D}, \infty}(P_n(G)),$$

$$(12\text{-}2.15)$$

where $\sigma_E = (M_E - m_E)/(2 - M_E - m_E)$. The right equality in (12-2.15) follows from (12-2.14) and the fact that $\sigma_E = [a/(1 - d)] = \sigma$. For the case when $M(G) = d + a$, we have equality throughout (12-2.15).

A difficult problem in the utilization of the Chebyshev-acceleration procedure given above is the determination of the bounding ellipse $\mathscr{D}$ which encloses the eigenvalues of G and which maximizes $\bar{R}_{\mathscr{D}, \infty}(P_n(G))$ with respect to all such ellipses. Numerical procedures to obtain such a bounding ellipse have been developed, for example, by Wrigley [1963], Wachspress [1966], and Manteuffel [1978]. (Also, see Section 11.2.)

We now discuss several special cases.

Chebyshev Acceleration When Bounding Ellipse Is a Circle

For this case, $a = b$ and $c = 0$. As $\rho_n = 1$ for all n when $a = b$, it follows that the Chebyshev procedure (12-2.10)–(12-2.12) with $a = b$ reduces to the basic method with extrapolation (see Eq. (2-2.13)). Manteuffel [1975] shows that this result for circles is compatible with the result for ellipses, i.e., he shows that

$$\lim_{b \to a} P_n(z) = [\gamma z + (1 - \gamma)]^n, \quad (12\text{-}2.16)$$

where, as in (12-2.9), $\gamma = 1/(1 - d)$. Note that with $z = G$, the right side of (12-2.16) is equal to $(G_{[\gamma]})^n$, where $G_{[\gamma]}$ is the iteration matrix of the extrapolated method (2-2.13). Thus if the domain containing the eigenvalues of G is a circle or nearly a circle, there will be little or no acceleration over that of extrapolation. This situation occurs, for example, when G is the SOR matrix $\mathscr{L}_\omega$ of Chapter 9 with ω close to ω_b (see Fig. 9-3.1).

Chebyshev Acceleration When the Eigenvalues of G Are Purely Imaginary

This situation occurs whenever G is similar to a skew-symmetric matrix.

For this case, with $b = S(G)$, the bounding ellipse (12-2.4) reduces to the interval $[-ib, ib]$ in the complex plane. Thus with $a = d = 0$, the Chebyshev procedure (12-2.10)–(12-2.12) becomes

$$u^{(n+1)} = \rho_{n+1}(Gu^{(n)} + k) + (1 - \rho_{n+1})u^{(n-1)}, \qquad (12\text{-}2.17)$$

where

$$\rho_1 = 1, \qquad \rho_2 = (1 + \tfrac{1}{2}b^2)^{-1},$$
$$\rho_{n+1} = (1 + \tfrac{1}{4}b^2\rho_n)^{-1} \qquad \text{if} \quad n \geq 2. \qquad (12\text{-}2.18)$$

Moreover, from (12-2.13), we have that

$$\bar{R}_{\mathscr{D}, \infty}(P_n(G)) = -\tfrac{1}{2}\log \bar{r}, \qquad (12\text{-}2.19)$$

where

$$\bar{r} = (\sqrt{1 + b^2} - 1)/(\sqrt{1 + b^2} + 1). \qquad (12\text{-}2.20)$$

When the eigenvalues of G are purely imaginary, we also can utilize the Chebyshev procedure given earlier in Chapter 4. To show this, consider the "double method" based on (12-1.1):

$$u^{(n+1)} = G^2u^{(n)} + Gk + k. \qquad (12\text{-}2.21)$$

Since $G^2u^{(n)} + Gk + k = G(Gu^{(n)} + k) + k$, it is evident that (12-2.21) is equivalent to two applications of (12-1.1). Clearly, the eigenvalues of G^2 are real and satisfy $M(G^2) \leq 0$ and $m(G^2) = -b^2$. This, coupled with the assumption that G has no eigenvector deficiency, implies that the double method (12-2.21) is symmetrizable. Hence the Chebyshev-acceleration method of Chapter 4 may be applied to (12-2.21). Using $M(G^2) = 0$ and $m(G^2) = -b^2$, we can express this Chebyshev procedure in the form

$$u^{(n+1)} = \bar{\rho}_{n+1}[\bar{\gamma}_{n+1}(G^2u^{(n)} + Gk + k) + (1 - \bar{\gamma}_{n+1})u^{(n)}] + (1 - \bar{\rho}_{n+1})u^{(n-1)}, \qquad (12\text{-}2.22)$$

where, from (4-2.12), $\bar{\gamma}_{n+1} = 2/(2 + b^2)$ and where the $\bar{\rho}_{n+1}$ are given by (4-2.15) with $\bar{\sigma} = b^2/(2 + b^2)$.

Hageman et al. [1980] show that the Chebyshev procedures (12-2.17) and (12-2.22) are, in a sense defined by them, equivalent. Moreover, since effective adaptive procedures can be developed† for both Chebyshev methods, there is no apparent reason to favor one method over the other.

---

† For each Chebyshev method, the results of Section 6.5 can be used to develop an adaptive procedure for estimating b. We remark that only pseudoresiduals corresponding to even iterations (i.e., $\delta^{(2n)}$) should be examined for the Chebyshev method of (12-2.17).

Chebyshev Accleration When Bounding Eigenvalue Region Is a Rectangle

We now consider the case when the eigenvalues μ of G satisfy

$$|\text{Re}(\mu - d)| \le \bar{a}, \qquad |\text{Im}(\mu)| \le \bar{b}, \qquad (12\text{-}2.23)$$

where, analogous to (12-2.5), we assume that $(\bar{a} + d) < 1$. Such bounds for the eigenvalues of G can be obtained, for example, by examining the eigenvalues of the symmetric matrix $H \equiv \frac{1}{2}(G + G^T)$ and the skew-symmetric matrix $K \equiv \frac{1}{2}(G - G^T)$ associated with G. For any eigenvalue μ of G, it can be shown (Manteuffel [1975]) that $\text{Re}(\mu)$ lies in the interval $[m(H), M(H)]$ and that $|\text{Im}(\mu)| \le S(K)$. Thus if the extreme eigenvalues of H and K can be determined, then the eigenvalues of G satisfy (12-2.23) with $d = \frac{1}{2}[M(H) + m(H)]$, $\bar{a} = \frac{1}{2}[M(H) - m(H)]$, and $\bar{b} = S(K)$. Another example where (12-2.23) holds is given in Section 12.6.

Without loss of generality, we can assume that $d = 0$. Otherwise, we simply consider the extrapolated method

$$u^{(n+1)} = (1/(1 - d))(Gu^{(n)} + k) - (d/(1 - d))u^{(n)}, \qquad (12\text{-}2.24)$$

whose iteration matrix has eigenvalues $\hat{\mu}$ which are bounded by

$$|\text{Re}(\hat{\mu})| \le \hat{a}, \qquad |\text{Im}(\hat{\mu})| \le \hat{b}, \qquad (12\text{-}2.25)$$

where $\hat{a} = \bar{a}/(1 - d)$ and $\hat{b} = \bar{b}/(1 - d)$.

There are many ellipses of the form

$$[(\text{Re } \mu)/a]^2 + [(\text{Im } \mu)/b]^2 = 1 \qquad (12\text{-}2.26)$$

which contain the rectangle (12-2.23) with $d = 0$. We choose the one for which $\bar{R}_{\mathscr{D}, \infty}(P_n(G))$ given by (12-2.13) (with $d = 0$) is as large as possible. From the analysis of Young [1971], see also Kjellberg [1958], it can be shown that the choice of a and b which minimizes $\bar{R}_{\mathscr{D}, \infty}(P_n(G))$ is given by

$$a = \frac{2\rho(1 + \rho^2)^{-1/3}\bar{a}^{2/3}}{\bar{b}^{2/3}(1 - \rho^2)^{2/3} + \bar{a}^{2/3}(1 + \rho^2)^{2/3}} \qquad (12\text{-}2.27)$$

$$b = \frac{2\rho(1 - \rho^2)^{-1/3}\bar{b}^{2/3}}{\bar{b}^{2/3}(1 - \rho^2)^{2/3} + \bar{a}^{2/3}(1 + \rho^2)^{2/3}}. \qquad (12\text{-}2.28)$$

Here ρ is the solution of the cubic equation

$$\left(\frac{1 + \rho^2}{2\rho}\right)^{2/3}\bar{a}^{2/3} + \left(\frac{1 - \rho^2}{2\rho}\right)^{2/3}\bar{b}^{2/3} = 1. \qquad (12\text{-}2.29)$$

We choose the root of (12-2.29) such that $0 \leq \rho < 1$. It can also be shown that

$$\rho = \frac{a + b}{1 + \sqrt{1 + b^2 - a^2}} \tag{12-2.30}$$

and that

$$\bar{R}_{\mathscr{D}, \infty}(P_n(G)) = -\log \rho. \tag{12-2.31}$$

Chebyshev Acceleration for Red/Black Partitioned Matrix Problems

We now consider Chebyshev acceleration of the Jacobi method when the matrix problem (12-1.2) is partitioned into the red/black form

$$\begin{bmatrix} D_R & H \\ K & D_B \end{bmatrix} \begin{bmatrix} u_R \\ u_B \end{bmatrix} = \begin{bmatrix} b_R \\ b_B \end{bmatrix}, \tag{12-2.32}$$

We assume that D_R and D_B are nonsingular. As in previous chapters, we let B denote the iteration matrix for the associated Jacobi method.

Since the coefficient matrix of (12-2.32) has Property $\mathscr{A}$ (see Section 9.2), relative to the red/black partitioning imposed, it can be shown (see, e.g., Varga [1962]) that the real parts of the eigenvalues μ of B occur in $\pm$ pairs. Thus a bounding ellipse for the eigenvalues of B can be given by (12-2.4) with $d = 0$. From (12-2.10)–(12-2.11), Chebyshev acceleration of the Jacobi method can then be defined by

$$u_R^{(n+1)} = \rho_{n+1}[F_R u_R^{(n)} + c_R] + (1 - \rho_{n+1})u_R^{(n-1)}$$
$$u_B^{(n+1)} = \rho_{n+1}[F_B u_R^{(n)} + c_B] + (1 - \rho_{n+1})u_B^{(n-1)}, \tag{12-2.33}$$

where ρ_{n+1} is given by (12-2.11) with $\sigma^2 = a^2 - b^2$. In (12-2.33), we let $F_R = -D_R^{-1}H$, $F_B = -D_B^{-1}K$, $c_R = D_R^{-1}b_R$, and $c_B = D_B^{-1}b_B$.

That the Golub–Varga cyclic procedure given in Chapter 8 is also applicable here can be shown, as before in Section 8.3. Thus if we let $U_R^{(n)} = u_R^{(2n-1)}$ and $U_B^{(n)} = u_B^{(2n)}$, the process (12-2.33) can be carried out by the cyclic formulation:

$$U_R^{(n)} = \rho_R^{(n)}[F_R U_B^{(n-1)} + c_R - U_R^{(n-1)}] + U_R^{(n-1)}$$
$$U_B^{(n)} = \rho_B^{(n)}[F_B U_R^{(n)} + c_B - U_B^{(n-1)}] + U_B^{(n-1)}, \tag{12-2.34}$$

where, with $\sigma^2 = a^2 - b^2$, we have $\rho_R^{(1)} = 1, \rho_B^{(1)} = (1 - \tfrac{1}{2}\sigma^2)^{-1}$, and for $n \geq 1$

$$\rho_R^{(n+1)} = [1 - \tfrac{1}{4}\sigma^2 \rho_B^{(n)}]^{-1}, \qquad \rho_B^{(n+1)} = [1 - \tfrac{1}{4}\sigma^2 \rho_R^{(n+1)}]^{-1}. \tag{12-2.35}$$

If $a < 1$, it follows from (12-2.13) that the asymptotic virtual rate of convergence for the cyclic accelerated method (12-2.34) is $-\log[(a + b)/(1 + \sqrt{1 - \sigma^2})]^2$.

We remark that the successive overrelaxation (SOR) method can also be used to solve the matrix problem (12-1.2) when A is not symmetric. To briefly discuss this application, let the system (12-1.2) be partitioned into the form (9-3.1) and let B denote the iteration matrix for the Jacobi method associated with this partitioning. We assume, as in Chapter 9, that the coefficient matrix A has Property $\mathscr{A}$ and is consistently ordered relative to the partioning imposed. Because A has Property $\mathscr{A}$, it again follows that the real parts of the eigenvalues of B occur in $\pm$ pairs. Thus a bounding ellipse for the eigenvalues of B can be given by (12-2.4) with $d = 0$, i.e.,

$$x^2/a^2 + y^2/b^2 = 1. \tag{12-2.36}$$

When a is less than unity, we have

Theorem 12-2.2.   Let the coefficient matrix A be consistently ordered relative to the partitioning imposed. If the eigenvalues of the associated Jacobi iteration matrix B lie in or on the ellipse (12-2.36) with $a < 1$, then the SOR method (9-3.2) converges for all ω satisfying $0 < \omega < 2/(1 + b)$. Moreover, if

$$\omega_b = 2/(1 + \sqrt{1 + b^2 - a^2}), \tag{12-2.37}$$

then the spectral radius of the SOR iteration matrix $\mathscr{L}_{\omega_b}$ satisfies

$$\mathsf{S}(\mathscr{L}_{\omega_b}) \le [(a + b)/(1 + \sqrt{1 + b^2 - a^2})]^2. \tag{12-2.38}$$

Further, if some eigenvalue of B lies on the ellipse (12-2.36), then equality holds in (12-2.38) and $\mathsf{S}(\mathscr{L}_\omega) \ge \mathsf{S}(\mathscr{L}_{\omega_b})$ for $\omega \ne \omega_b$.

Proof.   See, for example, Young [1971].

As for the case when the eigenvalues of B are real, the SOR method and the cyclic method (12-2.34) for the red/black partitioned problem (12-2.32) have identical asymptotic convergence rates.

12.3   GENERALIZED CONJUGATE GRADIENT ACCELERATION PROCEDURES

In this section, we consider three procedures for accelerating the convergence of the basic method (12-1.1) in the nonsymmetrizable case. One of the schemes, which we call ORTHOMIN, was developed by Vinsome [1976]. Each of the procedures we present can be derived by a method which yields conjugate gradient (CG) acceleration in the symmetrizable case. All the procedures are thus equivalent in the symmetrizable case. In general, we refer to the schemes of this section as "generalized CG-acceleration procedures."

If the basic method (12-1.1) is symmetrizable, there exists a nonsingular matrix W such that $W(I - G)W^{-1}$ is SPD. An equivalent condition for (12-1.1) to be symmetrizable is that $Z(I - G)$ is SPD for some matrix Z which is also SPD. This follows since if $W(I - G)W^{-1}$ is SPD, then $Z(I - G)$ is SPD for $Z = W^T W$. On the other hand, if $Z(I - G)$ is SPD and if Z is SPD, then $W(I - G)W^{-1}$ is SPD for $W = Z^{1/2}$. For two of the generalized CG-acceleration procedures, we will utilize a matrix Z, not necessarily SPD, such that $Z(I - G)$ is SPD. It is necessary and sufficient that such a matrix Z have the form

$$Z = (I - G)^T Y, \tag{12-3.1}$$

where Y is SPD. Obviously, if Z has the form (12-3.1), then $Z(I - G)$ is SPD. On the other hand, if $Z(I - G)$ is SPD, then, since $(I - G)$ is nonsingular, we can write $Z = (I - G)^T Y$, where $Y = (I - G)^{-T} Z$. But since $Z(I - G)$ is SPD, it then follows that $(I - G)^T Y (I - G)$ is SPD, and, hence, Y is SPD.

The generalized CG procedures which we consider are polynomial acceleration procedures as described in Chapter 3. From (3-2.5), the error vector associated with a polynomial method applied to (12-1.1) can be expressed in the form

$$\varepsilon^{(n)} = Q_n(G)\varepsilon^{(0)}, \tag{12-3.2}$$

where $Q_n(x) = \sum_{k=0}^n \alpha_{n,k} x^k$ with $Q_n(1) = 1$. Let $\delta^{(0)}$ denote, as before, the pseudoresidual vector

$$\delta^{(0)} = k - (I - G)u^{(0)} \tag{12-3.3}$$

and let $K_n(\delta^{(0)})$ denote the so-called *Krylov space*, i.e., the space spanned by the *Krylov sequence* of vectors $\delta^{(0)}, (I - G)\delta^{(0)}, \ldots, (I - G)^{n-1}\delta^{(0)}$. We now show for $n = 1, 2, \ldots$ that

$$u^{(n)} - u^{(0)} \in K_n(\delta^{(0)}). \tag{12-3.4}$$

Since $Q_n(1) = 1$, we have, by (12-3.2), that

$$u^{(n)} - u^{(0)} = (u^{(n)} - \bar{u}) - (u^{(0)} - \bar{u})$$

$$= \varepsilon^{(n)} - \varepsilon^{(0)} = (Q_n(G) - I)\varepsilon^{(0)}$$

$$= (Q_n(G) - Q_n(I))\varepsilon^{(0)}$$

$$= \sum_{k=0}^n \alpha_{n,k}(G^k - I)\varepsilon^{(0)} \tag{12-3.5}$$

Now, from (12-1.4), $G = I - Q^{-1}A$ and from (5-2.3), $\varepsilon^{(0)} = (G - I)^{-1}\delta^{(0)} = (-Q^{-1}A)^{-1}\delta^{(0)}$. Using the binomial expansion for $(I - Q^{-1}A)^k$, we then can express $(G^k - I)\varepsilon^{(0)}$ in the form

$$(G^k - I)\varepsilon^{(0)} = [kI + \tfrac{1}{2}k(k-1)(-Q^{-1}A) + \cdots + (-Q^{-1}A)^{k-1}]\delta^{(0)}. \tag{12-3.6}$$

Since $Q^{-1}A = I - G$, it now follows from (12-3.5) that $u^{(n)} - u^{(0)}$ can be expressed as a linear combination of the vector sequence $\{\delta^{(0)}, (I - G)\delta^{(0)}, \ldots, (I - G)^{n-1}\delta^{(0)}\}$ and (12-3.4) is proved.

Let $\varepsilon^{(0)}$ be given and suppose that the basic method (12-1.1) is symmetrizable, i.e., there exists an SPD matrix Z such that $Z(I - G)$ is SPD. It can be shown (see, for instance, Young et al. [1980]) that one can derive the CG-acceleration procedure of Chapter 7 by requiring for all n that $\|\varepsilon^{(n)}\|_{[Z(I-G)]^{1/2}}$ be minimized over all $\varepsilon^{(n)}$ such that $\varepsilon^{(n)} - \varepsilon^{(0)} \in K_n(\delta^{(0)})$. This minimization procedure can also be used when Z is not SPD but where $Z(I - G)$ is SPD. As we show below, this minimization procedure can lead to a generalized CG-acceleration method for the nonsymmetrizable case.

Our discussion in this section will be very brief and we will omit many details. These details can be found in Young et al. [1980] and in Young and Jea [1980a, 1980b].

ORTHODIR and ORTHOMIN

We now assume that $Z(I - G)$ is SPD but that Z is not necessarily SPD. We let $\bar{u}$ denote the true solution of (12-1.2) and let $u^{(n)}$ denote the iterates associated with the polynomial method (3-2.1)–(3-2.3). We assume the starting vector $u^{(0)}$ is arbitrary.

To derive formulas for a generalized CG-acceleration procedure as described above, it is convenient to choose a set of vectors $q^{(0)}, q^{(1)}, \ldots$ which are pairwise $(Z(I - G))^{1/2}$ orthogonal and are such that $q^{(i)} \in K_{i+1}(\delta^{(0)})$, $i = 0, 1, \ldots$. Such a set of vectors could be constructed by applying the Gram–Schmidt procedure to the set of vectors $\delta^{(0)}, (I - G)\delta^{(0)}, \ldots$. However, it is more convenient to use the following procedure to generate the $\{q^{(i)}\}$:

$$q^{(0)} = \delta^{(0)}$$

$$q^{(n)} = (I - G)q^{(n-1)} + \beta_{n, n-1}q^{(n-1)} + \cdots + \beta_{n, 0}q^{(0)}, \qquad n = 1, 2, \ldots,$$

$$(12\text{-}3.7)$$

where

$$\beta_{n, i} = -\frac{(Z(I - G)^2 q^{(n-1)}, q^{(i)})}{(Z(I - G)q^{(i)}, q^{(i)})}, \qquad i = 0, 1, \ldots, n - 1. \quad (12\text{-}3.8)$$

It is easy to see that $q^{(i)} \in K_{i+1}(\delta^{(0)})$ and that the $\{q^{(i)}\}$ are pairwise $(Z(I - G))^{1/2}$ orthogonal, i.e.,

$$(Z(I - G)q^{(i)}, q^{(j)}) = 0, \qquad i \neq j. \tag{12-3.9}$$

Since $q^{(i)} \in K_{i+1}(\delta^{(0)})$, we also have that

$$q^{(i)} = c_{i,0}\delta^{(0)} + c_{i,1}(I - G)\delta^{(0)} + \cdots + c_{i,i-1}(I - G)^{i-1}\delta^{(0)} + (I - G)^i\delta^{(0)}$$

$$(12\text{-}3.10)$$

and that

$$(I - G)^i\delta^{(0)} = e_{i,0}q^{(0)} + e_{i,1}q^{(1)} + \cdots + e_{i,i-1}q^{(i-1)} + q^{(i)} \quad (12\text{-}3.11)$$

for some coefficients $c_{i,0}, c_{i,1}, \ldots, c_{i,i-1}, e_{i,0}, e_{i,1}, \ldots, e_{i,i-1}$.

In the symmetrizable case, where Z and $Z(I - G)$ are SPD, it can be shown that $\beta_{n,i} = 0$ for $i < n - 2$; see, for instance, Young et al. [1980].

If we now require that $u^{(n)} - u^{(0)} \in K_n(\delta^{(0)})$ and that $\|u^{(n)} - \bar{u}\|_{(Z(I-G))^{1/2}}$ $\leq \|w - \bar{u}\|_{(Z(I-G))^{1/2}}$ for all w such that $w - u^{(0)} \in K_n(\delta^{(0)})$ we obtain, after some calculation, that

$$\begin{aligned} u^{(n+1)} &= u^{(0)} + \hat{\lambda}_0 q^{(0)} + \hat{\lambda}_1 q^{(1)} + \cdots + \hat{\lambda}_n q^{(n)} \\ &= u^{(n)} + \hat{\lambda}_n q^{(n)}, \end{aligned} \quad (12\text{-}3.12)$$

where

$$\hat{\lambda}_i = \frac{(Z\delta^{(i)}, q^{(i)})}{(Z(I - G)q^{(i)}, q^{(i)})}, \qquad i = 0, 1, 2, \ldots \quad (12\text{-}3.13)$$

$$\delta^{(i)} = k - (I - G)u^{(i)}. \quad (12\text{-}3.14)$$

We refer to the procedure defined by (12-3.7), (12-3.8), (12-3.12), and (12-3.13) as "ORTHODIR" or sometimes as "ORTHODIR(∞)."

It can be shown (Young and Jea [1980a, 1980b]) that the true solution $\bar{u}$ of (12-1.2) can be written in the form

$$\bar{u} = u^{(0)} + \hat{\lambda}_0 q^{(0)} + \hat{\lambda}_1 q^{(1)} + \cdots + \hat{\lambda}_{d-1} q^{(d-1)}, \quad (12\text{-}3.15)$$

where d is the smallest integer such that the vectors $\delta^{(0)}, (I - G)\delta^{(0)}, \ldots,$ $(I - G)^d\delta^{(0)}$ are linearly dependent. (We assume that $\delta^{(0)} \neq 0$ so that $d \geq 0$.) Evidently, $0 \leq d \leq N$.

From (12-3.12) and (12-3.15) it follows that, in the absence of rounding errors, ORTHODIR converges in d steps. It can also be shown (Eisenstat et al. [1979c]) that

$$(Z\delta^{(i)}, q^{(j)}) = 0, \qquad j < i, \quad i, j = 0, 1, \ldots, d - 1, \quad (12\text{-}3.16)$$

$$(Z\delta^{(i)}, \delta^{(j)}) = 0, \qquad i \neq j, \quad i, j = 0, 1, \ldots, d - 1. \quad (12\text{-}3.17)$$

If, in addition to the assumption that $Z(I - G)$ is SPD, we also assume that Z is positive real† (PR) then we can transform the formulas for ORTHODIR

---

† A matrix Z is PR if $Z + Z^T$ is SPD.

into a form which more closely resembles the two-term form of CG accelera-tion. To do this we introduce the vectors $p^{(0)}$, $p^{(1)}$, ... defined by

$$p^{(0)} = q^{(0)},$$

$$p^{(n)} = -\hat{\lambda}_{n-1} q^{(n)}, \qquad n = 1, 2, \ldots, d-1, \qquad (12\text{-}3.18)$$

where $\hat{\lambda}_{n-1}$ and $q^{(n)}$ are defined as in (12-3.7) and (12-3.13). It can be shown (Young and Jea [1980a, 1980b]) that since Z is PR the $\{\hat{\lambda}_i\}$ do not vanish, and hence the $\{p^{(i)}\}$ are linearly independent. It is also easy to show that the $\{p^{(i)}\}$ can alternatively be given by

$$p^{(n)} = \begin{cases} \delta^{(0)}, & n = 0, \\ \delta^{(n)} + \alpha_{n,n-1}p^{(n-1)} + \cdots + \alpha_{n,0}p^{(0)}, & n = 1, 2, \ldots, d-1, \end{cases}$$

$$(12\text{-}3.19)$$

where

$$\alpha_{n,i} = -\frac{(Z(I-G)\delta^{(n)}, p^{(i)})}{(Z(I-G)p^{(i)}, p^{(i)})}, \qquad i = 1, 2, \ldots, n-1, \quad n = 0, 1, \ldots, d-1.$$

$$(12\text{-}3.20)$$

From (12-3.12), (12-3.13), and (12-3.19) we obtain the following method:

$$u^{(n+1)} = u^{(n)} + \lambda_n p^{(n)}, \qquad (12\text{-}3.21)$$

where $p^{(0)}$, $p^{(1)}$, ... are given by (12-3.19) and (12-3.20) and where

$$\lambda_n = \frac{(Z\delta^{(n)}, p^{(n)})}{(Z(I-G)p^{(n)}, p^{(n)})}. \qquad (12\text{-}3.22)$$

We refer to the method defined by (12-3.19)–(12-3.22) as "ORTHOMIN." This method was given by Vinsome [1976]. Evidently, ORTHOMIN closely resembles the two-term form of CG acceleration given in Section 7.4. We remark that in the symmetrizable case $\alpha_{n,i} = 0$ for $i < n-1$ and ORTHO-MIN reduces to (7-4.7). For a proof see, for instance, Young and Jea [1980a, 1980b]. (We remark that $(Z\delta^{(n)}, p^{(n)})$ can be shown to equal $(Z\delta^{(n)}, \delta^{(n)})$ and that $Z = W^T W$.)

Because of the equivalence of ORTHOMIN and ORTHODIR, it follows that (12-3.9) and (12-3.16) hold with $q^{(j)}$ replaced by $p^{(j)}$, and that (12-3.17) holds.

Truncated Procedures

Neither ORTHODIR nor ORTHOMIN in their "idealized" forms described above are computationally feasible in general since the determina-tion of $q^{(n)}$ and $p^{(n)}$ requires the use of information from all previous iterations.

(See (12-3.7) and (12-3.19).) We now consider "truncated" versions of the above schemes. For each nonnegative integer s, we define ORTHODIR(s) and ORTHOMIN(s) as follows. For ORTHODIR(s) we use (12-3.7), (12-3.8), (12-3.12), and (12-3.13) but with

$$\beta_{n,i} = 0 \quad \text{if} \quad i < n - s. \tag{12-3.23}$$

For ORTHOMIN(s) we use (12-3.19)–(12-3.22), but with

$$\alpha_{n,i} = 0 \quad \text{if} \quad i < n - s. \tag{12-3.24}$$

Evidently, the idealized versions of ORTHODIR(s) and ORTHOMIN(s) are ORTHODIR(∞) and ORTHOMIN(∞), respectively. In the symmetrizable case, ORTHODIR(s) for $s \geq 2$ is equivalent to ORTHODIR(∞), while ORTHOMIN(s) for $s \geq 1$ is equivalent to ORTHOMIN(∞).

As stated above, ORTHODIR(∞) is more generally applicable than ORTHOMIN(∞) since for ORTHODIR(∞) we only require that $Z(I - G)$ be SPD, while for ORTHOMIN(∞) we also require that Z be PR. If Z is PR, then both methods are equivalent. However, this is not necessarily the case for ORTHODIR(s) and ORTHOMIN(s) for finite s. For these procedures we have no convergence theorems. We can, however, say something about "breakdown." We say that breakdown occurs for ORTHODIR(s) if $\delta^{(n)} \neq 0$ but $q^{(n)} = 0$ and that it occurs for ORTHOMIN if $\delta^{(n)} \neq 0$ but $p^{(n)} = 0$. It can be shown (Young and Jea [1980a, 1980b]) that breakdown cannot occur for ORTHODIR(s) for $n < d$, where d is defined in (12-3.15). It is assumed that $Z(I - G)$ is SPD but not that Z is PR. We have no proof that breakdown cannot occur for $n \geq d$ even if Z is PR. On the other hand, for ORTHOMIN(s), breakdown cannot occur provided that $Z(I - G)$ is SPD and Z is PR. This follows from results of Eisenstat et al., [1979c]; see also Young and Jea [1980a, 1980b].

So far there are relatively few theoretical results available concerning the behavior of the generalized CG-acceleration schemes defined above. However, numerical results reported by Eisenstat et al. [1979c] and by Eisenstat et al. [1979a] are encouraging. Encouraging numerical results have also been obtained by Young and Jea [1980a].

ORTHORES

We now seek to obtain a procedure which more closely resembles the three-term form of CG acceleration. To do this, we first observe that if $Z(I - G)$ is SPD and if (12-3.4) holds, then the condition that

$$\|u^{(n)} - \bar{u}\|_{(Z(I-G))^{1/2}} \leq \|w - \bar{u}\|_{(Z(I-G))^{1/2}}$$

for all w such that $w - u^{(0)} \in K_n(\delta^{(0)})$ is equivalent to the condition that

$$(Z\delta^{(n)}, v) = 0 \qquad (12\text{-}3.25)$$

for all $v \in K_n(\delta^{(0)})$. We now can weaken the condition that $Z(I - G)$ be SPD and only require that $Z(I - G)$ be PR. Using (12-3.25), we can derive a generalized version of ORTHODIR by choosing an ordered set of vectors $q^{(0)}, q^{(1)}, \ldots$ which are pairwise "semiorthogonal" with respect to $Z(I - G)$ in the sense that

$$(Z(I - G)q^{(i)}, q^{(j)}) = 0, \qquad j < i. \qquad (12\text{-}3.26)$$

To do this we use the procedure defined by (12-3.7) but with

$$\beta_{n,i} = -\frac{(Z(I - G)^2 q^{(n-1)}, q^{(i)}) + \sum_{j=0}^{i-1} \beta_{n,j}(Z(I - G)q^{(j)}, q^{(i)})}{(Z(I - G)q^{(i)}, q^{(i)})},$$

$$i = 0, 1, \ldots, n - 1. \quad (12\text{-}3.27)$$

The generalized ORTHODIR procedure is given by (12-3.7), (12-3.12), (12-3.13), and (12-3.27).

If we now assume that Z as well as $Z(I - G)$ is PR and if we define $\{p^{(i)}\}$ by (12-3.18), we can derive a generalized form of ORTHOMIN. Thus we obtain (12-3.19), (12-3.21), (12-3.22), and

$$\alpha_{n,i} = -\frac{(Z(I - G)\delta^{(n)}, p^{(i)}) + \sum_{j=0}^{i-1} \alpha_{n,j}(Z(I - G)p^{(j)}, p^{(i)})}{(Z(I - G)p^{(i)}, p^{(i)})},$$

$$i = 0, 1, \ldots, n - 1. \quad (12\text{-}3.28)$$

Let us now assume that Z is SPD. For the generalized ORTHODIR and ORTHOMIN procedures it can be shown that the $\{\delta^{(i)}\}$ are pairwise $Z^{1/2}$-orthogonal, i.e.,

$$(Z\delta^{(i)}, \delta^{(j)}) = 0, \qquad i \neq j. \qquad (12\text{-}3.29)$$

By (12-3.21) and (12-3.19) we have

$$u^{(n+1)} = u^{(n)} + \lambda_n p^{(n)}$$

$$= u^{(n)} + \lambda_n(\delta^{(n)} + \alpha_{n,n-1} p^{(n-1)} + \cdots + \alpha_{n,0} p^{(0)})$$

$$= u^{(n)} + \lambda_n \delta^{(n)} + \frac{\lambda_n}{\lambda_{n-1}} \alpha_{n,n-1}(u^{(n)} - u^{(n-1)}) + \cdots$$

$$+ \frac{\lambda_n}{\lambda_0} \alpha_{n,0}(u^{(1)} - u^{(0)}). \qquad (12\text{-}3.30)$$

Thus we have

$$u^{(n+1)} = \gamma_{n+1} f_{n+1,n} \delta^{(n)} + f_{n+1,n} u^{(n)} + \cdots + f_{n+1,0} u^{(0)}, \quad (12\text{-}3.31)$$

where $\gamma_{n+1} f_{n+1,n} = \lambda_n$ and

$$f_{n+1,1} + f_{n+1,2} + \cdots + f_{n+1,n} = 1. \quad (12\text{-}3.32)$$

Using (12-3.32) and (12-3.29), with $i = n + 1$, we obtain, by (12-3.31) that
$\delta^{(n+1)} = -\gamma_{n+1} f_{n+1,n}(I - G)\delta^{(n)} + f_{n+1,n}\delta^{(n)} + \cdots + f_{n+1,0}\delta^{(0)}$, where

$$\gamma_{n+1} = \sigma_{n+1,n}^{-1},$$

$$f_{n+1,n} = \left(1 + \gamma_{n+1} \sum_{i=0}^{n-1} \sigma_{n+1,i}\right)^{-1},$$

$$\sigma_{n+1,i} = \frac{(Z(I - G)\delta^{(n)}, \delta^{(i)})}{(Z\delta^{(i)}, \delta^{(i)})}, \quad i = 0, 1, \ldots, n, \quad (12\text{-}3.33)$$

$$f_{n+1,i} = \gamma_{n+1} f_{n+1,n} \sigma_{n+1,i} \quad i = 0, 1, \ldots, n - 1.$$

We refer to the procedure defined by (12-3.31) and (12-3.33) as "ORTHORES," or "ORTHORES(∞)." We now consider the truncated procedure ORTHORES(s) where we choose a nonnegative integer s and we let

$$\sigma_{n+1,i} = 0 \quad \text{for} \quad i < n - s, \quad (12\text{-}3.34)$$

In the symmetrizable case, where Z and $Z(I - G)$ are SPD, $\sigma_{n+1,i} = 0$ for $i < n - 1$ for ORTHORES(∞); hence ORTHORES(s) for $s \geq 1$ reduces to the three-term CG-acceleration scheme of Section 7.4. (Here, $\rho_{n+1} = f_{n+1,n}$.)

An important special case is ORTHORES(1). If we let $f_{n+1,n} = \rho_{n+1}$, then $f_{n+1,n-1} = 1 - \rho_{n+1}$ and we obtain

$$u^{(n+1)} = \rho_{n+1}[\gamma_{n+1}\delta^{(n)} + u^{(n)}] + (1 - \rho_{n+1})u^{(n-1)},$$

$$\gamma_{n+1} = \frac{(Z\delta^{(n)}, \delta^{(n)})}{(Z(I - G)\delta^{(n)}, \delta^{(n)})}, \quad (12\text{-}3.35)$$

$$\rho_{n+1} = \left[1 + \gamma_{n+1} \frac{(Z(I - G)\delta^{(n)}, \delta^{(n-1)})}{(Z\delta^{(n-1)}, \delta^{(n-1)})}\right]^{-1}, \quad n \geq 1 \quad (\rho_1 = 1).$$

In the symmetrizable case one could transform (12-3.35) into the alternative form (7-3.1)–(7-3.3). However, the two forms are not, in general, equivalent in the nonsymmetrizable case. Preliminary numerical experiments indicate that (12-3.35) is more effective than (7-3.1)–(7-3.3). These experiments also indicate that ORTHORES(s) is quite promising in spite of the fact that, except for the case $s = \infty$, we have no theoretical results concerning convergence or breakdown.

Choice of Auxiliary Matrix

Let us now discuss the choice of the auxiliary matrix Z corresponding to a given coefficient matrix A and to a given splitting matrix Q. In order to apply ORTHODIR or ORTHOMIN, we desire that $Z(I - G)$ should be SPD. But, as we have seen above, this implies that $Z = (I - G)^{1} Y$ for some SPD matrix Y. In order to apply ORTHORES, we desire that Z should be SPD. Thus $Z = Y$ for some SPD matrix Y. Thus for all three methods, we choose an SPD matrix Y and we use $Z = (I - G)^T Y$ for ORTHODIR and OR-THORES and $Z = Y$ for ORTHORES.

Let us assume we are dealing with a family $\mathscr{F}$ of problems each involving a coefficient matrix A and that for the basic iterative method there is a splitting matrix $Q(A)$ corresponding to each $A \in \mathscr{F}$ such that $Q(A)$ is a continuous function of A. We also assume that there is a matrix $\bar{A} \in \mathscr{F}$ such that the basic iterative method corresponding to $Q(\bar{A})$ is symmetrizable, i.e., there exists an SPD matrix H such that $H(I - G)$ is SPD, where $G = I - [Q(\bar{A})]^{-1}\bar{A}$.

As an example, one can consider the RF method or the SSOR method for the family of problems discussed in Section 12.6. The matrix $\bar{A}$ corresponds to the case $\beta = 0$ and is SPD. For the RF method $Q(\bar{A}) = I$ and $H = I$. For the SSOR method $Q(\bar{A})$ is SPD and $H = Q(\bar{A})$.

Our procedure is to choose, for each $A \in \mathscr{F}$, an SPD matrix $Y(A)$ such that $Y(A)$ depends continuously on A and such that $Y(\bar{A})Q(\bar{A})^{-1}\bar{A}$ is SPD. We indicate below several possible choices of Y.

(a)   If $Q(A)$ is PR and if $\bar{A}$ and $Q(\bar{A})$ are SPD, we can let

$$Y = \tfrac{1}{2}(Q + Q^T). \tag{12-3.36}$$

(b)   If A is PR and if $\bar{A}$ and $Q(\bar{A})$ are SPD, we can let

$$Y = \tfrac{1}{2}(A + A^T). \tag{12-3.37}$$

(c)   If $\bar{A}$ and $Q(\bar{A})$ are SPD, we can let

$$Y = Q_U^T |Q_D| Q_U, \tag{12-3.38}$$

where Q_D and Q_U are such that for some permutation matrix P we have $PQ(A) = Q_L Q_D Q_U$. Here Q_U is unit upper triangular, Q_D is diagonal, and Q_L is unit lower triangular. We let $|Q_D|$ be the matrix whose elements are the absolute values of Q_D. The above representation is valid since $Q(A)$ is nonsingular (see, e.g. Strang [1976]). Clearly, $Y = Y(A)$ is a continuous function of A.

(d)   In the general case we can let

$$Y = H. \tag{12-3.39}$$

Having chosen $Y = Y(A)$, we can apply ORTHODIR(s) or ORTHO-MIN(s) with $Z = Z(A) = (I - G(A))^{\mathrm{T}} Y(A)$ or we can apply ORTHORES(s) with $Z = Y(A)$. In the former case $Z(I - G)$ is SPD, while in the latter case Z is SPD. We now show that for A sufficiently close to $\bar{A}$ the condition that $Z(A)$ is PR holds for ORTHODIR(s) and ORTHOMIN(s) and the condition that $Z(A)(I - G(A))$ is PR holds for ORTHORES(s). Thus we seek to show that $L(A) \equiv Y(A)(I - G(A))$ is PR for A sufficiently close to $\bar{A}$. But since $Y(\bar{A})$ $(I - G(\bar{A})) = Y(\bar{A})Q(\bar{A})^{-1}\bar{A}$ is SPD, it follows that the eigenvalues of $L(\bar{A})$ are positive. Hence the eigenvalues of $K(\bar{A})$ are positive where, for any A, we let $K(A) = \frac{1}{2}(L(A) + L(A)^{\mathrm{T}})$. Hence for A sufficiently close to $\bar{A}$ the eigenvalues of $K(A)$ are positive and $L(A)$ is positive real.

From the above discussion one would expect that, for A sufficiently close to $\bar{A}$, ORTHODIR(s) and ORTHOMIN(s) would behave much like CG acceleration with $Z = (I - G(A))^{\mathrm{T}} Y(A)$. One would also expect that ORTHORES(s) would behave much like CG acceleration with $Z = Y(A)$. This analysis, however, says nothing about the case where A is not necessarily close to $\bar{A}$. If one could be sure that $Y(A)(I - G(A))$ were PR, then one could be sure that ORTHOMIN would not break down. In some cases this is possible. An example is the case where A is PR for all $A \in \mathscr{F}$ and where $Q(A)$ is SPD for all A. (See the example of Section 12.6.) A special case is the RF method where $Q(A) = I$. Here we can let $Y(A) = Q(A)$. Evidently, $Y(A)$ $(I - G(A)) = Y(A)Q(A)^{-1}A = A$ which is PR. Thus ORTHOMIN(s) will not break down for all s. Additional examples are given in Young and Jea [1980a, 1980b].

As stated earlier, numerical experiments carried out to date indicate that the methods are promising. However, a great deal of additional work, both theoretical and experimental, is needed. Based on the discussion in this section, it would seem that ORTHOMIN(s) should be used with a small value of s provided that one can choose $Y(A)$ such that $Y(A)(I - G(A))$ is PR.

Another possibility would be to use the idealized form of ORTHODIR or, if $Y(A)(I - G(A))$ is PR, the idealized forms of ORTHOMIN or ORTHORES and then "restart" every few iterations. This procedure would not break down and would not require an excessive amount of storage. Such a procedure for ORTHOMIN is discussed by Eisenstat et al. [1979c].

12.4   LANCZOS ACCELERATION

In this section we describe another procedure which appears very promising for accelerating basic iterative methods in the nonsymmetrizable case. The procedure is based on a method of Lanczos [1950, 1952].

We first describe the basic Lanczos procedure, which can be regarded as an acceleration process for the RF method. Instead of considering only the solution of the system

$$Au = b, \tag{12-4.1}$$

we also consider an auxiliary system

$$A^T\tilde{u} = \tilde{b}. \tag{12-4.2}$$

The choice of $\tilde{b}$ will be discussed later.

In Section 7.3 we presented a derivation of the three-term form of CG method for the case where A is SPD. We considered the procedure defined by

$$u^{(n+1)} = \rho_{n+1}\gamma_{n+1}r^{(n)} + \rho_{n+1}u^{(n)} + (1 - \rho_{n+1})u^{(n-1)}, \tag{12-4.3}$$

where $r^{(n)}$ is the residual vector

$$r^{(n)} \equiv b - Au^{(n)}. \tag{12-4.4}$$

The coefficients ρ_{n+1} and γ_{n+1} were chosen so that $(r^{(n+1)}, r^{(n)}) = (r^{(n+1)}, r^{(n-1)})$ $= 0$. If A is SPD, the method thus defined is the CG method. In particular, $(r^{(i)}, r^{(j)}) = 0$ for $i \neq j$.

For the Lanczos method we consider the procedure defined by (12-4.3) and also by

$$\tilde{u}^{(n+1)} = \rho_{n+1}\gamma_{n+1}R^{(n)} + \rho_{n+1}\tilde{u}^{(n)} + (1 - \rho_{n+1})\tilde{u}^{(n-1)}, \tag{12-4.5}$$

where

$$R^{(n)} = \tilde{b} - A^T\tilde{u}^{(n)} \tag{12-4.6}$$

is the residual vector for the auxiliary system. We now seek to choose the ρ_{n+1} and γ_{n+1} so that the sets $r^{(0)}, r^{(1)}, \ldots$ and $R^{(0)}, R^{(1)}, \ldots$ are *biorthogonal* in the sense that for $i \neq j$ we have

$$(r^{(i)}, R^{(j)}) = 0. \tag{12-4.7}$$

By an extension of the procedure used in Section 7.3 and by requiring that $(r^{(n+1)}, R^{(n)}) = (r^{(n+1)}, R^{(n-1)}) = 0$, we obtain

$$\gamma_{n+1} = \frac{(r^{(n)}, R^{(n)})}{(Ar^{(n)}, R^{(n)})}, \tag{12-4.8}$$

$$\rho_{n+1} = \left[1 - \frac{\gamma_{n+1}}{\gamma_n}\frac{(r^{(n)}, R^{(n)})}{(r^{(n-1)}, R^{(n-1)})}\frac{1}{\rho_n}\right]^{-1}, \quad \text{if } n \geq 1 \quad (\rho_1 = 1).$$

$$\tag{12-4.9}$$

It can then be shown that the biorthogonality condition (12-4.7) holds for all i and j.

To summarize, the Lanczos procedure is defined by

$u^{(0)}$ and $R^{(0)}$ are arbitrary,

$$r^{(0)} = b - Au^{(0)},$$

$$u^{(n+1)} = \rho_{n+1}\gamma_{n+1}r^{(n)} + \rho_{n+1}u^{(n)} + (1 - \rho_{n+1})u^{(n-1)},$$

$$r^{(n+1)} = -\rho_{n+1}\gamma_{n+1}Ar^{(n)} + \rho_{n+1}r^{(n)} + (1 - \rho_{n+1})r^{(n-1)},$$

$$R^{(n+1)} = -\rho_{n+1}\gamma_{n+1}A^{\mathsf{T}}R^{(n)} + \rho_{n+1}R^{(n)} + (1 - \rho_{n+1})R^{(n-1)},$$

$$\gamma_{n+1} = \frac{(r^{(n)}, R^{(n)})}{(Ar^{(n)}, R^{(n)})},$$

$$\rho_{n+1} = \left[1 - \frac{\gamma_{n+1}}{\gamma_n}\frac{(r^{(n)}, R^{(n)})}{(r^{(n-1)}, R^{(n-1)})}\frac{1}{\rho_n}\right]^{-1}, \qquad \text{if} \quad n \geq 1 \qquad (\rho_1 = 1).$$

$$(12\text{-}4.10)$$

We remark that rather than choosing $\tilde{u}^{(0)}$, we simply let $R^{(0)} = \tilde{b} - A^{\mathsf{T}}\tilde{u}^{(0)}$. This also avoids the need to choose $\tilde{b}$.

Evidently, if A is SPD and if $R^{(0)} = r^{(0)}$, the above procedure reduces to the CG method.

In the nonsymmetric case, the Lanczos process may fail if at some stage $r^{(n)} \neq 0$ but $(r^{(n)}, R^{(n)}) = 0$. This can happen if one makes an unfortunate choice of $R^{(0)}$ or if the matrix A has an eigenvector deficiency (see Section 2.3). On the other hand, if none of the numbers $(r^{(0)}, R^{(0)})$, $(r^{(1)}, R^{(1)})$, ... vanishes, then the process will converge in at most N iterations. This can be proved by showing that the vectors $R^{(0)}$, $R^{(1)}$, ... are linearly independent and that the condition $(r^{(N)}, R^{(i)}) = 0$, $i = 0, 1, \ldots, N - 1$ implies that $r^{(N)} = 0$.

Let us now consider the use of the Lanczos method to accelerate the convergence of a basic iterative method of the form

$$u^{(n+1)} = Gu^{(n)} + k, \qquad (12\text{-}4.11)$$

where

$$G = I - Q^{-1}A, \qquad k = Q^{-1}b. \qquad (12\text{-}4.12)$$

We consider the related equation corresponding to (12-4.11), namely,

$$(I - G)u = k. \qquad (12\text{-}4.13)$$

This can be regarded as a preconditioning of the system (12-4.1).

The residuals of (12-4.13) are the pseudoresiduals $\delta^{(n)} = Gu^{(n)} + k - u^{(n)}$. We let $\Delta^{(n)}$ denote the residuals of the corresponding auxiliary system $(I - G)^{\mathsf{T}}\tilde{u} = \tilde{k}$. If we replace A by $I - G$, A^{T} by $I - G^{\mathsf{T}}$, $r^{(n)}$ by $\delta^{(n)}$, and $R^{(n)}$ by $\Delta^{(n)}$ in (12-4.10), we obtain the more general procedure:

$u^{(0)}$ and $\Delta^{(0)}$ are arbitrary,

$$\delta^{(0)} = Gu^{(0)} + k - u^{(0)},$$

$$u^{(n+1)} = \rho_{n+1}\gamma_{n+1}\delta^{(n)} + \rho_{n+1}u^{(n)} + (1 - \rho_{n+1})u^{(n-1)},$$

$$\delta^{(n+1)} = -\rho_{n+1}\gamma_{n+1}(I - G)\delta^{(n)} + \rho_{n+1}\delta^{(n)} + (1 - \rho_{n+1})\delta^{(n-1)},$$

$$\Delta^{(n+1)} = -\rho_{n+1}\gamma_{n+1}(I - G^{T})\Delta^{(n)} + \rho_{n+1}\Delta^{(n)} + (1 - \rho_{n+1})\Delta^{(n-1)}, \quad (12\text{-}4.14)$$

$$\gamma_{n+1} = \frac{(\delta^{(n)}, \Delta^{(n)})}{((I - G)\delta^{(n)}, \Delta^{(n)})},$$

$$\rho_{n+1} = \left[1 - \frac{\gamma_{n+1}}{\gamma_{n}} \frac{(\delta^{(n)}, \Delta^{(n)})}{(\delta^{(n-1)}, \Delta^{(n-1)})} \frac{1}{\rho_{n}}\right]^{-1} \quad \text{if} \quad n \geq 1 \quad (\rho_{1} = 1).$$

Suppose now that (12-4.11) is symmetrizable and that Z is any SPD matrix such that $Z(I - G)$ is SPD. If we let $\Delta^{(0)} = Z\delta^{(0)}$, then the above procedure reduces to CG acceleration.

The choices of Z which are recommended for Lanczos acceleration are the same as for the ORTHOMIN and ORTHORES procedures described in Section 12.3.

Preliminary numerical experiments based on the Lanczos-acceleration procedure defined above have been very encouraging. However, there is a need for many more experiments and for a complete theoretical analysis of the method.

Fletcher [1976] introduced a method based on Lanczos' method which resembles the two-term form of the CG method. This method is referred to as the "bi-CG algorithm." Wong [1980] describes some numerical experiments together with a preconditioning technique involving "row-sums agreement factorization." The method was very effective for the class of problems considered.

12.5 ACCELERATION PROCEDURES FOR THE GCW METHOD

Let us now assume that the matrix A of (12-1.2) is "positive real" in the sense that $A + A^{T}$ is SPD. We consider the basic iterative method defined by

$$u^{(n+1)} = Gu^{(n)} + k, \quad (12\text{-}5.1)$$

where

$$G = I - Q^{-1}A = Q^{-1}R, \quad (12\text{-}5.2)$$

$$k = Q^{-1}b, \quad (12\text{-}5.3)$$

$$Q = \tfrac{1}{2}(A + A^{T}), \quad (12\text{-}5.4)$$

$$R = Q - A. \quad (12\text{-}5.5)$$

We will refer to this method as the "GCW method." Generalized CG-acceleration procedures for the GCW method were developed by Concus and Golub
[1976a] and by Widlund [1978].

In order to carry out a single iteration of the GCW method, it is necessary
to solve an auxiliary linear system whose matrix is the SPD matrix Q. This
can often be done by direct methods. Concus and Golub [1976a] and Widlund [1978] used fast direct methods for problems arising from elliptic partial
differential equations. One could also solve the auxiliary linear system by an
iterative method. Since Q is SPD, one can solve the auxiliary system iteratively,
for example, using one of the procedures discussed in previous chapters. We
note that good initial approximations to the auxiliary linear systems would be
available, especially at later stages of the overall procedure.

An important property of the GCW method is that the eigenvalues of the
iteration matrix G are purely imaginary. To show this, we first note that the
eigenvalues of G are the same as those of $\tilde{G}$, where

$$\tilde{G} = Q^{1/2}GQ^{-1/2} = Q^{-1/2}RQ^{-1/2}, \tag{12-5.6}$$

Moreover, since $R^{T} = -R$, we have

$$\begin{aligned}
\tilde{G}^2 &= Q^{-1/2}RQ^{-1/2}Q^{-1/2}RQ^{-1/2} \\
&= -Q^{-1/2}RQ^{-1/2}Q^{-1/2}R^{T}Q^{-1/2} \\
&= -[Q^{-1/2}RQ^{-1/2}][Q^{-1/2}RQ^{-1/2}]^{T}.
\end{aligned} \tag{12-5.7}$$

Thus $-\tilde{G}^2$ is symmetric and positive semidefinite. From this, it follows that
the eigenvalues of $\tilde{G}^2$ are nonpositive and hence that the eigenvalues of $\tilde{G}$
and those of G are purely imaginary.

Both CG and Chebyshev acceleration can effectively be applied to the
GCW method. For Chebyshev acceleration, we can use either the method
(12-2.17) or (12-2.22) with

$$b = S(G). \tag{12-5.8}$$

The asymptotic average rate of convergence is given by (12-2.19).

Concus and Golub [1976a] and Widlund [1978] developed a generalized
CG procedure based on the GCW method. Their method is defined by

$$u^{(n+1)} = \rho_{n+1}(Gu^{(n)} + k) + (1 - \rho_{n+1})u^{(n-1)}, \tag{12-5.9}$$

where G, k, Q, and R are given by (12-5.2)–(12-5.5) and where

$$\delta^{(n)} = Gu^{(n)} + k - u^{(n)}, \tag{12-5.10}$$

$$\rho_{n+1} = \left[1 + \frac{(Q^{1/2}\delta^{(n)}, Q^{1/2}\delta^{(n)})}{(Q^{1/2}\delta^{(n-1)}, Q^{1/2}\delta^{(n-1)})}\frac{1}{\rho_n}\right]^{-1} \quad \text{if} \quad n \geq 1 \quad (\rho_1 = 1). \tag{12-5.11}$$

We refer to the above method as GCW–GCG *method*.

It can be verified that the pseudoresiduals $\delta^{(0)}, \delta^{(1)}, \ldots$ are pairwise $Q^{1/2}$-orthogonal in the sense that

$$(Q^{1/2}\delta^{(i)}, Q^{1/2}\delta^{(j)}) = 0, \qquad i \neq j. \tag{12-5.12}$$

Thus the method is guaranteed to converge in at most N iterations.

We now give a derivation of (12-5.9)–(12-5.11) by using Lanczos acceleration. We simply let

$$\Delta^{(0)} = Q\delta^{(0)}. \tag{12-5.13}$$

We show by induction that

$$\Delta^{(n)} = (-1)^n Q\delta^{(n)}. \tag{12-5.14}$$

But by (12-4.14) we have

$$\gamma_{n+1} = \frac{(\delta^{(n)}, \Delta^{(n)})}{((I - G)\delta^{(n)}, \Delta^{(n)})}, \tag{12-5.15}$$

which equals unity if $(G\delta^{(n)}, \Delta^{(n)}) = 0$. But

$$(G\delta^{(n)}, \Delta^{(n)}) = (Q^{-1}R\delta^{(n)}, Q(-1)^n\delta^{(n)}) = (-1)^n(R\delta^{(n)}, \delta^{(n)}) = 0 \tag{12-5.16}$$

since R is skew symmetric. Moreover, since $\gamma_{n+1} = 1$, we have by (12-4.14) that

$$\begin{aligned}
\Delta^{(n+1)} &= \rho_{n+1}(-1)^n(Q\delta^{(n)} - \gamma_{n+1}(I - G^T)Q\delta^{(n)}) \\
&\quad + (1 - \rho_{n+1})(-1)^{n-1}Q\delta^{(n-1)}) \\
&= \rho_{n+1}(-1)^n(G^TQ\delta^{(n)}) + (1 - \rho_{n+1})(-1)^{n-1}Q\delta^{(n-1)} \\
&= \rho_{n+1}(-1)^n(R^T\delta^{(n)}) + (1 - \rho_{n+1})(-1)^{n-1}Q\delta^{(n-1)} \\
&= (-1)^{n+1}\{\rho_{n+1}R\delta^{(n)} + (1 - \rho_{n+1})Q\delta^{(n-1)}\} \\
&= (-1)^{n+1}Q\{\rho_{n+1}G\delta^{(n)} + (1 - \delta_{n+1})\delta^{(n-1)}\} = (-1)^{n+1}Q\delta^{(n+1)},
\end{aligned} \tag{12-5.17}$$

Thus (12-5.14) follows.

It can be shown (see Hageman et al. [1977]) that the GCW–GCG method is equivalent to the regular CG procedure of Section 7.4 applied to the "double" GCW method

$$u^{(n+1)} = G^2u^{(n)} + Gk + k. \tag{12-5.18}$$

This is true provided the symmetrization matrix $W_D = Q^{1/2}(I + G)^{-1}$ is used as the symmetrization matrix for the double method. The equivalence holds in the sense that if $v^{(0)} = u^{(0)}, v^{(1)}, v^{(2)}, \ldots$ are obtained by the accelerated double method and $u^{(0)}, u^{(1)}, u^{(2)}, \ldots$ are obtained by the GCW–GCG method, then

$$v^{(n)} = u^{(2n)}. \tag{12-5.19}$$

From this it can be shown† that

$$\|u^{(2n)} - \bar{u}\|_{Q^{1/2}} \le [2\bar{r}^{n/2}/(1 + \bar{r}^n)]\|u^{(0)} - \bar{u}\|_{Q^{1/2}}, \qquad (12\text{-}5.20)$$

where

$$\bar{r} = \frac{1 - \sqrt{1 - \{S(G)^2/[2 + S(G)^2]\}^2}}{1 + \sqrt{1 - \{S(G)^2/[2 + S(G)^2]\}^2}} = \left(\frac{\sqrt{1 + S(G)^2} - 1}{\sqrt{1 + S(G)^2} + 1}\right)^2. \qquad (12\text{-}5.21)$$

It should be noted that Widlund [1978] has reported that in some cases the sequence of odd iterants $u^{(1)}, u^{(3)}, u^{(5)}, \ldots$ for the GCW–GCG method converges much more rapidly than the sequence of even iterants $u^{(2)}, u^{(4)}, \ldots$.

12.6   AN EXAMPLE

Let us consider the linear system obtained by using the standard five-point finite difference discretization of the differential equation

$$u_{xx} + u_{yy} + \beta u_x = 0. \qquad (12\text{-}6.1)$$

The region considered is the unit square $0 \le x \le 1$, $0 \le y \le 1$. Dirichlet boundary conditions are assumed. The difference equation used is given by

$$h^{-2}\{u(x + h, y) + u(x - h, y) + u(x, y + h) + u(x, y - h) - 4u(x, y)\}$$
$$+ \tfrac{1}{2}\beta h^{-1}(u(x + h, y) - u(x - h, y)) = 0. \qquad (12\text{-}6.2)$$

It can be shown (see, e.g., Young and Jea [1980a]) that the eigenvalues of the RF method are given by

$$\lambda_{p,q} = \begin{cases} \tfrac{1}{2} \cos p\pi h + \tfrac{1}{2} \cos q\pi h \sqrt{1 - (\tfrac{1}{2}h\beta)^2}, & \text{if} \quad \tfrac{1}{2}h|\beta| \le 1, \\ \tfrac{1}{2} \cos p\pi h + [\tfrac{1}{2} \cos q\pi h \sqrt{(\tfrac{1}{2}\beta h)^2 - 1}]i, & \text{if} \quad \tfrac{1}{2}h|\beta| \ge 1, \end{cases} \qquad (12\text{-}6.3)$$

Here $p, q = 1, 2, \ldots, h^{-1} - 1$. It can also be shown that for the GCW method‡

$$\mu_{p,q} = \tfrac{1}{2}h\beta \, \frac{i}{\sqrt{[(2 - \cos q\pi h)/\cos p\pi h]^2 - 1}}, \qquad p, q = 1, 2, \ldots, h^{-1} - 1. \qquad (12\text{-}6.4)$$

† We compare the CG-accelerated error $v^{(n)} - \bar{u}$ with the error vector associated with Chebyshev acceleration applied to (12-5.18) with $M(G^2) = 0$ and $m(G^2) = -S(G)^2$. We also used the fact that $W_D^T W_D (I - G^2) = Q$.

‡ Widlund [1978] gave asymptotic values of the eigenvalues of the GCW method. These values were undoubtedly obtained from formulas similar to (12-6.4).

Thus we have

$$S(G) = |\mu_{1,1}| = \frac{h|\beta|}{4\sqrt{2}} \frac{\cos \pi h}{\sin(\pi h/2)}$$

$$\sim \frac{|\beta|}{2\sqrt{2\pi}}, \qquad h \to 0. \tag{12-6.5}$$

Let us apply the RF method with Chebyshev acceleration (RF–SI method) for the case $\beta = -100$, $h = 10^{-1}$. In this case, the eigenvalues μ of the RF method lie in the rectangle

$$|\mathrm{Re}\,\mu| \le \cos \pi h = 0.47553 = \bar{a},$$
$$|\mathrm{Im}\,\mu| \le \tfrac{1}{2}\sqrt{24} \cos \pi h = 2.32961 = \bar{b}. \tag{12-6.6}$$

Solving the cubic equation (12-2.29) for ρ, we obtain

$$\rho = 0.90141 \tag{12-6.7}$$

and, by (12-2.27), (12-2.28), and (12-2.12), with $d = 0$,

$$a = 0.6081, \qquad b = 3.7372, \qquad \sigma^2 = 13.5969. \tag{12-6.8}$$

Thus by (12-2.31), the asymptotic average rate of convergence of the RF–SI method is

$$\bar{R}_{\mathscr{D},\infty}(P_n(G)) = -\log \rho = 0.10380. \tag{12-6.9}$$

Based on (12-6.9), the number of iterations required for convergence with $\zeta = 10^{-6}$ is approximately 133. The corresponding numerical experiment for this case required 109 iterations for convergence. For the RF method with Lanczos acceleration with $Z = I$, 57 iterations were required. With ORTHO-MIN(s) for the RF method with $Z = A^T$, the following results were obtained:

s	0	1	2	∞
No. of iterations	185	133	108	55

For the GCW method, the eigenvalues of the iteration matrix are purely imaginary and lie in the interval $-iS(G)$ to $iS(G)$, where $S(G)$ is given by (12-6.5). We can then apply Chebyshev acceleration to the GCW method (GCW–SI method) as described in Section 12.2 with $b = S(G)$. By (12-2.19), the asymptotic average rate of convergence for small h and for large $S(G)^2$ is approximately

$$R_{\mathscr{D},\infty}(P_n(G)) = -\tfrac{1}{2}\log \frac{\sqrt{S(G)^2 + 1} - 1}{\sqrt{S(G)^2 + 1} + 1} \doteqdot \frac{1}{S(G)} \doteqdot \left(\frac{|\beta|}{2\sqrt{2\pi}}\right)^{-1}, \tag{12-6.10}$$

and the number of iterations needed for convergence with $\zeta = 10^{-6}$ is approximately, for small h and for $\beta = -100$,

$$\mathcal{N} \doteq (6 \log 10)\mathbf{S}(G) \doteq (6 \log 10)(|\beta|/2\sqrt{2\pi})$$
$$\doteq 155. \qquad (12\text{-}6.11)$$

Thus in this case, the GCW–SI method requires more iterations than the RF–SI method in spite of the fact that each iteration of the GCW–SI method requires solving an auxiliary linear system. The RF–SI method would have a much greater advantage for linear systems corresponding to the differential equation†

$$u_{xx} + u_{yy} + \beta(u_x + u_y) = 0, \qquad (12\text{-}6.12)$$

where h is small and $|\beta| h$ is large. For in this case, it can be shown that if $|\beta|h > 2$, then all eigenvalues of the RF method are purely imaginary and

$$\mathbf{S}(I - A) \sim \tfrac{1}{2}h|\beta|, \qquad h \to 0. \qquad (12\text{-}6.13)$$

On the other hand, for the GCW method we have

$$\mathbf{S}(G) \sim |\beta|/2\pi, \qquad h \to 0. \qquad (12\text{-}6.14)$$

The ratio of number of iterations required by the RF–SI method to the number required by the GCW–SI method is proportional to h. For small h, this would be a substantial improvement.

It should be noted, however, that for linear systems arising from (12-6.1) or (12-6.12) the number of iterations required with the GCW–SI method is nearly independent of h, while the number of iterations required with the RF–SI method behaves as h^{-1} for small h. For values of h considerably less than $2|\beta|^{-1}$, the GCW–SI method is much faster.

It should also be noted that the GCW–GCG method has been observed to converge more rapidly in certain cases than one would predict, based on the analysis of the GCW–SI method. See Widlund [1978].

A series of experiments based on the use of ORTHOMIN acceleration and other acceleration procedures applied to the SSOR method and other basic iterative methods for problems involving Eq. (12-6.1) are described in the paper of Eisenstat et al. [1979a]. They considered cases where $\beta < 0$ and used "upwind differences" for u_x rather then central differences. It was found that when the SSOR method was used, ORTHOMIN acceleration with $s = 0$ (which is essentially the method of steepest descent) was better than when values of $s = 1$ or $s = 6$ were used. Also, it was found that the convergence was faster as $|\beta|$ increased. The GCW–CG method was also used. The method was quite effective for small $|\beta|$ but was much less effective for large $|\beta|$.

† This was pointed out to us by T. Manteuffel, private communication.

LISTING OF A SUBROUTINE
DESIGNED TO PROVIDE ACCELERATION PARAMETERS
AND TO MEASURE THE ITERATION ERROR
FOR THE CHEBYSHEV ACCELERATION METHOD

ALGORITHM 6-5.1   CHEBYSHEV POLYNOMIAL ACCELERATION METHOD

```
      SUBROUTINE CHEBY(DELNP,DELNE,YUN,SIP,NPRT,          001
     X             RHO,GAM,ICONV,ILIMIT,ITP,              002
     X             XME,XLME,D,F,IE,IO,ILIM)               003
C                                                         004
C                                                         005
C  SUBROUTINE CHEBY IS AN IMPLEMENTATION OF ALGORITHM 6-5.1 GIVEN IN  006
C  *APPLIED ITERATIVE METHODS* BY LOUIS A. HAGEMAN AND DAVID M. YOUNG,  007
C  ACADEMIC PRESS(1981).  EQUATION NUMBERS AND SECTION NUMBERS GIVEN  008
C  BELOW REFER TO THIS BOOK.                             009
C                                                         010
C  THE CHEBY SUBROUTINE COMPUTES ACCELERATION PARAMETERS AND MEASURES  011
C  THE ITERATION ERROR VECTOR FOR THE CHEBYSHEV POLYNOMIAL METHOD  012
C  DEFINED BY EQUATION (6-3.1).  THIS SUBROUTINE MUST BE ENTERED BEFORE  013
C  EACH ITERATION.  ON EXIT, RHO AND GAM ARE THE CHEBYSHEV PARAMETERS TO  014
C  BE USED ON THE NEXT ITERATION.  ICONV INDICATES WHETHER OR NOT  015
C  CONVERGENCE WAS ACHIEVED ON THE PREVIOUS ITERATION.   016
C  WARNING:  SOME LOCAL VARIABLES IN THIS SUBROUTINE ARE ASSUMED TO  017
C  RETAIN THEIR VALUES BETWEEN CALLS.  THUS, IF THIS SUBROUTINE IS  018
C  RELOADED BETWEEN CALLS, SUCH VARIABLES MUST BE STORED IN COMMON.  019
C                                                         020
C                                                         021
C  ASSUMPTIONS:  THE BASIC ITERATION METHOD IS SYMMETRIZABLE.  SEE  022
C                DEFINITION 2-2.1.  IN THE FOLLOWING, WE LET G DENOTE  023
C                THE ITERATION MATRIX OF THE BASIC METHOD.  024
C                                                         025
C                                                         026
C  ALL VARIABLES IN THE CALLING SEQUENCE ARE REAL EXCEPT FOR  027
C  NPRT,ICONV,ILIMIT,ITP,IE,IO,ILIM.                     028
C                                                         029
C                                                         030
C  INITIALIZATION ENTRY(I.E., THE ENTRY BEFORE COMPUTING ITERATION 1).  031
C                                                         032
C     ITP    = 0                                         033
C     XME    = THE INITIAL ESTIMATE FOR MAX MU, THE LARGEST  034
C              EIGENVALUE OF G.  XME MUST LIE IN THE     035
C              INTERVAL  (XLME,1.0).                     036
C     XLME   = THE INITIAL APPROXIMATION FOR MIN MU, THE SMALLEST  037
C              EIGENVALUE OF G.  XLME MUST BE LESS THAN XME.  038
C     IE     = CONTROL WORD THAT CONVEYS MORE INFORMATION CONCERNING  039
C              THE ESTIMATE XLME.  (SEE THE DISCUSSION GIVEN IN  040
C              SECTION 5.3.)                             041
C              IE=-1 IMPLIES THAT THE INITIAL ESTIMATE FOR MIN MU IS TO  042
C                    BE CALCULATED BY THE SUBROUTINE.  IF LAM IS THE  043
C                    ESTIMATE CALCULATED, THEN THE INITIAL  044
C                    APPROXIMATION FOR MIN MU IS TAKEN TO BE THE  045
C                    MINIMUM OF (LAM,XLME,-1.0).  IF IE=-1 AND IF  046
C                    NOTHING IS KNOWN ABOUT MIN MU, SET XLME=-1.0.  047
C              IE=0  IMPLIES THAT XLME PROBABLY IS LESS THAN OR EQUAL  048
```

357

```
C                        TO MIN MU.  IF THIS INEQUALITY IS NOT SATISFIED,     049
C                        THE SUBROUTINE WILL UPDATE THE ESTIMATE FOR MIN MU   050
C                        IF NECESSARY.                                        051
C              IE=1      IMPLIES THAT THE INPUT ESTIMATE FOR MIN MU IS TO     052
C                        BE USED FOR ALL ITERATIONS.  THIS OPTION SHOULD BE   053
C                        USED ONLY IF A LOWER BOUND FOR MIN MU IS KNOWN.      054
C                        ESSENTIALLY, ALGORITHM 6-4.1 IS CARRIED OUT WHEN     055
C                        IE IS SET EQUAL TO ONE.  WARNING:  IF IE=1 AND IF    056
C                        XLME IS GREATER THAN MIN MU, ITERATIVE DIVERGENCE    057
C                        MAY OCCUR.                                           058
C         ILIM   = THE MAXIMUM NUMBER OF ITERATIONS WHICH MAY BE DONE IN      059
C                  OBTAINING AN INITIAL ESTIMATE FOR MIN MU.  ILIM IS         060
C                  USED ONLY IF IE=-1 AND SHOULD BE GREATER THAN 7.           061
C         D      = THE STRATEGY PARAMETER DEFINED BY EQUATION (6-3.19).       062
C                  THE DEFAULT VALUE D=.1 IS USED IF D IS OUTSIDE THE         063
C                  INTERVAL   [.001,.8].                                      064
C         F      = THE STRATEGY PARAMETER DEFINED BY EQUATION (6-3.21),       065
C                  THE DEFAULT VALUE F=.7 IS USED IF F IS OUTSIDE THE         066
C                  INTERVAL   [.1,.9].                                        067
C         IO     = OUTPUT UNIT USED FOR PRINTING                             068
C         ILIMIT = THE UPPER LIMIT ON THE NUMBER OF ITERATIONS WHICH          069
C                  MAY BE DONE.  ILIMIT MUST BE NON-ZERO.                     070
C         REMARK 1:  IF IT IS KNOWN ONLY THAT MAX MU .GT. 0.0 AND THAT        071
C                    MIN MU .LE. 0.0, SET XME=.01.                            072
C         REMARK 2:  IF BOTH MAX MU AND MIN MU ARE TO BE FIXED FOR ALL        073
C                    ITERATIONS, SET XME AND XLME TO THEIR DESIRED VALUES AND 074
C                    SET ILIMIT NEGATIVE.  WARNING:  WHEN THE VALUES FOR XME  075
C                    AND XLME ARE FIXED, THE ESTIMATE FOR THE ITERATION ERROR 076
C                    VECTOR MAY NOT BE ACCURATE.  WARNING:  IF XLME IS FIXED  077
C                    AND XLME IS GREATER THAN MIN MU, ITERATIVE DIVERGENCE    078
C                    MAY OCCUR.                                               079
C         REMARK 3:  AFTER THE INITIALIZATION ENTRY INTO CHEBY, THE USER      080
C                    SHOULD NOT MODIFY ANY VARIABLES EXCEPT                   081
C                    DELNP,DELNE,YUN,SIP,NPRT.                                082
C                                                                            083
C                                                                            084
C   ON SUCCEEDING ENTRIES, SAY AFTER COMPUTING ITERATION (N+1), THE USER      085
C   MUST SUPPLY THE FOLLOWING NUMBERS:                                        086
C                                                                            087
C         NPRT   = PRINT CONTROL WORD.                                        088
C                  NPRT=0 IMPLIES NO PRINTING OF ITERATION DATA EXCEPT FOR    089
C                         THAT ITERATION ON WHICH CONVERGENCE IS ACHIEVED.    090
C                  NPRT=1 IMPLIES PRINTING OF ITERATION DATA.                 091
C         DELNP  = THE 2-NORM(OR THE W-NORM) OF THE PSEUDO-RESIDUAL           092
C                  VECTOR DEFINED BY EQUATION (5-2.1).                        093
C         DELNE  = THE BETA NORM OF THE PSEUDO-RESIDUAL VECTOR.               094
C         YUN    = THE ETA-NORM OF A RECENT APPROXIMATION FOR THE             095
C                  SOLUTION VECTOR.                                           096
C         SIP    = THE STOPPING CRITERION NUMBER, ZETA, IN (6-3.23).          097
C                  IN ORDER THAT ACCURATE ESTIMATES BE OBTAINED FOR THE       098
C                  SPECTRAL RADIUS OF G AND FOR THE ITERATION ERROR, SIP      099
C                  SHOULD NOT BE TOO LARGE.                                   100
C         REMARK 4:  DELNP IS USED IN THE ADAPTIVE PARAMETER ESTIMATION       101
C                    PROCEDURE.                                               102
C         REMARK 5:  DELNE, YUN, AND SIP ARE USED IN THE STOPPING TEST OF     103
C                    (6-3.23).  FOR A DISCUSSION OF APPROPRIATE BETA AND ETA  104
C                    NORMS TO USE FOR DELNE AND YUN, SEE THE LAST FEW PAGES   105
C                    OF SECTION 5.4.  SET YUN=1.0 IF THE RELATIVE NORM IS     106
C                    USED FOR DELNE.                                          107
C         REMARK 6:  IF THE CHEBYSHEV POLYNOMIAL METHOD IS USED AS AN INNER   108
C                    ITERATION PROCEDURE, THE STOPPING TEST (6-3.23) SHOULD   109
C                    BE REPLACED BY A STOPPING TEST BASED ON THE TOTAL ERROR  110
C                    REDUCTION.  SEE COMMENTS GIVEN IN CHAPTER 11.            111
C                                                                            112
C                                                                            113
C   ON EXIT, THE CHEBY SUBROUTINE SUPPLIES THE FOLLOWING DATA(ASSUME          114
C   ITERATION N+1 HAS BEEN COMPUTED):                                         115
C                                                                            116
C         RHO    = THE CHEBYSHEV PARAMETER, RHO, OF EQUATION (6-3.1) TO       117
C                  BE USED ON THE NEXT ITERATION, WHICH IS ASSUMED HERE       118
C                  TO BE ITERATION N+2.                                       119
C         GAM    = THE CHEBYSHEV PARAMETER, GAMMA, OF EQUATION (6-3.1) TO     120
C                  BE USED ON THE NEXT ITERATION.                             121
C         ICONV  = THE CONVERGENCE INDICATOR.                                 122
C                  ICONV=-1 IMPLIES POSSIBLE DIVERGENCE AND THAT THE          123
```

```
C                            ITERATIVE PROCESS SHOULD BE TERMINATED.           124
C                 ICONV=0   IMPLIES THAT CONVERGENCE WAS NOT ACHIEVED ON        125
C                            ITERATION N+1.                                     126
C                 ICONV=1   IMPLIES THAT CONVERGENCE HAS BEEN ACHIEVED(I.E.     127
C                            THE STOPPING TEST (6-3.23) WAS SATISFIED ON        128
C                            ITERATION N+1).                                    129
C                 ICONV=10  IMPLIES THAT THE ITERATION COUNT EQUALS OR          130
C                            EXCEEDS ILIMIT.                                    131
C        ITP    = THE ITERATION COUNTER.                                        132
C        DELNE  = THE ESTIMATE FOR THE ITERATION ERROR VECTOR ON                133
C                 ITERATION N AS GIVEN BY THE LEFT SIDE OF                      134
C                 INEQUALITY (6-3.23).  DELNE=1.0 IF NO ESTIMATE WAS            135
C                 CALCULATED.                                                   136
C        XME    = THE ESTIMATE FOR MAX MU CURRENTLY BEING USED.                 137
C        XLME   = THE ESTIMATE FOR MIN MU CURRENTLY BEING USED.                 138
C                                                                               139
C                                                                               140
C     EDITS.  ASSUME ITERATION N+1 HAS BEEN COMPUTED.                           141
C                                                                               142
C        I(P)   = ITP(P), WHERE P IS THE DEGREE OF THE CHEBYSHEV                143
C                 POLYNOMIAL APPLIED ON ITERATION N+1.  ITP IS THE              144
C                 ITERATION COUNT FOR THE ITERATION JUST COMPLETED.             145
C        R      = THE RATIO OF SUCCESSIVE PSEUDO-RESIDUAL VECTOR NORMS           146
C                 DEFINED BY (6-5.11).                                          147
C        TER    = THE THEORETICAL ERROR REDUCTION FACTOR WHICH WOULD HAVE       148
C                 BEEN ACHIEVED ON ITERATION N HAD THE ACCELERATION             149
C                 PARAMETERS BEEN OPTIMAL.  WITH Q(P) DEFINED BY (5-4.11),      150
C                 THEN TER=Q(P)/Q(P-1).                                         151
C        CR     = THE RATIO OF ACTUAL TO THEORETICAL CONVERGENCE RATES.         152
C                 CR IS DEFINED BY (6-4.11).                                    153
C        EPS    = THE ESTIMATE FOR THE NORMALIZED ITERATION ERROR VECTOR        154
C                 ON ITERATION N AS GIVEN BY THE LEFT SIDE OF (6-3.23).         155
C        REMARK 7:  VALUES FOR TER, CR, AND EPS ARE NOT COMPUTED ON ALL         156
C                 ITERATIONS.  WHEN NO COMPUTATIONS ARE MADE, THE VALUE         157
C                 PRINTED IS UNITY.                                             158
C                                                                               159
C                                                                               160
C     WHENEVER ESTIMATES FOR MAX MU AND MIN MU ARE CHANGED, THE FOLLOWING       161
C     DATA IS PRINTED PROVIDED NPRT=1:                                          162
C                                                                               163
C        MAX MU   = THE ESTIMATE FOR MAX MU COMPUTED BY (5-4.24).               164
C        MAX MU U = THE ESTIMATE FOR MAX MU USED IN THE GENERATION              165
C                   OF THE CHEBYSHEV POLYNOMIAL.  MAX MU U WILL DIFFER          166
C                   FROM MAX MU ONLY BECAUSE OF THE UPPER BOUND RESTRICTION      167
C                   IMPOSED BY THE TAU CONSTANTS GIVEN BY (6-4.1).              168
C        MIN MU   = THE ESTIMATE FOR MIN MU COMPUTED BY (6-5.22)                169
C                   OR (6-5.23).                                                170
C        A.E.R.   = THE ASYMPTOTIC ERROR REDUCTION FACTOR FOR THE               171
C                   CHEBYSHEV POLYNOMIAL METHOD CORRESPONDING TO MAX MU U        172
C                   AND MIN MU.  A.E.R.=SQRT(R), WHERE R IS DEFINED BY          173
C                   (4-3.21).                                                   174
C                                                                               175
C                                                                               176
C     THANKS ARE DUE TO L. A. ONDIS, C. J. PFEIFER, AND C. J. SPITZ OF          177
C     WESTINGHOUSE WHO MADE MANY HELPFUL SUGGESTIONS CONCERNING THE             178
C     PROGRAMMING OF THIS SUBROUTINE.                                           179
C                                                                               180
C                                                                               181
      INTEGER P                                                                 182
      INTEGER PS                                                                183
      REAL TAU(7)                                                               184
      DATA TAU(1),TAU(2),TAU(3),TAU(4),TAU(5),TAU(6),TAU(7)/.948,.985,          185
     X.995,9975,.9990,.9995,.99995/                                            186
      ICONV=0                                                                   187
      IPWM=0                                                                    188
      C=1.0                                                                     189
      CRR=1.0                                                                   190
      EPS=1.0                                                                   191
      IF (ITP.GT.0) GO TO 130                                                   192
C                                                                               193
C INITIALIZE                                                                    194
      P=-1                                                                      195
      IDIV=0                                                                    196
      DELNP=1.0                                                                 197
      R=1.0                                                                     198
```

```
      ILIMI=ILIM                                                           199
      ICLME=0                                                             200
      INPRT=0                                                             201
      ILF=ILIMIT                                                          202
      IF (XME.LT.1.0) GO TO 10                                            203
      XME=0.0                                                             204
      ILF=IABS(ILF)                                                       205
   10 IF (XME.GT.XLME) GO TO 20                                           206
      XME=XLME+.1                                                         207
      ILF=IABS(ILF)                                                       208
      IF (XME.LT.1.0) GO TO 20                                            209
      XLME=0.0                                                            210
      XME=.1                                                              211
      IF (IE.GT.0) IE=0                                                   212
   20 IEE=IE                                                              213
      XMEP=XME                                                            214
      XLMT=XLME                                                           215
      XMET=XME                                                            216
      DD=D                                                                217
      FF=F                                                                218
      IF (D.LT..001.OR.D.GT..8) DD=.1                                     219
      IF (F.LT..1.OR.F.GT..9) FF=.7                                       220
      IS=0                                                                221
      IF (ILF.LT.0) IEE=1                                                 222
      IF (IEE.GE.0) GO TO 30                                              223
      XMEP=TAU(1)                                                         224
      XLME=-XMEP                                                          225
C                                                                         226
C NEXT ITERATION                                                          227
   30 ITP=ITP+1                                                           228
      P=P+1                                                               229
      DELNO=DELNP                                                         230
      ILIMIT=IABS(ILIMIT)                                                 231
      IF (P.GT.0) GO TO 70                                                232
C     INITIALIZE FOR START OF NEW CHEBYSHEV POLYNOMIAL                    233
      IS=IS+1                                                             234
      ICT=0                                                               235
      IF (IS.GT.7) IS=7                                                   236
      TAUS=TAU(IS)                                                        237
      XME=XMEP                                                            238
      IF (IEE.LT.0) GO TO 40                                              239
      IF (XME.GT.TAUS) XME=TAUS                                           240
   40 IF (ILF.GE.0) GO TO 50                                              241
      XME=XMEP                                                            242
      XLME=XLMT                                                           243
   50 GAM=2.0/(2.0-XME-XLME)                                              244
      SE=(XME-XLME)/(2.0-XME-XLME)                                        245
      SESF=SE*SE/4.0                                                      246
      RHO=1.0                                                             247
      Z=1.0-SE*SE                                                         248
      Z=SQRT(Z)                                                           249
      XLR=(1.0-Z)/(1.0+Z)                                                 250
      AERO=SQRT(XLR)                                                      251
      ACR=-ALOG(XLR)                                                      252
      Z=-ALOG(DD)/ACR                                                     253
      PS=INT(Z)                                                           254
      IF (PS.LT.6) PS=6                                                   255
      IF (ILF.LT.0) PS=10000                                             256
      IF (IEE.LT.0) PS=8                                                  257
      ACR=ACR/2.0                                                         258
      IPWM=1                                                              259
   60 FORMAT (1H0,7HMAX MU=,E11.4,12H   MAX MU U=,E11.4,10H   MIN MU=,E1  260
     11.4,10H   A.E.R.=,E11.4)                                            261
      GO TO 80                                                            262
C     CONTINUE POLYNOMIAL GENERATION                                     263
   70 RHO=1.0/(1.0-SESF*RHO)                                             264
      IF (P.EQ.1) RHO=1.0/(1.0-2.0*SESF)                                 265
C                                                                         266
C CLEAN UP BEFORE EXIT                                                    267
   80 DELNE=EPS                                                           268
      ITM=ITP-1                                                           269
      IF (ITM.EQ.0) WRITE(IO,85)                                          270
   85 FORMAT (1H0,28HINITIAL EIGENVALUE ESTIMATES)                        271
      IF (ITM.EQ.0) GO TO 110                                             272
      IF (ITM.GE.ILIMIT) ICONV=10                                        273
```

```
         IF (ICONV.NE.0) GO TO 250                                      274
         IF (NPRT.EQ.0) GO TO 120                                       275
      90 WRITE (IO,100) IPM,ITM,R,C,CRR,EPS                             276
     100 FORMAT (1H ,2HI(,I4,2H)=,I4,5H   R=,E11.4,7H   TER=,E11.4,6H   CR= 277
        1,E11.4,7H   EPS=,E11.4)                                        278
     110 IF (IPWM.EQ.1) WRITE (IO,60) XMEP,XME,XLME,AERO                279
     120 IPM=P+1                                                        280
         RETURN                                                         281
C                                                                       282
C CALCULATE NEW ESTIMATE XMEP                                           283
     130 RO=R                                                           284
         R=DELNP/DELNO                                                  285
         IF (P.EQ.0) DELNPI=DELNP                                       286
         IF(ITP.EQ.2) RESDS=DELNP                                       287
         IF (P.LE.2) GO TO 30                                           288
         XP=FLOAT(P)                                                    289
         XPOT=XP/2.0                                                    290
         Z=1.0+XLR**P                                                   291
         Q=(2.0*XLR**XPOT)/Z                                            292
         B=DELNP/DELNPI                                                 293
         IF (B.LT.1.0) GO TO 160                                        294
         ICT=ICT+1                                                      295
         IF (ICT.GE.5.AND.IEE.GT.0) GO TO 140                           296
         IF (IEE.LE.0) GO TO 180                                        297
         GO TO 30                                                       298
C        PRINT THAT ITERATIONS ARE POSSIBLY DIVERGENT.                  299
     140 IF (IDIV.EQ.1) GO TO 30                                        300
         WRITE (IO,150)                                                 301
     150 FORMAT (1H0,50HTHE QUANTITY B DEFINED BY (5-4.10) IS GREATER THAN, 302
        140H UNITY FIVE TIMES.  POSSIBLE DIVERGENCE.)                   303
         IDIV=1                                                         304
         ICONV=-1                                                       305
         GO TO 30                                                       306
     160 CRR=ALOG(Q)                                                    307
         CRR=ALOG(B)/CRR                                                308
         XPM1=XP-1.0                                                    309
         ZM1=1.0+XLR**XPM1                                              310
         C=AERO*ZM1/Z                                                   311
         XMEP=XME                                                       312
         IF (B.LE.Q) GO TO 170                                          313
         Z1=B*B-Q*Q                                                     314
         CX=B+SQRT(Z1)                                                  315
         Z1=1.0/XP                                                      316
         CX=(Z*CX/2.0)**Z1                                              317
         Z1=(2.0-XME-XLME)/(1.0+XLR)                                    318
         Z=(CX*CX+XLR)/CX                                               319
         XMEP=.5*(XME+XLME+Z1*Z)                                        320
C                                                                       321
C CONVERGENCE TEST                                                      322
     170 Z=1.0/(1.0-XMEP)                                               323
         IF (ICLME.NE.1) GO TO 175                                      324
C        THE FOLLOWING 2 CARDS ARE INCLUDED TO PREVENT POSSIBLE PSEUDO  325
C        CONVERGENCE ON THE ITERATIONS IMMEDIATELY FOLLOWING A CHANGE IN 326
C        THE XLME ESTIMATE.                                             327
         IF (P.LE.PS.OR.R.GE.1.) GO TO 180                             328
         IF(Z.LT.100.0) Z=100.0                                         329
     175 EPS=DELNE*Z/YUN                                                330
         IF (EPS.LE.SIP) ICONV=1                                        331
C                                                                       332
C PARAMETER CHANGE TEST                                                 333
     180 IC=MOD(P,2)                                                    334
         IF (IC.GT.0) GO TO 30                                          335
         IF (P.LT.PS) GO TO 30                                          336
         IF (IEE.LT.0) GO TO 200                                        337
         Z=Q**FF                                                        338
         IF (B.LE.Z) GO TO 30                                           339
         IF (ICT.GT.0) GO TO 190                                        340
         P=-1                                                           341
         ICLME=0                                                        342
         GO TO 30                                                       343
     190 IF (R.LT.1.0 .OR. IEE.GT.0) GO TO 30                           344
         IF (ICT.LT.5) GO TO 30                                         345
C                                                                       346
C UPDATE XLME                                                           347
     200 XLEP=0.0                                                       348
```

```
      XLEPP=0.0                                                              349
      IDR=0                                                                  350
      IF (R.LE.1.0) GO TO 210                                               351
      XP=FLOAT(P)                                                            352
      XPM1=XP-1.0                                                            353
      U=(1.0/SQRT(XLR))*R*(1.0+XLR**XP)/(1.0+XLR**XPM1)                     354
      XLEPP=.5*(XME+XLME-(XME-XLME)*(U*U+1.0)/(2.0*U))                      355
      Z=ABS(RO-R)                                                            356
      IF (Z.LT..1) IDR=1                                                    357
  210 IF (R.LE.1.0) GO TO 220                                               358
      Z=B/Q                                                                  359
      Z1=1.0/XP                                                              360
      Y=Z*Z-1.0                                                              361
      Y=(Z+SQRT(Y))**Z1                                                      362
      XLEP=.5*(XME+XLME-(XME-XLME)*(Y*Y+1.0)/(2.0*Y))                       363
      GO TO 230                                                              364
  220 IF (IEE.EQ.0) GO TO 30                                                365
  230 IF (XLEP.LT.XLEPP) XLEPP=XLEP                                         366
      XLEPP=1.1*XLEPP                                                        367
      Y=XLME                                                                 368
      IF (IEE.LT.0) GO TO 240                                               369
      IF (IDR.EQ.0) GO TO 30                                                370
      IF (XLEPP.LT.XLME) XLME=XLEPP                                         371
      XMEP=.1                                                                372
C     THE FOLLOWING 2 CARDS ARE A SLIGHT VARIATION FROM ALGORITHM 6-5.1.    373
      IF (Y.GT.XLME) XMEP=Y                                                 374
      ICLME=1                                                                375
      P=-1                                                                   376
      GO TO 30                                                               377
  240 IF (IDR.EQ.0.AND.ITP.LT.ILIMI) GO TO 30                              378
      IF (XLMT.GT.-1.0) XLMT=-1.0                                           379
      XLME=XLEPP                                                             380
      IF (XLMT.LT.XLME) XLME=XLMT                                           381
      XMEP=.1                                                                382
C     THE FOLLOWING 2 CARDS ARE A SLIGHT VARIATION FROM ALGORITHM 6-5.1.    383
      IF (Y.GT.XLME) XMEP=Y                                                 384
      ICLME=1                                                                385
      IEE=0                                                                  386
      P=-1                                                                   387
      IS=0                                                                   388
      GO TO 30                                                               389
C                                                                           390
C IF NPRT=0, PRINT FINAL ITERATION RESULTS.                                391
  250 IF (NPRT.EQ.1) GO TO 90                                              392
      IF (INPRT.NE.0) GO TO 90                                             393
      WRITE (IO,255)                                                        394
  255 FORMAT (1H0,17HITERATION SUMMARY)                                    395
      INPRT=1                                                               396
      WRITE (IO,60) XMEP,XME,XLME,AERO                                     397
      GO TO 90                                                              398
      END                                                                   399
```

APPENDIX

B

CCSI Subroutine

LISTING OF A SUBROUTINE
DESIGNED TO PROVIDE ACCELERATION PARAMETERS
AND TO MEASURE THE ITERATION ERROR
FOR THE CCSI ACCELERATION METHOD

ALGORITHM 8-3.1   CYCLIC CHEBYSHEV (CCSI) POLYNOMIAL METHOD

```
      SUBROUTINE CCSI (DELNP,DELNE,YUN,SIP,NPRT,RHOR,RHOB,ICONV,ILIMIT,    001
     1ITP,XME,D,F,IO)                                                      002
C                                                                         003
C                                                                         004
C     SUBROUTINE CCSI IS AN IMPLEMENTATION OF ALGORITHM 8-3.1 GIVEN IN     005
C     *APPLIED ITERATIVE METHODS* BY LOUIS A. HAGEMAN AND DAVID M. YOUNG,  006
C     ACADEMIC PRESS(1981).  EQUATION NUMBERS AND SECTION NUMBERS GIVEN    007
C     BELOW REFER TO THIS BOOK.                                           008
C                                                                         009
C     THE CCSI SUBROUTINE COMPUTES ACCELERATION PARAMETERS AND MEASURES THE 010
C     ITERATION ERROR VECTOR FOR THE CYCLIC CHEBYSHEV POLYNOMIAL METHOD    011
C     DEFINED BY EQUATION (8-1.11).  THIS SUBROUTINE MUST BE ENTERED BEFORE 012
C     EVERY CCSI ITERATION.  ON EXIT, RHOR AND RHOB ARE THE CHEBYSHEV      013
C     PARAMETERS TO BE USED ON THE NEXT ITERATION.  ICONV INDICATES WHETHER 014
C     OR NOT CONVERGENCE WAS ACHIEVED ON THE PREVIOUS ITERATION.          015
C     WARNING:  SOME LOCAL VARIABLES IN THIS SUBROUTINE ARE ASSUMED TO     016
C     RETAIN THEIR VALUES BETWEEN CALLS.  THUS, IF THIS SUBROUTINE IS      017
C     RELOADED BETWEEN CALLS, SUCH VARIABLES MUST BE STORED IN COMMON.     018
C                                                                         019
C                                                                         020
C     ASSUMPTIONS:  (1) THE COEFFICIENT MATRIX A IS PARTITIONED INTO A     021
C                       RED-BLACK FORM.  SEE SECTIONS 1.5, 8.1, AND 9.2.   022
C                   (2) THE COEFFICIENT MATRIX A IS SYMMETRIC AND POSITIVE 023
C                       DEFINITE;  OR ELSE THE ASSOCIATED JACOBI MATRIX    024
C                       IS SYMMETRIZABLE (SEE DEF. 2-2.1).                 025
C                                                                         026
C                                                                         027
C     ALL VARIABLES IN THE CALLING SEQUENCE ARE REAL EXCEPT FOR NPRT,      028
C     ICONV,ILIMIT,ITP,AND IO.                                            029
C                                                                         030
C                                                                         031
C     INITIALIZATION ENTRY(THAT IS, THE ENTRY BEFORE COMPUTING ITERATION 1) 032
C                                                                         033
C        ITP   = 0                                                        034
C        XME   = THE INITIAL ESTIMATE FOR THE SPECTRAL RADIUS OF THE       035
C                JACOBI ITERATION MATRIX B.  XME MUST LIE IN THE           036
C                INTERVAL [0.0,1.0).                                       037
C        D     = THE STRATEGY PARAMETER DEFINED BY EQUATION (6-3.19).      038
C                THE DEFAULT VALUE D=.1 IS USED IF D IS OUTSIDE  THE       039
C                INTERVAL [.001,.8].                                       040
C        F     = THE STRATEGY PARAMETER DEFINED BY EQUATION (8-3.36).      041
C                THE DEFAULT VALUE F = .7 IS USED IF F IS OUTSIDE THE      042
C                INTERVAL [.1,.9].                                        043
C        IO    = OUTPUT UNIT USED FOR PRINTING.                           044
C        ILIMIT = THE UPPER LIMIT ON THE NUMBER OF ITERATIONS WHICH        045
C                MAY BE DONE.  ILIMIT MUST BE NON-ZERO.                    046
C        REMARK 1:  SET XME = 0.0 UNLESS SOME A-PRIORI INFORMATION         047
C                CONCERNING THE SPECTRAL RADIUS OF B IS KNOWN.             048
```

363

```
C     REMARK 2:  IF A FIXED VALUE FOR THE SPECTRAL RADIUS OF B IS TO BE      049
C                USED FOR ALL ITERATIONS, SET XME TO THE DESIRED VALUE       050
C                AND SET ILIMIT NEGATIVE.  WARNING:  IF XME IS FIXED         051
C                AND IF XME IS GREATER THAN THE SPECTRAL RADIUS OF B, THE    052
C                ESTIMATE FOR THE ITERATION ERROR VECTOR MAY NOT BE          053
C                ACCURATE.  SEE FIGURE 8-4.4.                                054
C     REMARK 3:  AFTER THE INITIALIZATION ENTRY INTO CCSI, THE USER          055
C                SHOULD NOT MODIFY ANY VARIABLES EXCEPT DELNP, DELNE,         056
C                YUN, SIP, AND NPRT.                                         057
C                                                                            058
C                                                                            059
C     ON SUCCEEDING ENTRIES,SAY AFTER CALCULATING ITERATION (N+1), THE USER  060
C     MUST SUPPLY THE FOLLOWING NUMBERS:                                     061
C                                                                            062
C                                                                            063
C         NPRT  = PRINT CONTROL WORD.                                        064
C                 NPRT=0 IMPLIES NO PRINTING OF ITERATION DATA EXCEPT FOR    065
C                     THAT ITERATION ON WHICH CONVERGENCE IS ACHIEVED.       066
C                 NPRT=1 IMPLIES PRINTING OF ITERATION DATA.                 067
C         DELNP = THE 2-NORM(OR THE S-SUB-B NORM) OF THE DIFFERENCE          068
C                 VECTOR FOR THE BLACK UNKNOWNS.  THE DIFFERENCE             069
C                 VECTOR IS DEFINED BY EQUATION (8-3.19).                    070
C         DELNE = THE BETA-NORM OF THE DIFFERENCE VECTOR FOR THE             071
C                 BLACK UNKNOWNS.                                            072
C         YUN   = THE ETA-NORM OF A RECENT APPROXIMATION FOR THE  SOLUTION   073
C                 VECTOR OF BLACK UNKNOWNS.                                  074
C         SIP   = THE STOPPING CRITERION NUMBER, ZETA, IN (8-3.35).          075
C                 IN ORDER THAT ACCURATE ESTIMATES BE OBTAINED FOR THE       076
C                 SPECTRAL RADIUS OF B AND FOR THE ITERATION ERROR, SIP      077
C                 SHOULD NOT BE TOO LARGE.                                   078
C     REMARK 4:  DELNP IS USED IN THE ADAPTIVE PARAMETER ESTIMATION          079
C                PROCEDURE.                                                  080
C     REMARK 5:  DELNE, YUN, AND SIP ARE USED IN THE STOPPING TEST           081
C                OF (8-3.35).  FOR A DISCUSSION OF APPROPRIATE BETA AND       082
C                ETA NORMS TO USE FOR DELNE AND YUN, SEE THE LAST FEW        083
C                PAGES OF SECTION 5.4.  SET YUN = 1.0 IF THE RELATIVE        084
C                NORM IS USED FOR DELNE.                                     085
C     REMARK 6:  IF THE CCSI METHOD IS USED AS AN INNER ITERATION            086
C                PROCEDURE, THE STOPPING TEST (8-3.35) SHOULD BE REPLACED    087
C                BY A STOPPING TEST BASED ON THE TOTAL ERROR REDUCTION.      088
C                SEE COMMENTS GIVEN IN CHAPTER 11.                           089
C                                                                            090
C                                                                            091
C     ON EXIT, THE CCSI SUBROUTINE SUPPLIES THE FOLLOWING DATA(ASSUME        092
C     ITERATION N+1 HAS BEEN COMPUTED):                                      093
C                                                                            094
C         RHOR  = THE CHEBYSHEV PARAMETER FOR THE RED UNKNOWNS OF            095
C                 EQUATION (8-1.11) FOR THE NEXT ITERATION(WHICH IS          096
C                 ASSUMED HERE TO BE ITERATION N+2).                         097
C         RHOB  = THE CHEBYSHEV PARAMETER FOR THE BLACK UNKNOWNS OF          098
C                 EQUATION (8-1.11).                                         099
C         ICONV = THE CONVERGENCE INDICATOR.                                 100
C                 ICONV=1   IMPLIES CONVERGENCE HAS BEEN ACHIEVED(I.E.,THE   101
C                           STOPPING TEST (8-3.35) HAS BEEN SATISFIED ON     102
C                           ITERATION N+1).                                  103
C                 ICONV=0   IMPLIES THAT CONVERGENCE WAS NOT ACHIEVED ON     104
C                           ITERATION N+1.                                   105
C                 ICONV=10  IMPLIES THAT THE ITERATION COUNT EQUALS OR       106
C                           EXCEEDS ILIMIT.                                  107
C                 ICONV=-1  IMPLIES POSSIBLE DIVERGENCE AND THAT THE         108
C                           ITERATIVE PROCESS SHOULD BE TERMINATED.          109
C                           IF THE ITERATIONS ARE ALLOWED TO CONTINUE,       110
C                           THE GAUSS-SEIDEL METHOD WILL BE USED.            111
C         ITP   = THE ITERATION COUNTER.                                     112
C         DELNE = THE ESTIMATE FOR THE ITERATION ERROR VECTOR ON            113
C                 ITERATION N AS GIVEN BY THE LEFT SIDE OF                    114
C                 INEQUALITY (8-3.35).  DELNE=1.0 IF NO ESTIMATE IS MADE.    115
C         XME   = THE ESTIMATE FOR THE SPECTRAL RADIUS OF B                  116
C                 CURRENTLY BEING USED.                                       117
C                                                                            118
C                                                                            119
C     EDITS.  ASSUME ITERATION N+1 HAS BEEN COMPUTED.                        120
C                                                                            121
C         I(P) = ITP(P), WHERE P IS THE DEGREE OF THE CHEBYSHEV              122
C                POLYNOMIAL APPLIED TO THE BLACK UNKNOWNS ON ITERATION       123
C                N+1.  ITP IS THE ITERATION COUNT FOR THE ITERATION
```

```
C              JUST COMPLETED.                                        124
C        V   = RATIO OF SUCCESSIVE DIFFERENCE VECTOR NORMS.  THAT IS, 125
C              V=(DELNP FROM ITERATION N+1)/(DELNP FROM ITERATION N). 126
C        R   = THE R DEFINED BY EQUATION (8-3.33).  THE QUANTITY R    127
C              APPROXIMATES THE ERROR REDUCTION FACTOR WHICH WAS      128
C              ACTUALLY ACHIEVED ON ITERATION N.                      129
C        TER = THE THEORETICAL ERROR REDUCTION FACTOR WHICH WOULD HAVE 130
C              BEEN ACHIEVED ON ITERATION N HAD THE ACCELERATION      131
C              PARAMETERS BEEN OPTIMAL.  TER IS THE SECOND QUANTITY    132
C              GIVEN IN BRACKETS IN THE DEFINITION OF H WHICH IS       133
C              GIVEN FOLLOWING EQUATION (8-3.35).                     134
C        CR  = THE RATIO OF ACTUAL TO THEORETICAL CONVERGENCE RATES.  135
C              CR IS DEFINED BY (8-3.57).                             136
C        EPS = THE ESTIMATE FOR THE NORMALIZED ITERATION ERROR VECTOR 137
C              ON ITERATION N AS GIVEN BY THE LEFT SIDE OF (8-3.35).  138
C     REMARK 7: VALUES FOR R, TER, CR, AND EPS ARE NOT COMPUTED ON    139
C              ALL ITERATIONS.  WHEN NO COMPUTATIONS ARE MADE, THE    140
C              VALUE PRINTED IS UNITY.                                141
C                                                                     142
C  WHENEVER THE ESTIMATE FOR THE SPECTRAL RADIUS IS CHANGED, THE      143
C  FOLLOWING DATA IS PRINTED PROVIDED NPRT=1.                         144
C                                                                     145
C        MEP P. = THE ESTIMATE FOR THE SPECTRAL RADIUS OF THE JACOBI  146
C              ITERATION MATRIX B COMPUTED BY (8-3.44).               147
C        MEP U. = THE ESTIMATE FOR THE SPECTRAL RADIUS USED IN THE    148
C              GENERATION OF THE CHEBYSHEV POLYNOMIALS.  MEP U. WILL   149
C              DIFFER FROM MEP P. ONLY BECAUSE OF THE UPPER BOUND      150
C              RESTRICTION IMPOSED BY THE TAU CONSTANTS.  SEE         151
C              DISCUSSIONS GIVEN IN SECTION 6.4 AND ABOVE             152
C              EQUATION (8-3.53).                                     153
C        A.E.R. = THE ASYMPTOTIC ERROR REDUCTION FACTOR CORRESPONDING TO 154
C              MEP U.  A.E.R. IS THE SAME AS THE LOWER CASE R IN      155
C              EQUATION (8-3.44).                                     156
C                                                                     157
C  THANKS ARE DUE TO L. A. ONDIS, C. J. PFEIFER, AND C. J. SPITZ OF   158
C  WESTINGHOUSE WHO MADE MANY HELPFUL SUGGESTIONS CONCERNING THE      159
C  PROGRAMMING OF THIS SUBROUTINE.                                    160
C                                                                     161
C                                                                     162
      INTEGER P                                                       163
      INTEGER PST                                                     164
      INTEGER PS                                                      165
      REAL TAU(7)                                                     166
      DATA TAU(1),TAU(2),TAU(3),TAU(4),TAU(5),TAU(6),TAU(7)/.948,.985, 167
     X.995,.9975,.9990,.9995,.99995/                                  168
      ICONV=0                                                         169
      IPWM=0                                                          170
      AER=1.0                                                         171
      C=1.0                                                           172
      CRR=1.0                                                         173
      EPS=1.0                                                         174
      IF (ITP.GT.0) GO TO 100                                         175
C                                                                     176
C INITIALIZE                                                          177
      P=-1                                                            178
      IS=0                                                            179
      IF (XME.GE.0.0.AND.XME.LT.1.0) GO TO 5                          180
      XME=0.0                                                         181
      ILIMIT=IABS(ILIMIT)                                             182
    5 XMEP=XME                                                        183
      RESD=1.0                                                        184
      INPRT=0                                                         185
      ILF=ILIMIT                                                      186
      DD=D                                                            187
      FF=F                                                            188
      IF (D.LT..001.OR.D.GT..8) DD=.1                                 189
      IF (F.LT..1.OR.F.GT..9) FF=.7                                   190
C                                                                     191
C NEXT ITERATION                                                      192
   10 ITP=ITP+1                                                       193
      P=P+1                                                           194
      ILIMIT=IABS(ILIMIT)                                             195
      IF (P.GT.0) GO TO 40                                            196
C     INITIALIZE FOR START OF NEW POLYNOMIAL                          197
      IS=IS+1                                                         198
```

```
      ICT=0                                                               199
      ICS=0                                                               200
      IF ( IS.GT.7) IS=7                                                  201
      Z=TAU( IS)                                                          202
      XME=XMEP                                                            203
      IF ( XME.GT.Z) XME=Z                                               204
      IF ( ILF.LT.0) XME=XMEP                                            205
      XMESF=XME*XME                                                       206
      RHOR=1.0                                                            207
      RHOB=2.0/(2.0-XMESF)                                               208
      R=SQRT( 1.0-XMESF)                                                 209
      R=( 1.0-R)/( 1.0+R)                                                210
      IPWM=1                                                              211
   20 FORMAT ( 1H0,7HMEP P.=,E11.4,10H    MEP U.=,E11.4,10H    A.E.R.=,E11. 212
     14)                                                                  213
      XPS=8.0                                                             214
      IF ( XME.EQ.0.0) IS=IS-1                                           215
      IF ( XME.EQ.0.0) GO TO 30                                          216
      XPS=ALOG( DD)/ALOG( R)                                             217
   30 PS=INT( XPS)                                                       218
      IF ( ILF.LT.0) PS=10000                                           219
      ILF=0                                                               220
      IF ( PS.LT.8) PS=8                                                 221
      XMESF=XMESF/4.0                                                     222
      GO TO 50                                                            223
C       CONTINUE POLYNOMIAL GENERATION                                   224
   40 DELNO=DELNP                                                        225
      RHOBO=RHOB                                                          226
      IF .( RHOB.EQ.1.0) GO TO 50                                        227
      RHOR=1.0/( 1.0-XMESF*RHOB)                                        228
      RHOB=1.0/( 1.0-XMESF*RHOR)                                        229
C                                                                        230
C CLEAN UP BEFORE EXIT                                                   231
   50 DELNE=EPS                                                          232
      RESDO=RESD                                                         233
      IF ( ITP.EQ.1) WRITE( IO,55).                                     234
   55 FORMAT ( 1H0,27HINITIAL EIGENVALUE ESTIMATE)                      235
      IF ( ITP.EQ.1) GO TO 80                                           236
      IPT=2*P                                                            237
      IF ( XOMEGB.EQ.1.0) IPT=0                                         238
      ITM=ITP-1                                                          239
      IF ( ITM.GE.ILIMIT) ICONV=10                                     240
      IF ( ICONV.NE.0) GO TO 220                                       241
      IF ( NPRT.EQ.0) GO TO 90                                         242
   60 WRITE ( IO,70) IPT,ITM,V,AER,C,CRR,EPS                           243
   70 FORMAT ( 1H ,2HI( ,I4,2H)=,I4,5H    V=,E11.4,7H    R=,E11.4,7H    TER 244
     1=,E11.4,6H   CR=,E11.4,7H   EPS=,E11.4)                          245
   80 IF ( IPWM.EQ.1) WRITE ( IO,20) XMEP,XME,R                        246
   90 XOMEGB=RHOB                                                       247
      RETURN                                                            248
C                                                                        249
C CALCULATE NEW ESTIMATE XMEP                                            250
  100 RESD=DELNP                                                        251
      IF ( ITP.EQ.2) RESDS=DELNP                                       252
      V=RESD/RESDO                                                      253
      IF ( P.LT.3) GO TO 110                                           254
      IF ( XME.EQ.0.0) GO TO 200                                       255
      GO TO 120                                                         256
  110 IF ( P.EQ.0) DELNPI=DELNP                                        257
      GO TO 10                                                          258
  120 IF ( RHOB.GT.1.0) GO TO 170                                      259
      RP=R**P                                                           260
      XQ=( 2.0*RP)/( 1.0+RP*RP)                                        261
      B=( 2.0/( 2.0-XME*XME))*DELNP/DELNPI                             262
      BDQ=B/XQ                                                          263
      IF ( BDQ.GE.1.0) GO TO 130                                       264
      XMEP=XME                                                          265
      P=-1                                                              266
      GO TO 10                                                          267
  130 IF ( B.LT.1.0) GO TO 160                                         268
      ICONV=-1                                                          269
      WRITE ( IO,140)                                                   270
  140 FORMAT ( 1H0,45HTHE QUANTITY B DEFINED BY (8-3.43) IS GREATER,34H T 271
     1HAN UNITY.  POSSIBLE DIVERGENCE.)                                272
      WRITE ( IO,150)                                                   273
```

```
      150 FORMAT (1H ,46HIF CONTINUED, THE GAUSS-SEIDEL METHOD IS USED.)    274
          ILF=-1                                                            275
          XMEP=0.0                                                          276
          P=-1                                                              277
          GO TO 10                                                          278
      160 Z=SQRT(B*B-XQ*XQ)                                                 279
          Z=(1.0+RP*RP)*(B+Z)/2.0                                           280
          XP=FLOAT(P)                                                       281
          C3A=1.0/(2.0*XP)                                                  282
          Z=Z**C3A                                                          283
          XMEP=(Z+R/Z)/(1.0+R)                                              284
          P=-1                                                              285
          GO TO 10                                                          286
C                                                                           287
C STOPPING TEST                                                             288
      170 CONTINUE                                                          289
          AER=(2.0-RHOBO)/(2.0-RHOB)                                        290
          AER=AER*(DELNP/DELNO)                                             291
          RSP=R**P                                                          292
          RSP=RSP*RSP                                                       293
          RSPM=RSP/(R*R)                                                    294
          C=R*(1.0+RSPM)/(1.0+RSP)                                          295
          BET=(RHOB-1.0)/(RHOBO-1.0)                                        296
          H=BET*AER                                                         297
          IF (AER.LT.1.0) GO TO 180                                         298
          ICS=ICS+1                                                         299
          GO TO 10                                                          300
      180 IF (AER.GE.C) GO TO 190                                           301
          ICT=ICT+1                                                         302
          H=BET*C                                                           303
      190 EPS=DELNE/(YUN*(1.0-H))                                           304
          IF (EPS.LE.SIP) ICONV=1                                           305
          CRR=ALOG(C)                                                       306
          CRR=ALOG(AER)/CRR                                                 307
C                                                                           308
C PARAMETER CHANGE TEST                                                     309
          PST=PS/2                                                          310
          IF (P.LT.PST) GO TO 10                                            311
          PST=ICS*ICT                                                       312
          IF (PST.GT.0) GO TO 10                                            313
          Z=C**FF                                                           314
          IF (AER.LE.Z) GO TO 10                                            315
          ITP=ITP+1                                                         316
          P=P+1                                                             317
          RHOR=1.0                                                          318
          RHO3=1.0                                                          319
          GO TO 50                                                          320
C                                                                           321
C INITIAL GAUSS-SEIDEL ITERATIONS                                          322
      200 AER=DELNP/DELNO                                                   323
          IF (AER.GE.1.0) GO TO 10                                          324
          XMEP=SQRT(AER)                                                    325
          EPS=DELNE/(YUN*(1.0-AER))                                         326
          PST=(PS/2)-1                                                      327
          IF (P.LT.PST) GO TO 210                                           328
          P=-1                                                              329
      210 CONTINUE                                                          330
          IF (EPS.LE.SIP) ICONV=1                                           331
C THE TEST USING EPSS IS USED TO PREVENT PSEUDO CONVERGENCE                 332
C DURING ANY INITIAL GAUSS-SEIDEL ITERATIONS.                               333
          EPSS=RESD/RESDS                                                   334
          IF (EPSS.GE..10) ICONV=0                                          335
          GO TO 10                                                          336
C                                                                           337
C IF NPRT=0, PRINT EIGENVALUE ESTIMATES LAST USED.                          338
      220 IF(NPRT .EQ. 1) GO TO 60                                          339
          IF(INPRT .NE. 0) GO TO 60                                         340
          WRITE (IO,225)                                                    341
      225 FORMAT (1H0,17HITERATION SUMMARY)                                 342
          INPRT=1                                                           343
          WRITE(IO,20) XMEP,XME,R                                           344
          GO TO 60                                                          345
          END                                                              346
```

C

SOR Subroutine

LISTING OF A SUBROUTINE
DESIGNED TO PROVIDE ACCELERATION PARAMETERS
AND TO MEASURE THE ITERATION ERROR
FOR THE SOR METHOD

ALGORITHM 9-6.1   SUCCESSIVE OVERRELAXATION (SOR) METHOD

```
      SUBROUTINE SOR (DELNP,DELNE,YUN,SIP,NPRT,OM,ICONV,ILIMIT,ITP,OME,F   001
     1,PSP,RSP,IO)                                                         002
C                                                                          003
C                                                                          004
C     SUBROUTINE SOR IS AN IMPLEMENTATION OF ALGORITHM 9-6.1 GIVEN IN      005
C     *APPLIED ITERATIVE METHODS*  BY LOUIS A. HAGEMAN AND DAVID M. YOUNG, 006
C     ACADEMIC PRESS(1981).  EQUATION NUMBERS AND SECTION NUMBERS GIVEN    007
C     BELOW REFER TO THIS BOOK.                                            008
C                                                                          009
C     THE SOR SUBROUTINE COMPUTES THE OVERRELAXATION PARAMETER OMEGA AND   010
C     MEASURES THE ITERATION ERROR VECTOR FOR THE SUCCESSIVE OVERRELAXATION 011
C     METHOD DEFINED BY EQUATION (9-3.2).  THIS SUBROUTINE MUST BE ENTERED 012
C     BEFORE EVERY SOR ITERATION.  ON EXIT, OM IS THE VALUE OF OMEGA TO    013
C     USE ON THE NEXT ITERATION.  ICONV INDICATES WHETHER OR NOT           014
C     CONVERGENCE WAS ACHIEVED ON THE PREVIOUS ITERATION.                  015
C     WARNING:  SOME LOCAL VARIABLES IN THIS SUBROUTINE ARE ASSUMED TO     016
C     RETAIN THEIR VALUES BETWEEN CALLS.  THUS, IF THIS SUBROUTINE IS      017
C     RELOADED BETWEEN CALLS, SUCH VARIABLES MUST BE STORED IN COMMON.     018
C                                                                          019
C                                                                          020
C     ASSUMPTIONS:   (1) THE COEFFICIENT MATRIX A HAS  PROPERTY A  (OR IS  021
C                        2-CYCLIC) AND IS CONSISTENTLY ORDERED RELATIVE    022
C                        TO THE PARTITIONING IMPOSED.  SEE SECTIONS 9.2    023
C                        AND 9.3.                                          024
C                    (2) THE COEFFICIENT MATRIX A IS SYMMETRIC AND POSITIVE 025
C                        DEFINITE;  OR ELSE THE ASSOCIATED JACOBI METHOD   026
C                        IS SYMMETRIZABLE (SEE DEF. 2-2.1).                027
C                                                                          028
C                                                                          029
C     THE VARIABLES IN THE CALLING SEQUENCE ARE REAL EXCEPT FOR NPRT,      030
C     ICONV,ILIMIT,ITP,AND IO.                                             031
C                                                                          032
C                                                                          033
C     REQUIREMENTS FOR INITIALIZATION ENTRY(THAT IS, THE ENTRY BEFORE      034
C     COMPUTING ITERATION 1).                                              035
C                                                                          036
C         ITP    = 0                                                       037
C         OME    = THE INITIAL ESTIMATE FOR OMEGA.  OME MUST LIE IN THE    038
C                  INTERVAL [1.0,2.0).                                     039
C         F      = THE STRATEGY PARAMETER DEFINED BY EQUATION (9-5.10).    040
C                  THE DEFAULT VALUE F = .7 IS USED IF F IS OUTSIDE THE    041
C                  INTERVAL  [.1,.9].                                      042
C         PSP    = THE STRATEGY PARAMETER DEFINED BY (9-5.20).  THE        043
C                  DEFAULT VALUE PSP = .5 IS USED IF PSP IS OUTSIDE THE    044
C                  INTERVAL  [.001,.9].                                    045
C         RSP    = THE STRATEGY PARAMETER DEFINED BY (9-5.21).  THE        046
C                  DEFAULT VALUE RSP = .0001 IS USED IF RSP IS OUTSIDE     047
C                  THE INTERVAL  (1.0E-6,.01].                             048
```

```
      ILIMIT =  THE UPPER LIMIT ON THE NUMBER OF ITERATIONS WHICH MAY    049
                BE DONE.                                                 050
      IO     =  OUTPUT UNIT USED FOR PRINTING.                           051
      REMARK 1:  SET OME=1.0 UNLESS SOME A-PRIORI INFORMATION IS KNOWN   052
                CONCERNING THE OPTIMAL VALUE OF OMEGA.                   053
      REMARK 2:  IF A FIXED VALUE FOR OMEGA IS TO BE USED FOR ALL        054
                ITERATIONS, SET OME TO THE NEGATIVE OF THE DESIRED       055
                FIXED VALUE.  WARNING:  IF OMEGA IS FIXED FOR ALL        056
                ITERATIONS AND IF OME IS GREATER THAN THE OPTIMAL VALUE  057
                OF OMEGA, THE ESTIMATE FOR THE ITERATION ERROR MAY NOT   058
                BE ACCURATE.  SEE FIGURE 9-8.3.                          059
      REMARK 3:  AFTER THE INITIALIZATION ENTRY INTO SOR, THE USER       060
                SHOULD NOT MODIFY ANY VARIABLES EXCEPT DELNE, DELNP,     061
                YUN, SIP, AND NPRT.                                      062
                                                                        063
                                                                        064
ON SUCCEEDING ENTRIES, SAY AFTER CALCULATING ITERATION (N+1), THE        065
USER MUST SUPPLY THE FOLLOWING NUMBERS:                                  066
                                                                        067
      NPRT   =  PRINT CONTROL WORD.                                     068
                NPRT=0 IMPLIES NO PRINTING OF ITERATION DATA EXCEPT FOR  069
                       THAT ITERATION ON WHICH CONVERGENCE IS ACHIEVED.  070
                NPRT=1 IMPLIES PRINTING OF ITERATION DATA.              071
      DELNP  =  THE 2-NORM OF THE DIFFERENCE VECTOR.  THE DIFFERENCE     072
                VECTOR IS DEFINED BY EQUATION (9-4.1).                  073
      DELNE  =  THE BETA-NORM OF THE DIFFERENCE VECTOR.                 074
      YUN    =  THE ETA-NORM OF A RECENT APPROXIMATION FOR THE SOLUTION  075
                VECTOR.                                                 076
      SIP    =  THE STOPPING CRITERION NUMBER, ZETA, IN (9-5.17).       077
                IN ORDER THAT ACCURATE ESTIMATES BE OBTAINED FOR THE     078
                SPECTRAL RADIUS OF B AND FOR THE ITERATION ERROR, SIP    079
                SHOULD NOT BE TOO LARGE.                                080
      REMARK 4:  DELNP IS USED IN THE ADAPTIVE PARAMETER ESTIMATION      081
                PROCEDURE.                                              082
      REMARK 5:  DELNE, YUN, AND SIP ARE USED IN THE STOPPING TEST       083
                (9-5.17).  FOR A DISCUSSION OF APPROPRIATE BETA AND ETA  084
                NORMS TO USE FOR DELNE AND YUN, SEE THE LAST FEW PAGES   085
                OF SECTION 5.4.  IF THE RELATIVE NORM IS USED TO COMPUTE 086
                DELNE, SET YUN=1.0.                                     087
      REMARK 6:  IF THE SOR METHOD IS USED AS AN INNER ITERATION         088
                PROCEDURE, THE STOPPING TEST (9-5.17) SHOULD BE REPLACED 089
                BY A STOPPING TEST BASED ON THE TOTAL ERROR REDUCTION.   090
                SEE COMMENTS GIVEN IN CHAPTER 11.                       091
                                                                        092
                                                                        093
ON EXIT, THE SOR SUBROUTINE SUPPLIES THE FOLLOWING DATA(ASSUME           094
ITERATION (N+1) HAS BEEN COMPUTED):                                      095
                                                                        096
      OM     =  THE VALUE OF OMEGA TO BE USED ON THE NEXT ITERATION,     097
                WHICH IS ASSUMED HERE TO BE ITERATION N+2.              098
      ICONV  =  THE CONVERGENCE INDICATOR.                             099
                ICONV=0 IMPLIES THE STOPPING TEST (9-5.17) IS NOT        100
                        SATISFIED.                                     101
                ICONV=1 IMPLIES THE STOPPING TEST (9-5.17) IS SATISFIED. 102
                ICONV=10 IMPLIES THE ITERATION LIMIT, ILIMIT, HAS BEEN   103
                         EXCEEDED.                                     104
      ITP    =  THE ITERATION COUNTER.                                 105
      DELNE  =  THE ESTIMATE FOR THE ITERATION ERROR ON ITERATION N      106
                AS GIVEN BY THE LEFT SIDE OF (9-5.17).                 107
                                                                        108
                                                                        109
EDITS.  ASSUME ITERATION (N+1) HAS BEEN COMPUTED.                        110
                                                                        111
      I      =  THE ITERATION COUNTER.                                 112
      R      =  RATIO OF DIFFERENCE VECTOR NORMS(SEE EQUATION (9-5.9)).  113
      CR     =  THE RATIO OF ACTUAL TO THEORETICAL CONVERGENCE RATES FOR 114
                ITERATION N.  SEE (9-6.5).                             115
      EPS    =  THE ESTIMATE FOR THE ITERATION ERROR VECTOR FOR          116
                ITERATION N AS DEFINED BY THE LEFT SIDE OF (9-5.17).    117
      REMARK 7:  VALUES FOR CR AND EPS ARE NOT COMPUTED FOR ALL          118
                ITERATIONS.  WHEN NO COMPUTATIONS ARE MADE, THE         119
                VALUE PRINTED IS UNITY.                                120
                                                                        121
WHENEVER OMEGA IS CHANGED AND NPRT=1, THE FOLLOWING IS PRINTED:          122
                                                                        123
```

```
C      OMEG P. = THE ESTIMATE FOR OMEGA COMPUTED BY (9-5.13),           124
C      OMEG U. = THE ESTIMATE FOR OMEGA TO BE USED.  OMEG U. WILL DIFFER 125
C                FROM OMEG P. ONLY BECAUSE OF THE TAU RESTRICTIONS.  SEE 126
C                (9-5.22).                                              127
C      P-HAT   = THE VALUE DEFINED BY (9-6.2).                          128
C      P-STAR  = THE VALUE DEFINED BY (9-5.20).                         129
C      MU      = THE ESTIMATE FOR THE SPECTRAL RADIUS OF THE JACOBI     130
C                ITERATION MATRIX B.                                    131
C                                                                       132
C   THANKS ARE DUE TO L. A. ONDIS, C. J. PFEIFER, AND C. J. SPITZ OF    133
C   WESTINGHOUSE WHO MADE MANY HELPFUL SUGGESTIONS CONCERNING THE       134
C   PROGRAMMING OF THIS SUBROUTINE.                                     135
C                                                                       136
C                                                                       137
       INTEGER P                                                        138
       INTEGER PS                                                       139
       INTEGER PH                                                       140
       REAL TAU(8)                                                      141
       DATA TAU(1),TAU(2),TAU(3),TAU(4),TAU(5),TAU(6),TAU(7),TAU(8)/1.6, 142
      X1.8,1.90,1.95,1.975,1.985,1.990,1.995/                          143
       ICONV=0                                                          144
       EPS=1.0                                                          145
       XC=1.0                                                           146
       IPWM=0                                                           147
       IF (ITP.GT.0) GO TO 130                                          148
C                                                                       149
C INITIALIZE                                                            150
       P=-1                                                             151
       IS=0                                                             152
       Z=ABS(OME)                                                       153
       IF (Z.LT.1.0.OR.Z.GE.2.0) OME=1.0                               154
       XMUP=0.0                                                         155
       IF (Z.GT.1.0) XMUP=2.0*SQRT(Z-1.0)/Z                            156
       OMEP=OME                                                         157
       DELNP=1.0                                                        158
       INPRT=0                                                          159
       R=1.0                                                            160
       FF=F                                                             161
       IF (F.LT..1.OR.F.GT..9) FF=.7                                   162
       PSPP=PSP                                                         163
       IF (PSP.LT..001.OR.PSP.GT..9) PSPP=.5                           164
       RSPP=RSP                                                         165
       IF (RSP.LT.1.0E-6.OR.RSP.GT..01) RSPP=.0001                     166
C                                                                       167
C NEXT ITERATION                                                        168
   10  ITP=ITP+1                                                        169
       P=P+1                                                            170
       DELNO=DELNP                                                      171
       IF (P.GT.0) GO TO 80                                             172
C      INITIALIZE FOR NEW ESTIMATE FOR OMEGA                            173
       ICS=0                                                            174
       IS=IS+1                                                          175
       PH=3                                                             176
       PS=6                                                             177
       OME=OMEP                                                         178
       IF (IS.GT.8) IS=8                                                179
       TAUS=TAU(IS)                                                     180
       IF (OME.GT.TAUS) OME=TAUS                                        181
       IF (OMEP.LT.0.0) OME=-OMEP                                       182
       XOM1=OME-1.0                                                     183
       XC1=XOM1/(2.0-OME)                                               184
       IPH=INT(XC1)                                                     185
       IF (IPH.GT.PH) PH=IPH                                            186
       XOCR=1.0                                                         187
       IF (XOM1.GT.0.0) XOCR=ALOG(XOM1)                                188
       DO 20 I=PS,5000                                                  189
       Z=FLOAT(I)                                                       190
       J=I-1                                                            191
       IF (Z*XOM1**J.LT.PSPP) GO TO 30                                  192
   20  CONTINUE                                                         193
       I=5000                                                           194
   30  PS=I                                                             195
       IF (OMEP.GT.0.0) GO TO 40                                        196
       PS=10000                                                         197
       GO TO 50                                                         198
```

```
 40 IF (XOM1.GT.0.0) GO TO 50                                        199
    PH=2                                                             200
    PS=3                                                             201
    IS=IS-1                                                          202
 50 XWDEL=RSPP                                                       203
    IF (OME.LE.1.9) GO TO 60                                         204
    Z=((2.0-OME)**2)*.01                                            205
    IF (Z.LT.RSPP) XWDEL=Z                                           206
 60 IPWM=1                                                           207
 70 FORMAT (1H0,8HOMEG P.=,E11.4,11H   OMEG U.=,E11.4,9H   P-HAT=,I3,1 208
   10H   P-STAR=,I5,6H   MU=,E11.4)                                  209
C                                                                    210
C CLEAN UP BEFORE EXIT                                               211
 80 DELNE=EPS                                                        212
    ITPP=ITP-1                                                       213
    IF (ITPP.GE.ILIMIT) ICONV=10                                    214
    IF (ICONV.GT.0) GO TO 190                                        215
    IF (ITP.EQ.1) WRITE (IO,85)                                      216
 85 FORMAT (1H0,31HINITIAL ESTIMATE USED FOR OMEGA)                  217
    IF (ITP.EQ.1) GO TO 110                                          218
    IF(NPRT.EQ.0) GO TO 120                                          219
 90 WRITE (IO,100) ITPP,R,XC,EPS                                     220
100 FORMAT (1H ,2HI=,I4,5H   R=,E11.4,6H   CR=,E11.4,7H   EPS=,E11.4) 221
110 IF (IPWM.EQ.1) WRITE (IO,70) OMEP,OME,PH,PS,XMUP                 222
    IF(PS.GE.10000 .AND. ICONV.GT.0) GO TO 200                       223
120 OM=OME                                                           224
    RETURN                                                           225
C                                                                    226
C STOPPING TEST                                                      227
130 RO=R                                                             228
    R=DELNP/DELNO                                                    229
    H=R                                                              230
    IF (ITP.EQ.2) RESDS=DELNP                                        231
    IF (P.LT.PH) GO TO 10                                            232
    IF (R.GE.1.0) GO TO 10                                           233
    IF (R.GT.XOM1) GO TO 140                                         234
    ICIS=ITP                                                         235
    H=XOM1                                                           236
    ICS=ICS+1                                                        237
140 EPS=DELNE/(YUN*(1.0-H))                                          238
    IF (XOM1.GT.0.0) XC=ALOG(R)/XOCR                                 239
    IF (EPS.LE.SIP) ICONV=1                                          240
C                                                                    241
C PARAMETER CHANGE TEST                                              242
150 IF (P.LT.PS) GO TO 10                                            243
    IF (ICS.EQ.0) GO TO 160                                          244
    ITPD=ITP-ICIS                                                    245
C THIS TEST INVOLVING ICS AND ITPD IS NOT INCLUDED IN ALGORITHM 9-6.1. 246
    IF (ICS.LT.3.AND.ITPD.GT.40) GO TO 160                           247
    GO TO 10                                                         248
160 IF (OME.GT.1.0) GO TO 170                                        249
    Z=1.0-R                                                          250
    OMEP=2.0/(1.0+SQRT(Z))                                           251
    P=-1                                                             252
    XMUP=SQRT(R)                                                     253
    GO TO 10                                                         254
170 Z=XOM1**FF                                                       255
    IF (R.LT.Z) GO TO 10                                             256
    DELR=RO-R                                                        257
    Z=-10.0*XWDEL                                                    258
    IF (DELR.LE.XWDEL.AND.DELR.GE.Z) GO TO 180                       259
    IF (IS.GT.2) GO TO 10                                            260
    Z=XOM1**.1                                                       261
    IF (R.LT.Z) GO TO 10                                             262
180 XMUP=(R+XOM1)/(OME*SQRT(R))                                      263
    Z=1.0-XMUP*XMUP                                                  264
    OMEP=2.0/(1.0+SQRT(Z))                                           265
    P=-1                                                             266
    GO TO 10                                                         267
C                                                                    268
C PRINT FINAL ESTIMATE FOR OMEGA IF NPRT=0                           269
190 IF(INPRT .NE. 0) GO TO 90                                        270
    IF(PS .GE. 10000) GO TO 90                                       271
    IF(NPRT .EQ. 1) GO TO 90                                         272
    WRITE (IO,195)                                                   273
```

```
195 FORMAT ( 1H0,17HITERATION SUMMARY)                              274
    INPRT=1                                                         275
    WRITE( IO,70) OMEP,OME,PH,PS,XMUP                               276
    GO TO 90                                                        277
C                                                                   278
C IF OMEGA IS FIXED, GIVE ESTIMATE OF ITS WORTH.                    279
200 IF( INPRT.NE.0) GO TO 120                                       280
    INPRT=1                                                         281
    DELNP=-1.0                                                      282
    IF(R .LT. 1.0) GO TO 250                                        283
210 WRITE( IO,220) OME                                              284
220 FORMAT( 1H0,35HNUMERICAL RESULTS IMPLY THAT OMEGA= ,E11.4,47H IS GRE  285
   XATER THAN OR EQUAL TO THE OPTIMAL VALUE.)                       286
230 WRITE( IO,240)                                                  287
240 FORMAT( 1H0)                                                    288
    GO TO 120                                                       289
250 IPRED=1                                                         290
    IF( ICS .EQ. 0) GO TO 260                                       291
    ITPD=ITP - ICIS -1                                              292
    IF( ICS .LT. 3 .AND. ITPD .GT. 40) GO TO 260                    293
    GO TO 210                                                       294
260 Z=.0001                                                         295
    ZZ=-.001                                                        296
    DELR=RO-R                                                       297
    IF(DELR .LE. Z .AND. DELR .GE. ZZ) IPRED=2                      298
    IF(OME .EQ. 1.0) GO TO 290                                      299
    XMUP=(R+XOM1)/(OMF*SQRT(R))                                     300
270 Z=1.0-XMUP*XMUP                                                 301
    DELNP=2.0/(1.0+SQRT(Z))                                         302
    WRITE( IO,280) DELNP                                            303
280 FORMAT( 1H0,59HNUMERICAL RESULTS INDICATE THAT THE OPTIMAL OMEGA IS  304
   X ABOUT ,E11.4)                                                  305
    IF( IPRED.EQ.1) WRITE( IO,300)                                  306
    IF( IPRED.EQ.2) WRITE( IO,310)                                  307
    Z=1.0                                                           308
    IF( IPRED .EQ. 1) Z=-1.0                                        309
    DELNP=Z*DELNP                                                   310
    GO TO 230                                                       311
290 XMUP=SQRT(R)                                                    312
    GO TO 270                                                       313
300 FORMAT( 1H ,30HTHIS ESTIMATE IS QUESTIONABLE.)                  314
310 FORMAT( 1H ,28HTHIS ESTIMATE IS REASONABLE.)                    315
    END                                                             316
```

Bibliography

Abu-Shumays, I. K. [1977]. Compatible product angular quadrature for neutron transport in x-y geometry. *Nucl. Sci. Eng.* **64**, 299–316.

Abu-Shumays, I. K., and Hageman, L. A. [1975]. Development and comparison of practical discretization methods for the neutron diffusion equation over general convex quadrilateral partitions. *Proc. Conf. Comput. Methods Nucl. Eng.* **1** (Conf.-750413), I-117–I-165.

Arms, R. J., Gates, L. D., and Zondek, B. [1956]. A method of block iteration. *J. Soc. Ind. Appl. Math.* **4**, 220–229.

Axelsson, O. [1972]. A generalized SSOR method. *BIT* **13**, 443–467.

Axelsson, O. [1974]. On Preconditioning and Convergence Acceleration in Sparse Matrix Problems, CERN 74–10. European Organization for Nuclear Research (CERN), Data Handling Division.

Axelsson, O. [1977]. Solution of linear systems of equations: iterative methods. *In* "Sparse Matrix Techniques, Copenhagen (1976)" (V. A. Barker, ed.), pp. 1–51. Lecture Notes in Mathematics, 572. Springer-Verlag, Berlin and New York.

Axelsson, O. [1978]. An Iterative Solution of Large Sparse Systems of Equations with Particular Emphasis on Boundary Value Problems. Rep. 78–4, Texas Institute of Computational Mechanics, Univ. of Texas at Austin.

Babuska, I., and Kellogg, R. B. [1972]. Numerical solution of the neutron diffusion equation in presence of corners and interfaces. *Proc. Seminar Numer. Reactor Calculat.* pp. 473–485. International Atomic Energy Agency, Vienna.

Bank, R. E. [1975]. Marching Algorithms for Elliptic Boundary Value Problems. Doctoral Thesis, Harvard Univ., Cambridge, Massachusetts.

Bank, R. E. [1976]. Marching algorithms and block Gaussian elimination. *In* "Sparse Matrix Computation" (J. R. Bunch and D. J. Rose, eds.), pp. 293–307. Academic Press, New York.

Bank, R. E., and Rose, D. J. [1975]. An $O(n^2)$ method for solving constant coefficient boundary value problems in two dimensions. *SIAM J. Numer. Anal.* **12**, 529–540.

Barnard, S., and Child, J. M. [1952]. "Higher Algebra." Macmillan, New York.

Bartels, R., and Daniel, J. W. [1974]. A conjugate gradient approach to nonlinear elliptic boundary value problems in irregular regions. *Proc. Conf. Numerical Solution of Differential Equations, Dundee, Scotland, 1973.* Springer-Verlag, Berlin and New York.

Beauwens, R. [1979]. Factorization iterative methods, M-operators and H-operators. *Numer. Math.* **31**, 335–357.

Beckman, F. S. [1960]. The solution of linear equations by the conjugate gradient method. *In* "Mathematical Methods for Digital Computers" (A. Ralston and H. S. Wilf. eds.), Vol. I, pp. 62–72. Wiley, New York.

Bell, G. I., and Glasstone, S. [1970]. "Nuclear Reactor Theory." Van Nostrand-Reinhold, New York.

Benokraitis, V. J. [1974]. On the Adaptive Acceleration of Symmetric Successive Over-relaxation. Doctoral thesis, Univ. of Texas at Austin.

Birkhoff, G., and Kellogg, R. B. [1966]. Solution of equilibrium equations in thermal networks. *Proc. Symp. Generalized Networks*, (J. Fox, ed.), Vol. 16, pp. 443–452. Microwave Research Institute Symposia Series, Brooklyn Polytechnic Press, New York.

Birkhoff, G., and Varga, R. S. [1958]. Reactor criticality and nonnegative matrices. *SIAM J. Appl. Math.* **6**, 354–377.

Birkhoff, G., and Varga, R. S. [1959]. Implicit alternating direction methods. *Trans. Amer. Math. Soc.* **92**, 13–24.

Birkhoff, G., Varga, R. S., and Young, D. M. [1962]. Alternating direction implicit methods. *In* "Advances in Computers," (F. Alt and M. Rubinoff, eds.), Vol. 3, pp. 189–273. Academic Press, New York.

Birkhoff, G., Cavendish, J. C., and Gordon, W. J. [1974]. Multivariate approximation by locally blended univariate interpolants. *Proc. Nat. Acad. Sci. U.S.A.* **71**, 3423–3425.

Bracha-Barak, A., and Saylor, P. E. [1973]. A symmetric factorization procedure for the solution of elliptic boundary value problems. *SIAM J. Numer. Anal.* **10**, 190–206.

Brandt, A. [1977]. Multi-level adaptive solution to boundary value problems. *Math. Comp.* **31**, 333–391.

Buzbee, B. L., Golub, G. H., and Nielson, C. W. [1970]. The method of odd/even reduction and factorization with application to Poisson's equation. *SIAM J. Numer. Anal.* **7**, 617–656.

Buzbee, B. L., Dorr, F. W., George, J. A., and Golub, G. H. [1971]. The direct solution of the discrete Poisson equation on irregular regions. *SIAM J. Numer. Anal.* **8**, 722–736.

Buzbee, B. L., Golub, G. H., and Howell, J. A. [1977]. Vectorization for the CRAY-1 of some methods for solving elliptic difference equations. *Proc. Symp. High-speed Comput. and Algorithm Organization* (D. J. Kuck, D. H. Lawrie, and A. H. Semah, eds.), pp. 255–272. Academic Press, New York.

Byrne, G. D., and Hindmarsh, A. C. [1975]. A polyalgorithm for the numerical solution of ordinary differential equations. *ACM Trans. Math. Software* **1**, 71–96.

Carlson, B. G. [1963]. The numerical theory of neutron transport. *In* "Methods of Computational Physics," Vol. I, Statistical Physics, pp. 1–42. Academic Press, New York.

Carré, B. A. [1961]. The determination of the optimum acceleration factor for successive over-relaxation. *Comput. J.* **4**, 73–78.

Chandra, R. [1978]. Conjugate Gradient Methods for Partial Differential Equations. Research Rep. 129, Department of Computer Science, Yale Univ., New Haven, Connecticut.

Chandra, R., Eisenstat, S. C., and Schultz, M. H. [1977]. The modified conjugate residual method for partial differential equations. *In* "Advances in Computer Methods for Partial Differential Equations" (R. Vichnevetsky, ed.), Vol. II. IMACS.

Clayton, A. J. [1963]. Further Results on Polynomials Having Least Maximum Modulus Over an Ellipse in the Complex Plane. AEEW-M348, United Kingdom Atomic Energy Authority, Winfrith, Dorchester.

Concus, P., and Golub, G. H. [1973]. Use of fast direct methods for the effective numerical solution of nonseparable elliptic equations. *SIAM J. Numer. Anal.* **10**, 1103–1120.

Concus, P. and Golub, G. H. [1976a]. A Generalized Conjugate Gradient Method for Nonsymmetric Systems of Linear Equations. Rep. Stan-CS-76-646, Computer Science Department, Stanford Univ.

Concus, P., Golub, G. H., and O'Leary, D. P. [1976b]. A generalized conjugate gradient method for the numerical solution of elliptic partial differential equations. *In* "Sparse Matrix Computation" (J. R. Bunch and D. J. Rose, eds.), pp. 309–332. Academic Press, New York.

Cuthill, E. [1964]. Digital computers in nuclear reactor design. *In* "Advances in Computers" (F. L. Alt and M. Rubinoff, eds.), Vol. 5, pp. 289–347. Academic Press, New York.

Cuthill, E., and McKee, J. [1969]. Reducing the bandwidth of sparse symmetric matrices. *Proc. Nat. Conf. ACM, 24th* (S. L. Pollack, ed.), pp. 157–172.

Cuthill, E. H., and Varga, R. S. [1959]. A method of normalized block iteration. *J. Assoc. Comput. Mach.* **6**, 236–244.

Daniel, J. W. [1965]. The Conjugate Gradient Method for Linear and Nonlinear Operator Equations. Doctoral thesis, Stanford Univ.

Daniel, J. W. [1967]. The conjugate gradient method for linear and nonlinear operator equations. *SIAM J. Numer. Anal.* **4**, 10–26.

Davis, J. A., and Hageman, L. A. [1969]. An iterative method for solving the neutron transport equation in x–y geometry. *SIAM J. Appl. Math.* **17**, 149–161.

Davis, J. A., Hageman, L. A., and Kellogg, R. B. [1967]. Singular difference approximations for the discrete ordinate equations in x–y geometry. *Nucl. Sci. Eng.* **29**, 237–243.

Davis, P. J. [1963]. "Interpolation and Approximation." Ginn (Blaisdell), New York.

Davison, B., and Sykes, J. B. [1957]. "Neutron Transport Theory." Oxford Univ. Press (Clarendon), London and New York.

de la Vallee Poussin, F. [1968]. An accelerated relaxation algorithm for iterative solution of elliptic equations. *SIAM J. Numer. Anal.* **5**, 340–351.

Diamond, M. A. [1971]. An Economical Algorithm for the Solution of Finite Difference Equations. Doctoral thesis, Univ. of Illinois, Urbana, Illinois.

Dorr, F. W. [1970]. The direct solution of the discrete Poisson equation on a rectangle. *SIAM Rev.* **12**, 248–263.

Douglas, J., Jr. [1955]. On the numerical integration of $\partial^2 u/\partial x^2 + \partial^2 u/\partial y^2 = \partial u/\partial t$ by implicit methods. *J. SIAM* **3**, 42–65.

Douglas, J., Jr., and Rachford, H. [1956]. On the numerical solution of heat conduction problems in two and three space variables. *Trans. Amer. Math. Soc.* **82**, 421–439.

Dupont, T., Kendall, R., and Rachford, H. H., Jr. [1968]. An approximate factorization procedure for solving self-adjoint elliptic difference equations. *SIAM J. Numer. Anal.* **5**, 559–573.

Ehrlich, L. W. [1963]. The Block Symmetric Successive Overrelaxation Method. Doctoral thesis, Univ. of Texas, Austin.

Ehrlich, L. W. [1964]. The block symmetric successive overrelaxation method. *J. Soc. Indust. Appl. Math.* **12**, 807–826.

Eisenstat, S. C., Elman, H., Schultz, M. H., and Sherman, A. H. [1979a]. Solving approximations to the convection diffusion equation. *Proc. Soc. Petroleum Eng. AIME Symp. Reservoir Simulation, 5th, Denver, Colorado.*

Eisenstat, S., George, A., Grimes, R., Kincaid, D., and Sherman, A. [1979b]. Some comparisons of software packages for large sparse linear systems. *In* "Advances in Computer Methods for Partial Differential Equations" (R. Vichnevetsky and R. S. Stepleman, eds.), Vol. III. IMACS, Rutgers Univ., New Brunswick, New Jersey.

Eisenstat, S. C., Elman, H., and Schultz, M. H. [1979c]. Variational Iterative Methods for Nonsymmetric Systems of Linear Equations. Unpublished manuscript.

Engeli, M., Ginsburg, M., Rutishauser, H., and Stiefel, E. [1959]. Refined iterative methods for the computation of the solution and the eigenvalues of self-adjoint boundary value problems. *Mitt. Inst. Angew. Math. ETH, Zürich*, Nr. 8, Basel-Stuttgart.

Evans, D. J. [1967]. The use of preconditioning in iterative methods for solving linear equations with symmetric positive definite matrices. *J. Inst. Math. Appl.* **4**, 295–314.

Faddeev, D. K., and Faddeeva, V. N. [1963]. "Computational Methods of Linear Algebra." Freeman, San Francisco, California.

Flanders, D., and Shortley, G. [1950]. Numerical determination of fundamental modes. *J. Appl. Phys.* **21**, 1326–1332.

Fletcher, R. [1976]. "Conjugate Gradient Methods for Indefinite Systems." Lecture Notes in Mathematics, 506. Springer-Verlag, Berlin and New York.

Gear, W. C. [1971]. "Numerical Initial Value Problems in Ordinary Differential Equations." Prentice-Hall, Englewood Cliffs, New Jersey.

Gelbard, E. M., and Hageman, L. A. [1968]. The synthetic method as applied to the S_n equations. *Nucl. Sci. Eng.* **39**, 288–298.

George, A. [1973]. Nested dissection of a regular finite element mesh. *SIAM. J. Numer. Anal.* **10**, 345–363.

Golub, G. H., and Varga, R. S. [1961]. Chebyshev semi-iterative methods, successive over-relaxation iterative methods, and second-order Richardson iterative methods. *Numer. Math. Parts I and II* **3**, 147–168.

Grimes, R. C., Kincaid, D. R., MacGregor, W. I., and Young, D. M. [1978]. ITPACK Report: Adaptive Iterative Algorithms Using Symmetric Sparse Storage. Rep. CNA-139, Center for Numerical Analysis, Univ. of Texas, Austin.

Grimes, R. G., Kincaid, D. R., and Young, D. M. [1979]. ITPACK 2.0 User's Guide, CNA-150. Center for Numerical Analysis, Univ. of Texas, Austin.

Habetler, G., and Martino, M. [1961]. Existence theorems and spectral theory for the multi-group diffusion model. *Proc. Symp. Appl. Math.* **11**, 127–139.

Habetler, G. J., and Wachspress, E. L. [1961]. Symmetric successive overrelaxation in solving diffusion difference equations. *Math. Comp.* **15**, 356–362.

Hackbusch, W. [1977]. On the Computation of Approximate Eigenvalue and Eigenfunctions of Elliptic Operators by Means of Multi-grid Methods. Rep. 77–10, Univ. zu Köln, West Germany.

Hageman, L. A. [1975]. Computational Aspects in the Numerical Solution of Partial Differential Equations. Invited talk at SIAM 1975 National Meeting in Troy, New York.

Hageman, L. A., and Kellogg, R. B. [1968]. Estimating optimum overrelaxation parameters. *Math. Comp.* **22**, 60–68.

Hageman, L. A., and Porsching, T. A. [1975]. Aspects of nonlinear block successive over-relaxation. *SIAM J. Numer. Anal.* **12**, 316–335.

Hageman, L. A., and Varga, R. S. [1964]. Block iterative methods for cyclically reduced matrix equations. *Numer. Math.* **6**, 106–119.

Hageman, L. A., Luk, F., and Young, D. M. [1977]. On the Acceleration of Iterative Methods: Preliminary Report. Rep. CNA-129, Center for Numerical Analysis, University of Texas at Austin.

Hageman, L. A., Luk, F. T., and Young, D. M. [1980]. On the equivalence of certain iterative acceleration methods. *SIAM. J. Numer. Anal.* **17**, 852–872.

Hayes, L. J., and Young, D. M. [1977]. The Accelerated SSOR Method for Solving Large Linear Systems. CNA-123, Center for Numerical Analysis, University of Texas at Austin.

Hestenes, M. R. [1956]. The conjugate-gradient method for solving linear systems. *In* "Numerical Analysis" (J. Curtiss, ed.), Vol. VI. McGraw-Hill, New York.

Hestenes, M. R., and Stiefel, E. L. [1952]. Methods of conjugate gradients for solving linear systems. *Nat. Bur. Std. J. Res.* **49**, 409–436.

Hindmarsh, A. C. [1974]. Numerical Solution of Ordinary Differential Equations: Lecture Notes. UCID-16558, Lawrence Livermore Laboratory, Livermore, California.

Hockney, R. W. [1965]. A fast direct solution of Poisson's equation using Fourier analysis. *J. Assoc. Comp. Mach.* **12**, 95–113.

Irons, B. [1970]. A frontal solution program of finite element analysis. *Int. J. Numer. Methods Eng.* **2**, 5–32.

Kahan, W. [1958]. Gauss–Seidel Methods of Solving Large Systems of Linear Equations. Doctoral thesis, Univ. of Toronto, Toronto, Canada.

Kaniel, S. [1966]. Estimate for some computational techniques in linear algebra. *Math. Comp.* **20**, 369–378.

Kellogg, R. B. [1969]. On the spectrum of an operator associated with the neutron transport equation. *SIAM J. Appl. Math.* **17**, 162–171.

Kellogg, R. B., and Noderer, L. C. [1960]. Iterations and linear equations. *J. Soc. Ind. Appl. Math.* **8**, 634–661.

Kellogg, R. B., and Spanier, J. [1965]. On optimal alternating direction parameters for singular matrices. *Math. Comp.* **19**, 448–452.

Kershaw, D. S. [1978]. The incomplete Cholesky-conjugate gradient method for the iterative solution of systems of linear equations. *J. Comp. Phys.* **26**, 43–65.

Kjellberg, G. [1958]. On the convergence of successive over-relaxation applied to a class of linear systems of equations with complex eigenvalues. *Ericsson Tech. Stockholm* **2**, 245–258.

Kopp, H. J. [1963]. Synthetic method solution of the transport equation. *Nucl. Sci. Eng.* **17**, 65–74.

Kristiansen, G. K. [1963]. Description of DC-2, a Two-dimensional Cylindrical Geometry, Two-Group Diffusion Theory Code for DASK, and a Discussion of the Theory for Such Codes. Risö Rep. No. 55, Danish Atomic Energy Commission, Risö, Roskilde, Denmark.

Kulsrud, H. E. [1961]. A practical technique for the determination of the optimum relaxation factor of the successive overrelaxation method. *Comm. ACM* **4**, 184–187.

Lanczos, C. [1950]. An iteration method for the solution of the eigenvalue problem of linear differential and integral operators. *Nat. Bur. Std. J. Res.* **45**, 255–282.

Lanczos, C. [1952]. Solution of systems of linear equations by minimized iterations. *Nat. Bur. Std. J. Res.* **49**, 33–53.

Lathrop, K. D. [1969a]. User's Guide for the TWOTRAN(x, y) Program. Los Alamos Scientific Laboratory Rep. LA-4058.

Lathrop, K. D. [1969b]. Spatial differentiating of the transport equation: positivity vs. accuracy. *J. Comp. Phys.* **4**, 475–498.

Lewis, E. E. [1977]. Progress in multidimensional neutron computation. *Nucl. Sci. Eng.* **64**, 279–293.

Luenberger, D. G. [1973]. "Introduction to Linear and Nonlinear Programming." Addison-Wesley, Reading, Massachusetts.

Madsen, N. K. [1971]. Pointwise convergence of the three-dimensional discrete ordinate method. *SIAM J. Numer. Anal.* **8**, 266–269.

Manteuffel, T. A. [1975]. An Iterative Method for Solving Nonsymmetric Linear Systems with Dynamic Estimation of Parameters. Rep. UTUCDS-R-758, Department of Computer Science, Univ. of Illinois, Urbana, Illinois.

Manteuffel, T. A. [1977]. The Tchebychev iteration for nonsymmetric linear systems. *Numer. Math.* **28**, 307–327.

Manteuffel, T. A. [1978]. Adaptive procedure for estimating parameters for the nonsymmetric Tchebyshev iteration. *Numer. Math.* **31**, 183–208.

Meijerink, J. A., and van der Vorst, H. A. [1974]. "Iterative Solution of Linear Systems Arising from Discrete Approximations to Partial Differential Equations." Academisch Computer Centrum, Utrecht, The Netherlands.

Meijerink, J. A., and van der Vorst, H. A. [1977]. An iterative solution method for linear systems of which the coefficient matrix is a symmetric M-matrix. *Math. Comp.* **31**, 148–162.

Miles, G. A., Stewart, K. L., and Tee, G. T. [1964]. "Elementary divisors of the Liebmann Process. *Comput. J.* **6**, 353–355.

Nakamura, S. [1974]. Effect of weighting functions on the coarse mesh rebalancing acceleration. *Proc. Conf. Math. Models and Comput. Tech. for Anal. Nucl. Syst.* Vol. II, pp. 120–135. U.S. Atomic Energy Commission, CONF, 730414.

Nakamura, S. [1977]. "Computational Methods in Engineering and Science." Wiley, New York.

Nicolaides, R. A. [1975]. On multiple grid and related techniques for solving discrete elliptic systems. *J. Comp. Phys.* **19**, 418–431.

Nicolaides, R. A. [1977]. On the l^2 convergence of an algorithm for solving finite element equations. *Math. Comp.* **31**, 892–906.

Nichols, N. [1973]. On the convergence of two-stage iterative processes for solving linear equations. *SIAM. J. Numer. Anal.* **10**, 460–469.

Noble, B. and Daniel, W. [1977]. "Applied Linear Algebra." Prentice-Hall, Englewood Cliffs, New Jersey.

O'Carroll, M. J. [1973]. Inconsistencies and S.O.R. convergence for the discrete Neumann problem. *J. Inst. Math. Appl.* **11**, 343–350.

O'Leary, D. P. [1975]. Hybrid Conjugate Gradient Algorithms. Doctoral thesis, Stanford Univ., Palo Alto, California.

Ortega, J., and Rheinboldt, W. [1970]. "Iterative Solution of Nonlinear Equations in Several Variables." Academic Press, New York.

Ortega, J., and Rockoff, M. [1966]. Nonlinear difference equations and Gauss–Seidel type iterative methods. *SIAM J.* **3**, 497–513.

Paige, C. C. [1971]. The Computation of Eigenvalues and Eigenvalues of Very Large Sparse Matrices. Doctoral thesis, London Univ., Inst. of Computer Science.

Parter, S. [1961]. Multi-line iterative methods for elliptic difference equations and fundamental frequencies. *Numer. Math.* **3**, 305–319.

Peaceman, D. W., and Rachford, H. H., Jr. [1955]. The numerical solution of parabolic and elliptic differential equations, *J. Soc. Ind. Appl. Math.* **3**, 28–41.

Pfeifer, C. [1963]. Data flow and storage allocation for the PDQ-5 program on the Philco-2000. *Comm. ACM* **6**, 365–366.

Porsching, T. A. [1971]. On rates of convergence of Jacobi and Gauss–Seidel methods for M-functions. *J. Soc. Ind. Appl. Math.* **8**, 575–582.

Price, H., and Varga, R. S. [1962]. Recent Numerical Experiments Comparing Successive Overrelaxation Iterative Methods with Implicit Alternating Direction Methods. Rep. 91, Gulf Research and Development Co., Pittsburg, Pennsylvania.

Proskurowski, W., and Widlund, O. [1976]. On the numerical solution of Helmholtz equation by the capacitance matrix method. *Math. Comp.* **30**, 433–468.

Reid, J. K. [1966]. A method for finding the optimum successive overrelaxation factor. *Comput. J.* **9**, 200–204.

Reid, J. K. [1971]. On the method of conjugate gradients for the solution of large sparse systems of linear equations. *Proc. Conf. Large Sparse Sets of Linear Equations*, pp. 231–254. Academic Press, New York.

Reid, J. K. [1972]. The use of conjugate gradients for systems of linear equations possessing Property A. *SIAM J. Numer. Anal.* **9**, 325–332.

Reid, J. K. [1977]. Sparse matrices. *In* "The State of the Art in Numerical Analysis" (D. Jacobs, ed.), pp. 85–146, Academic Press, New York.

Rheinboldt, W. [1970a]. On M-functions and their application to nonlinear Gauss–Seidel iterations and network flows. *J. Math. Anal. Appl.* **32**, 274–307.

Rheinboldt, W. [1970b]. On classes on n-dimensional nonlinear mappings generalizing several types of matrices. *Proc. Symp. Numer. Solut. Partial Differential Equations* (B. Hubbard, ed.), Vol. II, pp. 501–546. Academic Press, New York.

Richardson, L. F. [1910]. The approximate arithmetical solution by finite differences of physical problems involving differential equations with an application to the stresses in a masonry dam. *Philos. Trans. Roy. Soc. London A* **210**, 307–357.

Rose, D. J., and Willoughby, R. A. eds. [1972]. "Sparse Matrices and Their Applications." Plenum Press, New York.

Setturi, A., and Aziz, K. [1973]. A generalization of the additive correction methods for the iterative solution of matrix equations. *SIAM J. Numer. Anal.* **10**, 506–521.

Sheldon, J. [1955]. On the numerical solution of elliptic difference equations. *Math Tables Aids Comput.* **9**, 101–112.

Stone, H. L. [1968]. Iterative solutions of implicit approximations of multi-dimensional partial differential equations. *SIAM J. Numer. Anal.* **5**, 530–538.

Strang, G. [1976]. "Linear Algebra and Its Applications." Academic Press, New York.

Strang, W. G., and Fix, G. J. [1973]. "An Analysis of the Finite Element Method." Prentice-Hall, Englewood Cliffs, New Jersey.

Swartztrauber, P. M. [1974]. A direct method for the discrete solution of separable elliptic equations. *SIAM J. Numer. Anal.* **11**, 1136–1150.

Sweet, R. A. [1973]. Direct methods for the solution of Poisson's equation on a staggered grid. *J. Comp. Phys.* **12**, 422–428.

Sweet, R. A. [1974]. A generalized cyclic reduction algorithm. *SIAM J. Numer. Anal.* **11**, 506–620.

Tee, G. J. [1963]. Eigenvectors of the successive overrelaxation process, and its combination with Chebyshev semi-iteration. *Comput. J.* **6**, 250–263.

Varga, R. S. [1960]. Factorization and normalized iterative methods. *In* "Boundary Problems in Differential Equations" (R. E. Langer, ed.), pp. 121–142. Univ. of Wisconsin Press, Madison, Wisconsin.

Varga, R. S. [1961]. Numerical methods for solving multi-dimensional multigroup diffusion equations. *In Proc. Symp. Appl. Math.* (G. Birkhoff and E. P. Wigner, eds.), Vol. XI, pp. 164–189. American Mathematical Society, Providence, Rhode Island.

Varga, R. S. [1962]. "Matrix Iterative Analysis." Prentice-Hall, Englewood Cliffs, New Jersey.

Vinsome, P. K. W. [1976]. ORTHOMIN, an iterative method for solving sparse sets of simultaneous linear equations. Paper SPE 5739, *Symp. Numer. Simulation of Reservoir Performance of SPE of AIME, 4th, Los Angeles, California, February 19-20, 1976.*

Wachspress, E. L. [1963]. Extended application of alternating direction implicit iteration model problem theory. *SIAM J. Appl. Math.* **11**, 994–1016.

Wachspress, E. L. [1966]. "Iterative Solution of Elliptic Systems and Applications to the Neutron Diffusion Equations of Reactor Physics." Prentice-Hall, Englewood Cliffs, New Jersey.

Wachspress, E. L. [1977]. Two-level finite element calculation. *In* "Formulations and Computational Algorithms in Finite Element Analysis" (K. J. Bathe, J. T. Oden, and W. Wunderlick, eds.), pp. 877–913. Halliday Lithograph Corp., Boston, Massachusetts.

Wang, H. H. [1977]. The Application of the Symmetric SOR and the Symmetric SIP Methods for the Numerical Solution of the Neutron Diffusion Equation. Rep. G320–3358. IBM Palo Alto Scientific Center, Palo Alto, California.

Widlund, O. [1966]. On the rate of convergence of an alternating direction implicit method in a noncommutative case. *Math. Comp.* **20**, 500–515.

Widlund, O. [1969]. On the effects of scaling of the Peaceman–Rachford method. *Proc. Conf. Numer. Solution of Differential Equations, Dundee, Scotland, June, 23-27.* pp. 113–132. Springer-Verlag, Berlin and New York.

Widlund, O. [1978]. A Lanczos method for a class of nonsymmetric systems of linear equations. *SIAM J. Numer. Anal.* **15**, 801–812.

Wilkinson, J. H. [1965]. "The Algebraic Eigenvalue Problem." Oxford Univ. Press, London and New York.

Wong, Yau Shu [1979]. Conjugate Gradient Type Methods for Unsymmetric Matrix Problems. Rep. TR 79–36, Department of Computer Science, Univ. of British Columbia, Vancouver, Canada.

Wrigley, E. E. [1963]. Accelerating the Jacobi method for solving simultaneous equations by Chebyshev extrapolation when the eigenvalues of the iteration matrix are complex. *Comput. J.* **6**, 169–176.

Young, D. M. [1950]. Iterative Methods for Solving Partial Difference Equations of Elliptic Type. Doctoral thesis, Harvard Univ.

Young, D. M. [1954]. Iterative methods for solving partial difference equations of elliptic type. *Trans. Amer. Math. Soc.* **76**, 92–111.

Young, D. M. [1971]. "Iterative Solution of Large Linear Systems." Academic Press, New York.

Young, D. M. [1972]. Second-degree iterative methods for the solution of large linear systems. *J. of Approx. Theory* **5**, 137–148.

Young, D. M. [1977]. On the accelerated SSOR method for solving large linear systems. *Adv. in Mathematics* **23**, 215–271.

Young, D. M., and Gregory, R. T. [1972]. "A Survey of Numerical Mathematics." Vol. I. Addison-Wesley, Reading, Massachusetts.

Young, D. M., and Jea, K. C. [1980a]. Conjugate Gradient Acceleration of Iterative Methods: Part II, The Nonsymmetrizable Case. Rep. CNA-163, Center for Numerical Analysis, Univ. of Texas at Austin.

Young, D. M., and Jea, K. C. [1980b]. Generalized Conjugate Gradient Acceleration of Non-symmetrizable Iterative Methods. *Linear Algebra and Appl.* **34**, 159–194.

Young, D. M., Hayes, L., and Jea, K. C. [1980]. Conjugate Gradient Acceleration of Iterative Methods: Part I, The Symmetrizable Case. Rep. CNA-162, Center for Numerical Analysis, University of Texas at Austin.

Zienkiewicz, O. C. [1973]. Finite elements—the background story, *In* "The Mathematics of Finite Elements and Applications" (J. R. Whiteman, ed.), pp. 1–35. Academic Press, New York.

Index

Computer Science and Applied Mathematics

A SERIES OF MONOGRAPHS AND TEXTBOOKS

Editor
Werner Rheinboldt
University of Pittsburgh

HANS P. KÜNZI, H. G. TZSCHACH, and C. A. ZEHNDER. Numerical Methods of Mathematical Optimization: With ALGOL and FORTRAN Programs, Corrected and Augmented Edition

AZRIEL ROSENFELD. Picture Processing by Computer

JAMES ORTEGA AND WERNER RHEINBOLDT. Iterative Solution of Nonlinear Equations in Several Variables

AZARIA PAZ. Introduction to Probabilistic Automata

DAVID YOUNG. Iterative Solution of Large Linear Systems

ANN YASUHARA. Recursive Function Theory and Logic

JAMES M. ORTEGA. Numerical Analysis: A Second Course

G. W. STEWART. Introduction to Matrix Computations

CHIN-LIANG CHANG AND RICHARD CHAR-TUNG LEE. Symbolic Logic and Mechanical Theorem Proving

C. C. GOTLIEB AND A. BORODIN. Social Issues in Computing

ERWIN ENGELER. Introduction to the Theory of Computation

F. W. J. OLVER. Asymptotics and Special Functions

DIONYSIOS C. TSICHRITZIS AND PHILIP A. BERNSTEIN. Operating Systems

ROBERT R. KORFHAGE. Discrete Computational Structures

PHILIP J. DAVIS AND PHILIP RABINOWITZ. Methods of Numerical Integration

A. T. BERZTISS. Data Structures: Theory and Practice, Second Edition

N. CHRISTOPHIDES. Graph Theory: An Algorithmic Approach

ALBERT NIJENHUIS AND HERBERT S. WILF. Combinatorial Algorithms

AZRIEL ROSENFELD AND AVINASH C. KAK. Digital Picture Processing

SAKTI P. GHOSH. Data Base Organization for Data Management

DIONYSIOS C. TSICHRITZIS AND FREDERICK H. LOCHOVSKY. Data Base Management Systems

JAMES L. PETERSON. Computer Organization and Assembly Language Programming

WILLIAM F. AMES. Numerical Methods for Partial Differential Equations, Second Edition

ARNOLD O. ALLEN. Probability, Statistics, and Queueing Theory: With Computer Science Applications

ELLIOTT I. ORGANICK, ALEXANDRA I. FORSYTHE, AND ROBERT P. PLUMMER. Programming Language Structures

ALBERT NIJENHUIS AND HERBERT S. WILF. Combinatorial Algorithms. Second edition.

JAMES S. VANDERGRAFT. Introduction to Numerical Computations